ANALYSIS OF QUEUEING SYSTEMS

OPERATIONS RESEARCH AND INDUSTRIAL ENGINEERING

Consulting Editor: J. William Schmidt
Virginia Polytechnic Institute and State University
Blacksburg, Virginia

Applied Statistical Methods, *I. W. Burr*

Mathematical Foundations of Management Science and Systems Analysis, *J. William Schmidt*

Urban Systems Models, *Walter Helly*

Introduction to Discrete Linear Controls: Theory and Application, *Albert B. Bishop*

Integer Programming: Theory, Applications, and Computations, *Hamdy A. Taha*

Transform Techniques for Probability Modeling, *Walter C. Giffin*

Analysis of Queueing Systems, *J. A. White, J. W. Schmidt, and G. K. Bennett*

Models for Public Systems Analysis, *Edward J. Beltrami*

Computer Methods in Operations Research, *Arne Thesen*

Cost-Benefit Analysis: A Handbook, *Peter G. Sassone and William A. Schaffer*

Modeling of Complex Systems, *V. Vemuri*

Applied Linear Programming: For The Socioeconomic and Environmental Sciences, *Michael R. Greenberg*

In preparation:

Analysis of Queueing Systems

J. A. White
School of Industrial and Systems Engineering
Georgia Institute of Technology
Atlanta, Georgia

J. W. Schmidt
CBM, Inc.
Cleveland, Ohio

G. K. Bennett
Department of Industrial Systems Engineering
University of South Florida
Tampa, Florida

ACADEMIC PRESS   New York   San Francisco   London   1975
A Subsidiary of Harcourt Brace Jovanovich, Publishers

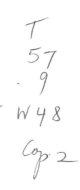

COPYRIGHT © 1975, BY ACADEMIC PRESS, INC.
ALL RIGHTS RESERVED.
NO PART OF THIS PUBLICATION MAY BE REPRODUCED OR
TRANSMITTED IN ANY FORM OR BY ANY MEANS, ELECTRONIC
OR MECHANICAL, INCLUDING PHOTOCOPY, RECORDING, OR ANY
INFORMATION STORAGE AND RETRIEVAL SYSTEM, WITHOUT
PERMISSION IN WRITING FROM THE PUBLISHER.

ACADEMIC PRESS, INC.
111 Fifth Avenue, New York, New York 10003

United Kingdom Edition published by
ACADEMIC PRESS, INC. (LONDON) LTD.
24/28 Oval Road, London NW1

Library of Congress Cataloging in Publication Data

White, John A   (date)
    Analysis of queueing systems.

    (Industrial engineering and operations research
series)
    Includes bibliographies and index.
    1. Queuing theory.   I. Schmidt, Joseph William,
joint author.   II. Bennett, George Kemble, joint author.
III.   Title.
T57.9.W48   519.8'2   75-6860
ISBN 0–12–746950–8

PRINTED IN THE UNITED STATES OF AMERICA

80 81 82   9 8 7 6 5 4 3 2

CONTENTS

Preface ix

1. Introduction

Why This Book? 1
Conducting a Systems Analysis 2
Where to from Here? 16
References 17

2. Probability Theory and Transform Methods

Probability Theory 18
Transform Methods 50
Summary 78
References 78
Problems 78

3. Poisson Queues

Introduction 85
Infinite Population Models: $(M\,|\,M\,|\,c):(GD\,|\,N\,|\,\infty)$ Queue 89
Finite Population Models: $(M\,|\,M\,|\,c):(GD\,|\,K\,|\,K)$ Queue 113
Bulk Arrivals: $(M^{(b)}\,|\,M\,|\,c):(GD\,|\,\infty\,|\,\infty)$ Queue 120
Network of Poisson Queues 124
Summary 133
References 134
Problems 134

4. Non-Poisson Queues

Introduction	143
Pollaczek–Khintchine Formula	144
Method of Stages	150
Numerical Solution of Steady-State Balance Equations	174
Summary	177
References	178
Problems	178

5. Decision Models

Introduction	182
Classical Optimization	184
Search Techniques	193
Cost Models	207
Aspiration Level Models	226
Cost Determination	231
Summary	238
References	239
Problems	240

6. Transient Analysis and Special Topics

Introduction	245
$(M\mid G\mid 1):(GD\mid \infty \mid \infty)$ Queue	246
Busy Period	252
$(GI\mid M\mid 1):(GD\mid \infty \mid \infty)$ Queue	255
Priority Service Disciplines	257
Transient Analysis	265
$(M\mid M\mid \infty):(GD\mid \infty \mid \infty)$ Queue	272
Summary	291
References	292
Problems	292

7. Data Analysis—Estimation

Introduction	297
Identifying the Distribution	298
Point Estimation	302
Goodness-of-Fit Tests	330
Interval Estimation	344
Summary	355
References	355
Problems	355

Contents   vii

8. Data Analysis—Hypothesis Testing

Introduction	360
Null and Alternative Hypotheses	361
Type I and Type II Errors	361
Sample Size	362
Tests for a Single Parameter	363
Tests for the Comparison of Two Parameters	380
Effects of Nonnormality	393
Other Statistical Tests	394
Charting Techniques	395
Summary	409
References	409
Problems	410

9. Simulation of Queueing Systems

Introduction	414
Simulation Modeling	415
Monte Carlo Method	418
Generation of Uniformly Distributed Random Numbers	421
Process Generation of Continuous Random Variables with Known Density Functions	422
Process Generators for Discrete Random Variables with Known Probability Mass Functions	426
Empirical Process Generators	428
Simulation of a Single-Channel Queueing System $(G\,|\,G\,|\,1):(FCFS\,|\,\infty\,|\,\infty)$	435
Multiple Channels in Parallel $(G\,|\,G\,|\,C):(FCFS\,|\,\infty\,|\,\infty)$	450
Simulation of Networks of Queues	490
Summary	506
References	506
Problems	506

Appendix. Tables   515

Index   529

PREFACE

This book is designed for the individual who is interested in the analysis of management problems. While introductory in nature, it requires a background of quantitatively oriented programs within business administration, engineering, management science, operations research, systems engineering, computer science, and transportation. A prerequisite knowledge of differential and integral calculus is assumed. The reader having a background in probability theory can omit Chapter 2, with the exception of the material on Laplace transforms and probability generating functions.

Our interest is in the *applied* problem of analyzing queueing systems. We set as our objective the development of the reader's ability to carry out such analyses from problem definition to implementation and follow-up of the solution. As a secondary objective we wish to achieve a general improvement in the reader's ability to build models of physical systems.

We believe the major contribution of this book to be a unified treatment of the analysis of queueing systems. Our presentations on data analysis, cost modeling, and simulation are not normally encountered in a book on queues. We believe that queueing problems are more plentiful than and offer as much potential for improvement as, say, inventory problems. Although there exist a number of excellent texts dealing with inventory *systems*, queueing texts are concerned primarily with the theory of queues, as opposed to the analysis of queueing *systems*. We have concentrated on applied queueing problems, and have presented systems analysis as it relates to the study of queueing systems.

If used in a course, the instructor will find that this book is designed in such a way that he can select that material consistent with the objectives of the course and background of the student. The contents can be covered in either one or two courses. If two courses are to be taught, we suggest that the coverage be as follows: Chapters 1–5 for the first course and Chapters 6–9 for the second course. If only one course is to be taught, we suggest Chapters 1, 3, 4, 5, and 6 or 9 be covered.

This book has evolved from our combined experiences in both solving real-world queueing problems and teaching the subject in both undergraduate and graduate courses. Undoubtedly, the resulting manuscript has been influenced by the writings of and/or the associations with U. Narayan Bhat, Robert B. Cooper, D. R. Cox, Ralph L. Disney, Walter C. Giffin, Frederick S. Hillier, A. M. Lee, Philip M. Morse, Marcel F. Neuts, A. Alan B. Pritsker, Thomas L. Saaty, Walter L. Smith, and Hamdy A. Taha. We gratefully acknowledge the contributions they have made to our understanding of the study of queues. Additionally, the many graduate students who, during the past five years, studied from various versions of the manuscript are acknowledged for their many helpful suggestions. A special note of appreciation is given to Marvin H. Agee at Virginia Polytechnic Institute and State University, who provided the support and encouragement necessary for the completion of the manuscript when we were teaching there. Finally, we wish to express our appreciation to our wives Mary Lib, Joan, and Cindy for not balking and/or reneging while waiting in line for us to complete the manuscript.

Chapter 1

INTRODUCTION

WHY THIS BOOK?

Waiting lines, or queues, seem as inevitable a consequence of modern-day life as death and taxes. In fact, it is quite likely that you had to wait in line during the process of purchasing this book. It is our purpose to organize and present some of the methods of analysis appropriate for the study of systems involving queues. How well we achieve our objective will, in large measure, determine whether your wait in line (and purchase price) was a worthwhile investment.

Queueing problems have received the attention of management and operations analysts for a number of years. In fact, there exists an overwhelming body of literature devoted to the study of queues. Much of this literature involves analytical treatments of very narrowly defined problems. As an example, one can find literally hundreds of papers dealing with some mathematical aspect of a single-server queue having random arrivals and random service. On the other hand, there has been very little emphasis given to systems analysis of real-world queueing problems. We feel there exists a need for such a treatment. It is to this problem that this book is addressed. We are concerned not only with the development of mathematical models, but with the associated problems of *modeling*. Specifically, how does one go about the business of analyzing queueing systems? How should the data be collected and analyzed? How is the model developed? How is the system to be improved? Should an analytic or simulation model be used? How sensitive is the model? These are some of the questions that we hope you can answer after reading this book. Hopefully, in the text we have provided you with the materials necessary to solve your particular queueing problem.

At the outset it should be made perfectly clear that this book is designed for the analyst (or analyst-to-be). We omit some of the more sophisticated models that are available, since this is not a review of the queueing literature. On the other hand, we have made a judicious selection of those models that we feel will extend your capability and give you the confidence to address real-world problems of significance. Furthermore, once you have mastered the contents of this book, you should be equipped to read a large portion of the more sophisticated queueing literature. For the most part, we have presented models that have withstood the test of time. Thus, we feel that these models offer considerable potential for application.

CONDUCTING A SYSTEMS ANALYSIS

In conducting an analysis of a queueing system we recommend a sequence of steps similar to the following:

1. Identify the system
2. Formulate the problem
3. Analyze the problem
4. Specify alternate solutions
5. Construct models
6. Solve the models
7. Test and validate the models
8. Specify the solution
9. Implement the solution
10. Follow up

1. Identify the System

The first step in the analysis of queueing systems is to identify the system under study. More specifically, we wish to place boundaries on the problem. A queueing system consists of three basic elements: customers, servers, and randomness. As depicted in Fig. 1.1, a queueing system is defined as the collection of activities and events associated with providing service to an arriving customer. Depending on the particular application, a customer may require service from a number of servers and may enter a number of different waiting lines before finally departing from the system.

The terms *customers, server*, and *service facility* can take on very general interpretations. A customer can be a person, machine, part, book, requisition, etc. The customer is the item that requires the attention of the server. Likewise, a server can be a person, machine, or other less obvious mechanism, which satisfies the demands of the customer. The service facility is the interface between the customer and the server; obviously, there need not be a physical service facility.

During the course of a day, if you attempted to list all of the queueing problems you encountered, it is quite likely that you would either be hoarse or have writer's cramp. Obviously, not all of these problems are worthy of

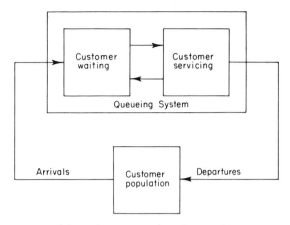

Fig. 1.1 Schematic representation of a queueing system.

analysis. In most cases, the potential payoff is not sufficient to offset the cost of the analysis. Such payoffs become substantial when either the number of similar problems is large or the dollar expenditures are large for a single problem.

Let us consider a system with which you are, no doubt, familiar—a university. Symptomatic of queueing systems are waiting lines. Think for a moment about the waiting lines that exist within a university. The number is overwhelming. Let us reduce the system to that of a university library. How might we analyze a library system? Customers arrive to check out books, books are returned for shelving, requests for new books arrive, new books arrive for cataloging, customers arrive at the card catalogs, new cards arrive to be filed in the card catalog, requests arrive to place books on reserve, customers arrive to use reserve books, ... The list of queueing problems seems endless. If we further restrict our specification of the system, we might arrive at a study of the check-out desk within a university library.

The identification of the queueing system is extremely important. The decisions made and actions taken in subsequent steps of the analysis are critically dependent on this initial decision.

2. Formulate the Problem

System identification and problem formulation go hand in hand. Careful consideration must be given to the formulation of the problem. As an example, a major food processor was faced with the problem of having a large number of fields of peas ripening at a rate that exceeded the processing capacity. Under the existing system, it was not physically possible to harvest the peas, transport them to the processing plant, and process the peas in the time necessary to maintain the required quality standards for the product. A number of unsuccessful attempts were made to speed up the process of harvesting and processing of the peas. Finally, rather than concentrate on shortening the service time, the problem was formulated as determining the optimum arrival rate for the customers (peas). The problem was solved by staggering the planting to achieve a staggered ripening of the peas. Thus, the problem was not with the service rate, but with the arrival rate.

In some cases the formulation of the problem is obvious. In fact, it may be formulated for you; e.g., "Go over to the storeroom and determine the number of attendants needed to man the counter." In such a case, care should be taken to assure that the proper problem is formulated. Indeed, rather than having the problem formulated correctly for you, you will often be confronted with a situation and you must proceed from there. Typically, the situation is the present solution to the problem.

The formulation of the problem is aided by taking a "black-box" approach (Krick, 1965). Specifically, there is an originating state of affairs (state A) and a desired state of affairs (state B). There is a transformation that must take place in going from state A to state B, as shown in Fig. 1.2.

Fig. 1.2 Black-box approach.

There is more than one method of accomplishing the transformation from A to B, and there is unequal preferability of these methods (otherwise no problem exists). The solution to the problem is visualized as a black box of unknown, unspecified contents, having input A and output B.

The black-box approach facilitates a proper identification of states A and B during problem formulation. As an illustration, recall the check-out counter in the university library. The problem could be formulated in terms of states A and B as follows:

State A	State B
1. Customer at counter desiring book	Customer leaving with book
2. Customer entering library desiring book	Customer leaving with book
3. Customer at office (home) desiring book	Customer at office (home) with book
4. Customer desiring information	Customer with information

Conducting a Systems Analysis

The first formulation is very narrow. Using it, we would concentrate our attention on the improvement of the check-out process and the number of servers at the counter. In the second formulation, we would broaden our thinking and also consider alternate ways of getting the desired book for the customer. The third formulation allows consideration of ways of satisfying the borrower's demands without his necessarily coming to the library. Finally, the fourth formulation allows us to eliminate the book as the method of transferring information and the library as the storage location for the information. No doubt there are other formulations having varying degrees of breadth.

This discussion should have motivated the question of just how broadly the problem should be formulated. In general, a problem should be formulated as broadly as the economics of the situation and organizational boundaries permit. Specifically, as suggested by Krick (1965) the formulation of the problem should involve a brief, general description of the characteristics of the problem. It should be free of detail and restrictions, and should concern at least states A and B, and perhaps the major criteria, volume of the activity anticipated and a time limit within which the problem is to be solved.

It is pointless to spend time and money on the improvement of a queueing system that should be eliminated. The analyst leaves himself open to this possibility when he immerses himself in the details of the present solution. By spending some thought on problem formulation at the outset of the analysis, the analyst can save himself possible embarrassment in the future.

Rarely will the true problem be placed before you. Rather, you must attempt to see through the present solution, tradition, and opinions of others, to determine for yourself what the problem actually is. This process has been made more difficult by the practice of presenting problems to students in an unrealistically pure form. Thus, when the student finally confronts "real-world" queueing problems, which are not so pure, he often forces the problem to fit one of the pure forms encountered in school. In this book, we have tried to avoid this by emphasizing alternate approaches and by offering a number of problems of varying degrees of complexity and clarity at the end of each chapter.

It is interesting to note that the very presence of waiting lines is sometimes interpreted as an indication of the existence of a problem. Such an assumption may well be in error. We will find in subsequent discussions that the *absence* of waiting lines may be an indication of the existence of a problem. What is often desired is a proper balance between server idleness and customer idleness.

3. Analyze the Problem

The analysis of the problem, which has been previously formulated, consists of a relatively detailed phrasing of the characteristics of the problem, including restrictions. This phase of the system study involves considerable "fact gathering." You must be extremely careful in making decisions as to what constitutes real restrictions of the system and what constitutes fictitious restrictions. Furthermore, you must identify the appropriate criteria to be used in evaluating alternative solutions to the problem.

The process of fact gathering will bring you face to face with the present solution to the problem. Consequently, you should make every effort to avoid becoming biased in your thinking. Preoccupation with the present solution often results in solutions to the problem that are merely slight modifications of the present solution. You must cultivate the ability to free your thinking from the present solution, to see through it, and to identify the problem properly.

In analyzing queueing problems we are concerned with the following four elements: arrival population, service population, queue discipline, service discipline.

Arrival population. The size of the customer population must be specified. When the number of customers in the population is sufficiently large such that the probability of a customer arriving for service is not significantly affected by the number of customers already at the service facility, we refer to the population size as being infinite. Otherwise, a finite population is assumed and its size specified.

In subsequent discussions we will find that when the population size is assumed to be infinitely large, we are interested in developing the probability distribution for the time between consecutive arrivals of customers (demands for service). A finite population of customers gives rise to the development of the probability distribution for the time between service $k - 1$ being completed and service demand k occurring *for a particular customer*. We expand on this in Chapter 3.

On analyzing the arrival population, we may find that arrival times are constant, rather than random. If arrival times are random, the probability distribution developed from our analysis will be either discrete valued or continuous. We might find that the arrival-time distribution is changeable over time. For example, more customers may arrive, on the average, on Mondays and Fridays than on Tuesdays, Wednesdays, and Thursdays.

In addition to determining the size of the population we wish to know if each customer in the population can be considered to be identical, or if some customers are significantly different from others. We will find that different customer classes often arise because of different service-time and/or arrival-

Conducting a Systems Analysis

time distributions, different priorities of service, different acceptances of waiting lines (balking and reneging), etc. Explicit consideration of significant differences in the customers within a population is quite important. Some of these differences will be examined in discussions of other elements of the queueing process.

We must also determine whether arrivals occur singly or in bulk. If bulk arrivals occur, we must decide if the number is fixed or dependent on, say, the number of servers available or the number of customers already in the system. In the case of bulk arrivals, we must also determine if the number of customers per arrival is deterministic or probabilistic.

Service population. The service population concerns the number of servers and their arrangement. The number of servers can also be considered finite or infinite. If finite, one must know if there is a single server or if there are multiple servers. If there are multiple servers, are they arranged in parallel or in series? Additional questions to be answered are:

1. Is the number of servers fixed or is it a function of the number in the system?
2. Is the number of servers deterministic or random?
3. Are the servers identical?
4. Do all *customers* have the same service-time distribution?
5. Do all *servers* have the same service-time distribution for the same *customer*?
6. Is the service-time distribution a known distribution?
7. Is the service-time distribution stationary or nonstationary?

Queue discipline. The queue discipline concerns the behavior of customers in the waiting line. Questions to be answered concerning the queue discipline are:

1. Do some customers refuse to join the waiting line due to its length?
2. Do customers join a common waiting line from which customers go directly to be served or do customers form separate queues in front of each server?
3. Do customers jockey back and forth between lines?
4. Do some customers renege, i.e., leave the system after entering the line, but before being served?
5. If customers renege, is such behavior deterministic or random, and can it be reasonably predicted?
6. Is there a limit on the number of customers that can be in the system? If so, is this limit dependent on the number of servers, and is it deterministic or random?

Service discipline. The manner in which customers are served is referred to as the service discipline for the queueing problem. Typical questions to be asked in analyzing the service discipline are:

1. Are customers served singly or in bulk?
2. If bulk service is used, is the number served deterministic or random, and is the number dependent on the number of customers in the system?
3. Are customers processed on a first come–first served, a random, or a priority basis?
4. If priorities are used, is a preemptive or nonpreemptive priority employed? If preemptive, is service resumed or repeated?
5. If priorities are used, how many priorities are there?

In obtaining answers to the above questions, you must distinguish between "what is" and "what must be." You may find that customers arrive randomly when scheduled arrivals are possible. You may observe that service time is constant and, say, equal to 8 minutes. However, by improving the method of providing service, it might be possible to reduce service time to, say, 5 minutes. You must keep in mind that a large number of the characteristics of the present system are probably due to evolution, rather than design. Replacement of human servers with mechanized servers (or vice versa) might be desirable. Additionally, it might be beneficial to adopt a priority service scheme for meeting customer demands.

The above discussion emphasizes the fact that a large number of different variations of queueing problems exists. As Wagner has put it so succinctly, "What is said about birds is also true of waiting line models: their variety and number seem infinite" (Wagner, 1969, p. 838). Even though the number of distinct problems is large, we will attempt to bring order to the discussion of queueing problems by adopting the classification scheme initially proposed by Kendall (1953), and subsequently extended by Lee (1966) and Taha (1971). The classification scheme is given as

$$(x \mid y \mid z) : (u \mid v \mid w)$$

where the symbols are as follows:

- x arrival (or interarrival) distribution
- y departure (or service-time) distribution
- z number of parallel service channels in the system
- u service discipline
- v maximum number allowed in the system (in service plus waiting)
- w size of the population

The following codes are commonly used to replace the symbols x and y:

- M Poisson arrival or departure distributions (or equivalently exponential interarrival or service-time distributions; M refers to the Markov property of the exponential distribution)
- GI general independent distribution of arrivals (or interarrival times)
- G general distribution of departures (or service times)
- D deterministic interarrival or service times
- E_k Erlangian or gamma interarrival or service-time distribution with k phases
- K_n chi-square interarrival or service-time distribution with n degrees of freedom
- HE_k hyperexponential interarrival or service-time distribution with k phases.

The symbols z, v, and w are replaced by the appropriate numerical designations. The symbol u is replaced by a code similar to the following:

- $FCFS$ first come–first served
- $LCFS$ last come–first served
- $SIRO$ service in random order
- SPT shortest processing (service) time
- GD general service discipline

Additionally, a superscript is attached to the first symbol if bulk arrivals exist and to the second symbol if bulk service is used. To illustrate the use of the notation, consider $(M^{(b)} \mid G \mid c) : (FCFS \mid N \mid \infty)$. This denotes exponential interarrival times with b customers per arrival, general service times, c parallel servers, first come–first served service discipline, a maximum allowable number of N in the system, and an infinite population.

Although the designation does not allow us to classify series and networks of queues, balking of customers, state-dependent service times, and a number of other variations of queueing systems, it does prove helpful in the majority of cases considered in this book. Furthermore, the designation is commonly used in the queueing literature. So, it is a good idea to become familiar with its usage.

4. Specify Alternate Solutions

On the basis of a detailed analysis of the problem, you should be in a position to generate a number of alternate solutions to the problem. The desired solution is a system design. At this phase of the analysis alternate contents of the black box employed in formulating the problem are to be specified.

Every effort should be made to be creative and to divorce one's thinking from the present solution. A number of aids to creativity exist, which improve one's ability to generate more and better ideas. The following are given by Krick (1962):

1. Exert the necessary effort.
2. Do not get bogged down in details too soon.
3. Make liberal use of the questioning attitude.
4. Seek many alternatives.
5. Avoid conservatism.
6. Avoid premature rejection.
7. Avoid premature satisfaction.
8. Refer to analogous problems for ideas.
9. Consult others.
10. Attempt to divorce your thinking from the existing solution.
11. Try the group approach.
12. Remain conscious of the limitations of the mind in this process of idea generating.

The success of your analysis is dependent on the successful generation of alternate solutions. Since the final solution to the problem must come from the set of alternatives generated, you should make every attempt to maximize the number, quality, and variety of alternate solutions.

Among the many alternate solutions to a queueing problem there are usually some that are relatively complex and some that are refreshingly simple, but no less effective than their complex counterparts. Since the preferred solution must be implemented, simple solutions are usually favored. A simple solution is especially important when the servers are people, rather than machines.

You should keep in mind that a given system might consist of several subsystems. In such a situation, an aid in specifying alternate solutions that has proven successful is to consider eliminating, combining, and simplifying subsystems. Additionally, consideration should be given to controlling arrivals and reducing the service time.

5. Construct Models

Once a set of alternate solutions has been generated, the alternatives must be evaluated. In the case of queueing systems, we frequently find that alternate designs can be best evaluated by constructing a model of each of the proposed systems and evaluating the performance of the models.

In evaluating alternate system designs we will be interested in gaining insights into cause–effect relationships of the system. Since the present solution (if such exists) will probably be contained in the set of alternate solu-

Conducting a Systems Analysis

tions, modifications could be made to the existing system and the performance of the system evaluated. However, such modifications may be very expensive and may have disruptive influences on the organization. Furthermore, there may not exist a present solution. In this case, such manipulation of a nonexistent physical system is not possible. Therefore, we find it desirable to develop a model of the system and analyze the model. In doing this, we must keep in mind that the results of our analysis must be tempered with judgment since a *model* of the system, rather than the system, is studied.

Models can be classified as being either *iconic*, *analog*, or *symbolic* (Elmaghraby, 1968). Iconic models are scalar representations of the system elements in that they look like the system represented. As an example, a doll house is an iconic model of a house. Thus, if a service station is to be analyzed using an iconic model, a toy service station could be used with toy cars arriving for service. Iconic models are used primarily to assist the analyst in visualizing the interactions within the system, as well as presenting alternate solutions to a problem.

Analog models substitute one property for another. After the problem is solved in the substituted state, the solution is translated back to the original dimensions or properties. Some queueing systems can be analyzed accurately by developing an electrical, mechanical, or hydraulic analog of the system.

Symbolic or mathematical models are an abstract representation of a system and are more general than other types of models. We will be concerned primarily with the use of mathematical models of queueing systems.

The classification of models as being iconic, analog, and symbolic is neither mutually exclusive nor collectively exhaustive. Some models have the properties of both analog and symbolic models. Additionally, we find that some models are referred to as *simulation* and *heuristic* models.

A simulation model is a synthetic representation of a physical system. It yields the output of a system, given the input to its interacting subsystems. A simulation model can be characterized as an "if . . . , then . . ." device, i.e., *if* a certain input is specified, *then* the output can be determined. Some of the major reasons for employing simulation are (Schmidt and Taylor, 1970):

1. Simulation is useful either when the problem cannot be solved mathematically or when it cannot be solved mathematically without great difficulty.
2. Simulation is useful in selling a system modification to management.
3. Simulation allows complete control over time; i.e., the simulation can "speed up" or "slow down" the phenomenon under study, as desired.

4. One simulation model can supply the information that can only be obtained from several analytical models.

5. Simulation can be used as a verification of analytical solutions.

6. Simulation may be the only possible way of experimentation due to the difficulty of experimentation and observation with the real-world system.

A heuristic model is one that uses rules-of-thumb or strategies that are intuitively appealing, but are not guaranteed to produce optimum solutions. It differs from an optimum producing model, since no claim of optimality can be made. Heuristic models can be incorporated in a simulation model. The search techniques we describe in Chapter 5 are examples of heuristic models.

Models may be further classified as being *descriptive* or *prescriptive* (*normative*). A descriptive model is used to describe the behavior of the system. Queueing models are descriptive models. Prescriptive models are used to prescribe a course of action that, in some sense, is optimal. Linear and nonlinear programming models are examples of prescriptive models. In a subsequent discussion we will use descriptive queueing models to obtain inputs for a prescriptive model. The choice between descriptive and prescriptive models is dependent on the criterion specified in the analysis of the problem.

In analyzing a complex queueing system a combination of two or more models will often be required. The models developed can be simple or complex, depending on the objectives of the analysis. There exist tradeoffs between (a) the degree to which the model represents the system under study, (b) the ease of manipulation of the model, (c) the cost of obtaining a solution from the model, and (d) the clarity of cause–effect relationships. If a model is to have a very high degree of realism, it will probably be difficult to manipulate, costly to solve, and difficult to ascertain cause–effect relationships. On the other hand, if the model is simple to solve and manipulate and has high clarity of cause–effect relationships, it may have a low degree of realism.

Because of the low cost of solution, simple models are usually developed in the initial stages of systems analysis. Furthermore, simple models will serve to capture a number of the interesting aspects of queueing systems. As we will see in later discussions, many interesting problems can be solved using relatively simple queueing models. Additionally, a simple model can be used to determine the feasibility of more detailed analysis. Thus, we see that there are many practical arguments for developing simple representations of queueing systems.

Many times the system studied will consist of a number of interacting subsystems. Each subsystem might be a separate queueing system. However, the subsystems might not be independent and suboptimization might result if the subsystems are analyzed without regard to their system interactions.

Conducting a Systems Analysis

We are interested in improvement of the total system, not just components of the system. Thus, we should take a total systems approach in solving the problem. However, the overall system might be so complex as to preclude solution. Therefore, we often resort to suboptimization and use simple models to analyze subsystems as though they were independent of the other components of the system. This approach is referred to as a "component" approach, rather than a systems approach.

Based on the insight gained from the simple model, subsequent analysis might involve the use of more complex models. Our treatment of queueing models will follow the component approach. We will be concerned initially with some very simple models. As we build our confidence in developing simple models, we will gradually enrich the analysis until our models have reached a fairly high level of sophistication. In order to extend our analysis further, we move from a discussion of analytic models to a study of simulation models.

Even though our discussion will be oriented toward a discussion of queueing models, you should keep in mind the important distinction between *models* and *modeling*. We are not suggesting that you take a model that we describe and use it to solve a particular queueing problem. Rather, you should analyze the queueing system, identify the problem, and then develop a model to solve the problem. The "development" of the model may well involve the application of an existing model. The distinction to be made is that the model should be selected based on the characteristics of the problem, rather than the reverse. We do not want to be guilty of slotting the heads of nails so we can use our new screwdriver, when a hammer is the appropriate tool.

6. Solve the Models

Once the models have been constructed, they must be solved. The method of solution is, of course, dependent on the type of model formulated. If a simulation model has been constructed, a number of simulation runs are performed and the performance of the system noted. Detailed discussion is given to simulation models in Chapter 9. If a descriptive mathematical model is developed, it can normally be solved using standard mathematical techniques. Some prescriptive mathematical models can be solved analytically; others must be solved numerically. Quite likely you will find that the solution of the models is the simplest step to perform in the analysis of queueing systems.

In the case of prescriptive models, we are interested in obtaining an optimum or best solution. However, since the model is necessarily an idealization rather than an exact representation of the real system, we can make no claim of optimality for the system. We do expect that the solution obtained will be a "good" solution and, using the terminology of March and Simon (1958), will "satisfice." Indeed, our reliance on model construction is based on the pragmatic view that a better answer can be obtained through analysis than would otherwise be obtained. This is the true test of our thesis.

7. Test and Validate the Models

A critical step in the use of models is model verification. Due to the type of problems we are studying we can never develop a model that is an *exact* representation of the physical system. Thus, there will be some deviation in the response of the model to a stimulus and the response of the physical system to that same stimulus. The important question is whether or not that deviation is significant. The answer depends, of course, on how we decide what is and is not significant. Indeed, what system response is the appropriate one to be measured and compared? The question of validation is not an easy one to answer. One can be led into some very involved philosophical issues in model verification.

Since we are directing our discussion to the systems analyst, we will adopt an operational approach in testing model validity. Specifically, we will subject the model to certain input conditions and compare the resulting output with the output from the actual system. If a physical system does not exist, we will compare the outputs from two separate models of the system. Although the system response(s) we select may be arbitrarily chosen, we tend to accept the model as a valid representation of the system if the difference in the response between the model and system (or other model) is not greater than what might occur due to chance variation.

Even though model validation is a difficult undertaking, it should not be ignored. Since it is the model that is being solved, not the problem, it is critical that the model be a reasonable representation of the system under study.

Once you are satisfied that the model is a valid representation of the system, it should be tested. In particular, the sensitivity of the model to variations in the values of the parameters of the system is to be tested. We are interested in the sensitivity of the model for at least two reasons. First, the parameter values may have been incorrectly estimated. If the model is relatively insensitive, we will not be overly concerned about making such errors. Thus, we can assign values to the parameters without the expense of obtaining more accurate parameter values. Second, the values of the parameters may undergo changes in the future. If the model is insensitive, it will not be necessary to modify the solution to the problem every time a parameter value changes.

Sensitivity analysis is an important step in the use of any model. It is especially important in the evaluation of alternate solutions to queueing problems. We will have more to say about sensitivity analysis in later discussions.

8. Specify the Solution

Once the models have been tested and validated it remains to choose the solution from among the set of alternate solutions that is preferred. This step may not be yours to perform. Rather, you may choose to present two or more alternate solutions to the appropriate decision maker and rely on him to judge the solutions and designate the one that is preferred.

The specification of the solution is not a simple undertaking. There are both long- and short-term considerations to be reckoned with. Also, there are many nonquantifiables to be dealt with. The alternatives must be evaluated with respect to the amount of resistance each will encounter in implementation. Remember, no useful purpose is served by specifying a solution that will not be implemented.

Another important consideration in specifying the solution has been mentioned previously. The solutions that have been obtained are solutions to models, not the actual system. Subjective judgment must be employed to "massage" the solutions obtained into workable solutions of the real-world problem. Recall, our objective is to find a solution that "satisfices." Thus, we wish to find an answer that is good from the viewpoint of the overall organization.

9. Implement the Solution

For many, the study has ended: That is truly unfortunate. There have been many excellent analyses performed that were never implemented. Undoubtedly, there are a number of reasons for this. Systems analysts tend to be problem solvers. As such, many of them bring a great deal of enthusiasm to the analysis until, for them, the problem has been solved. Typically, these analysts turn in their reports to management with some statement to the effect that they will be available to answer any questions concerning the recommendations of the study.

Another reason solutions are not implemented is a poor job of selling. Not only must top, middle, and lower management be sold on the solution, but more critically, the individual worker must be sold. Management may *make* the decisions, but the worker can *break* them if he is not sold.

A third reason implementation fails is that a good systems analyst is in high demand. He will have a backlog of requests to be met. As such, management may pressure him to let someone else handle the implementation so that he will be free to tackle other problems. This is very shortsighted and may result in dissatisfaction with the work of the systems analyst.

Implementation consists of *selling* all persons involved on the solution, *documenting* the procedures to be followed, *training* personnel in the performance of new duties, and *monitoring* the operation of the newly designed system. Remember, a solution not implemented is really not a solution.

We have painted a negative picture concerning implementation. Hopefully, this will serve as motivation to perform all ten steps in the analysis of queueing systems, rather than just the first eight steps.

10. Follow Up

In addition to implementing the preferred solution, control procedures should be devised to monitor the system. Management information systems should be designed to indicate when revisions should be made to the solution. The use of control charts to indicate when arrival rates or service rates have changed significantly is one example of the follow-up action that should be taken. Additionally, cost values might change over time, necessitating modifications in the system.

It is also important for a follow-up study to be conducted after the system has stabilized and adjusted to any modifications that might have been made. The results of this study should be compared with those that were predicted. If serious discrepancies exist between the two, it is an indication that the model is not an accurate representation of the system. The model should be appropriately modified so that it can be used in subsequent analyses with more confidence in its validity.

WHERE TO FROM HERE?

This book is about the analysis of queueing systems. We have outlined a sequence of ten steps, which if properly executed should yield an improved design of the system. In the remaining chapters of this book, we will concentrate mainly on the fifth, sixth, and seventh steps in the sequence. It is our belief that the major ingredients for successful execution of the first four steps are an open mind and creativity. The last three steps require an adherence to purpose and ability to work closely with people. The steps we have selected for emphasis are those that rely on your having certain quantitative skills. We set as our objective the development of these skills in you. So, from here we go into quantitative aspects of queueing systems analysis.

In Chapter 2 we develop the necessary background in probability theory and transform methods. The development of queueing models begins by treating some very simple cases in Chapter 3. In Chapter 4, we show how the simple models developed in Chapter 3 can be applied in more complex situations. The development of prescriptive models is the subject of Chapter 5. Although the emphasis is on the development of cost models, other objectives are also considered. Chapter 6 presents a treatment of transient analysis, as opposed to the steady-state treatment in previous chapters. Also presented in Chapter 6 are modeling techniques for use in analyzing more complex queueing systems than are treated in Chapter 4. The design of data collection systems and the analysis of data are discussed in Chapters 7 and 8. The development of simulation models is treated in Chapter 9.

REFERENCES

Elmaghraby, S. E., The role of modeling in I.E. design, *J. Industr. Eng.* **19**, 292, 1968.
Kendall, D. G., Stochastic processes occurring in the theory of queues and their analysis by means of the imbedded Markov chain, *Ann. Math. Statist.* **24**, 338–354, 1953.
Krick, E. V., *Methods Engineering.* New York: Wiley, 1962.
Krick, E. V., *An Introduction to Engineering and Engineering Design.* New York: Wiley, 1965.
Lee, A., *Applied Queueing Theory.* New York: Macmillan, 1966.
March, J. G., and Simon, H. A., *Organizations.* New York: Wiley, 1958.
Schmidt, J. W., and Taylor, R. E., *Simulation and Analysis of Industrial Systems.* Homewood, Illinois: Irwin, 1970.
Taha, H. A., *Operations Research: An Introduction.* New York: Macmillan, 1971.
Wagner, H. M., *Principles of Operations Research.* Englewood Cliffs, New Jersey: Prentice-Hall, 1969.

Chapter 2

PROBABILITY THEORY AND TRANSFORM METHODS

PROBABILITY THEORY

The discussion in Chapter 1 indicated that, by their nature, queueing systems involve arrivals that demand a service. In most systems of this type, the precise time at which any arrival will occur is unpredictable. Service times may be similarly unpredictable. Consider, for example, a barber shop with three barbers. Suppose that the barbers wish to determine the number of chairs to be provided for waiting customers. To resolve this problem it would be useful to know something about the time between customer arrivals and the time required per haircut. However, neither of these time intervals can be predicted with certainty and this uncertainty should be reflected in the analysis of the problem. In this chapter we discuss ways to describe these uncertainties.

Basic Concepts of Probability

Suppose that we were to place a counter at a given point on a highway. The counter records the number of vehicles passing that point during intervals of length one minute. Each one-minute interval may be viewed as the period of time required to conduct an experiment that consists of counting the number of vehicles passing the point at which the counter is located. We will refer to the result or findings of an experiment as an *outcome*. In this case the outcome of any experiment is a nonnegative integer representing the

Probability Theory

number of vehicles passing the specified point on the highway. Thus the result of every experiment is an outcome. The set of all outcomes that can occur at any single trial of an experiment is called the sample space. In the example we are considering here, the sample space is the collection of nonnegative integers.

The experiment described above was such that we could define the set of outcomes, any one of which might occur at a single trial of the experiment. However, it is unlikely that we would be able to predict with certainty the outcome of a specific trial of the experiment and, in this sense, the experiment may be considered one of chance. In this case it would be natural to associate a nonnegative integer with each outcome. That is, an x (integer, ≥ 0) recorded at a given trial of the experiment would indicate that x vehicles passed the location of the counter during that trial. When we associate a real number with each outcome or with sets of outcomes of an experiment, this number is called a random variable. More specifically, a *random variable* is a real-valued function defined on the sample space such that to each possible outcome of the experiment a real number is assigned. Let X be the random variable denoting the number of vehicles passing the counter. Then $X = 0$ if no vehicles pass the counter at a given trial, $X = 1$ if one vehicle passes the counter at a given trial, and so forth.

Suppose that instead of the number of vehicles passing a point on a highway in a one-minute interval, we were interested in the time taken for a vehicle to traverse a segment of the highway. In this case, an experiment starts when a vehicle enters the highway segment and terminates when the vehicle leaves the segment. The sample space for the experiment consists of all time intervals $t > 0$. The random variable of interest is a positive real number that corresponds to the time taken to traverse the highway segment measured in whatever units are desired.

In the two examples discussed above, there was a natural correspondence between the outcome of the experiment and the value of the number associated with that outcome. Such an obvious association does not always occur. To illustrate, suppose that an experiment consists of inspecting a unit of product. As a result of the experiment we simply record "good" or "bad" to designate the acceptability of the unit. Therefore, the sample space consists of the outcomes "good" and "bad." However, we have not assigned a real number to these outcomes, thus defining an associated random variable. We might assign the number -5 to "bad" and $+5$ to "good," or 137.2 to "bad" and -8.1 to "good," or any other pair of numbers to "bad" and "good" such that the distinction between the outcomes "bad" and "good" is maintained. The usual approach would be to assign 0 to one outcome and 1 to the other. Trials in which there can only be two distinct outcomes are often referred to as Bernoulli trials.

Example 2.1 An experiment consists of recording the source of demands for a particular service during a period of one day. Demands may come from three sources, A, B, or C. Define the sample space and a random variable corresponding to the possible outcomes.

Let A, B, and C represent the occurrence of demands from the three sources, respectively, and $\bar{A}$, $\bar{B}$, $\bar{C}$ the absence of each. The set of possible outcomes—the sample space—is defined by

Outcome definition	Demand condition
$\bar{A}\bar{B}\bar{C}$	None
$A\bar{B}\bar{C}$	From A only
$\bar{A}B\bar{C}$	From B only
$AB\bar{C}$	From A and B
$\bar{A}\bar{B}C$	From C only
$A\bar{B}C$	From A and C
$\bar{A}BC$	From B and C
ABC	From A, B, and C

Let X be the random variable associated with the sample space described in the above table. The function associating each outcome in the sample space with a set of real numbers is given by

X	Outcome definition
0	$\bar{A}\bar{B}\bar{C}$
1	$A\bar{B}\bar{C}$
2	$\bar{A}B\bar{C}$
3	$AB\bar{C}$
4	$\bar{A}\bar{B}C$
5	$A\bar{B}C$
6	$\bar{A}BC$
7	ABC

The reader will note that any other definition of the random variable X could have been used, provided the distinction among outcomes was maintained.

In Example 2.1 suppose that we are interested in the number of sources from which demands emanated, rather than the specific source of the demand. In this case we are concerned with four sets of outcomes of the experiment: $\bar{A}\bar{B}\bar{C}$ (0 demands); $A\bar{B}\bar{C}$, $\bar{A}B\bar{C}$, $\bar{A}\bar{B}C$ (1 source of demand); $AB\bar{C}$, $A\bar{B}C$, $\bar{A}BC$ (2 sources of demands); and ABC (3 sources of demands). Each of these subsets of the possible outcomes is called an *event*. The grouping of outcomes into events is dependent on the experimental results of interest to the analyst. In the case just considered, we were concerned with the number of sources demanding service, and four events resulted: 0 demands ($\bar{A}\bar{B}\bar{C}$), 1 source of demand ($A\bar{B}\bar{C}$, $\bar{A}B\bar{C}$, $\bar{A}\bar{B}C$), 2 sources of demands ($AB\bar{C}$, $A\bar{B}C$, $\bar{A}BC$), and 3 sources of demands (ABC). On the other hand, we might be concerned merely with whether or not any demand for the service occurred at all. In this case two events may occur as a result of the experiment: no demand ($\bar{A}\bar{B}\bar{C}$), and one or more demands ($A\bar{B}\bar{C}$, $\bar{A}B\bar{C}$, $\bar{A}\bar{B}C$, $AB\bar{C}$, $A\bar{B}C$, $\bar{A}BC$, ABC). Of course, the events of interest may correspond to the individual experimental outcomes themselves, rather than a grouping of these outcomes.

The experiments we have discussed have been referred to as experiments of chance since the outcome of each is uncertain. Although the outcome of a given chance experiment cannot be predicted with certainty, the likelihood of occurrence of each of the possible outcomes can be measured or at least estimated. To illustrate, suppose that the same experiment is conducted N times. At each trial of the experiment any one of m outcomes may occur. Let n_i be the number of times the ith outcome occurred during the N trials of the experiment, where

$$\sum_{i=1}^{m} n_i = N \tag{2.1}$$

The likelihood or *probability* of occurrence of the ith outcome at a given trial, $P(i)$, can be estimated by n_i/N. The true probability of occurrence of the ith outcome can be determined by repeating the experiment an infinite number of times, that is,

$$P(i) = \lim_{N \to \infty} \frac{n_i}{N} \tag{2.2}$$

and $0 \leq P(i) \leq 1$. Furthermore,

$$\sum_{i=1}^{m} P(i) = \sum_{i=1}^{m} \lim_{N \to \infty} \frac{n_i}{N} = \lim_{N \to \infty} \sum_{i=1}^{m} \frac{n_i}{N} = \lim_{N \to \infty} \frac{N}{N} = 1 \tag{2.3}$$

Now assume that event A occurs whenever outcomes i, j, or k occur. The probability of occurrence of A, $P(A)$, is given by

$$P(A) = P(i) + P(j) + P(k) \tag{2.4}$$

that is, the probability of occurrence of any event is given by the sum of the probabilities of the individual outcomes comprising that event.

Example 2.2 Two dice, one red and one green, are thrown simultaneously. At any throw, the number of dots showing on each die after it comes to rest lies between one and six inclusive. If i is the number of dots showing on the red die and j the number of dots showing on the green die, then an outcome of the experiment consists of the combination of i and j resulting when the dice are cast. Let $P(i, j)$ be the probability that an i appears on the red die and a j on the green die. Then

$$P(i, j) = \tfrac{1}{36}, \quad i = 1, 2, \ldots, 6, \quad j = 1, 2, \ldots, 6$$

Event A occurs whenever the sum of the dots showing is seven, and event B occurs whenever the sum of the dots showing is eleven. Find $P(A)$ and $P(B)$.

Event A occurs whenever the outcome of the experiment is such that $i + j = 7$. Therefore, A occurs when the outcome of the experiment is (1, 6), (2, 5), (3, 4), (4, 3), (5, 2), or (6, 1), and

$$P(A) = P(1, 6) + P(2, 5) + P(3, 4) + P(4, 3) + P(5, 2) + P(6, 1)$$
$$= \tfrac{1}{36} + \tfrac{1}{36} + \tfrac{1}{36} + \tfrac{1}{36} + \tfrac{1}{36} + \tfrac{1}{36} = \tfrac{1}{6}$$

Event B occurs whenever $i + j = 11$. Hence

$$P(B) = P(5, 6) + P(6, 5) = \tfrac{1}{36} + \tfrac{1}{36} = \tfrac{1}{18}$$

Two events A and B are said to be *mutually exclusive* if the occurrence of one excludes the possibility of the other occurring. Hence, two events are not mutually exclusive if both may occur simultaneously. If A and B are mutually exclusive events with individual probabilities of occurrence of $P(A)$ and $P(B)$, then the probability $P(A + B)$ that either A or B occurs is given by

$$P(A + B) = P(A) + P(B) \qquad (2.5)$$

If two events are such that the occurrence of one in no way affects the occurrence of the other, then the events are said to be *independent*. If events A and B are independent, then the probability $P(AB)$ that both occur at a given trial of an experiment is given by

$$P(AB) = P(A)P(B) \qquad (2.6)$$

Events that are not independent are said to be *dependent*. For example, events A and B of Example 2.2 are dependent since the occurrence of one eliminates the possibility of the other occurring. If two events A and B are mutually exclusive, then

$$P(AB) = 0 \qquad (2.7)$$

To consider dependent events further, suppose that two cards are drawn, one at a time, from a standard deck of playing cards, the second being drawn without replacement of the first. Let us calculate the probability that the second card drawn is an ace. Since there are only four aces in a standard deck of playing cards, we might anticipate that the likelihood of drawing an ace on the second card would depend to some extent upon whether or not an ace was drawn on the first. If an ace was drawn on the first, the probability of

Probability Theory

drawing an ace on the second is 3/51. If an ace was not drawn on the first card, then the probability of selecting an ace on the second becomes 4/51. Therefore, the probability of selection of an ace on the second draw is dependent or conditioned on whether an ace was drawn on the first. To generalize the concept of conditional probability, let A and B be two events. The probability that A occurs if B is known to have occurred is expressed by $P(A|B)$ and is read "probability of A given B." In general,

$$P(A|B) = \frac{P(AB)}{P(B)} \tag{2.8}$$

If, in fact, A and B are independent, we have

$$P(A|B) = \frac{P(AB)}{P(B)} = \frac{P(A)P(B)}{P(B)} = P(A) \tag{2.9}$$

Consider now the probability $P(A + B)$ that either A or B or both occur:

$$P(A + B) = P(A) + P(B) - P(AB) \tag{2.10}$$

When A and B are mutually exclusive, $P(AB) = 0$ and (2.10) reduces to (2.5).

Example 2.3 A particular product is partially processed on one of two production lines A or B, and finished on production line X and Y. However it may be completely processed on a third production line C. If it is partially processed on line A, it is completed on line X with probability 0.20 and completed on line Y with probability 0.80. If it is partially processed on line B, it is completed on line X with probability 0.60 and line Y with probability 0.40. The probability that a unit is initially processed on A is 0.30 and is 0.20 for line B. A completed unit is randomly selected.
 (a) What is the probability that it was processed on line C?
 (b) What is the probability that it was processed on line X?
 (c) What is the probability that it was processed on line Y?

$$P(\text{processing on line } C) = 0.50$$

$$P(\text{processing on line } X) = (0.20)(0.30) + (0.60)(0.20)$$

$$= 0.06 + 0.12 = 0.18$$

$$P(\text{processing on line } Y) = (0.80)(0.30) + (0.40)(0.20)$$

$$= 0.24 + 0.08 = 0.32$$

Let $B_1, B_2, \ldots, B_n$ be a set of mutually exclusive events and let A be an event that must be accompanied by some B_i, $i = 1, 2, \ldots, n$. That is,

$$P(A) = \sum_{i=1}^{n} P(AB_i) \tag{2.11}$$

Then

$$P(B_i \mid A) = \frac{P(A \mid B_i)P(B_i)}{\sum_{i=1}^{n} P(A \mid B_i)P(B_i)} \qquad (2.12)$$

The result given in (2.12) is known as *Bayes's theorem*.

Discrete Random Variables

Let X be a random variable representing the number of failures of a piece of machinery per week. Let $p(x)$ be the probability that X assumes the value x in a particular week. Then

$$p(x) = P(X = x) \qquad (2.13)$$

In general, a capital letter will be used to denote the name of a random variable and the corresponding lowercase letter will be used to denote the value the random variable assumes. For the example considered here, the number of failures per week could assume only integer values between zero and some upper limit N, inclusive. Thus X is a discrete random variable and

$$\sum_{x=0}^{N} p(x) = 1 \qquad (2.14)$$

from (2.3). The function $p(x)$ is called the *probability mass function* of X.

Quite often it is necessary to calculate the probability that the value of a random variable falls between two specified limits. For example, suppose that X is a discrete random variable with probability mass function $p(x)$. The probability that X has a value between a and b inclusive, $a \leq b$, is given by

$$P(a \leq x \leq b) = \sum_{x=a}^{b} p(x) \qquad (2.15)$$

To calculate a probability such as that given in (2.15), it is sometimes convenient to use the *cumulative distribution function*, often referred to simply as the *distribution function* and given by

$$F(x) = \sum_{y=-\infty}^{x} p(y) \qquad (2.16)$$

where the summation is over specified discrete values of y, usually integers. It should be noted that $p(y)$ may be zero for some set of values of y. We can now represent $P(a \leq x \leq b)$ by

$$P(a \leq x \leq b) = F(b) - F(a - 1) \qquad (2.17)$$

where the difference between successive values of x is 1.

Probability Theory

The probability that x is greater than a but less than or equal to b is given by

$$P(a < x \leq b) = F(b) - F(a) \qquad (2.18)$$

The definition of a distribution function suggests several properties that it possesses. Since $0 \leq p(x) \leq 1$ for all x, $0 \leq F(x) \leq 1$ for all x. Further, since

$$\sum_{x=-\infty}^{\infty} p(x) = 1 \qquad (2.19)$$

we have

$$F(-\infty) = 0 \qquad (2.20)$$

and

$$F(\infty) = 1 \qquad (2.21)$$

Finally, since $F(x)$ is a partial sum of $p(x)$, $F(x)$ is a nondecreasing function of x.

Example 2.4 Let X be a discrete random variable with probability mass function given by

$$p(x) = \begin{cases} 0, & x < 1 \\ 0.1, & x = 1 \\ 0.2, & x = 2 \\ 0.3, & x = 3 \\ 0.4, & x = 4 \\ 0, & x > 4 \end{cases}$$

Find and graph the distribution function of X.

Since $p(x) = 0$, $x < 1$, we have

$$F(x) = \sum_{y=-\infty}^{x} p(y) = 0, \qquad x < 1$$

For $1 \leq x \leq 4$, we have

$$F(x) = \begin{cases} 0.1, & x = 1 \\ 0.3, & x = 2 \\ 0.6, & x = 3 \\ 1.0, & x = 4 \end{cases}$$

Since $p(x) = 0$, $x > 4$, we have

$$F(x) = \sum_{y=-\infty}^{x} p(y) = \sum_{y=-\infty}^{4} p(y) + \sum_{y=5}^{x} p(y) = 1.0 + 0, \quad x > 4$$

To summarize then,

$$F(x) = \begin{cases} 0, & x < 1 \\ 0.1, & 1 \leq x < 2 \\ 0.3, & 2 \leq x < 3 \\ 0.6, & 3 \leq x < 4 \\ 1.0, & x \geq 4 \end{cases}$$

The graph of $F(x)$ is shown in Fig. 2.1.

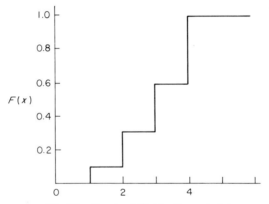

Fig. 2.1 Graph of $F(x)$ for Example 2.4.

There are several discrete probability distributions that are of general interest in operations research and probability theory. Some of these distributions are of particular interest in queueing theory. For example, as we shall see in later chapters, the Poisson distribution is a fundamental building block in the theory of queues. Specifically, the initial treatment of most queueing systems usually entails the assumption that the number of arrivals to the system in a fixed interval of time is Poisson distributed. In addition, the Bernoulli distribution is often used to describe the operation of a queueing system when an arriving unit chooses to enter one of two service facilities with probability p and the other with probability $(1 - p)$. The geometric distribution sometimes has application when units arrive in bulk, that is, the number of units arriving at one time is a random variable X, and this random variable is described by a geometric distribution.

Other discrete random variables treated in this chapter include the binomial, hypergeometric, rectangular, and negative binomial. Although these random variables are not of fundamental interest in queueing theory, they are of general interest in such areas as quality control, inventory theory, reliability theory, and traffic flow theory. This, however, is not intended to imply that these random variables are never encountered in queueing theory. A summary of the salient characteristics of several important discrete random variables is given in Table 2.1.

Bernoulli distribution. Suppose that a given experiment of chance may result in one of two outcomes. To the first outcome we assign the number zero and to the second we assign the number one. If X is the random variable used to describe the outcome of the experiment, then X is a discrete random variable that may assume the values zero or one. Let the probability of the first outcome (zero) be $1 - p$ and let p be the probability of the second outcome (one). Then the probability mass function of X is given by

$$p(x) = p^x(1-p)^{1-x}, \quad x = 0, 1, \quad 0 < p < 1 \qquad (2.22)$$

Binomial distribution. Let an experiment consist of n independent Bernoulli subexperiments; that is, the result of each subexperiment is one of two outcomes to which we assign either a zero or a one. After conducting n such subexperiments, we count the number of ones occurring. Let X be the random variable representing the number of ones occurring in the n Bernoulli subexperiments. If $Y_1, Y_2, \ldots, Y_n$ are random variables representing the outcomes of the n subexperiments, $Y_i = 0, 1, i = 1, 2, \ldots, n$, then

$$X = \sum_{i=1}^{n} Y_i \qquad (2.23)$$

and the probability mass function of X is given by

$$p(x) = \frac{n!}{x!(n-x)!} p^x(1-p)^{n-x}, \quad x = 0, 1, 2, \ldots, n, \quad 0 < p < 1 \qquad (2.24)$$

where X is called a binomial random variable.

Poisson distribution. Suppose that we were to count the number of events occurring in successive nonoverlapping time intervals each of length T. Let X be the random variable representing the number of events in an interval of time T. If

1. the number of events occurring in an interval of length T is independent of the number occurring in any other nonoverlapping interval of length T,

TABLE 2.1

Random variable name	Probability mass function	Generating function
Bernoulli	$p^x(1-p)^{1-x}$	$pz + (1-p)$
Binomial	$\dfrac{n!}{x!(n-x)!} p^x(1-p)^{n-x}$	$[pz + (1-p)]^n$
Poisson	$\dfrac{\lambda^x}{x!} e^{-\lambda}$	$e^{-\lambda(1-z)}$
Geometric	$p(1-p)^{x-1}$	$\dfrac{pz}{1-z(1-p)}$
Negative binomial	$\dfrac{(x-1)!}{(n-1)!(x-n)!} p^n(1-p)^{x-n}$	$\left[\dfrac{pz}{1-z(1-p)}\right]^n$
Rectangular	$\dfrac{1}{b-a+1}$	$\dfrac{z^a - z^{b+1}}{(b-a+1)(1-z)}$
Hypergeometric	$\dfrac{\dfrac{M!}{x!(M-x)!} \dfrac{(N-M)!}{(n-x)!(N-M-n+x)!}}{\dfrac{N!}{n!(N-n)!}}$	

2. the distribution of X is the same for all intervals of length T, no matter when the interval began,
3. two events cannot occur simultaneously,
4. no matter how small T is, there is a positive probability that an event will occur provided $T > 0$,
5. $P(X > 0 \mid T = 0) = 0$,

then X has a Poisson distribution with probability mass function given by

$$p(x) = \frac{(\lambda T)^x}{x!} e^{-\lambda T}, \quad x = 0, 1, 2, \ldots, \quad \lambda T > 0 \qquad (2.25)$$

where λ is the mean number of events occurring in a unit time interval.

The Poisson distribution has been successfully used to describe such diverse phenomena as the number of busy channels in a telephone system, the demand for certain products, machine failures, radioactive decay, and customer demand for service. In addition, the Poisson distribution can frequently be used to approximate the binomial distribution.

Probability Theory

Discrete distributions

Mean	Variance	Range of variable	Parameter values
p	$p(1-p)$	$x = 0, 1$	$0 < p < 1$
np	$np(1-p)$	$x = 0, 1, \ldots, n$	$0 < p < 1$ $n = 1, 2, 3, \ldots$
λ	λ	$x = 0, 1, 2, \ldots$	$\lambda > 0$
$\dfrac{1}{p}$	$\dfrac{1-p}{p^2}$	$x = 1, 2, 3, \ldots$	$0 < p < 1$
$\dfrac{n}{p}$	$\dfrac{n(1-p)}{p^2}$	$x = n, n+1, \ldots$	$0 < p < 1$ $n = 1, 2, \ldots$
$\dfrac{a+b}{2}$	$\dfrac{(b-a)^2}{12} + \dfrac{b-a}{6}$	$x = a, a+1, \ldots, b$	$a < b$
$\dfrac{Mn}{N}$	$\dfrac{(N-n)Mn(N-M)}{(N-1)N^2}$	$x = 0, 1, \ldots, n$	$N > M > n$

Geometric distribution. As in the case of the binomial random variable, the geometric random variable is related to the Bernoulli random variable. Let X be the random variable representing the number of Bernoulli trials necessary to obtain the first one (1). Again, p is the probability of obtaining a one on any Bernoulli trial and $1 - p$ is the probability of obtaining a zero. If one first appears at the xth trial, then it must have been preceded by $x - 1$ successive zeros. Therefore,

$$P(X = x) = P[(x-1) \text{ zeros}]P(1) = (1-p)^{x-1}p$$

Hence the probability mass function for the geometric random variable is given by

$$p(x) = p(1-p)^{x-1}, \qquad x = 1, 2, 3, \ldots \qquad (2.26)$$

The geometric random variable is said to be without memory. A random variable X is without memory if and only if

$$P(X > x_1 + x_2 \mid X > x_1) = P(X > x_2) \qquad (2.27)$$

For the geometric random variable

$$P(X > x_1 + x_2 \mid X > x_1) = \frac{P(X > x_1 + x_2)}{P(X > x_1)}$$

$$= \frac{\sum_{y=x_1+x_2+1}^{\infty} p(1-p)^y}{\sum_{y=x_1+1}^{\infty} p(1-p)^y} \quad (2.28)$$

But

$$\sum_{y=a+1}^{\infty} p(1-p)^y = 1 - \sum_{y=0}^{a} p(1-p)^{y-1}$$

$$= 1 - p \frac{1-(1-p)^a}{1-(1-p)} = (1-p)^a \quad (2.29)$$

Therefore,

$$P(X > x_1 + x_2 \mid X > x_1) = \frac{(1-p)^{x_1+x_2}}{(1-p)^{x_1}}$$

$$= (1-p)^{x_2} = P(X > x_2) \quad (2.30)$$

As we shall see later, there is a continuous random variable, the exponential, that is also without memory.

Negative binomial distribution. In the discussion of the geometric distribution, the geometric random variable was defined as the number of Bernoulli trials necessary to obtain the first one (1). Let $Y_1, Y_2, \ldots, Y_n$ be n identically distributed geometric random variables; that is, Y_i is the number of Bernoulli trials necessary to obtain the first one on the ith experiment. If we let X be the random variable defined by

$$X = \sum_{i=1}^{n} Y_i \quad (2.31)$$

then X is a negative binomial random variable with probability mass function given by

$$p(x) = \frac{(x-1)!}{(n-1)!(x-n)!} p^n (1-p)^{x-n}, \quad x = n, n+1, n+2, \ldots \quad (2.32)$$

Based on the definition of the negative binomial random variable given in (2.32), this random variable can be interpreted as representing the number of Bernoulli trials required to achieve a total of n ones.

Continuous Random Variables

If X is a continuous random variable, then it may assume any value on a real line segment of nonzero length. For example, the time between customer arrivals for service is treated as a continuous random variable except in

Probability Theory

certain rare cases. Similarly, the time taken to provide the customer with the service he desires is normally a continuous random variable. Other continuous random variables include the time to satisfy a customer demand, the spatial distribution of particles in a gas, the spatial distribution of plant life, the weight of a human being, the volume of rainfall in a day, and the velocity of automobiles on a freeway. A summary of the properties of several important continuous random variables is given in Table 2.2.

In treating continuous random variables and their distributions it is convenient to start with a discussion of the distribution function $F(x)$. The characteristics of the distribution function of a continuous random variable are the same as those for a discrete random variable:

1. $F(-\infty) = 0$
2. $F(\infty) = 1$
3. $0 \leq F(x) \leq 1$ for $-\infty < x < \infty$
4. $F(x)$ is nondecreasing

The continuous analog of the probability mass function is the *probability density function*, or simply the density function. The density function $f(x)$ of a continuous random variable X is given by

$$f(x) = \frac{d}{dx} F(x) \tag{2.33}$$

Conversely, we have

$$F(x) = \int_{-\infty}^{x} f(y)\, dy \tag{2.34}$$

As in the case of discrete random variables, the distribution function can be used to calculate the probability that the value of a random variable falls between two limits a and b, where $a < b$:

$$P(a \leq x \leq b) = F(b) - F(a) = \int_{a}^{b} f(y)\, dy \tag{2.35}$$

Now let us calculate the probability that a continuous random variable assumes a specific value a. From (2.35),

$$P(X = a) = P(a \leq x \leq a) = 0 \tag{2.36}$$

Thus, unlike discrete random variables, the probability that a continuous random variable assumes a specific value is zero. Further, if X is a continuous random variable, then

$$P(a \leq x \leq b) = P(a < x \leq b) = P(a \leq x < b) = P(a < x < b) \tag{2.37}$$

TABLE 2.2

Random variable name	Probability density function	Laplace transform (one-sided)
Normal	$\dfrac{1}{\sigma(2\pi)^{1/2}} \exp - \dfrac{(x-\mu)^2}{2\sigma^2}$	$\exp\left(-\mu s + \dfrac{\sigma^2 s^2}{2}\right)$†
Exponential	$\lambda e^{-\lambda x}$	$\left(1 + \dfrac{s}{\lambda}\right)^{-1}$
Erlang	$\dfrac{(k\lambda)^k}{(k-1)!} x^{k-1} e^{-k\lambda x}$	$\left(1 + \dfrac{s}{k\lambda}\right)^{-k}$
Gamma	$\dfrac{\lambda^n}{\Gamma(n)} x^{n-1} e^{-\lambda x}$	$\left(1 + \dfrac{s}{\lambda}\right)^{-n}$
Uniform	$\dfrac{1}{b-a}$	$\dfrac{e^{-as} - e^{-bs}}{s(b-a)}, a > 0$
Beta	$\dfrac{\Gamma(a+b)}{\Gamma(a)\Gamma(b)} x^{a-1}(1-x)^{b-1}$	
Hyperexponential	$p\lambda_1 e^{-\lambda_1 x} + (1-p)\lambda_2 e^{-\lambda_2 x}$	$\dfrac{(\lambda_1 - \lambda_2)ps + \lambda_2(\lambda_1 + s)}{(s+\lambda_1)(s+\lambda_2)}$

† Two-sided Laplace transform.

However these relationships do not necessarily hold if X is a discrete random variable.

Several properties of the probability density function can be deduced from the distribution function and its relationship to the density function. First, since $F(-\infty) = 0$, $f(-\infty) = 0$. Second, since $F(\infty) = 1$, $f(\infty) = 0$. Finally, since $F(x)$ is a nondecreasing function of x, $f(x) \geq 0$ for $-\infty < x < \infty$.

There are three continuous random variables that are of particular interest in queueing theory. These random variables are the exponential, Erlang, and hyperexponential, all of which are related. Other random variables of general interest in operations research and probability theory include the normal, gamma, beta, t, F, chi-square, and uniform. Actually the exponential, Erlang, and chi-square random variables are special cases of the gamma random variable.

Probability Theory

Continuous distributions

Mean	Variance	Range of variable	Parameter values
μ	σ^2	$-\infty < x < \infty$	$-\infty < \mu < \infty$, $\sigma > 0$
$\dfrac{1}{\lambda}$	$\dfrac{1}{\lambda^2}$	$0 < x < \infty$	$\lambda > 0$
$\dfrac{1}{\lambda}$	$\dfrac{1}{k\lambda^2}$	$0 < x < \infty$	$\lambda > 0$, integer $k > 0$
$\dfrac{n}{\lambda}$	$\dfrac{n}{\lambda^2}$	$0 < x < \infty$	$\lambda > 0$, $n > 0$
$\dfrac{a+b}{2}$	$\dfrac{(b-a)^2}{12}$	$a < x < b$	$-\infty < a < \infty$, $-\infty < b < \infty$
$\dfrac{a}{a+b}$	$\dfrac{ab}{(a+b)^2(a+b+1)}$	$0 < x < 1$	$a > 0$, $b > 0$
$\dfrac{\lambda_1 + p(\lambda_2 - \lambda_1)}{\lambda_1 \lambda_2}$	$\dfrac{p\lambda_2^2 + (1-p)\lambda_1^2 + p(1-p)(\lambda_2 - \lambda_1)^2}{\lambda_1^2 \lambda_2^2}$	$0 < x < \infty$	$0 \le p \le 1$, $\lambda_1 > 0$, $\lambda_2 > 0$

Exponential distribution. Let the random variable X represent the time between successive arrivals to a given system with density function $f(x)$. If an arrival occurs at time 0, then the probability that the next arrival occurs after time t, measured from time 0, is given by

$$P(X > t_1) = 1 - F(t_1)$$

where $F(t)$ is the cumulative distribution function of X. Now we move forward in time to t_1 and start to measure the time of the next arrival from this point in time. The probability that the arrival does not occur until after $t_1 + t_2$, given that we know it did not occur in the interval t_1, is

$$P(X > t_1 + t_2 \mid X > t_1) = \frac{1 - F(t_1 + t_2)}{1 - F(t_1)} \qquad (2.38)$$

X can be shown to be exponentially distributed with density function

$$f(x) = \lambda e^{-\lambda x}, \quad 0 < x < \infty, \quad \lambda > 0 \quad (2.39)$$

if

$$P(X > t_1 + t_2 \mid X > t_1) = P(X > t_2) \quad (2.40)$$

The implication of (2.40) is that the density function of time until the next event, in this case an arrival, is the same whether or not we start measuring time from the occurrence of the last event; that is, the exponential random variable is without memory in the sense that the time until the next event occurs, measured from an arbitrary point in time, is independent of when the last event took place. This property of the exponential random variable makes it useful, particularly from an analytic point of view, in describing interarrival times and service times in queueing systems. As we shall show later, the exponential random variable is integrally related to the Poisson random variable and this relationship will prove to be of paramount importance throughout our discussion of queueing theory.

Gamma distribution. Let $Y_1, Y_2, \ldots, Y_n$ be the times between the occurrence of n successive events, and define X as

$$X = \sum_{i=1}^{n} Y_i \quad (2.41)$$

If Y_i is an exponential random variable with parameter λ, $i = 1, 2, \ldots, n$, then X is gamma distributed with density function

$$f(x) = \frac{\lambda^n}{(n-1)!} x^{n-1} e^{-\lambda x}, \quad 0 < x < \infty, \quad \lambda > 0 \quad (2.42)$$

A more general form of the gamma density function is

$$f(x) = \frac{\lambda^n}{\Gamma(n)} x^{n-1} e^{-\lambda x}, \quad 0 < x < \infty, \quad \lambda > 0, \quad n > 0 \quad (2.43)$$

where

$$\Gamma(n) = (n-1)!$$

when n is an integer. If, however, n is not an integer, the gamma random variable cannot be represented by the sum of identically distributed exponential random variables.

When n is unity the gamma density function reduces to the exponential density function. Thus, the exponential random variable is a special case of the gamma random variable. Furthermore, if $n = r/2$, where r is an integer,

Probability Theory

and $\lambda = 1/2$, the resulting gamma random variable is frequently called a chi-square random variable.

Erlang distribution. Closely related to both the exponential and gamma random variables is the Erlang random variable. To illustrate how this random variable arises in the context of queueing theory, consider a service facility that services units at a rate μ. However, service is performed in k phases, where the distribution of service time in each phase is exponential with rate $k\mu$. If Y_i is the time the unit spends in the ith phase, $i = 1, 2, \ldots, k$, then the density function of Y_i is given by

$$g(y_i) = k\mu e^{-k\mu y_i}, \quad 0 < y_i < \infty \tag{2.44}$$

The total time X spent in service is then

$$X = \sum_{i=1}^{k} Y_i \tag{2.45}$$

From (2.41) and (2.42) the density function of X is

$$f(x) = \frac{(k\mu)^k}{(k-1)!} x^{k-1} e^{-k\mu x}, \quad 0 < x < \infty \tag{2.46}$$

The relationship between the Erlang and gamma density functions should be apparent: if $\lambda = k\mu$ in (2.42), then the gamma density function becomes the Erlang.

Hyperexponential distribution. Suppose that a driver may take one of two routes to get to work. The time Y_1 taken over the first route is exponentially distributed with parameter λ_1 and the time Y_2 taken over the second route is also exponentially distributed but with parameter λ_2. Now suppose that he selects route one with probability p and route two with probability $(1 - p)$. Let X be the total time to get to work on an arbitrary day. Then

$$P(X > x) = P(Y_1 > x \,|\, \text{first route selected})p$$
$$+ P(Y_2 > x \,|\, \text{second route selected})(1 - p) \tag{2.47}$$

or

$$P(X > x) = p\lambda_1 \int_x^\infty \exp(-\lambda_1 y_1)\, dy_1$$
$$+ (1 - p)\lambda_2 \int_x^\infty \exp(-\lambda_2 y_2)\, dy_2 \tag{2.48}$$

Since

$$1 - F(x) = P(X > x)$$

we have, by (2.33),

$$f(x) = \frac{d}{dx}\left[1 - p\lambda_1 \int_x^\infty \exp(-\lambda_1 y_1)\, dy_1\right.$$

$$\left. - (1-p)\lambda_2 \int_x^\infty \exp(-\lambda_2 y_2)\, dy_2\right]$$

$$= p\lambda_1 \exp(-\lambda_1 x) + (1-p)\lambda_2 \exp(-\lambda_2 x), \quad 0 < x < \infty \quad (2.49)$$

which is the density function of the hyperexponential random variable.

The hyperexponential random variable occurs in queueing theory when the time to service a unit is exponentially distributed with a parameter that may assume one of two values. For example, suppose that a switchboard is controlled by one operator. However, two operators are available. The probability that the first is at the switchboard is p and is $(1-p)$ for the second. The first operator processes calls in an exponential manner with parameter λ_1, and the second in an exponential manner with parameter λ_2. The unconditional distribution of processing time per call is then hyperexponential with parameters p, λ_1, and λ_2.

Suppose, in the above example, that three operators were available for switchboard duty, each processing calls in an exponential fashion but with different parameters, λ_1, λ_2, and λ_3. If the first operator is on duty with probability p_1, the second with probability p_2, and the third with probability p_3, where $p_1 + p_2 + p_3 = 1$, then the density function of time X to process a call becomes

$$f(x) = \sum_{i=1}^{3} p_i \lambda_i \exp(-\lambda_i x), \quad 0 < x < \infty$$

Obviously, we can extend the number of alternatives implied by the hyperexponential random variable to n, which leads to

$$f(x) = \sum_{i=1}^{n} p_i \lambda_i \exp(-\lambda_i x), \quad 0 < x < \infty \quad (2.50)$$

where

$$\sum_{i=1}^{n} p_i = 1, \quad 0 \leq p_i \leq 1, \quad i = 1, 2, \ldots, n \quad (2.51)$$

The exponential, Erlang, and hyperexponential random variables are of paramount importance in the analysis of queueing systems. As we have already seen, the exponential random variable falls within the family of Erlang random variables; that is, the Erlang density function reduces to the exponential when $k = 1$. Similarly, the exponential random variable can be considered to fall within the family of hyperexponential random variables in that the hyperexponential density reduces to the exponential density when $p = 0$ or $p = 1$. Thus, the hyperexponential random variable can be considered a mixture of exponential random variables, while the Erlang can be considered a sum of exponential random variables.

An interesting and useful relationship exists among the variances of the hyperexponential, exponential, and Erlang random variables in the case where the mean of each is identical. In order to demonstrate this relationship, we will express the hyperexponential density function as

$$f(x) = 2\alpha^2 \mu e^{-\alpha\mu x} + 2(1-\alpha)^2 \mu e^{-2(1-\alpha)\mu x}, \qquad 0 < x < \infty \qquad (2.52)$$

where $0 < \alpha < 0.5$. The mean and variance of x are given by

$$E(x) = \frac{1}{\mu} \qquad (2.53)$$

$$\text{Var}(x) = \frac{1}{\mu^2}\left[\frac{1}{2\alpha(1-\alpha)} - 1\right] \qquad (2.54)$$

Now, if y is an exponential random variable and z an Erlang random variable each with mean $1/\mu$, then

$$\text{Var}(y) = \frac{1}{\mu^2} \qquad (2.55)$$

and

$$\text{Var}(z) = \frac{1}{k\mu^2} \qquad (2.56)$$

Hence if x, y, and z are, respectively, hyperexponential, exponential, and gamma random variables each with mean $1/\mu$, then

$$\text{Var}(z) \leq \text{Var}(x) \leq \text{Var}(y) \qquad (2.57)$$

The relationship established in (2.57) can be quite useful in identifying the distribution of random variables involved in queueing systems. For example, if the sample mean and standard deviation are nearly equal, then one might suspect that the random variable under study is exponentially distributed. On the other hand, if the sample mean is substantially larger than the sample standard deviation, then the random variable may well be Erlang distributed. Finally, if the sample mean is less in value than the sample standard deviation, then it may be appropriate to test the fit of a hyperexponential distribution to the observed data.

Normal distribution. The normal distribution is undoubtedly the best-known distribution in probability theory and statistics. This is due, in part, to the frequency with which it is encountered in attempting to describe random physical phenomena. The normal density function is "bell shaped" with mean at its center. Thus, if u is the mean of the normal random variable X, then

$$P(u - a \leq X \leq u) = P(u \leq X \leq u + a) \qquad (2.58)$$

Quite often physical phenomena such as human measurement error, product dimension, repetitive task completion time, height and weight of humans, ballistic measurements, and molecular velocity can be considered to be normally distributed. In addition the normal distribution has extensive application in statistical inference. Its use in statistics is a result of the *central limit theorem*, which states that, under appropriate conditions, the distribution of the sum of random variables approaches the normal distribution as the number of random variables included in the sum increases. If the sum of random variables has an approximate normal distribution, then the sample mean can also be shown to have an approximate normal distribution. The latter result has important implications in statistical hypothesis testing, where a conclusion is to be drawn about the mean of one or more populations. In fact, the normal distribution forms the basis for most parametric tests of hypotheses.

The density function of the normal random variable is given by

$$f(x) = \frac{1}{\sigma(2\pi)^{1/2}} \exp{-\frac{(x-\mu)^2}{2\sigma^2}}, \qquad -\infty < x < \infty, \quad \sigma > 0 \qquad (2.59)$$

When $\mu = 0$, $\sigma = 1$, the resulting normal random variable is called a standard normal random variable.

Uniform distribution. Consider a random variable X, which may take on values between a and b, $a < b$. Now let us divide the interval (a, b) into n subintervals each of width $(b - a)/n$. Then the ith subinterval, $1 \leq i \leq n$, begins at $a + (b - a)(i - 1)/n$ and ends at $a + (b - a)i/n$. If

$$P\left[a + (b-a)\frac{i-1}{n} < X < a + (b-a)\frac{i}{n}\right]$$

$$= P\left[a + (b-a)\frac{j-1}{n} < X < a + (b-a)\frac{j}{n}\right] \qquad (2.60)$$

for all i and j and all $n > 1$, then X has a uniform distribution with density function

$$f(x) = \frac{1}{b-a}, \qquad a < x < b \qquad (2.61)$$

Probability Theory

The uniform distribution is of paramount importance in simulation, since the uniform random variable provides the basis for generating numerous other random variables. This application of the uniform distribution will be illustrated when we discuss the simulation of queueing systems.

The t, F, and chi-square distributions are used extensively in statistical hypothesis testing. However, their applications to queueing theory are limited. Even so, they are of use in analyzing queueing systems through simulation and in data analysis prior to the development of queueing models. These applications will be demonstrated when we discuss data collection and analysis and in the chapter on the simulation of queueing systems.

Example 2.5 An income tax consultant works by himself. He spends an average of about 30 minutes with each client. However, the time spent with each client is exponentially distributed. The parameter λ of the exponential can be expressed by

$$\lambda = 1/\text{mean time per client}$$

Therefore, the density function of time X spent per client is given by

$$f(x) = 0.033e^{-0.033x}, \quad 0 < x < \infty$$

where x is measured in minutes. If he schedules appointments every 30 minutes, what is the probability that a client will have to wait to see the tax consultant, if he arrives for his appointment on time, and if the preceding client saw the tax consultant on schedule?

$$P(\text{client waits}) = P(\text{preceding client spent over 30 min}$$
$$\text{with tax consultant})$$

$$= 1 - F(30) = 1 - \int_0^{30} 0.033e^{-0.033x}\, dx$$

$$= e^{-1} = 0.368$$

Example 2.6 A certain part is produced on one of two manufacturing lines. The time, in hours, to produce a part on the first line is exponentially distributed with parameter 10. The time to produce a part on the second line is also exponentially distributed but with parameter 20. The probability that a unit is manufactured on the first line is 0.60. What is the probability that an arbitrary unit will be produced in 0.10 hr or less?

The conditions given in this problem are identical to those given for the hyperexponential random variable. Therefore, if X is the time to process a part, in hours, then

$$f(x) = (0.6)(10)e^{-10x} + (0.4)(20)e^{-20x}, \quad 0 < x < \infty$$

and

$$P(X \leq 0.10) = 0.6 \int_0^{0.1} 10e^{-10x}\, dx + 0.4 \int_0^{0.1} 20e^{-20x}\, dx$$
$$= 0.6(1 - e^{-1}) + 0.4(1 - e^{-2})$$
$$= 0.6(0.632) + 0.4(0.865) = 0.725$$

Joint Distributions

Quite often we are concerned with a random variable that is related to or dependent upon other random variables. For example, suppose that we were interested in the time a customer waits for service. Waiting time is, to some extent, dependent upon the number of customers in the store upon his arrival and the service time required for these customers. If we know the number of customers in the store at a given time and the density function or functions of service time, then we can determine the density function of waiting time. If, in addition, we know the probability mass function of the number of customers in the store at a given time, we can determine the joint density function of waiting time and the number of customers in the store. The joint density, probability mass, and distribution functions allow us to calculate the probability of occurrence of a set of events.

The joint distribution function of the random variables $X_1, X_2, \ldots, X_n$ is given by

$$F(x_1, x_2, \ldots, x_n) = P(X_1 \leq x_1, X_2 \leq x_2, \ldots, X_n \leq x_n) \quad (2.62)$$

In words, (2.62) expresses the probability that the events $X_1 \leq x_1, X_2 \leq x_2, \ldots, X_n \leq x_n$ all occur. The joint density function is expressed by $f(x_1, x_2, \ldots, x_n)$, and $p(x_1, x_2, \ldots, x_n)$ is generally used to denote a joint probability mass function.

If $X_1, X_2, \ldots, X_n$ are independent random variables, then

$$F(x_1, x_2, \ldots, x_n) = F(x_1)F(x_2) \cdots F(x_n) \quad (2.63)$$
$$f(x_1, x_2, \ldots, x_n) = f(x_1)f(x_2) \cdots f(x_n) \quad (2.64)$$

and

$$p(x_1, x_2, \ldots, x_n) = p(x_1)p(x_2) \cdots p(x_n) \quad (2.65)$$

Equations (2.63)–(2.65) can be used to determine whether a set of random variables is independent or not. Before continuing, we must first introduce

Probability Theory

the notion of the marginal density and probability mass functions. The *marginal density function* of the random variable X_i is given by

$$f(x_i) = \int_{x_n} \cdots \int_{x_{i+1}} \int_{x_{i-1}} \cdots \int_{x_1} f(x_1, x_2, \ldots, x_n)$$
$$\times dx_1 \cdots dx_{i-1}\, dx_{i+1} \cdots dx_n \qquad (2.66)$$

whereas the marginal probability mass function of X_i is given by

$$p(x_i) = \sum_{X_n} \cdots \sum_{X_{i+1}} \sum_{X_{i-1}} \cdots \sum_{X_1} p(x_1, \ldots, x_{i-1}, x_i, \ldots, x_n) \qquad (2.67)$$

Therefore, if the joint density function is equal to the product of the marginal density functions of the associated random variables, then the random variables are said to be independently distributed. Similarly, if the joint probability mass function is equal to the product of the individual marginal probability mass functions, the random variables are independently distributed.

Example 2.7 Let the joint density function of X_1 and X_2 be

$$f(x_1, x_2) = \frac{1}{x_1 T}, \qquad 0 < x_1 < T, \quad 0 < x_2 < x_1$$

Determine whether X_1 and X_2 are independently distributed. From (2.66) the marginal density function of X_1 is given by

$$f(x_1) = \int_0^{x_1} \frac{1}{x_1 T}\, dx_2 = \frac{1}{T}, \qquad 0 < x_1 < T$$

and the marginal density function of X_2 is

$$f(x_2) = \int_{x_2}^{T} \frac{1}{x_1 T}\, dx_1 = \frac{\log(T/x_2)}{T}, \qquad 0 < x_2 < T$$

Then

$$f(x_1)f(x_2) = \frac{\log(T/x_2)}{T^2}$$

Since

$$f(x_1, x_2) \neq f(x_1)f(x_2)$$

X_1 and X_2 are not independently distributed.

If two random variables are dependent upon one another, one would expect that the density or probability mass function of one would be in some way dependent upon the value of the other. This dependence is expressed through the *conditional density function* or the *conditional probability mass function.* The conditional density function of X_1 given X_2, $f(x_1 \mid x_2)$, is given by

$$f(x_1 \mid x_2) = \frac{f(x_1, x_2)}{f(x_2)}$$

and the conditional probability mass function of X_1 given X_2, $p(x_1 \mid x_2)$, is defined as

$$p(x_1 \mid x_2) = \frac{p(x_1, x_2)}{p(x_2)} \tag{2.68}$$

Example 2.8 The time T required to provide a given service is an exponential random variable with parameter λ. However, the parameter λ is also a random variable with density function

$$g(\lambda) = \mu e^{-\mu \lambda}, \qquad 0 < \lambda < \infty$$

Find the joint density function of T and λ and the conditional distribution of λ given T.

Since T is dependent upon λ, and λ is a random variable, the density function given for T must be considered to be conditioned on λ, that is,

$$f(t \mid \lambda) = \lambda e^{-\lambda t}, \qquad 0 < t < \infty$$

From (2.67),

$$f(t, \lambda) = f(t \mid \lambda) g(\lambda) = \mu \lambda e^{-\lambda(\mu + t)}, \qquad 0 < t < \infty, \ 0 < \lambda < \infty, \ \mu > 0$$

The conditional density function of λ given T is defined by

$$g(\lambda \mid t) = \frac{f(t, \lambda)}{h(t)}$$

where $h(t)$ is the marginal density function of T. By (2.66),

$$h(t) = \int_0^\infty \mu \lambda e^{-\lambda(\mu + t)} \, d\lambda = \frac{\mu}{(\mu + t)^2}, \qquad 0 < t < \infty$$

and

$$g(\lambda \mid t) = (\mu + t)^2 \lambda e^{-\lambda(\mu + t)}, \qquad 0 < \lambda < \infty$$

Expectation

Let $\phi(X)$ be a function of the random variable X. The expected value of $\phi(X)$, $E[\phi(x)]$, is defined by

$$E[\phi(x)] = \begin{cases} \int_{-\infty}^{\infty} \phi(x) f(x) \, dx, & x \text{ continuous} \\ \sum_{x = -\infty}^{\infty} \phi(x) p(x), & x \text{ discrete} \end{cases} \tag{2.69}$$

where $f(x)$ and $p(x)$ are density and probability mass functions, respectively. Expected values play a significant role in queueing theory and operations research as well as probability theory. There are two expected values that are particularly important in queueing theory. The first is the *mean*, defined by $E(X)$, and the second is the *variance*, given by $E[X - E(X)]^2$ or $E(X^2) - [E(X)]^2$. The mean can usually be considered a measure of central tendency or a point about which values of the random variable tend to cluster. The variance is a measure of dispersion or a measure of the degree to which values of the random variable cluster about the mean; that is, the smaller the variance the more values of the random variable tend to cluster about the mean. The variance is usually denoted by σ^2 and the mean by μ. However, since μ will be used otherwise in this book, the mean will be denoted by m in this chapter.

Example 2.9 Find the mean and variance of the gamma random variable. The density function of the gamma random variable is defined by

$$f(x) = \frac{\lambda^n}{\Gamma(n)} x^{n-1} e^{-\lambda x}, \quad 0 < x < \infty$$

For the mean, $\phi(x) = x$ in (2.69). Therefore,

$$m = \int_0^\infty x f(x)\, dx = \int_0^\infty \frac{\lambda^n}{\Gamma(n)} x^n e^{-\lambda x}\, dx = \frac{n}{\lambda}$$

For the variance, $\phi(X) = x^2 - [E(x)]^2$ and

$$\sigma^2 = \int_0^\infty \{x^2 - [E(x)]^2\} f(x)\, dx$$

$$= \int_0^\infty x^2 f(x)\, dx - \int_0^\infty [E(x)]^2 f(x)\, dx$$

$$= \int_0^\infty x^2 f(x)\, dx - m^2 \int_0^\infty f(x)\, dx$$

$$= \int_0^\infty \frac{\lambda^n}{\Gamma(n)} x^{n+1} e^{-\lambda x}\, dx - m^2$$

$$= \frac{n(n+1)}{\lambda^2} - \frac{n^2}{\lambda^2} = \frac{n}{\lambda^2}$$

Example 2.10 Find the mean and variance of the Poisson random variable. The probability mass function of the Poisson random variable is

$$p(x) = \frac{(\lambda T)^x}{x!} e^{-\lambda T}, \quad x = 0, 1, 2, \ldots$$

Then by (2.69)

$$m = \sum_{x=0}^{\infty} xp(x) = \sum_{x=0}^{\infty} x \frac{(\lambda T)^x}{x!} e^{-\lambda T} = \sum_{x=1}^{\infty} \frac{(\lambda T)^x}{(x-1)!} e^{-\lambda T}$$

Let $y = x - 1$. Then

$$m = \sum_{y=0}^{\infty} \frac{(\lambda T)^{y+1}}{y!} e^{-\lambda T} = \lambda T \sum_{y=0}^{\infty} \frac{(\lambda T)^y}{y!} e^{-\lambda T} = \lambda T$$

since, by definition of the probability mass function,

$$\sum_{y=0}^{\infty} \frac{(\lambda T)^y}{y!} e^{-\lambda T} = 1$$

For the variance,

$$\sigma^2 = \sum_{x=0}^{\infty} (x^2 - m^2) p(x) = \sum_{x=0}^{\infty} x^2 p(x) - m^2 \sum_{x=0}^{\infty} p(x)$$

$$= \sum_{x=0}^{\infty} x^2 \frac{(\lambda T)^x}{x!} e^{-\lambda T} - (\lambda T)^2$$

Let

$$x^2 = x(x-1) + x$$

Then

$$\sigma^2 = \sum_{x=0}^{\infty} x(x-1) \frac{(\lambda T)^x}{x!} e^{-\lambda T} + \sum_{x=0}^{\infty} x \frac{(\lambda T)^x}{x!} e^{-\lambda T} - (\lambda T)^2$$

$$= \sum_{x=2}^{\infty} \frac{(\lambda T)^x}{(x-2)!} e^{-\lambda T} + \lambda T - (\lambda T)^2$$

$$= (\lambda T)^2 \sum_{y=0}^{\infty} \frac{(\lambda T)^y}{y!} e^{-\lambda T} + \lambda T - (\lambda T)^2 = \lambda T$$

Thus the mean and variance of the Poisson random variable are the same, as indicated in Table 2.1.

Example 2.11 A company places orders for raw materials t weeks prior to scheduled production. The time X taken to receive an order after it has been placed is exponentially distributed with parameter λ. If the order arrives prior to production, $x \leq t$, an inventory carrying cost of $1500(t - x)$ is

Probability Theory

incurred. If the order arrives after the scheduled production startup, $x > t$, a cost of lost production of $5000(x - t)$ results. Therefore,

$$\text{cost} = \begin{cases} 1500(t - x), & x \leq t \\ 5000(x - t), & x > t \end{cases}$$

Find the expected cost of the ordering policy as a function of t and the value of t that minimizes expected cost. Let

$$c(t) = \begin{cases} 1500(t - x), & x \leq t \\ 5000(x - t), & x > t \end{cases}$$

Then

$$E[c(t)] = \int_0^\infty c(t)\lambda e^{-\lambda x}\, dx$$

But $c(t)$ is defined differently for $x \leq t$ and $x > t$. Thus

$$E[c(t)] = \int_0^t 1500(t - x)\lambda e^{-\lambda x}\, dx + \int_t^\infty 5000(x - t)\lambda e^{-\lambda x}\, dx$$

$$= 1500 e^{-\lambda t}\left[te^{\lambda t} + \frac{1}{\lambda}(1 - e^{\lambda t})\right] + \frac{5000}{\lambda}e^{-\lambda t}$$

$$= \frac{6500}{\lambda} e^{-\lambda t} + 1500 t - \frac{1500}{\lambda}$$

To find the value of t that minimizes $E[c(t)]$, we take the first derivative of $E[c(t)]$ with respect to t, set the derivative equal to zero, and solve for t. Now

$$\frac{d}{dt} E[c(t)] = -6500 e^{-\lambda t} + 1500 = 0$$

Therefore,

$$-\lambda t = \log(0.23) \quad \text{or} \quad t = \frac{1.47}{\lambda}$$

It is frequently necessary to find the expected value of a function of several random variables. For example, suppose that we wished to find the mean value of waiting time for an arrival that occurred when there were n units in the system ahead of it. If $Y_1, Y_2, \ldots, Y_n$ are the service times of the n units already in the system and X is waiting time for the arriving unit, then

$$E(X) = E(Y_1 + Y_2 + \cdots + Y_n)$$

Let $h(X_1, X_2, \ldots, X_n)$ be a function of the random variables $X_1, X_2, \ldots, X_n$. The expected value of $h(X_1, X_2, \ldots, X_n)$ is given by

$$E[h(X_1, X_2, \ldots, X_n)] = \begin{cases} \int_{x_n} \cdots \int_{x_1} h(x_1, \ldots, x_n) f(x_1, \ldots, x_n) \, dx_1 \cdots dx_n, \\ \qquad\qquad\qquad\qquad\qquad\qquad\qquad \text{continuous case} \\ \sum_{x_n} \cdots \sum_{x_1} h(x_1, \ldots, x_n) p(x_1, \ldots, x_n), \\ \qquad\qquad\qquad\qquad\qquad\qquad\qquad \text{discrete case} \end{cases} \quad (2.70)$$

For any set of random variables $X_1, X_2, \ldots, X_n$ and constants $a_1, a_2, \ldots, a_n$,

$$E(a_1 X_1 + a_2 X_2 + \cdots + a_n X_n) = \sum_{i=1}^{n} a_i E(X_i) \qquad (2.71)$$

If X_1 and X_2 are independent random variables and Y the product of these random variables, then

$$E(Y) = E(X_1 X_2) = E(X_1) E(X_2) \qquad (2.72)$$

Conditional expectation. Let X_1 and X_2 be dependent random variables. Suppose that we wish to find the expected value of X_1. We could accomplish this by finding the marginal density or probability mass function of X_1 and then using (2.69). However, it is often more convenient to work with the conditional density or probability mass function of X_1. For example, consider the problem of finding the expected waiting time for a random arrival to a system. Let X_1 be waiting time and X_2 the number of units in the waiting line at a random point in time. We can sometimes define $f(x_1 \mid x_2)$ and $f(x_2)$ without great difficulty. We could then define $f(x_1)$ by (2.66). However, we can define $E(X_1)$ as

$$E(X_1) = \sum_{x_2=0}^{\infty} E(X_1 \mid X_2) p(x_2)$$

where $p(x_2)$ is the probability mass function of X_2, and

$$E(X_1 \mid X_2) = \int_0^{\infty} x_1 f(x_1 \mid x_2) \, dx_1$$

where $f(x_1 \mid x_2)$ is the conditional density function of X_1 given X_2. In general,

$$E(X_1) = \begin{cases} \int_{X_2} E(X_1 \mid X_2) f(x_2) \, dx_2, & X_2 \text{ continuous} \\ \sum_{X_2} E(X_1 \mid X_2) p(x_2), & X_2 \text{ discrete} \end{cases} \qquad (2.73)$$

Example 2.12 Units of a perishable product are manufactured in large lots. The proportion of units that go bad before they are sold is Q, where Q is a random variable having a uniform distribution with density function

$$f(q) = 10, \quad 0 < q < 0.1$$

A unit is selected at random. If the unit is good, a profit P is realized. If the unit is discarded, a loss C is incurred. What is the expected profit on the unit?

Define the random variable X such that

$$X = \begin{cases} 0, & \text{unit sold} \\ 1, & \text{unit discarded} \end{cases}$$

Then X is dependent upon the proportion of bad items in the lot from which it was drawn, and

$$p(x \mid q) = q^x (1-q)^{1-x}, \quad x = 0, 1$$

Therefore,

$$E(\text{profit} \mid q) = P(1-q) - Cq$$

and

$$E(\text{profit}) = \int_0^{0.1} 10[P(1-q) - Cq]\, dq = 0.95P - 0.05C$$

Poisson Process

When we discussed the Poisson random variable we defined several properties that it possesses. We have also indicated an important relationship between the Poisson and exponential random variables. If the number of events occurring in an arbitrary fixed interval of time T is Poisson distributed, then the time between successive events is exponentially distributed. Let us now examine these properties of the Poisson random variable in detail since they play an important role in the theory of queues. For convenience, let us reiterate these properties:

1. The number of events occurring in an interval of length T is independent of the number occurring in any other nonoverlapping interval of length T.

2. The distribution of X is the same for all intervals of length T, no matter when the interval began.

3. Two events cannot occur simultaneously.

4. No matter how small T is, there is a positive probability that an event will occur provided $T > 0$.

5. $P(X > 0 \mid T = 0) = 0$.

Let us assume that X is a Poisson random variable with probability mass function

$$p(x) = \frac{(\lambda T)^x}{x!} e^{-\lambda T}, \qquad x = 0, 1, 2, \ldots \qquad (2.74)$$

The first property specified for X was that the number of events occurring in any two nonoverlapping intervals are independent random variables. Let X_1 and X_2 be the number of events occurring in the nonoverlapping intervals T_1 and T_2, respectively. Then

$$p(x_1, x_2) = \frac{(\lambda T_1)^{x_1}(\lambda T_2)^{x_2}}{x_1! x_2!} \exp[-\lambda(T_1 + T_2)],$$

$$x_1 = 0, 1, 2, \ldots, \quad x_2 = 0, 1, 2, \ldots \qquad (2.75)$$

The second assumption is implied by the definition of $p(x)$. The fourth assumption associated with the Poisson random variable is that (2.74) holds no matter how small or large T is, provided $T > 0$. For $T = 0$,

$$p(x) = 0, \qquad x = 1, 2, \ldots \qquad (2.76)$$

and

$$p(x) = e^{-\lambda(0)} = 1, \qquad x = 0 \qquad (2.77)$$

and the fifth assumption is verified.

The third assumption states that if X has a Poisson distribution, then no two events may occur at the same point in time. Suppose we consider the occurrence of x events in a time interval of length Δt. Then

$$p(x) = \frac{(\lambda \Delta t)^x}{x!} e^{-\lambda \Delta t}, \qquad x = 0, 1, 2, \ldots$$

Now

$$e^{-\lambda \Delta t} = \sum_{i=0}^{\infty} \frac{(-\lambda \Delta t)^i}{i!} \qquad (2.78)$$

Therefore,

$$p(x) = \frac{(\lambda \Delta t)^x}{x!} \sum_{i=0}^{\infty} \frac{(-\lambda \Delta t)^i}{i!} = \sum_{i=0}^{\infty} (-1)^i \frac{(\lambda \Delta t)^{x+i}}{x! i!} \qquad (2.79)$$

Now let us define a number $0(\Delta t) > 0$ such that if $|k| < 0(\Delta t)$, k may be considered zero for all practical purposes. Choose Δt such that

$$\frac{(\lambda \Delta t)^{x+i}}{x! i!} < 0(\Delta t), \qquad x + i > 1 \qquad (2.80)$$

Then

$$p(x) = \sum_{i=0}^{1}(-1)^i \frac{(\lambda \Delta t)^{x+i}}{x!\,i!} = \begin{cases} 0, & x > 1 \\ \lambda \Delta t, & x = 1 \\ 1 - \lambda \Delta t, & x = 0 \end{cases} \quad (2.81)$$

Hence, if we choose Δt small enough, then the probability of two or more events in that interval is zero.

We have already mentioned the unique relationship between the Poisson and exponential random variables. Specifically, if the number of events occurring in an interval of length T is a Poisson random variable, then the time between successive events is exponentially distributed. The converse is also easily shown (see Problem 23). To prove this result let t be the time between successive events. Then

$$P(T > t) = P(\text{no events in the interval } t) = e^{-\lambda t} \quad (2.82)$$

since events are assumed to occur in a Poisson fashion. If $F(t)$ is the distribution function of t, then

$$P(T > t) = 1 - F(t) \quad \text{and} \quad F(t) = 1 - e^{-\lambda t} \quad (2.83)$$

Therefore,

$$f(t) = \frac{d}{dt} F(t) = \lambda e^{-\lambda t}, \quad 0 < t < \infty \quad (2.84)$$

We have already demonstrated the relationship between the Poisson and exponential random variables. There is also an interesting relationship between the Poisson and uniform random variables. Suppose that k Poisson events have occurred in time t. Let the times of occurrence of the first through kth events be $v_1, v_2, \ldots, v_k$. Let $[t_i, t_i + dv_i]$, $i = 1, 2, \ldots, k$, be nonoverlapping intervals. Then

$$P(t < v_1 < t_1 + dv_1, \ldots, t_k < v_k < t_k + dv_k \mid k \text{ events in } t)$$

$$= \frac{P(1 \text{ event in } dv_1) P(1 \text{ event in } dv_2) \cdots P(1 \text{ event in } dv_k)}{P(k \text{ events in } t)}$$

$$\times P(0 \text{ events in } t - dv_1 - dv_2 - \cdots - dv_k)$$

$$= \frac{[\prod_{j=1}^{k} \lambda \, dv_j \exp(-\lambda \, dv_j)] \exp[-\lambda(t - dv_1 - dv_2 - \cdots - dv_k)]}{(1/k!)(\lambda t)^k e^{-\lambda t}}$$

$$= \frac{k!}{t^k} dv_1 \, dv_2 \cdots dv_k \quad (2.85)$$

Hence the joint density function of $v_1, v_2, \ldots, v_k$ given k events in t is given by

$$f(v_1, v_2, \ldots, v_k \mid k) = \frac{k!}{t^k}, \qquad 0 < v_1 < v_2 < \cdots < v_k < t$$

Now consider k independent uniformly distributed random variables x_i, $i = 1, 2, \ldots, k$, on the interval $(0, t)$. The probability density function of x_i is

$$f(x_i) = \frac{1}{t}, \qquad 0 < x_i < t, \quad i = 1, 2, \ldots, k \tag{2.86}$$

Now let us arrange these random variables in the order of their occurrence; that is, let

$$y_1 = \text{smallest of } x_1, x_2, \ldots, x_k$$
$$y_2 = \text{second smallest of } x_1, x_2, \ldots, x_k$$
$$\vdots$$
$$y_k = \text{largest of } x_1, x_2, \ldots, x_k$$

The random variables $y_1, y_2, \ldots, y_k$ are called the *order statistics* of $x_1, x_2, \ldots, x_k$. The joint density function of $y_1, y_2, \ldots, y_k$ is given by

$$f(y_1, y_2, \ldots, y_k) = \frac{k!}{t^k}, \qquad 0 < y_1 < y_2 < \cdots < y_k \tag{2.87}$$

Thus, given a Poisson process and the occurrence of k events in time t, then the times of occurrence of these k events are uniformly distributed over the interval $(0, t)$.

TRANSFORM METHODS

Functional transformations play an important role in nearly all of the physical and engineering sciences. Transforms commonly encountered include the Fourier (Giffin, 1971; LePage, 1961; Schmidt, 1974), Laplace (Giffin, 1971; Hall et al., 1959; LePage, 1961; Schmidt, 1974), Mellin and Z transforms (Brown, 1963; Giffin, 1971; Schmidt, 1974), the characteristic function (Lukacs, 1960; Parzen, 1960, 1965; Schmidt, 1974), moment generating function (Freeman, 1963; Giffin, 1971; Parzen, 1960; Wilks, 1962), and generating function (sometimes called the geometric transform) (Brown, 1963; Giffin, 1971). The characteristic, moment generating, and probability generating functions are frequently encountered in probability theory and its applications and bear a close relationship to the Fourier, Laplace, and Z transforms, respectively. The transforms with which we shall be concerned in

Transform Methods

this chapter and the remainder of this book are the Laplace transform and the generating function.

In general the usefulness of transform methods derives from the fact that many problems can be more readily solved after the application of a transform rather than in their initial form. For example, consider the problem of finding the distribution of the sum of identically distributed Erlang random variables. Conceptually this problem is one of simple convolution, although the mathematics involved is quite tedious. However, this problem can be solved without difficulty and with much less tedium by taking the transform of the density function of each random variable and taking the product of the transforms. The result is the transform of the density function of the sum desired. By inverting this transform (usually by referring to a table), the density of the sum of the random variables is obtained.

Probably the most widely used application of transforms—particularly the Laplace transform—lies in the solution of linear differential equations. In this case, one merely takes the transform of the linear differential equation and through rather simple algebraic manipulation obtains the transform of the function desired. Inversion of this transform yields the function that is the solution of the differential equation.

It should be emphasized that transform methods are basically a convenience; that is, they frequently offer a simplified approach to the solution of systems problems. On the other hand, the application of transform techniques can complicate the solution process unnecessarily. This is often the case, for example, when attempting to find the moments of a random variable with known probability density or mass function. Thus, transform methods should not be used indiscriminately, but rather where they will facilitate the solution process.

In the context of queueing theory, the Laplace transform is used principally to solve certain differential equations, to identify the distributions of continuous random variables, and in some cases to determine the moments of continuous random variables. The generating function serves a similar function in the case of discrete variables; that is, the probability generating function is used to solve certain difference equations, to identify the distributions of discrete random variables, and to determine the moments of discrete random variables.

Laplace Transform

Let $f(t)$ be a piecewise continuous function defined for real $t \geq 0$. The Laplace transform is defined by

$$\mathscr{L}[f(t)] = \int_0^\infty f(t)e^{-st}\,dt \qquad (2.88)$$

where the integral on the right exists, and s is a complex variable given by

$$s = \sigma + i\omega, \quad i = \sqrt{-1} \tag{2.89}$$

Thus, the Laplace transform converts a function $f(t)$ of a real variable to a function of a complex variable. Let

$$L(s) = \mathscr{L}[f(t)] \tag{2.90}$$

Then $f(t)$ and $L(s)$ form a unique transform pair; that is, $f(t)$ has only one Laplace transform, if it exists, and corresponding to each Laplace transform $L(s)$ there is only one function $f(t)$. Therefore, just as we can define $L(s)$ given $f(t)$ by (2.88), we should be able to define $f(t)$ given $L(s)$. This relationship is given by

$$f(t) = \frac{1}{2\pi i} \int_{a-i\infty}^{a+i\infty} L(s) e^{st}\, ds \tag{2.91}$$

Equation (2.91) defines the inverse Laplace transform, sometimes denoted by $\mathscr{L}^{-1}(s)$. Although (2.91) can be used to determine the inverse Laplace transform, its use requires a knowledge of contour integration (LePage, 1961; Schmidt, 1974). To circumvent this problem, extensive tables of Laplace and inverse Laplace transforms have been developed. (See Appendix, Table E, for a short list of Laplace transforms.)

Example 2.13 Find the Laplace transform of each of the following functions:

(a) $f(t) = \lambda e^{-\lambda t}, 0 < t < \infty, \lambda > 0$

(b) $f(t) = \dfrac{\lambda^n}{\Gamma(n)} t^{n-1} e^{-\lambda t}, 0 < t < \infty, \lambda > 0$

(c) $f(t) = \dfrac{1}{b-a}, a < t < b, 0 < a < b$

(d) $f(t) = a + bt + ct^2, 0 < t < \infty$

The function $f(t)$ defined in (a) is the density function of the exponential random variable, with Laplace transform given by

$$\mathscr{L}[f(t)] = \int_0^\infty \lambda e^{-\lambda t} e^{-st}\, dt = \int_0^\infty \lambda e^{-(\lambda+s)t}\, dt = \left(1 + \frac{s}{\lambda}\right)^{-1}$$

For (b),

$$\mathscr{L}[f(t)] = \int_0^\infty \frac{\lambda^n}{\Gamma(n)} t^{n-1} e^{-(\lambda+s)t}\, dt$$

by (2.88). Hence

$$\mathscr{L}[f(t)] = \frac{\lambda^n}{\Gamma(n)} \int_0^\infty t^{n-1} e^{-(\lambda+s)t} \, dt = \frac{\lambda^n}{\Gamma(n)} \frac{\Gamma(n)}{(\lambda + s)^n} = \left(1 + \frac{s}{\lambda}\right)^{-n}$$

For (c),

$$\mathscr{L}[f(t)] = \int_0^\infty f(t) e^{-st} \, dt$$

$$= \int_0^a 0 e^{-st} \, dt + \int_a^b \frac{e^{-st}}{b-a} \, dt + \int_b^\infty 0 e^{-st} \, dt$$

$$= \int_a^b \frac{e^{-st}}{b-a} \, dt = -\frac{e^{-st}}{s(b-a)} \bigg|_a^b = \frac{e^{-as} - e^{-bs}}{s(b-a)}$$

For (d),

$$\mathscr{L}[f(t)] = \int_0^\infty (a + bt + ct^2) e^{-st} \, dt$$

$$= a \int_0^\infty e^{-st} \, dt + b \int_0^\infty t e^{-st} + c \int_0^\infty t^2 e^{-st} \, dt$$

$$= \frac{a}{s} + \frac{b}{s^2} + \frac{2c}{s^3}$$

Laplace Transform of Special Functions

In subsequent chapters we will encounter situations requiring the development of the Laplace transform of special functions. In particular, we will require the Laplace transform of linear combinations of functions, derivatives and integrals of functions, convolutions, the unit step function, the Dirac delta function, and differential equations. In this section, we develop the Laplace transform for each of these special functions.

Linear combinations of functions. Let $f(t)$ be a linear combination of the functions $f_1(t), f_2(t), \ldots, f_n(t)$; that is,

$$f(t) = \sum_{i=1}^n a_i f_i(t) \tag{2.92}$$

Then the Laplace transform of $f(t)$ is given by

$$\mathscr{L}[f(t)] = \sum_{i=1}^n a_i \mathscr{L}[f_i(t)] \tag{2.93}$$

where $\mathscr{L}[f_i(t)]$, $i = 1, 2, \ldots, n$, are assumed to exist. We prove this result as follows:

$$\mathscr{L}[f(t)] = \int_0^\infty f(t)e^{-st}\,dt = \int_0^\infty \sum_{i=1}^n a_i f_i(t)e^{-st}\,dt$$

$$= \sum_{i=1}^n a_i \int_0^\infty f_i(t)e^{-st}\,dt$$

Since

$$\int_0^\infty f_i(t)e^{-st}\,dt = \mathscr{L}[f_i(t)]$$

the proof is complete. In words, (2.93) states that the Laplace transform is a linear operator.

Derivatives and integrals. In solving certain problems using Laplace transforms it sometimes becomes necessary to find the Laplace transform of the integral or derivative. Such problems may arise when attempting to solve integral or differential equations, when dealing with the distribution function of a continuous random variable, or when attempting to find the density function of the sum of continuous random variables. Let us first consider the problem of finding the Laplace transform of the derivative $d^n f(t)/dt^n$. For $n = 1$, we have

$$\mathscr{L}\left[\frac{d}{dt}f(t)\right] = \int_0^\infty f'(t)e^{-st}\,dt \qquad (2.94)$$

Integrating by parts leads to

$$\mathscr{L}\left[\frac{d}{dt}f(t)\right] = \int_0^\infty sf(t)e^{-st}\,dt + f(t)e^{-st}\Big|_0^\infty = s\mathscr{L}[f(t)] - f(0) \qquad (2.95)$$

assuming that the Laplace transform of $f(t)$ exists. For $n = 2$, let

$$G(t) = \frac{d}{dt}f(t) \qquad (2.96)$$

Then

$$\mathscr{L}\left[\frac{d^2}{dt^2}f(t)\right] = \mathscr{L}[G'(t)] = s\mathscr{L}[G(t)] - G(0)$$

$$= s^2\mathscr{L}[f(t)] - sf(0) - f'(0) \qquad (2.97)$$

by (2.95). Assume that

$$\mathscr{L}\left[\frac{d^{n-1}}{dt^{n-1}}f(t)\right] = s^{n-1}\mathscr{L}[f(t)] - \sum_{i=0}^{n-2} s^{n-2-i}f^{(i)}(0) \quad (2.98)$$

To find $\mathscr{L}[d^n f(t)/dt^n]$ let

$$G(t) = \frac{d^{n-1}}{dt^{n-1}}f(t) \quad (2.99)$$

Then

$$\mathscr{L}\left[\frac{d^n}{dt^n}f(t)\right] = \mathscr{L}[G'(t)] = s\mathscr{L}[G(t)] - G(0)$$

$$= s\left\{s^{n-1}\mathscr{L}[f(t)] - \sum_{i=0}^{n-2} s^{n-2-i}f^{(i)}(0)\right\} - f^{(n-1)}(0)$$

$$= s^n \mathscr{L}[f(t)] - \sum_{i=0}^{n-2} s^{n-1-i}f^{(i)}(0) - f^{(n-1)}(0)$$

$$= s^n \mathscr{L}[f(t)] - \sum_{i=0}^{n-1} s^{n-1-i}f^{(i)}(0) \quad (2.100)$$

where $f^{(i)}(0)$ are assumed to exist for $i = 1, 2, \ldots, n$.

Let us now turn our attention to the problem of finding the Laplace transform of an integral of the form $\int_0^t f(x)\,dx$. Let

$$G(t) = \int_0^t f(x)\,dx \quad (2.101)$$

Then $G'(t) = f(t)$ and

$$\mathscr{L}\left[\int_0^t f(x)\,dx\right] = \mathscr{L}[G(t)] = \int_0^\infty \int_0^t f(x)e^{-st}\,dx\,dt$$

By (2.95)

$$\mathscr{L}[G'(t)] = s\mathscr{L}[G(t)] - G(0)$$

but by definition

$$G(0) = 0$$

Therefore,

$$\mathscr{L}[G(t)] = \frac{1}{s}\mathscr{L}[G'(t)] \quad (2.102)$$

or

$$\mathscr{L}\left[\int_0^t f(x)\,dx\right] = \frac{1}{s}\mathscr{L}[f(t)] \qquad (2.103)$$

Convolutions. Let Y_1 and Y_2 be two independent continuous random variables with density functions $f(y_1)$ and $g(y_2)$, $y_1 > 0$, $y_2 > 0$, and assume that we wish to find the density function of T, where

$$T = Y_1 + Y_2$$

Now

$$F(t) = P(Y_1 + Y_2 \le t) = P(Y_1 \le t - y_2, Y_2 \le t)$$

$$= \int_0^t \int_0^{t-y_2} f(y_1)g(y_2)\,dy_1\,dy_2, \qquad t > 0 \qquad (2.104)$$

Then the density function of T, $h(t)$, is given by

$$h(t) = \frac{d}{dt}F(t) = \int_0^t f(t - y_2)g(y_2)\,dy_2 \qquad (2.105)$$

Equation (2.105) defines the convolution of the functions $f(y_1)$ and $g(y_2)$ and can be extended to define the convolution of an arbitrary number of functions. Although the analysis required to determine the convolution of functions is not conceptually complex, it may prove quite tedious. To simplify this problem we will use the Laplace transform. Assume that $h(t)$ and $g(t)$ are functions for which $\mathscr{L}[h(t)]$ and $\mathscr{L}[g(t)]$ exist. It should be noted that these functions are not necessarily density functions. If

$$f(t) = \int_0^t h(x)g(t - x)\,dx \qquad (2.106)$$

then

$$\mathscr{L}\left[\int_0^t h(x)g(t - x)\,dx\right] = \mathscr{L}[h(t)]\mathscr{L}[g(t)] \qquad (2.107)$$

The function $f(t)$ can then be found by finding the inverse transform of $\mathscr{L}[h(t)]\mathscr{L}[g(t)]$.

Example 2.14 Let Y_1 and Y_2 be independent, identically distributed exponential random variables each with parameter λ. Find the density function of $Y_1 + Y_2$ using (a) Eq. (2.106), (b) Eq. (2.107).

Let $g(y_1)$ and $g(y_2)$ be the density functions of Y_1 and Y_2, $X = Y_1 + Y_2$, and $f(x)$ the density function of X. Then

$$g(y_i) = \lambda \exp(-\lambda y_i), \qquad 0 < y_i < \infty, \quad i = 1, 2$$

Transform Methods

For (a), we have

$$f(x) = \int_0^x g(y_1)g(x - y_1)\,dy_1$$

$$= \lambda^2 \int_0^x \exp(-\lambda y_1) \exp[-\lambda(x - y_1)]\,dy_1$$

$$= \lambda^2 x e^{-\lambda x}, \quad 0 < x < \infty$$

For (b), we use (2.107), where

$$\mathscr{L}[g(x)] = \int_0^\infty \lambda e^{-\lambda x} e^{-sx}\,dx = \lambda \int_0^\infty e^{-(\lambda+s)x}\,dx = \left(1 + \frac{s}{\lambda}\right)^{-1}$$

Then

$$[f(x)] = \left(1 + \frac{s}{\lambda}\right)^{-2}$$

By Example 2.13(b),

$$f(x) = \lambda^2 x e^{-\lambda x}, \quad 0 < x < \infty$$

Unit step function. The unit step function, often called the Heaviside unit function, is defined by

$$U(t - a) = \begin{cases} 0, & t \le a \\ 1, & t > a \end{cases} \qquad (2.108)$$

and is used in the analysis of certain queueing systems. To illustrate, suppose that customer arrivals to a chain food store in an interval T are Poisson distributed with parameters $\lambda_1 T$ during the week and $\lambda_2 T$ on Saturday. Then the probability mass function for the number of arrivals X in any arbitrarily chosen interval T is given by

$$p(x) = [1 - U(t - 5)]\frac{(\lambda_1 T)^x}{x!} \exp(-\lambda_1 T)$$

$$+ U(t - 5)\frac{(\lambda_2 T)^x}{x!} \exp(-\lambda_2 T), \quad x = 0, 1, 2, \ldots \qquad (2.109)$$

where t is measured in days.

The Laplace transform of $U(t - a)$ is

$$\mathscr{L}[U(t - a)] = \int_0^\infty U(t - a)e^{-st}\,dt = \int_a^\infty e^{-st}\,dt = \frac{e^{-sa}}{s} \qquad (2.110)$$

Dirac delta function. The Dirac delta function, sometimes referred to as the unit impulse function or simply the delta function, is denoted by $\delta(t)$ and is defined by

$$\delta(t) = \lim_{\varepsilon \to 0} \begin{cases} \dfrac{1}{\varepsilon}, & 0 \leq t \leq \varepsilon \\ 0, & t > \varepsilon \end{cases} \quad (2.111)$$

Let

$$F_\varepsilon(t) = \begin{cases} \dfrac{1}{\varepsilon}, & 0 \leq t \leq \varepsilon \\ 0, & t > \varepsilon \end{cases} \quad (2.112)$$

Then

$$\int_0^\infty \delta(t)\, dt = \lim_{\varepsilon \to 0} \int_0^\infty F_\varepsilon(t)\, dt = \lim_{\varepsilon \to 0} \int_0^\varepsilon \frac{1}{\varepsilon}\, dt = 1 \quad (2.113)$$

Although (2.111) indicates that $\delta(t)$ does not exist, its integral as given in (2.113) does exist and is equal to unity. This and the following property make $\delta(t)$ useful in deriving waiting-time distributions for many queueing systems. Let $g(t)$ be any continuous function and let the delta function $\delta(t - a)$ be defined by

$$\delta(t - a) = \lim_{\varepsilon \to 0} \begin{cases} 0, & t < a \\ \dfrac{1}{\varepsilon}, & a \leq t \leq a + \varepsilon \\ 0, & t > a + \varepsilon \end{cases} \quad (2.114)$$

Then the convolution of $\delta(t)$ and $g(t)$ is

$$\int_0^\infty \delta(t - a) g(t)\, dt = \lim_{\varepsilon \to 0} \int_a^{a+\varepsilon} \frac{g(t)}{\varepsilon}\, dt = \lim_{\varepsilon \to 0} \frac{G(a + \varepsilon) - G(a)}{\varepsilon}$$

where

$$G(t) = \int_0^t g(x)\, dx, \quad \text{and} \quad \int_0^\infty \delta(t - a) g(t)\, dt = g(a) \quad (2.115)$$

since

$$\lim_{\varepsilon \to 0} \frac{G(a + \varepsilon) - G(a)}{\varepsilon} = \frac{d}{da} G(a) = g(a)$$

Transform Methods

The Laplace transform of $\delta(t)$ is given by

$$\mathscr{L}[\delta(t)] = \int_0^\infty \delta(t) e^{-st}\, dt = \lim_{\varepsilon \to 0} \int_0^\varepsilon \frac{1}{\varepsilon} e^{-st}\, dt = \lim_{\varepsilon \to 0} \frac{1 - e^{-s\varepsilon}}{s\varepsilon}$$

Letting

$$e^{-s\varepsilon} = \sum_{i=0}^\infty \frac{(-s\varepsilon)^i}{i!}$$

we have

$$\mathscr{L}[\delta(t)] = \lim_{\varepsilon \to 0} \frac{1 - \sum_{i=0}^\infty (-s\varepsilon)^i/i!}{s\varepsilon} = \lim_{\varepsilon \to 0}\left[1 - \sum_{i=2}^\infty (-1)^i \frac{(s\varepsilon)^{i-1}}{i!}\right] = 1 \tag{2.116}$$

To illustrate the use of the Dirac delta function, consider the following example.

Example 2.15 The probability mass function $p(x)$ of the number of units in a service system having one service facility is given by

$$p(x) = (1 - \rho)\rho^x, \qquad x = 0, 1, \ldots, \quad 0 < \rho < 1$$

The conditional density function of time spent waiting for service by a unit entering the system when there are x units already in the system is

$$f(t\mid x) = \begin{cases} 1, & t = 0, \quad x = 0 \\ 0, & t > 0, \quad x = 0 \\ \dfrac{u^x}{(x-1)!} t^{x-1} e^{-ut}, & 0 < t < \infty, \quad x = 1, 2, \ldots \end{cases}$$

Find the density function of t, $g(t)$.

The density function $g(t)$ is defined by

$$g(t) = \sum_{x=0}^\infty f(t\mid x) p(x)$$

As defined above, $f(t\mid x)$ is continuous for $x > 0$, but has a discontinuity at $t = 0$ for $x = 0$. For $x = 0$, the probability of an arriving unit having to wait is obviously zero. Thus, $f(0\mid 0)$ is the probability that an arriving unit does not wait for service, and is unity as we might expect.

We will define $g(t)$ by taking its Laplace transform and inverting:

$$\mathscr{L}[g(t)] = \int_0^\infty \sum_{x=0}^\infty f(t\,|\,x)p(x)e^{-st}\,dt$$

$$= \int_0^\infty f(t\,|\,0)p(0)e^{-st}\,dt + \int_0^\infty \sum_{x=1}^\infty f(t\,|\,x)p(x)e^{-st}\,dt$$

Since we cannot evaluate $\int_0^\infty f(t\,|\,0)p(0)e^{-st}\,dt$ directly, due to the definition of $f(t\,|\,0)$, we will use the Dirac delta function as an artifice to aid us in this evaluation. Let us redefine $\mathscr{L}[g(t)]$ as

$$\mathscr{L}[g(t)] = \int_0^\infty \delta(t)p(0)e^{-st}\,dt + \int_0^\infty \sum_{x=1}^\infty f(t\,|\,x)p(x)e^{-st}\,dt$$

where the coefficient of $\delta(t)$ here, and hereafter, is taken as the value of a time function at a discontinuity at $t = 0$. Now

$$\int_0^\infty \delta(t)p(0)e^{-st}\,dt = (1 - \rho)$$

since

$$\mathscr{L}[\delta(t)] = 1$$

Furthermore,

$$\int_0^\infty \sum_{x=1}^\infty f(t\,|\,x)p(x)e^{-st}\,dt = \int_0^\infty \sum_{x=1}^\infty (1 - \rho)\frac{(\rho u)^x}{(x-1)!}t^{x-1}e^{-ut}e^{-st}\,dt$$

$$= \sum_{x=1}^\infty (1 - \rho)\left(\frac{\rho u}{u + s}\right)^x = \frac{u\rho(1 - \rho)}{u(1 - \rho) + s}$$

Hence

$$\mathscr{L}[g(t)] = (1 - \rho) + \frac{u\rho(1 - \rho)}{u(1 - \rho) + s}$$

and

$$g(t) = (1 - \rho)\,\delta(t) + u\rho(1 - \rho)e^{-u(1-\rho)t}$$

or

$$g(t) = \begin{cases} (1 - \rho), & t = 0 \\ u\rho(1 - \rho)e^{-u(1-\rho)t}, & 0 < t < \infty \end{cases}$$

since the coefficient of $\delta(t)$, $1 - \rho$, is treated as the value of the time function $g(t)$ at the discontinuity at $t = 0$.

Transform Methods

Differential equations. The Laplace transform can be useful in solving ordinary linear differential equations with constant coefficients and some ordinary differential equations with variable coefficients. Our attention will be restricted to the case of constant coefficients. Consider the differential equation

$$a \frac{d^2}{dt^2} f(t) + b \frac{d}{dt} f(t) + cf(t) = g(t) \tag{2.117}$$

where a, b, and c are constants, $f(t)$ an unknown function of t, and $g(t)$ a known function of t. To solve this differential equation we must determine $f(t)$. Let

$$L(s) = \mathscr{L}[f(t)] \quad \text{and} \quad G(s) = \mathscr{L}[g(t)]$$

Then taking the Laplace transform of both sides of (2.117) yields

$$a[s^2 L(s) - sf(0) - f'(0)] + b[sL(s) - f(0)] + cL(s) = G(s) \tag{2.118}$$

or

$$L(s)[as^2 + bs + c] = G(s) + f(0)[as + b] + af'(0)$$

and

$$L(s) = \frac{G(s) + f(0)[as + b] + af'(0)}{as^2 + bs + c} \tag{2.119}$$

Assuming that $G(s)$ can be determined, if $f(0)$ and $f'(0)$, initial conditions, are known, then $L(s)$ is explicitly defined. Inverting $L(s)$ we have the function $f(t)$ satisfying (2.119).

As implied in the above discussion, the function $f(t)$ satisfying a given differential equation cannot be explicitly determined until initial conditions for the problem are defined. Of course, it is also assumed that the Laplace transforms of $f(t)$ and $g(t)$ exist. To illustrate the use of Laplace transforms for solving differential equations consider the following examples.

Example 2.16 Solve the following differential equation using Laplace transforms:

$$\frac{d^2}{dt^2} f(t) + 3f(t) = t$$

where $f'(0) = f(0) = 0$.

Taking the Laplace transform of both sides of the differential equation yields

$$\mathscr{L}\left[\frac{d^2}{dt^2} f(t) + 3f(t)\right] = s^2 L(s) - sf(0) - f'(0) + 3L(s)$$

and

$$\mathcal{L}[t] = \frac{1}{s^2}$$

from Table E in the Appendix. Since $f'(0) = f(0) = 0$, we have

$$L(s)[s^2 + 3] = \frac{1}{s^2} \quad \text{or} \quad L(s) = \frac{1}{s^2(s^2 + 3)}$$

Now $L(s)$ can be expressed by

$$L(s) = \frac{1}{3}\left[\frac{1}{s^2} - \frac{1}{(s^2 + 3)}\right]$$

Let $1/s^2$ be the Laplace transform of the function $g(t)$ and $1/(s^2 + 3)$ the Laplace transform of $h(t)$. Then

$$f(t) = \tfrac{1}{3}[g(t) - h(t)]$$

From Table E,

$$g(t) = t, \quad \text{and} \quad h(t) = \frac{1}{\sqrt{3}}\sin(\sqrt{3}\,t)$$

Therefore,

$$f(t) = \frac{1}{3}\left[t - \frac{1}{\sqrt{3}}\sin(\sqrt{3}\,t)\right]$$

To check this solution, let us substitute $f(t)$ into the differential equation:

$$\frac{d^2}{dt^2}f(t) = \frac{1}{\sqrt{3}}\sin(\sqrt{3}\,t)$$

Then

$$\frac{d^2}{dt^2}f(t) + 3f(t) = \frac{1}{\sqrt{3}}\sin(\sqrt{3}\,t) + t - \frac{1}{\sqrt{3}}\sin(\sqrt{3}\,t) = t$$

and $f(t)$ is the desired solution.

Example 2.17 The rate of production per hour of defective parts as a function of operating time $f(t)$ satisfies the differential equation

$$\frac{d^2}{dt^2}f(t) + \frac{d}{dt}f(t) = 0$$

where

$$f(0) = 0.15, \quad f'(0) = 0.05$$

Find the function $f(t)$.

The Laplace transform of the differential equation is given by

$$\mathscr{L}\left[\frac{d^2}{dt^2}f(t) + \frac{d}{dt}f(t)\right] = [s^2 L(s) - sf(0) - f'(0)] + [sL(s) - f(0)]$$

Therefore,

$$L(s)[s^2 + s] = sf(0) + f'(0) + f(0),$$

or

$$L(s) = \frac{sf(0) + f'(0) + f(0)}{s(s+1)}$$

Substituting the initial conditions into the expression for $L(s)$ yields

$$L(s) = \frac{0.15(s+1) + 0.05}{s(s+1)}, \quad \text{or} \quad L(s) = \frac{0.15}{s} + \frac{0.05}{s(s+1)}$$

From Table E in the Appendix, the inverse transform of $0.15/s$ is 0.15, and is $0.05(1 - e^{-t})$ for $0.05/s(s+1)$. Therefore,

$$f(t) = 0.20 - 0.05e^{-t}$$

Substitution of $f(t)$ into the differential equation will verify that it is the required solution.

In Example 2.16 the inverse Laplace transform of $L(s)$ could not be identified using Table E in the Appendix alone. To resolve this problem we expressed the original Laplace transform as a linear function of a Laplace transform that could be found in Table E in the Appendix. This reformulation of the Laplace transform is called partial fraction expansion and is possible in certain cases where the Laplace transform is a ratio of polynomials in s.

Partial Fraction Expansion

As demonstrated in Example 2.16, the Laplace transform is sometimes expressed as the ratio of polynomials; that is,

$$\mathscr{L}[f(t)] = \frac{I(s)}{H(s)} \tag{2.120}$$

where $I(s)$ and $H(s)$ are polynomials in s, of degree m_1 and m_2, respectively. If $H(s)$ can be expressed as

$$H(s) = \prod_{i=1}^{m_2} h_i(s) \tag{2.121}$$

where $h_i(s)$ is a linear function of s for all i, and if $m_1 < m_2$ then

$$\mathscr{L}[f(t)] = \sum_{i=1}^{m_2} \frac{a_i}{h_i(s)} \tag{2.122}$$

where a_i are constants for all i, and $h_i(s) \neq h_j(s)$ for $i \neq j$. Therefore, under the conditions given, one can express a Laplace transform as the sum of partial fractions where the numerator of each partial fraction is a constant and where the denominator of each partial fraction is a linear function of the transform variable s.

Partial fraction expansions are particularly useful in attempting to find the inverse of a Laplace transform, since each partial fraction is of the form $a/(s + b)$. Therefore, $\mathscr{L}^{-1}[f(t)]$ can be expressed as

$$\mathscr{L}^{-1}[f(t)] = \sum_{i=1}^{m_2} \mathscr{L}^{-1}\left(\frac{a_i}{s + b_i}\right) \tag{2.123}$$

or

$$f(t) = \sum_{i=1}^{m_2} a_i \exp(-b_i t) \tag{2.124}$$

To illustrate the use of the partial fraction expansion, consider the following example.

Example 2.18 Let

$$\mathscr{L}[f(t)] = \frac{s^2 - 3}{s^3 + 4s^2 + s - 6}$$

Find $f(t)$.

The denominator of $\mathscr{L}[f(t)]$ can be expressed as

$$s^3 + 4s^2 + s - 6 = (s - 1)(s + 2)(s + 3)$$

Since the degree of the polynomial in the numerator is less than that in the denominator, $\mathscr{L}[f(t)]$ can be defined as

$$\begin{aligned} \mathscr{L}[f(t)] &= \frac{a_1}{s-1} + \frac{a_2}{s+2} + \frac{a_3}{s+3} \\ &= \frac{a_1(s+2)(s+3) + a_2(s-1)(s+3) + a_3(s-1)(s+2)}{(s-1)(s+2)(s+3)} \\ &= \frac{a_1(s^2 + 5s + 6) + a_2(s^2 + 2s - 3) + a_3(s^2 + s - 2)}{(s-1)(s+2)(s+3)} \end{aligned}$$

But

$$s^2 \sum_{i=1}^{3} a_i = s^2, \qquad s(5a_1 + 2a_2 + a_3) = 0,$$

$$(6a_1 - 3a_2 - 2a_3) = -3$$

Solving for a_1, a_2, and a_3 yields

$$a_1 = -\tfrac{1}{6}, \qquad a_2 = -\tfrac{1}{3}, \qquad a_3 = \tfrac{3}{2}$$

and

$$\mathscr{L}[f(t)] = -\frac{1}{6}\left(\frac{1}{s-1}\right) - \frac{1}{3}\left(\frac{1}{s+2}\right) + \frac{3}{2}\left(\frac{1}{s+3}\right)$$

Therefore,

$$f(t) = -\tfrac{1}{6}e^{t} - \tfrac{1}{3}e^{-2t} + \tfrac{3}{2}e^{-3t}$$

Thus far we have assumed that the polynomial $H(s)$ in (2.120) can be expressed as the product of distinct linear functions of s. However, it is not always possible to express $H(s)$ in this manner. Consider, for example, the following Laplace transform:

$$\mathscr{L}[f(t)] = \frac{6s^2 + 3s + 1}{s^3 - s^2 - s + 1}$$

The expression in the denominator can be factored as $(s+1)(s-1)^2$ or $(s+1)(s-1)(s-1)$. If we let

$$h_1(s) = s + 1, \qquad h_2(s) = s - 1, \qquad h_3(s) = s - 1$$

then $H(s)$ can be expressed as a product of linear functions of s; but the linear functions are not distinct.

For generality, assume that $H(s)$ can be expressed as

$$H(s) = \prod_{i=1}^{m_2} [h_i(s)]^{k_i}$$

where $h_i(s)$ is a linear function of s, k_i is a positive integer, and $h_i(s) \neq h_j(s)$ for $i \neq j$. We can then express $\mathscr{L}[f(t)]$ as

$$\mathscr{L}[f(t)] = \frac{I(s)}{H(s)} = \sum_{j=1}^{k_1} \frac{a_{1j}}{[h_1(s)]^j} + \sum_{j=1}^{k_2} \frac{a_{2j}}{[h_2(s)]^j} + \cdots + \sum_{j=1}^{k_{m.}} \frac{a_{m_2 j}}{[h_{m_2}(s)]^j}$$

The coefficients $a_{1j}, a_{2j}, \ldots, a_{m_2 j}$ can now be evaluated in the manner described previously.

Example 2.19 Let

$$\mathscr{L}[f(t)] = \frac{6s^2 - 3s + 1}{s^3 - s^2 - s + 1}$$

Find $f(t)$.

$\mathscr{L}[f(t)]$ can be expressed as

$$\mathscr{L}[f(t)] = \frac{6s^2 - 3s + 1}{(s+1)(s-1)^2} = \frac{a_{11}}{s+1} + \frac{a_{21}}{s-1} + \frac{a_{22}}{(s-1)^2}$$

$$= \frac{a_{11}(s-1)^3 + a_{21}(s+1)(s-1)^2 + a_{22}(s+1)(s-1)}{(s+1)(s-1)^3}$$

$$= \frac{a_{11}(s-1)^2 + a_{21}(s+1)(s-1) + a_{22}(s+1)}{(s+1)(s-2)^2}$$

Now

$$a_{11}(s-1)^2 + a_{21}(s+1)(s-1) + a_{22}(s+1)$$
$$= s^2(a_{11} + a_{21}) + s(-2a_{11} + a_{22}) + (a_{11} - a_{21} + a_{22})$$

Therefore,

$$a_{11} + a_{21} = 6, \quad -2a_{11} + a_{22} = -3, \quad a_{11} - a_{21} + a_{22} = 1$$

and

$$a_{11} = \tfrac{5}{2}, \quad a_{21} = \tfrac{7}{2}, \quad a_{22} = 2$$

Hence

$$\mathscr{L}[f(t)] = \frac{5/2}{s+1} + \frac{7/2}{s-1} + \frac{2}{(s-1)^2}$$

From Table E in the Appendix,

$$\mathscr{L}^{-1}\left[\frac{1}{s+1}\right] = e^{-t} \quad \text{and} \quad \mathscr{L}^{-1}\left[\frac{1}{s-1}\right] = e^{t}$$

From Example 2.13(b), the Laplace transform of $[\lambda^n/\Gamma(n)]t^{n-1}e^{-\lambda t}$ is $\lambda^n/(\lambda + s)^n$. Letting $n = 2$ and $\lambda = -1$, we have

$$\mathscr{L}^{-1}\left[\frac{1}{(s-1)^2}\right] = \frac{1}{\Gamma(2)} t e^t$$

Finally,

$$f(t) = \tfrac{5}{2}e^{-t} + \tfrac{7}{2}e^t + 2te^t$$

Transform Methods

Moments and the Laplace Transform

Earlier in this chapter we discussed the expected value of a function of a random variable. Special attention was given to the mean $E(X)$ and the variance $E(X^2) - [E(X)]^2$, because of their importance in probability theory and its applications. $E(X)$ and $E(X^2)$ are referred to, respectively, as the first and second moments of the random variable X. Higher moments are also of interest, particularly in parameter estimation. The nth moment of X is defined by

$$E(X^n) = \int_0^\infty x^n f(x)\, dx \tag{2.125}$$

when X is positive valued and continuous with density function $f(x)$. Now the Laplace transform of $f(x)$ is given by

$$L(s) = \int_0^\infty e^{-sx} f(x)\, dx \tag{2.126}$$

Let us take the first derivative of $L(s)$ with respect to s:

$$\frac{d}{ds} L(s) = -\int_0^\infty x e^{-sx} f(x)\, dx \tag{2.127}$$

If we evaluate (2.127) at $s = 0$, we have

$$\left.\frac{d}{ds} L(s)\right|_{s=0} = -\int_0^\infty x f(x)\, dx \tag{2.128}$$

or

$$E(X) = -\left.\frac{d}{ds} L(s)\right|_{s=0} \tag{2.129}$$

If we evaluate $d^n L(s)/ds^n \big|_{s=0}$, we are led to

$$\left.\frac{d^n}{ds^n} L(s)\right|_{s=0} = (-1)^n \left.\int_0^\infty x^n e^{-sx} f(x)\, dx\right|_{s=0}$$

$$= (-1)^n \int_0^\infty x^n f(x)\, dx \tag{2.130}$$

or

$$E(X^n) = (-1)^n \left.\frac{d^n}{ds^n} L(s)\right|_{s=0} \tag{2.131}$$

2. Probability Theory and Transform Methods

Example 2.20 Let X be a continuous random variable having a uniform distribution with density function

$$f(x) = \frac{1}{b-a}, \qquad 0 < a < x < b$$

Find the mean and variance of X using the Laplace transform.

The Laplace transform of X is given by

$$L(s) = \int_0^\infty e^{-sx} f(x)\, dx = \int_a^b \frac{e^{-sx}}{b-a}\, dx = \frac{e^{-as} - e^{-bs}}{s(b-a)}$$

The first derivative of $L(s)$ is given by

$$\frac{d}{ds} L(s) = \frac{1}{b-a} \left[\frac{e^{-bs}(sb+1) - e^{-as}(sa+1)}{s^2} \right]$$

From (2.128),

$$E(X) = -\frac{d}{ds} L(s) \bigg|_{s=0}$$

However, $dL(s)/ds \big|_{s=0}$ is indeterminate. Applying L'Hospital's rule, we have

$$\left[\frac{e^{-bs}(sb+1) - e^{-as}(sa+1)}{s^2} \right] \bigg|_{s=0}$$

$$= \left[\frac{-be^{-bs}(sb+1) + be^{-bs} + ae^{-as}(sa+1) - ae^{-as}}{2s} \right] \bigg|_{s=0}$$

$$= \left[\frac{b^2 e^{-bs}(sb+1) - b^2 e^{-bs} - b^2 e^{-bs} - a^2 e^{-as}(sa+1)}{2} \right.$$

$$\left. + \frac{a^2 e^{-as} + a^2 e^{-as}}{2} \right] \bigg|_{s=0}$$

$$= \frac{a^2 - b^2}{2}$$

Hence

$$E(X) = -\frac{a^2 - b^2}{2(b-a)} = \frac{a+b}{2}$$

The variance of X is

$$\sigma_x^2 = E(X^2) - [E(X)]^2$$

Transform Methods

Therefore, we must determine $E(X^2)$ using (2.131). The second derivative of $L(s)$ is

$$\frac{d^2}{ds^2} L(s) = \frac{1}{b-a}\left[\frac{a^2 e^{-as} - b^2 e^{-bs}}{s} - 2\frac{be^{-bs} - ae^{-as}}{s^2} - 2\frac{e^{-bs} - e^{-as}}{s^3}\right]$$

Applying L'Hospital's rule as before yields

$$\sigma_x^2 = \frac{b^3 - a^3}{3(b-a)} - \frac{(b+a)^2}{4} = \frac{(b-a)^2}{12}$$

Generating Function

The generating function, sometimes called the probability generating function, geometric transform, or z transform (Beightler et al., 1961; Elmaghraby, 1966; Feller, 1957; Giffin, 1971; Hadley, 1967; Hadley and Whitin, 1963) is used in a manner similar to the Laplace transform, but for functions of a discrete variable. In particular, we shall apply this transform to positive, integer-valued random variables. Let $p(x)$ be the probability mass function of the random variable X, where $x = 0, 1, 2, \ldots$. The generating function of x is given by

$$\mathscr{G}[p(x)] = G(z) = \sum_{x=0}^{\infty} z^x p(x) \tag{2.132}$$

As in the case of the Laplace transform, the correspondence between $p(x)$ and $G(z)$ is unique. Given the generating function, the corresponding probability mass function can be obtained by successive differentiation of $G(z)$, which can be shown by expanding $G(z)$:

$$G(z) = p(0) + zp(1) + z^2 p(2) + \cdots \tag{2.133}$$

Therefore,

$$p(0) = G(z)\bigg|_{z=0} \tag{2.134}$$

Now taking the first derivative of $G(z)$,

$$\frac{d}{dz} G(z) = p(1) + 2zp(2) + 3z^2 p(3) + \cdots \tag{2.135}$$

we have

$$p(1) = \frac{d}{dz} G(z)\bigg|_{z=0} \tag{2.136}$$

2. Probability Theory and Transform Methods

Extending this analysis to the xth derivative of $G(z)$, we are led to

$$\frac{d^x}{dz^x} G(z) = x!\, p(x) + [(x+1)x \cdots 2]zp(x+1) \cdots \quad (2.137)$$

and

$$p(x) = \frac{1}{x!} \frac{d^x}{dz^x} G(z) \bigg|_{z=0} \quad (2.138)$$

Thus, unlike the Laplace transform, the inverse transform of the generating function can be found in a rather straightforward manner. Even so, it is generally convenient to use tables of generating functions to identify inverse transforms. Table F in the Appendix provides a collection of generating functions for this purpose.

Example 2.21 Find the generating function of the binomial random variable.

The probability mass function of the binomial random variable is given by

$$p(x) = \frac{n!}{x!\,(n-x)!} p^x (1-p)^{n-x}, \quad x = 0, 1, 2, \ldots, n$$

Therefore,

$$G(z) = \sum_{x=0}^{\infty} z^x p(x) = \sum_{x=0}^{n} z^x \frac{n!}{x!\,(n-x)!} p^x (1-p)^{n-x}$$

Noting that

$$\sum_{y=0}^{k} \frac{k!}{y!\,(k-y)!} a^y b^{k-y} = (a+b)^k$$

we have

$$G(z) = \sum_{x=0}^{n} \frac{n!}{x!\,(n-x)!} (zp)^x (1-p)^{n-x} = [1 + p(z-1)]^n$$

Example 2.22 Let

$$G(z) = e^{-\lambda(1-z)}$$

be the generating function of the random variable X. Find the probability mass function of X.

From (2.138),

$$p(x) = \frac{1}{x!} \frac{d^x}{dz^x} G(z) \bigg|_{z=0}$$

Transform Methods

Now

$$\frac{d^x}{dz^x} G(z) = \lambda^x e^{-\lambda(1-z)}$$

Therefore,

$$p(x) = \frac{\lambda^x}{x!} e^{-\lambda}, \qquad x = 0, 1, 2, \ldots$$

and X is Poisson distributed with parameter λ.

The first and second moments can be obtained from the generating function without great difficulty. However, higher-order moments are not so easily obtained. Consider the nth derivative of $G(z)$ with respect to z:

$$\frac{d^n}{dz^n} G(z) = \sum_{x=0}^{\infty} x^{(n)} z^{x-n} p(x) \qquad (2.139)$$

where $x^{(n)}$ is called a factorial polynomial and is defined by

$$x^{(n)} = x(x-1)(x-2) \cdots (x-n+1) = \begin{cases} 0, & x < n \\ \dfrac{x!}{(x-n)!}, & x \geq n \end{cases} \qquad (2.140)$$

Evaluating (2.139) at $z = 1$ leads to

$$\left. \frac{d^n}{dz^n} G(z) \right|_{z=1} = \sum_{x=0}^{\infty} x^{(n)} p(x) = E[X^{(n)}] \qquad (2.141)$$

For $n = 1$,

$$E[X^{(1)}] = E(X)$$

Therefore,

$$E(X) = \left. \frac{d}{dz} G(z) \right|_{z=1} \qquad (2.142)$$

For $n = 2$,

$$E[X^{(2)}] = E(X^2 - X) = E(X^2) - E(X) \qquad (2.143)$$

or

$$E(X^2) = \left. \frac{d^2}{dz^2} G(z) \right|_{z=1} + \left. \frac{d}{dz} G(z) \right|_{z=1} \qquad (2.144)$$

Since
$$\sigma_X^2 = E(X^2) - [E(X)]^2$$
we have
$$\sigma_X^2 = \frac{d^2}{dz^2}G(z)\bigg|_{z=1} + \frac{d}{dz}G(z)\bigg|_{z=1} - \left[\frac{d}{dz}G(z)\bigg|_{z=1}\right]^2 \quad (2.145)$$

Example 2.23 Find the mean and variance of the Poisson random variable using the generating function.
From Example 2.22,
$$G(z) = e^{-\lambda(1-z)} \quad \text{and} \quad \frac{d^n}{dz^n}G(z) = \lambda^n e^{-\lambda(1-z)}$$

From (2.142),
$$E(X) = \lambda$$

$E(X^2)$ can be found from (2.144):
$$E(X^2) = \lambda^2 + \lambda$$

Therefore,
$$\sigma_X^2 = \lambda^2 + \lambda - \lambda^2 = \lambda$$

Example 2.24 Let $f(x, t)$ be a function of the discrete variable x and the continuous variable t. In particular, $f(x, t)$ is defined for $x = 0, 1, 2, \ldots$ and $0 < t < \infty$. Show that the generating function of the Laplace transform of $f(x + 1, t)$ is given by

$$\mathcal{G}\{\mathcal{L}[f(x + 1, t)]\} = z^{-1}\mathcal{G}\{\mathcal{L}[f(x, t)]\} - z^{-1}\mathcal{L}[f(0, t)]$$

The Laplace transform of $f(x + 1, t)$ is

$$\mathcal{L}[f(x + 1, t)] = \int_0^\infty f(x + 1, t)e^{-st}\, dt$$

and the generating function of $\mathcal{L}[f(x + 1, t)]$ is given by

$$\mathcal{G}\{\mathcal{L}[f(x + 1, t)]\} = \sum_{x=0}^\infty z^x \int_0^\infty f(x + 1, t)e^{-st}\, dt$$
$$= \int_0^\infty \left[\sum_{x=0}^\infty z^x f(x + 1, t)\right] e^{-st}\, dt$$

Transform Methods

Now
$$\sum_{x=0}^{\infty} z^x f(x+1, t) = z^{-1} \sum_{x=0}^{\infty} z^{x+1} f(x+1, t)$$

To $\sum_{x=0}^{\infty} z^{x+1} f(x+1, t)$ add and subtract $f(0, t)$, yielding

$$\sum_{x=0}^{\infty} z^{x+1} f(x+1, t) = \sum_{y=0}^{\infty} z^y f(y, t) - f(0, t)$$

Then

$$\mathscr{G}\{\mathscr{L}[f(x+1, t)]\} = \int_0^{\infty} \left[z^{-1} \sum_{y=0}^{\infty} z^y f(y, t) - z^{-1} f(0, t) \right] e^{-st} dt$$

$$= z^{-1} \sum_{y=0}^{\infty} z^y \int_0^{\infty} f(y, t) e^{-st} dt - z^{-1} \int_0^{\infty} f(0, t) e^{-st} dt$$

But

$$\sum_{y=0}^{\infty} z^y \int_0^{\infty} f(y, t) e^{-st} dt = \sum_{y=0}^{\infty} z^y \mathscr{L}[f(y, t)] = \mathscr{G}\{\mathscr{L}[f(y, t)]\}$$

and

$$\int_0^{\infty} f(0, t) e^{-st} dt = \mathscr{L}[f(0, t)]$$

Therefore,

$$\mathscr{G}\{\mathscr{L}[f(x+1, t)]\} = z^{-1} \mathscr{G}\{\mathscr{L}[f(x, t)]\} - z^{-1} \mathscr{L}[f(0, t)]$$

Sum of Random Variables

One of the more important applications of the generating function is its use in identifying the distribution of the sum of discrete random variables. For example, suppose that demand for a certain product comes from n different sources. Let $Y_1, Y_2, \ldots, Y_n$ be independent random variables representing weekly demand from these sources with associated probability mass functions $p(y_1), p(y_2), \ldots, p(y_n)$. What then is the probability mass function of total weekly demand X? X is defined as

$$X = Y_1 + Y_2 + \cdots + Y_n$$

To resolve this problem, we first identify the generating functions of $Y_1, Y_2, \ldots, Y_n$. Let $G_i(z)$ be the generating function of Y_i, $i = 1, 2, \ldots, n$, and let $G(z)$ be the generating function of X. Then it can be shown that

$$G(z) = \prod_{i=1}^{n} G_i(z) \qquad (2.146)$$

Since $G_i(z)$ are assumed to be known, $G(z)$ is easily defined. To find the probability mass function of X we can invert $G(z)$ using (2.138). However, we are sometimes able to recognize the derived generating function and thus identify the corresponding random variable visually.

Example 2.25 Let $Y_1, Y_2, \ldots, Y_n$ be independent Poisson random variables with parameters $\lambda_1, \lambda_2, \ldots, \lambda_n$, respectively. If

$$X = Y_1 + Y_2 + \cdots + Y_n$$

what is the probability mass function of X?

From Example 2.22,

$$G_i(z) = \exp[-\lambda_i(1-z)], \quad i = 1, 2, \ldots, n$$

Then by (2.146),

$$G(z) = \prod_{i=1}^{n} \exp[-\lambda_i(1-z)] = \exp\left[-(1-z)\sum_{i=1}^{n} \lambda_i\right]$$

Let

$$\gamma = \sum_{i=1}^{n} \lambda_i$$

Then

$$G(z) = e^{-\gamma(1-z)}$$

Since the generating function uniquely characterizes a random variable, X is Poisson distributed with parameter $\gamma = \sum_{i=1}^{n} \lambda_i$ and

$$p(x) = \frac{(\sum_{i=1}^{n} \lambda_i)^x}{x!} \exp\left(-\sum_{i=1}^{n} \lambda_i\right)$$

Therefore, the sum of independent Poisson random variables is Poisson distributed with parameter equal to the sum of the parameters of the individual random variables.

In the preceding discussion we considered the number of random variables included in the sum $Y_1 + Y_2 + \cdots + Y_n$ to be a constant; that is, n is known. Let us now consider the case where n is also a random variable. Thus we wish to determine the probability mass function of a random sum of random variables. Let

$$X = Y_1 + Y_2 + \cdots + Y_n$$

For a given value of n, the generating function of X is given by

$$G(z \mid n) = \prod_{i=1}^{n} G_i(z) \qquad (2.147)$$

Transform Methods

from (2.146); that is, $G(z|n)$ is the conditional generating function of X given that there are n random variables included in the sum. If $p(n)$ is the probability mass function of the number of random variables in the sum, then the unconditional generating function of n is given by

$$G(z) = \sum_{n=1}^{\infty} G(z|n)p(n) = \sum_{n=1}^{\infty} \prod_{i=1}^{n} G_i(z)p(n) \qquad (2.148)$$

Example 2.26 A closed-loop belt conveyor is used to transfer parts from a loading station to an unloading station. The number of parts placed in a tote bin at the loading station is a discrete random variable having the following probability mass function:

$$p(x) = \binom{10}{x}(0.8)^x(0.2)^{10-x}, \qquad x = 0, 1, \ldots, 10$$

The time between placements of tote bins on the conveyor at the loading station is a random variable with the following probability density function:

$$f(t) = 0.25e^{-0.25t}, \qquad t > 0$$

where t is measured in minutes. The conveyor has a constant speed of 2 ft/min. Develop the transform for the number of parts per foot located on the conveyor.

Since the conveyor is moving at a constant speed and the time between placements of tote bins is exponentially distributed, N, the number of tote bins placed on the conveyor per foot, will follow the Poisson distribution with parameter 0.125. To obtain this result, notice that the probability mass function for N, the number of tote bins placed on the conveyor during a 1-min interval, is Poisson distributed with parameter 0.25. Furthermore, the elapsed time for 1 ft of conveyor to pass the loading station is 0.5 min. Consequently, the number of tote bins placed on the conveyor per foot of conveyor is the same as the number of tote bins placed on the conveyor during a 0.5-min interval. The latter is Poisson distributed with a mean of 0.125. Therefore,

$$p(n) = \frac{e^{-0.125}(0.125)^n}{n!}, \qquad n = 0, 1, 2, \ldots$$

Since X denotes the number of parts in a tote bin and N the number of tote bins per foot of conveyor, then Y, the number of parts per foot of conveyor, is given by

$$Y = \sum_{j=1}^{N} X_j$$

From (2.166) we see that $G_Y(z)$, the generating function for Y, can be expressed as

$$G_Y(z) = \sum_{n=0}^{\infty} \prod_{j=1}^{n} G_j(z) p(n)$$

where $G_j(z)$ is the generating function for X. However, since $G_1(z) = G_2(z) = \cdots = G_n(z) = G_X(z)$, we see that $G_Y(z)$ can be given as

$$G_Y(z) = \sum_{n=0}^{\infty} [G_X(z)]^n p(n)$$

Letting Θ equal $G_X(z)$,

$$G_Y(z) = \sum_{n=0}^{\infty} \Theta^n p(n)$$

or, from the definition of the generating function,

$$G_Y(z) = G_N(\Theta)$$

where $G_N(\Theta)$ is the generating function for N with z replaced by $G_X(z)$. Consequently, the generating function for Y is given by the generating function of X nested within the generating function for N, or

$$G_Y(z) = G_N[G_X(z)]$$

Now, since

$$G_N(z) = e^{-0.125(1-z)} \quad \text{and} \quad G_X(z) = (0.2 + 0.8z)^{10}$$

then

$$G_Y(z) = \exp\{-0.125[1 - (0.2 + 0.8z)^{10}]\}$$

Example 2.27 Determine the mean and variance for the number of parts per foot of conveyor in the previous example.

From Problem 34,

$$E(Y) = E(N)E(X_j) = 0.125(8) = 1.0$$

and

$$\sigma_Y^2 = \sigma_{X_j}^2 E(N) + \sigma_N^2 [E(X_j)]^2 = 1.6(0.125) + 0.125(64) = 8.2$$

Example 2.28 Cars arrive at a police roadblock at a Poisson rate of 10 cars per hour. Cars are chosen at random with probability of 0.6 for an intensive search. The time required to search a car is a random variable, with density function

$$f(t) = 0.1 e^{-0.1t}, \quad t > 0$$

where t is measured in minutes.

Transform Methods

(a) Develop the generating function and probability mass function for the number of cars searched per hour.

(b) Develop the Laplace transform for the total amount of time spent searching cars during a one-hour interval.

(c) Determine the expected value and variance for the total amount of time spent searching cars during a one-hour interval.

(a) Let X be a Bernoulli random variable, taking on the value one if a car is searched and zero otherwise. Let N be the number of cars arriving at the roadblock during a one-hour interval. If Y is the number of cars to be searched per hour, then

$$Y = \sum_{j=1}^{N} X_j$$

and the generating function for Y is

$$G_Y(z) = G_N[G_X(z)] = e^{-10(1-G_X(z))} = \exp\{-10[1-(0.4+0.6z)]\} = e^{-6(1-z)}$$

Consequently, Y is Poisson distributed with a mean of 6 cars searched per hour.

(b) If W is the total amount of time spent searching cars during a one-hour interval, then

$$W = \sum_{j=1}^{Y} T_j$$

Therefore, the Laplace transform for W, given a value of Y, is

$$L_W(s|y) = \prod_{j=1}^{y} L_j(s)$$

and the unconditional Laplace transform is

$$L_W(s) = \sum_{y=0}^{\infty} \prod_{j=1}^{y} L_j(s) p(y) = \sum_{y=0}^{\infty} [L_T(s)]^y p(y) = G_Y(z) \bigg|_{z=L_T(s)} = G_Y[L_T(s)]$$

Consequently, the Laplace transform of W is obtained by nesting the Laplace transform of T within the generating function for Y. For the data given,

$$L_W(s) = \exp\{-6[1 - 0.1/(0.1+s)]\}$$

(c) To determine $E(W)$ and σ_W^2, we make use of the results given in Problem 34. Specifically, we see that

$$E(W) = E(Y)E(T) = 6(10) = 60 \text{ min}$$

$$\sigma_W^2 = \sigma_T^2 E(Y) + \sigma_Y^2[E(T)]^2 = 100(6) + 6(100) = 1200$$

SUMMARY

In this chapter we have presented the basic concepts of probability theory with an introduction to transform methods. The treatment of these topics is intended to be neither comprehensive nor in depth. The emphasis of the chapter is on those areas of probability theory and transform methods that are most useful in the development and application of queueing models. The reader will find this material used repeatedly throughout the remainder of the book. For a more complete discussion of probability theory the reader should see Parzen (1960) or Wilks (1962). An excellent discussion of the Poisson process is presented in Parzen (1965). A discussion of the transform methods and their applications may be found in Griffin (1971) or Schmidt (1974). For an in-depth treatment of the Laplace transform and its applications the reader should see Hall *et al.* (1959) or LaPage (1961).

REFERENCES

Beightler, C. S., Mitten, L. G., and Nemhauser, G. L., A short table of z-transforms and generating functions, *Operations Res.* **9** (4), 574–578, 1961.

Brown, G. B., *Smoothing, Forecasting, and Prediction of Discrete Time Series.* Englewood Cliffs, New Jersey: Prentice-Hall, 1963.

Elmaghraby, S. E., *The Design of Production Systems.* New York: Reinhold, 1966.

Feller, W., *An Introduction to Probability Theory and Its Applications*, Volume I. New York: Wiley, 1957.

Freeman, H., *Introduction to Statistical Inference.* Reading, Massachusetts: Addison-Wesley, 1963.

Giffin, W. C., *Introduction to Operations Engineering.* Homewood, Illinois: Irwin, 1971.

Hadley, G., *Introduction to Probability and Statistical Decision Theory.* San Francisco: Holden-Day, 1967.

Hadley, G., and Whitin, T. M., *Analysis of Inventory Systems.* Englewood Cliffs, New Jersey: Prentice-Hall, 1963.

Hall, D. L., Maple, C. G., and Vinograde, B., *Introduction to the Laplace Transform.* New York: Appleton, 1959.

LePage, W. R., *Complex Variables and the Laplace Transform for Engineers.* New York: McGraw-Hill, 1961.

Lukacs, E., *Characteristic Functions.* London: Griffin, 1960.

Parzen, E., *Modern Probability Theory and Its Applications.* New York: Wiley, 1960.

Parzen, E., *Stochastic Processes.* San Francisco: Holden-Day, 1965.

Schmidt, J. W., *Mathematical Foundations for Management Science and Systems Analysis.* New York: Academic Press, 1974.

Wilks, S. S., *Mathematical Statistics.* New York: Wiley, 1962.

PROBLEMS

1. A production lot contains 1000 items, of which 300 are defective. A sample of three units is drawn from the lot at random. Let X be the number of defective units found in the sample. Find $P(X = 0)$, $P(X = 1)$, $P(X = 2)$, $P(X = 3)$ if:

(a) X is hypergeometrically distributed with $M = 300$, $N = 700$, $n = 3$.
(b) X is binomially distributed with $p = M/N$, where M and N are given in (a).
(c) X is Poisson distributed with $\lambda = np$, where n and p are given in (a) and (b) and $T = 1$.

2. Machining time for a particular operation is exponentially distributed. The average machining time has been found by time study to be 2 min. What is the probability that a particular part will require (a) less than 2 min for machining? (b) less than 2 min, but more than 1 min for machining?

3. The probability of a machine running at any given time is 0.80. If there are 4 machines in a department, what is the probability that at least 2 machines are running at the same time?

4. Along a turnpike cars pass an observation point at a rate of 30 cars per hour. If the cars arrive in a Poisson fashion, determine:
  (a) the probability no cars arrive in a 15-min interval,
  (b) the probability that the time between consecutive arrivals is greater than 15 min,
  (c) the probability that at most 10 cars arrive in a 15-min interval, given that at least 5 cars arrive in a 15-min interval.

5. In a particular residential area of a city it is found that the number of persons living in a dwelling has the probability distribution $p(x)$ and the number of dwellings in a residential section has the distribution $q(y)$, with

$$p(x) = \begin{cases} 0.2, & x = 1 \\ 0.6, & x = 2 \\ 0.2, & x = 3 \end{cases} \quad \text{and} \quad q(y) = \begin{cases} 0.25, & y = 1 \\ 0.50, & y = 2 \\ 0.25, & y = 3 \end{cases}$$

What is the probability that no more than 3 persons live in a residential section?

6. The proportion P of customers forced to wait for service has the probability density function

$$f(p) = ke^{-up}, \quad 0 \le p \le 1$$

Define k.

7. The distribution function of the Weibull random variable is given by

$$F(x) = 1 - \exp\left\{-\left[\frac{(x-a)}{(b-a)}\right]^n\right\}, \quad a \le x < \infty, \ b \ge a, \ n > 1$$

Find the probability density function of X.

8. Find the distribution function of the (a) rectangular random variable, (b) uniform random variable, (c) geometric random variable, (d) hyperexponential random variable.

9. Find the mean and variance of the beta random variable.

10. The mode of a continuous random variable is defined as the value of the random variable that maximizes its probability density function. Find the mode of (a) the gamma random variable, (b) the exponential random variable, (c) the hyperexponential random variable, (d) the beta random variable.

11. The median of a continuous random variable X is that value of X such that

$$F(x) = 0.5$$

Find the median of (a) the normal random variable, and (b) the exponential random variable.

12. The random variable X has the density function shown graphically below. Define a mathematical expression for $f(x)$.

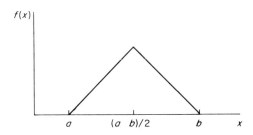

13. A continuous random variable is unimodal if its density function has only one local maximum. Are the random variables defined by the following probability density functions unimodal?

(a) $f(x) = \dfrac{1}{b-a}, \quad a < x < b$

(b) $f(x) = \dfrac{p}{\sigma_1 (2\pi)^{1/2}} \exp\left[-\dfrac{(x-m_1)^2}{2\sigma_1^2}\right] + \dfrac{(1-p)}{\sigma_2 (2\pi)^{1/2}} \exp\left[-\dfrac{(x-m_2)^2}{2\sigma_2^2}\right]$

$-\infty < x < \infty, \quad 0 \leq p \leq 1$

14. The number of units of product sold per customer sale is geometrically distributed with parameter $p = 0.25$. The selling price of a unit is $6.00. It costs $3.00 to make a unit and $0.50 to process a sale. What is the expected profit per sale?

15. A certain product is sold at a rate of 10,000 items per year. The proportion P of dissatisfied customers per year is beta distributed with parameters a and b. Each satisfied customer yields a net profit of $5.00 and each dissatisfied customer results in a net loss of $6.00. What is the expected net profit per year? What should $E(P)$ be to insure an expected net profit of $20,000 per year?

16. A man is involved in a game of chance. His probability of winning is $1/38$. If he bets A dollars and wins he is paid $36A$ dollars. Otherwise he loses the amount of his bet. What is his expected gain per play? How much should he bet to maximize his expected gain?

17. The time T until failure of a given product is exponentially distributed with parameter λ. Each unit of product is guaranteed for a period τ and costs C_m to manufacture. The sale price per unit sold is S and each unit that fails during the guarantee period is replaced at a cost C. What is the expected profit per unit sold?

18. Show that the exponential random variable is without memory.

19. Let

$$f(x) = kx^2, \quad 0 < x < a$$

Find (a) the value of k such that $f(x)$ is a probability density function, and (b) the Laplace transform of $f(x)$.

Problems

20. Let X_1 and X_2 be jointly distributed normal random variables with joint probability density function

$$f(x_1, x_2) = \frac{1}{2\pi(1-\rho^2)^{1/2}} \exp\left[-\frac{(x_1^2 - 2\rho x_1 x_2 + x_2^2)}{2(1-\rho^2)}\right], \qquad -\infty < x_1, x_2 < \infty$$

where $0 \leq \rho \leq 1$. Determine the value of ρ for which X_1 and X_2 are independent.

21. Let $p(x)$ be the probability mass function of the number of arrivals in time t. Let $F_n(t)$ be the distribution function of time until the nth arrival. Show that

$$p(x) = F_x(t) - F_{x+1}(t), \qquad x = 1, 2, 3, \ldots$$

$$p(0) = 1 - F_1(t)$$

22. If X is a gamma random variable with probability density function

$$f(x) = \frac{\lambda^n}{(n-1)!} x^{n-1} e^{-\lambda x}, \qquad 0 < x < \infty$$

show that the distribution function of X is given by

$$F(x) = 1 - \sum_{k=0}^{n-1} \frac{(\lambda x)^k}{k!} e^{-\lambda x}, \qquad 0 < x < \infty$$

where n is a positive integer.

23. The time between successive arrivals to a given system is exponentially distributed with parameter λ. Thus the time until the occurrence of n successive arrivals is gamma distributed with parameters n and λ. Using the results of Problems 21 and 22 show that the number of arrivals in a fixed period of time T is Poisson distributed with parameter λ.

24. The probability density function of the Cauchy random variable is given by

$$f(x) = \frac{1}{\pi} \frac{\beta}{\beta^2 + (x-\alpha)^2}, \qquad -\infty < x < \infty$$

Show that the mean of this random variable does not exist.

25. Let Y_1 and Y_2 be independent, identically distributed normal random variables. Show that

$$X = Y_1 + Y_2$$

is also normally distributed.

26. Show that the sum of independent identically distributed geometric random variables has a negative binomial distribution.

27. Using the Laplace transform, find the mean and variance of (a) the hyperexponential random variable, and (b) the gamma random variable.

28. Using the generating function, find the mean and variance of (a) the binomial random variable, (b) the negative binomial random variable, and (c) the rectangular random variable.

29. Solve Problem 23 using Eq. (2.82).

30. The moment-generating function of a discrete or continuous random variable X is defined by $E(e^{tx})$. Define an expression for $E(X^k)$ in terms of the moment-generating function, where $k = 0, 1, 2, \ldots$.

31. Let X_1 and X_2 be independently distributed Poisson random variables with parameters λ_1 and λ_2, respectively. Find $P(X_1 = m \mid X_1 + X_2 = n)$, where $m = 0, 1, 2, \ldots, n$.

32. Find the function $f(t)$ that satisfies the following equation:

$$\int_0^t f(x)\,dx + \frac{1}{u}f(t) = 1$$

33. Let

$$X = \sum_{i=1}^n a_i Y_i$$

where $Y_1, Y_2, \ldots, Y_n$ are independent, identically distributed random variables. Show that

$$\sigma_X^2 = \sum_{i=1}^n a_i^2 \sigma_{Y_i}^2$$

34. Let X be a random variable composed of the random sum of the discrete, independent, identically distributed random variables $Y_1, Y_2, \ldots, Y_n$. If the generating function of X is given by

$$G(z) = \sum_{n=1}^{\infty} [G_i(z)]^n p(n)$$

where $G_i(z)$ is the generating function of Y_i, $i = 1, 2, \ldots, n$, and $p(n)$ is the probability mass function of the number N of random variables comprising the sum, show that

$$E(X) = E(N)E(Y_i), \qquad \sigma_X^2 = \sigma_{Y_i}^2 E(N) + \sigma_N^2 [E(Y_i)]^2$$

35. Let $Y_1, Y_2, \ldots, Y_n$ be continuous, independent, identically distributed random variables with density functions $f(y_i)$, $i = 1, 2, \ldots, n$. If N is a random variable with probability mass function $p(n)$, find the Laplace transform of X, where

$$X = \sum_{i=1}^n Y_i$$

36. Let X be the random variable defined in Problem 19. Find the mean and variance of X.

37. Derive the Laplace transforms of the following functions:

   (a) $\dfrac{e^{at} - e^{bt}}{a - b}$, $\quad 0 < t < \infty$

   (b) $e^{at} \sin(at)$, $\quad 0 < t < \infty$

   (c) $\begin{cases} e^{at}, & 0 < t < T \\ 0, & t > T \end{cases}$

   (d) $e^{at} f(t)$, $\quad 0 < t < \infty$

38. Define the function $f(t)$ corresponding to the following Laplace transforms:

   (a) $\mathscr{L}[f(t)] = \dfrac{1}{(s - c_1)} + \dfrac{1}{(s - c_2)(s - c_3)}$

   (b) $\mathscr{L}[f(t)] = \dfrac{1}{(s - c_1)^2 (s - c_2)}$

   (c) $\mathscr{L}[f(t)] = \dfrac{1}{(s - 5)^3 (s - 4)^2}$

Problems

39. Solve the following differential equations:

(a) $\dfrac{d^2}{dt^2}f(t) + \dfrac{d}{dt}f(t) = -2\sin(t)$, $\quad \dfrac{d}{dt}f(t)\bigg|_{t=0} = 1$, $\quad f(0) = 1$

(b) $\dfrac{d^4}{dt^4}f(t) - 2\dfrac{d^2}{dt^2}f(t) - 3f(t) = 0$, $\quad \dfrac{d^3}{dt^3}f(t)\bigg|_{t=0} = -1$, $\quad \dfrac{d^2}{dt^2}f(t)\bigg|_{t=0} = 1$, $\quad f(0) = 1$

(c) $\dfrac{d^2}{dt^2}f(t) + \mu\dfrac{d}{dt}f(t) + f(t) = \mu e^{-\mu t}$, $\quad \dfrac{d}{dt}f(t)\bigg|_{t=0} = -\mu^2$, $\quad f(0) = \mu$

40. In a service station the probability distribution for the number of hours between the arrival of customers is

$$f(t) = 10e^{-10t}, \quad t > 0$$

On the average, 80% of the customers drive cars and the remaining 20% drive motorcycles. The number of hours the station is open per day is distributed as follows:

$$p(h) = \begin{cases} 0.3, & h = 12 \\ 0.5, & h = 14 \\ 0.2, & h = 16 \end{cases}$$

The probability distribution for the number of gallons of gasoline demanded for a customer driving a car is the normal distribution with a mean of 18 and a standard deviation of 5 gal. The number of gallons demanded for a customer driving a motorcycle is normally distributed with a mean of 4 gal and a standard deviation of 1 gal. The time between deliveries of gasoline to the station, measured in days, is Poisson distributed with a variance of 10.

(a) Determine the transform for the number of gallons of gasoline sold to cars per hour. Is the transform a Laplace transform or a generating function?

(b) Determine the transform for the number of gallons of gasoline sold per hour.

(c) Determine the transform for the number of gallons sold between deliveries to the station.

(d) Determine the mean and variance for the number of gallons sold between deliveries to the station.

41. Based on recent events, it is estimated that the probability of a building take-over occurring at VIP University within t days following the most recent take-over is given by

$$1 - e^{-0.2t}, \quad t \geq 0$$

Furthermore, the number x of students involved in a building take-over has the following probability mass function:

$$p(x) = (0.5)^{x+1}, \quad x = 0, 1, 2, \ldots$$

Also, the number y of protest signs carried per student is equally likely to be 1, 2, or 3. The length of a school term can be estimated by the continuous random variable z (measured in days) with probability density function given by

$$h(z) = \dfrac{e^{-0.1z}}{10}, \quad z \geq 0$$

A local protest sign shop needs to determine its demand distribution for the coming school term. Students at VIP only carry protest signs during a take-over.

(a) Determine the appropriate transforms for the probability distributions of each of the random variables z, y, and x.

(b) Give the probability generating function for the number of protest signs per building take-over.

(c) What is the transform for the number of protest signs per school term?

(d) What is the expected number of protest signs per school term?

42. Consider a situation in which arrivals are discrete and service time is a continuous random variable with density function $f(t)$. Let $p(x \mid t)$ be the probability that x customers arrive in time t and let $G(z; t)$ be the generating function for this mass function. Let y be the random variable representing the number of arrivals during service time, and let $G^*(z)$ be its generating function. Show that

$$G^*(z) = \int_0^\infty G(z; t) f(t) \, dt.$$

Find $G^*(z)$ in the case where a Poisson process generates arrivals, arrivals occurring one at a time, and the service time has a gamma distribution. In this way show that the number of arrivals during service time has a negative binomial distribution.

43. Let $p(x \mid t)$ be the probability that x arrivals occur in a time t. Assume that the mean of this distribution is λt, where λ is the mean arrival rate and its variance is $V(x)t$. Let $f(t)$ be the density function for service time with mean $E(t)$ and variance $V(t)$. Assume that x is either discrete or continuous and that t is continuous. If y is the random variable representing the number of arrivals during service time, and $E(y)$ and $V(y)$ are the mean and variance of y, respectively, show that

$$E(y) = \lambda E(t) \quad \text{and} \quad V(y) = E(t)V(x) + \lambda^2 V(t).$$

Chapter 3

POISSON QUEUES

INTRODUCTION

In Chapter 1 we emphasized the importance of simple models in the analysis of queueing systems. Of the many different varieties of queueing problems, a queue that is relatively simple to model is the $(M \mid M \mid c)$ queue. Recalling the identifying scheme described in Chapter 1, we are referring to a problem in which the arrivals follow a Poisson distribution, the service time is exponentially distributed, and there are c parallel service channels. In this chapter we will develop a number of descriptive models for variations of the $(M \mid M \mid c)$ queue. Specifically, we will consider queueing systems for both finite and infinite populations, as well as those involving single- and multiple-service facilities. Additional topics to be considered include bulk arrivals and balking customers. In addition to the variety of situations discussed in detail in the text, several other situations are treated in the problems at the end of the chapter.

To facilitate the treatment of Poisson queues, it is helpful to review the following properties of a queueing system having Poisson distributed arrivals and exponentially distributed service times:

1. If λ is the expected number of arrivals per unit time, then the probability of x arrivals occurring in the interval $(0, t)$ is

$$p_x(t) = e^{-\lambda t} \frac{(\lambda t)^x}{x!}, \qquad x = 0, 1, 2, \ldots, \infty$$

and the probability density function for the time between consecutive arrivals is

$$f(t) = \lambda e^{-\lambda t}, \quad t > 0$$

2. If μ is the reciprocal of expected service time, then the probability mass function for the number of departures (services) x in time t, given that customers are available to be served throughout time t, is

$$p_x(t) = e^{-\mu t} \frac{(\mu t)^x}{x!}, \quad x = 0, 1, 2, \ldots, \infty$$

and the probability density function for service time is

$$f(t) = \mu e^{-\mu t}, \quad t > 0$$

3. The probability of service time (interarrival time) T being greater than $\alpha + \gamma$, given $t > \alpha$, is simply the probability that service time (interarrival time) is greater than γ. This is called the *forgetfulness* or *memoryless property* of a Poisson process.

4. Arrivals and departures are completely random. Thus, an arrival is equally likely to occur in any interval of time Δt, and during the time customers are being served a departure is equally likely to occur in any interval of time Δt.

5. The number of arrivals (services) per unit time is Poisson distributed if and only if the interarrival (service) times are exponentially distributed.

At this point, it is instructive to consider how we obtain the Poisson distribution for, say, arrivals. To show this, let us assume:

1. The probability of exactly one arrival in an interval of length Δt is directly proportional to Δt, with λ the constant of proportionality.

2. The probability of k arrivals in an interval Δt is $\alpha_k(\Delta t)$ for all $k > 1$, where $\alpha_k(\Delta t)$ is an infinitesimal of higher order than Δt, such that

$$\lim_{\Delta t \to 0} \frac{\alpha_k(\Delta t)}{\Delta t} = 0$$

Now let $P_n(t)$ denote the probability of exactly n arrivals in the interval $(0, t)$. Thus, $P_n(t + \Delta t)$ represents the probability of n arrivals during the interval $(0, t + \Delta t)$. Note that there can be n arrivals in time $t + \Delta t$ only if one of the following events has occurred:

1. n arrivals in time t and no arrivals in time Δt,
2. $n - 1$ arrivals in time t and one arrival in time Δt,
3. $n - k$ arrivals in time t and k arrivals in time Δt, where $k > 1$.

Introduction

Assuming arrivals occur independently, we see that

$$P_n(t + \Delta t) = P_n(t)\left[1 - \lambda \Delta t - \sum_{k=2}^{n}\alpha_k(\Delta t)\right]$$
$$+ P_{n-1}(t)\lambda \Delta t + \sum_{k=2}^{n} P_{n-k}(t)\alpha_k(\Delta t)$$

We are interested in measuring the change in the probability mass function for n as a function of the time increment Δt. Consequently, we subtract $P_n(t)$ from both sides and divide by Δt to obtain

$$\frac{P_n(t + \Delta t) - P_n(t)}{\Delta t} = -P_n(t)\left[\lambda + \sum_{k=2}^{n}\frac{\alpha_k(\Delta t)}{\Delta t}\right]$$
$$+ P_{n-1}(t)\lambda + \sum_{k=2}^{n} P_{n-k}(t)\frac{\alpha_k(\Delta t)}{\Delta t} \quad (3.1)$$

Since we wish to model the system under continuous rather than discrete time, we let Δt approach zero and obtain a differential equation in time and a difference equation in the number of arrivals. Consequently, the instantaneous rate of change in $P_n(t)$ can be given as

$$\frac{dP_n(t)}{dt} = -\lambda P_n(t) + \lambda P_{n-1}(t), \quad n > 0 \quad (3.2)$$

Notice that $P_{n-1}(t)$ is not defined for $n = 0$. Therefore,

$$\frac{dP_0(t)}{dt} = -\lambda P_0(t) \quad (3.3)$$

Taking the Laplace transform of (3.2) and (3.3) gives

$$s\mathscr{L}[P_n(t)] - P_n(0) = -\lambda\mathscr{L}[P_n(t)] + \lambda\mathscr{L}[P_{n-1}(t)] \quad (3.4)$$

and

$$s\mathscr{L}[P_0(t)] - P_0(0) = -\lambda\mathscr{L}[P_0(t)] \quad (3.5)$$

If we assume no arrivals have occurred at time zero, then (3.5) becomes

$$s\mathscr{L}[P_0(t)] - 1 = -\lambda\mathscr{L}[P_0(t)]$$

or

$$\mathscr{L}[P_0(t)] = \frac{1}{s + \lambda} \quad (3.6)$$

Since $P_n(0) = 0$ for $n \neq 0$, then (3.4) reduces to

$$(s + \lambda)\mathscr{L}[P_n(t)] = \lambda\mathscr{L}[P_{n-1}(t)], \quad n = 1, 2, \ldots$$

or

$$(s + \lambda)\mathscr{L}[P_{n+1}(t)] = \lambda\mathscr{L}[P_n(t)], \qquad n = 0, 1, \ldots \qquad (3.7)$$

Taking the generating function of both sides of (3.7) gives

$$(s + \lambda)\mathscr{G}\{\mathscr{L}[P_{n+1}(t)]\} = \lambda\mathscr{G}\{\mathscr{L}[P_n(t)]\} \qquad (3.8)$$

but, from Chapter 2,

$$\mathscr{G}\{\mathscr{L}[P_{n+1}(t)]\} = z^{-1}\mathscr{G}\{\mathscr{L}[P_n(t)]\} - z^{-1}\mathscr{L}[P_0(t)] \qquad (3.9)$$

Therefore, on combining (3.6), (3.8), and (3.9) and solving for $\mathscr{G}\{\mathscr{L}[P_n(t)]\}$, we obtain

$$\mathscr{G}\{\mathscr{L}[P_n(t)]\} = \frac{1}{s + \lambda(1 - z)} \qquad (3.10)$$

On obtaining the inverse Laplace transform of (3.10), we have

$$\mathscr{G}[P_n(t)] = e^{-\lambda t(1 - z)}$$

which, from Chapter 2, is seen to be the generating function for the Poisson distribution. Therefore,

$$P_n(t) = \frac{(\lambda t)^n e^{-\lambda t}}{n!}, \qquad n = 0, 1, \ldots \qquad (3.11)$$

as stated in Property 1.

We have succeeded in our objective of deriving the Poisson arrival distribution. In like manner we can obtain the Poisson departure distribution. The point to be made by all of this is that in many practical situations we can expect arrivals to be Poisson distributed if arrivals are independent and have a probability of occurrence in any small interval Δt approximately equal to $\lambda \Delta t$.

We can easily demonstrate for the Poisson distribution that as Δt becomes small, the probability of more than one arrival occurring approaches zero. Suppose that on the average one arrival occurs per minute. Figure 3.1 shows the values for $P_0(\Delta t) + P_1(\Delta t)$, and $1 - [P_0(\Delta t) + P_1(\Delta t)]$, where

$$P_n(\Delta t) = \frac{e^{-\Delta t}(\Delta t)^n}{n!}, \qquad n = 0, 1, 2, \ldots$$

Notice also that we can utilize (3.11) to determine the number of customers who have arrived during t time units of operation. In some cases, there exists a buildup of customers before service begins. If customers arrive in a completely random manner and service begins at time t, then the number of customers in the system at the time service begins is given by (3.11).

Infinite Population Models

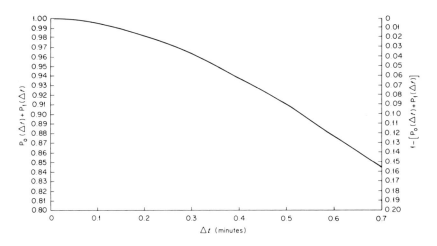

Fig. 3.1 Representation of the probability of more than one arrival during time Δt.

At this point we suspect it is possible to develop a descriptive model of the $(M\,|\,M\,|\,c)$ queue using a differential–difference equation approach. Knowing the probability of there being n customers present in the system, we can determine a number of characteristics of the system. Typical operating characteristics of the system are the expected number of customers waiting to be served and the expected amount of time a customer must wait before service begins.

INFINITE POPULATION MODELS: $(M\,|\,M\,|\,c):(GD\,|\,N\,|\,\infty)$ QUEUE

We will begin our analysis of a queueing system by considering the $(M\,|\,M\,|\,c):(GD\,|\,N\,|\,\infty)$ queue. From Chapter 1 recall that the notation $(M\,|\,M\,|\,c):(GD\,|\,N\,|\,\infty)$ means that arrivals and services are generated by a Poisson process, there are c parallel servers, a general service discipline is used, a maximum of N customers can be in the system at any given point in time, and an infinite population of customers exists. Of course, since a general service discipline is assumed, the results to be obtained will hold for any other service discipline, such as first come–first served and last come–first served. The capacity of N customers in the system at any given time can be due either to a physical limitation on the amount of space available or to a level of waiting that is intolerable to arriving customers. In the latter case, we refer to N as a balking level, since an arriving customer who sees N customers in the system ahead of him balks and refuses to enter the system. (In a problem at the end of the chapter, we let the balking level be a random variable.) Finally, we consider the population of customers to be infinite.

It should also be emphasized that the notation we employ does not completely identify the system since, for example, no indication is given concerning the jockeying, reneging, and balking behavior of customers or the influence of the number of customers in the system on the service rate for the servers. However, the notation does serve as an aid in developing a general classification of the type of problems being considered.

The $(M\,|\,M\,|\,c):(GD\,|\,N\,|\,\infty)$ queueing situation can be considered a general model, allowing us to consider $(M\,|\,M\,|\,1):(GD\,|\,N\,|\,\infty)$, $(M\,|\,M\,|\,c):(GD\,|\,\infty\,|\,\infty)$, and $(M\,|\,M\,|\,1):(GD\,|\,\infty\,|\,\infty)$ queues as special cases. Therefore, the operating characteristics that we will develop for the

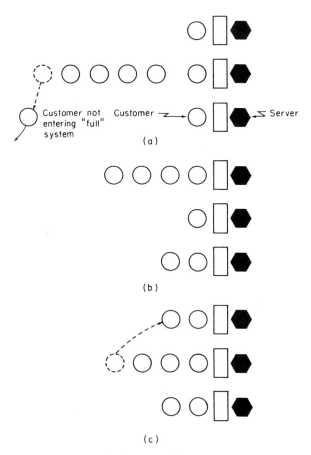

Fig. 3.2 Representation of $(M\,|\,M\,|\,3):(GD\,|\,7\,|\,\infty)$ queue. (a) Case 1: single waiting line; (b) Case 2: multiple waiting line without jockeying; (c) Case 3: multiple waiting line with jockeying.

Infinite Population Models

$(M \mid M \mid c) : (GD \mid N \mid \infty)$ can be easily reduced to the particular special case desired.

The assumption is made for the $(M \mid M \mid c) : (GD \mid N \mid \infty)$ queue that a single waiting line, or its equivalent, exists. As depicted in Fig. 3.2, three waiting-line behaviors can occur: first, a single waiting line can exist; second, a waiting line can form in front of each server and customers cannot change waiting lines; third, jockeying of customers is allowed. If either the first or third condition exists, then the results to be derived are valid; otherwise, an alternate approach is required.

General Birth–Death Process

Using an approach similar to that employed in developing the Poisson arrival distribution, let $P_n(t)$ denote the probability of n units in the *system* at time t. The *system* includes units in service, as well as those units in the queue. Let λ_n and μ_n denote the arrival and service rates, respectively, when n units are in the system.

For a general birth–death process, with a sufficiently small time increment Δt, we know that given n in the system

$\lambda_n \Delta t =$ probability of one arrival during Δt

$1 - \lambda_n \Delta t =$ probability of no arrival during Δt

$\mu_n \Delta t =$ probability of one service during Δt

$1 - \mu_n \Delta t =$ probability of no service during Δt

For Δt sufficiently small, both an arrival and a service during Δt cannot occur. Thus,

$$P_n(t + \Delta t) = P_n(t)(1 - \lambda_n \Delta t)(1 - \mu_n \Delta t) + P_{n-1}(t)\lambda_{n-1} \Delta t(1 - \mu_{n-1} \Delta t)$$
$$+ P_{n+1}(t)(1 - \lambda_{n+1} \Delta t)\mu_{n+1} \Delta t \qquad (3.12)$$

From (3.12) the probability of n units in the system at $t + \Delta t$ equals the sum of the probabilities of the three collectively exhaustive and mutually exclusive events:

1. n units in the system at t, no arrival or service during Δt,
2. $n - 1$ units in the system at t, one arrival and no service during Δt, and
3. $n + 1$ units in the system at t, no arrival and one service during Δt.

Any other event will involve either multiple arrivals, multiple services, or both arrivals and services during Δt and cannot occur.

Expanding the right-hand side of (3.12), taking $P_n(t)$ to the left-hand side, dividing by Δt, and letting $\Delta t \to 0$ gives the following differential–difference equation:

$$\frac{dP_n(t)}{dt} = -P_n(t)(\lambda_n + \mu_n) + P_{n-1}(t)\lambda_{n-1} + P_{n+1}(t)\mu_{n+1} \quad (3.13)$$

Equation (3.13) is a differential equation in t and a difference equation in n. For $n = 0$, $P_{n-1}(t)$, λ_{n-1}, and μ_n equal zero (there must be a zero departure rate if there are no units present).

It is possible, in some cases, for (3.13) to be solved directly. However, even for the simplest cases a direct solution for $P_n(t)$ is quite complex. For the present, only a *steady-state solution* will be obtained. Steady state means that $t \to \infty$, or $dP_n(t)/dt = 0$. In steady state, $P_n(t)$ is no longer a function of t. Thus t can be excluded, and (3.13) becomes

$$0 = -P_n(\lambda_n + \mu_n) + P_{n-1}\lambda_{n-1} + P_{n+1}\mu_{n+1}, \quad 0 < n < N \quad (3.14)$$

$$0 = -P_0\lambda_0 + P_1\mu_1, \quad 0 = -P_N\mu_N + P_{N-1}\lambda_{N-1} \quad (3.15)$$

Equations (3.14) and (3.15) are referred to as the steady-state, *general balance equations* for the $(M\,|\,M\,|\,c)$ queue. In order to solve the steady-state general balance equations, we begin by noting that (3.15) gives

$$P_1 = \frac{\lambda_0}{\mu_1} P_0 \quad (3.16)$$

For $n = 1$, from (3.14) we see that

$$P_2 = \frac{\lambda_1 + \mu_1}{\mu_2} P_1 - \frac{\lambda_0}{\mu_2} P_0 \quad (3.17)$$

Substituting (3.16) in (3.17) and solving for P_2 gives

$$P_2 = \frac{\lambda_1 \lambda_0}{\mu_2 \mu_1} P_0$$

We make the following inductive claim,

$$P_n = \prod_{k=1}^{n} \frac{\lambda_{k-1}}{\mu_k} P_0, \quad 1 \leq n \leq N \quad (3.18)$$

To establish the claim, let us assume (3.18) true for n. Therefore, from (3.14) we see that

$$P_{n+1} = \frac{P_n(\lambda_n + \mu_n)}{\mu_{n+1}} - \frac{P_{n-1}\lambda_{n-1}}{\mu_{n+1}}$$

Infinite Population Models

or, on applying (3.18),

$$P_{n+1} = \prod_{k=1}^{n} \frac{\lambda_{k-1}}{\mu_k} \frac{\lambda_n}{\mu_{n+1}} P_0 + \prod_{k=1}^{n} \frac{\lambda_{k-1}}{\mu_k} \frac{\mu_n}{\mu_{n+1}} P_0 - \prod_{k=1}^{n-1} \frac{\lambda_{k-1}}{\mu_k} \frac{\lambda_{n-1}}{\mu_{n+1}} P$$

or

$$P_{n+1} = \prod_{k=1}^{n+1} \frac{\lambda_{k-1}}{\mu_k} P_0 + \frac{\mu_n}{\lambda_n} \prod_{k=1}^{n+1} \frac{\lambda_{k-1}}{\mu_k} P_0 - \frac{\mu_n}{\lambda_n} \prod_{k=1}^{n+1} \frac{\lambda_{k-1}}{\mu_k} P_0$$

which gives

$$P_{n+1} = \prod_{k=1}^{n+1} \frac{\lambda_{k-1}}{\mu_k} P_0$$

Therefore, we know that if (3.18) holds for n, it also holds for $n + 1$. It remains to show that (3.18) holds for $n = 1$. Letting $n = 1$ in (3.18) gives (3.16). Thus, we can conclude that P_n is given by (3.18).

At this point we know the value of P_n if the value for P_0 is known. To obtain the value of P_0, we note that

$$\sum_{n=0}^{N} P_n = 1$$

must be true. So from (3.18)

$$P_0 \left[1 + \sum_{n=1}^{N} \prod_{k=1}^{n} \frac{\lambda_{k-1}}{\mu_k} \right] = 1 \quad \text{or} \quad P_0 = \left[1 + \sum_{n=1}^{N} \prod_{k=1}^{n} \frac{\lambda_{k-1}}{\mu_k} \right]^{-1}$$

(3.19)

Substituting (3.19) in (3.18) gives, for $n = 1, 2, \ldots, N$,

$$P_n = \left[\prod_{k=1}^{n} \frac{\lambda_{k-1}}{\mu_k} \right] \Big/ \left[1 + \sum_{n=1}^{N} \prod_{k=1}^{n} \frac{\lambda_{k-1}}{\mu_k} \right] \qquad (3.20)$$

The steady-state, general balance equations for the $(M \mid M \mid c)$ queue are often referred to as birth–death equations. The analogy is drawn between births and arrivals and between deaths and services (departures). Thus when only arrivals occur, the process is called a pure birth process. Likewise, when only services occur, we have a pure death process.

Example 3.1 Based on (3.20) we can analyze a number of interesting queueing systems. To illustrate, suppose we are analyzing a barber shop having two barbers ($c = 2$). The data for the system indicate that customers arrive in a Poisson fashion at a rate of 12/hr so long as no more than one customer is waiting to be served. If two customers are waiting, the arrival rate is reduced to 8/hr. If three customers are waiting, customers arrive at a rate of 4/hr. If four customers are waiting, no additional customers arrive

($N = 6$). The time required to cut a customer's hair is exponentially distributed with a mean of 10 min.

To determine the probability distribution for the number of customers in the barber shop, we first enter the values for λ_n and μ_n from Table 3.1. Second, we use (3.18) to compute the value of P_n in terms of P_0. The third step is to sum the values of P_n, set the sum equal to one, and determine the value of P_0. The last step is to compute the value of P_n.

TABLE 3.1 Computation of P_n values for Example 3.1

n	λ_n	μ_n	P_n	P_n
0	12	0	$1 P_0$	9/97
1	12	6	$\dfrac{\lambda_0}{\mu_1} P_0 = 2 P_0$	18/97
2	12	12	$\dfrac{\lambda_1}{\mu_2} P_1 = 2 P_0$	18/97
3	12	12	$\dfrac{\lambda_2}{\mu_3} P_2 = 2 P_0$	18/97
4	8	12	$\dfrac{\lambda_3}{\mu_4} P_3 = 2 P_0$	18/97
5	4	12	$\dfrac{\lambda_4}{\mu_5} P_4 = (4/3) P_0$	12/97
6	0	12	$\dfrac{\lambda_5}{\mu_6} P_5 = (4/9) P_0$	4/97
			$(97/9) P_0 = 1$	97/97

Note that μ_n equals 6 customers/hr when one barber is busy and 12 customers/hr when both barbers are busy. Also, there will never be more than 6 customers in the system at the same time.

Just how realistic is our model? We did not assume anything about the way in which customers are served, except that there will not be an idle barber if there are two or more customers in the shop. Thus, as soon as a barber becomes available, service begins on the next customer if there is a customer available.

Operating Characteristics

From Example 3.1, several questions come to mind. In particular, how might the system be improved? The owner of the shop might consider hiring an additional barber. He might also purchase new equipment to

Infinite Population Models

shorten service time. By getting more comfortable chairs, installing television and stereo systems, and advertising, he could possibly increase the arrival rate and encourage customers to enter the shop, regardless of the number of customers waiting. No doubt there are a number of alternative ways of improving the system. However, each of these probably involves some added cost. How might we compare these alternatives? At this point, all we know is the probability mass function P_n. It would be difficult to compare probability distributions for different systems. Therefore, we are probably more interested in obtaining measurements of the system's operating characteristics. The following operating characteristics are typically used to compare systems:

1. L, expected number in the system (waiting and service),
2. L_q, expected number waiting,
3. W, expected time spent in the system,
4. W_q, expected time spent waiting,
5. D, probability of delay.

The values for L and L_q are obtained as follows:

$$L = \sum_{n=0}^{N} nP_n \qquad (3.21)$$

$$L_q = \sum_{n=c}^{N} (n-c)P_n \qquad (3.22)$$

where c is the number of servers. Thus, for Example 3.1,

$$L = 0(9/97) + 1(18/97) + 2(18/97) + 3(18/97) + 4(18/97)$$
$$+ 5(12/97) + 6(4/97)$$
$$= 264/97 = 2.72 \text{ customers}$$
$$L_q = 1(18/97) + 2(18/97) + 3(12/97) + 4(4/97)$$
$$= 106/97 = 1.09 \text{ customers}$$

Therefore, on the average there are 2.72 customers in the shop. Of these, 1.09 are waiting and 1.63 are being served. Consequently, the expected number of busy barbers is 1.63, which indicates that the barbers are busy 81.5% of the time.

In order to compute the value of W, we note that W equals W_q plus the expected service time. The expected service time $E(t_s)$ is 10 min. Now how can we determine the expected waiting time? One approach is to consider an amount of time T during which arrivals occur according to some arbitrary distribution with an average interarrival time of $(1/\tilde{\lambda})$.† Now let the total waiting time during T be ξ_T. Therefore, the average amount of time spent waiting per customer is

$$W_q = \xi_T \div \frac{T}{1/\tilde{\lambda}} = \frac{\xi_T}{\tilde{\lambda} T} \quad \text{or} \quad \tilde{\lambda} W_q = \frac{\xi_T}{T}$$

Suppose that during T the average number of persons waiting is L_q. Therefore, the total waiting time for all customers arriving during T is $L_q T$, and

$$\xi_T = L_q T \quad \text{or} \quad L_q = \frac{\xi_T}{T}$$

Consequently,

$$L_q = \tilde{\lambda} W_q \tag{3.23}$$

where $\tilde{\lambda}$ denotes the effective arrival rate for the system. By similar arguments we can establish that

$$L = \tilde{\lambda} W \tag{3.24}$$

A more rigorous justification for (3.23) and (3.24) is given by Stidham (1974), among others.

From the fact that

$$W = W_q + E(t_s) \tag{3.25}$$

and that

$$L = L_q + \tilde{\lambda} E(t_s) \tag{3.26}$$

we see that

$$\tilde{\lambda} = (L - L_q)\mu \tag{3.27}$$

Therefore, for Example 3.1

$$\tilde{\lambda} = (158/97)(6) = 948/97 = 9.77 \text{ customers/hr}$$

or

$$\tilde{\lambda} = 0.163 \text{ customers/min}$$

---

† $\tilde{\lambda}$ is defined as the *effective* arrival rate for the system and can differ from λ, the arrival rate for the system. $\tilde{\lambda}$ is the average rate at which customers *enter* the system, whereas λ is the average rate at which customers *arrive* at the system. In the case of a system capacity of N, not all arriving customers enter the system. Thus, $\tilde{\lambda}$ and λ need not be the same. We will find subsequently that $\tilde{\lambda}$ is very valuable in developing the operating characteristics of the system.

Consequently,

$$W_q = \frac{L_q}{\tilde{\lambda}} = 6.71 \text{ min/customer}$$

and

$$W = W_q + E(t_s) = 16.71 \text{ min/customer}$$

Notice that once the values of L and L_q have been obtained using (3.21) and (3.22), then the values of $\tilde{\lambda}$, W, and W_q can be found using (3.27), (3.23), and (3.24), respectively.

To compute the probability that a customer is delayed, or has to wait, we note that not all customers enter the system. Therefore, to be meaningful, we express delay probability as the probability that a customer who enters the system will be delayed. In order to determine this value we must develop the probability mass function for the number of customers in the system ahead of an *entering* customer. This probability is denoted by $\hat{P}_n$ and is often referred to as the *entering customer's distribution*, whereas P_n is referred to as the *outside observer's distribution*. Therefore,

$$\hat{P}_n = P(n \text{ in system} \mid \text{arriving customer enters})$$

or

$$\hat{P}_n = \frac{P(\text{arriving customer enters} \mid n \text{ in system})P_n}{P(\text{arriving customer enters})} \quad (3.28)$$

For Example 3.1, the probability that an arriving customer enters the shop is

$$\sum_{n=0}^{6} P(\text{enters} \mid n)P_n$$

or

$$(1)(9/97) + (1)(18/97) + (1)(18/97) + (1)(18/97)$$
$$+ (2/3)(18/97) + (1/3)(12/97) + (0)(4/97) = 79/97$$

Therefore,

$$\hat{P}_0 = \frac{(1)(9/97)}{79/97} = 9/79, \qquad \hat{P}_1 = \frac{(1)(18/97)}{79/97} = 18/79$$

$$\hat{P}_2 = \frac{(1)(18/97)}{79/97} = 18/79, \qquad \hat{P}_3 = \frac{(1)(18/97)}{79/97} = 18/79$$

$$\hat{P}_4 = \frac{(2/3)(18/97)}{79/97} = 12/79, \qquad \hat{P}_5 = \frac{(1/3)(12/97)}{79/97} = 4/79$$

Thus, the probability that an entering customer will be delayed is given by the probability that both servers are busy, or

$$D = 1 - \hat{P}_0 - \hat{P}_1 = 52/79 = 0.657$$

Notice that we can also obtain the delay probability in the following alternative manner. Let delay probability be defined as

$$D = P(\text{entering customer is delayed})$$

or

$$D = P(\text{customer is delayed} \,|\, \text{customer enters})$$

Therefore, by the definition of conditional probability,

$$D = \frac{P(\text{customer enters and is delayed})}{P(\text{customer enters})}$$

or

$$D = \frac{\sum_n P(\text{customer enters and is delayed} \,|\, n)P_n}{\sum_n P(\text{customer enters} \,|\, n)P_n}$$

Alternatively,

$$D = \frac{\sum_n P(\text{customer is delayed} \,|\, \text{enters and } n \text{ in system})P(\text{enters} \,|\, n)P_n}{\sum_n P(\text{customer enters} \,|\, n)P_n}$$

For the barber shop example, the delay probability is given by

$$D = \frac{0(1)P_0 + 0(1)P_1 + 1(1)P_2 + 1(1)P_3 + 1(2/3)P_4 + 1(1/3)P_5 + 1(0)P_6}{79/97}$$

$$= 52/79 = 0.658.$$

Knowing the probability that an arriving customer will enter the shop allows us to compute the effective arrival rate in an alternate way. Specifically, since the probability that an arriving customer will enter the shop is 79/97 and customers arrive at an average rate of 12/hr, then the effective arrival rate is also given as

$$\tilde{\lambda} = (79/97)(12) = 948/97$$

Infinite Population Models

Let us summarize what we have learned thus far about the present system:

$$\text{expected number in the shop} = 2.72$$
$$\text{expected number waiting} = 1.09$$
$$\text{expected number being served} = 1.63$$
$$\text{barber utilization} = 81.5\%$$
$$\text{expected service time (min)} = 10.00$$
$$\text{expected waiting time (min)} = 6.71$$
$$\text{expected time in the shop (min)} = 16.71$$
$$\text{effective arrival rate (arrivals/hr)} = 9.77$$
$$\text{probability of delay for entering customer} = 0.658$$
$$\text{probability of customer entering shop} = 0.814$$

If we observe the barber shop over a long period of time and obtain values of the operating characteristics comparable to those given above, we would conclude that our model of the barber shop is a valid representation of the system. Furthermore, we can use the model to predict the behavior of the system under a variety of conditions.

$(M\,|\,M\,|\,c):(GD\,|\,N\,|\,\infty)$ *results.* We have presented a very general procedure for analyzing variations of the $(M\,|\,M\,|\,c):(GD\,|\,N\,|\,\infty)$ queueing system. Even though the procedure can be used to analyze a large number of queueing problems, there are some situations that allow a closed-form solution for P_n, L, L_q, W, and W_q. Since these situations occur quite often, it is worthwhile to consider them at this time.

In performing the analysis, we will find that the following summations of series arise:

$$S_1 = \sum_{n=0}^{N} \rho^n, \qquad S_2 = \sum_{n=0}^{N} n\rho^n$$

The series having the sum S_1 is the sum of the first $N + 1$ terms of the geometric series and has the value

$$S_1 = \begin{cases} N + 1, & \rho = 1 \\ \dfrac{1 - \rho^{N+1}}{1 - \rho}, & \rho \neq 1, \ \rho > 0 \end{cases}$$

The series S_2 is related to S_1 as follows:

$$S_2 = \rho \frac{dS_1}{d\rho}$$

Therefore, when $\rho \neq 1$ and $\rho > 0$, S_2 has the value

$$S_2 = \rho \frac{d}{d\rho} \frac{1 - \rho^{N+1}}{1 - \rho}$$

Consequently,

$$S_2 = \begin{cases} \frac{1}{2}N(N + 1), & \rho = 1 \\ \dfrac{\rho[1 - \rho^{N+1} - (1 - \rho)(N + 1)\rho^N]}{(1 - \rho)^2}, & \rho \neq 1, \; \rho > 0 \end{cases}$$

We will consider first a variation of the $(M \mid M \mid c) : (GD \mid N \mid \infty)$ queue that occurs frequently in practice. Specifically, c identical servers, each having a mean service rate of μ, operate in parallel to provide service to identical customers who arrive at a mean rate of λ. When all servers are busy, a single waiting line forms as shown in Fig. 3.3. Assuming $N \geq c$, the specific arrival and service rates to be considered are written as follows:

$$\lambda_n = \begin{cases} \lambda, & n = 0, 1, \ldots, N - 1 \\ 0, & n = N \end{cases}$$

and

$$\mu_n = \begin{cases} n\mu, & n = 0, 1, \ldots, c \\ c\mu, & n = c + 1, \ldots, N \end{cases}$$

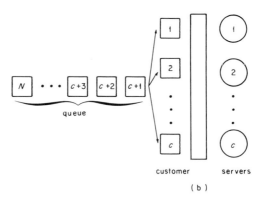

Fig. 3.3 Representation of an $(M \mid M \mid c) : (GD \mid N \mid \infty)$ queue that is full.

Infinite Population Models

Substituting λ_n and μ_n into (3.18) and letting $\rho = (\lambda/c\mu)$ gives

$$P_n = \begin{cases} \dfrac{(c\rho)^n}{n!} P_0, & n = 0, 1, \ldots, c-1 \\ \dfrac{c^c \rho^n}{c!} P_0, & n = c, \ldots, N \\ 0, & n > N \end{cases} \qquad (3.29)$$

Using the fact that $\sum_{n=0}^{N} P_n = 1$, we find that

$$P_0 = \left[\sum_{n=0}^{c-1} \frac{(c\rho)^n}{n!} + \sum_{n=c}^{N} \frac{c^c \rho^n}{c!} \right]^{-1}$$

which reduces to

$$P_0 = \begin{cases} \left[\dfrac{(c\rho)^c (1 - \rho^{N-c+1})}{c!(1-\rho)} + \sum_{n=0}^{c-1} \dfrac{(c\rho)^n}{n!} \right]^{-1}, & \rho \neq 1 \\ \left[\dfrac{c^c}{c!}(N - c + 1) + \sum_{n=0}^{c-1} \dfrac{c^n}{n!} \right]^{-1}, & \rho = 1 \end{cases} \qquad (3.30)$$

The quantity ρ is referred to as the *traffic intensity* for the system. Other terminology encountered in the queueing literature includes the *offered load*, *carried load*, and *server utilization* or *server occupancy*. These terms are defined as follows:

$$\text{offered load} \equiv \text{mean number of arrivals that occur during a service time}$$
$$= \frac{\lambda}{\mu}$$

$$\text{carried load} \equiv \text{mean number of arrivals that occur and enter the system during a service time, as well as the mean number of busy servers}$$
$$= \frac{\tilde{\lambda}}{\mu}$$

$$\text{server occupancy} \equiv \text{carried load per server, or average percent utilization of the group of servers}$$
$$= \frac{\tilde{\lambda}}{c\mu}$$

Thus, the traffic intensity can be defined as the offered load per server. We will see that ρ plays an important role in our analysis of queueing systems. The expected number of customers waiting for service to begin is given as

$$L_q = \sum_{n=c}^{N} (n-c) P_n$$

or

$$L_q = \begin{cases} \dfrac{\rho(c\rho)^c P_0}{c!(1-\rho)^2} [1 - \rho^{N-c+1} - (N-c+1)\rho^{N-c}(1-\rho)], & \rho \neq 1 \\ \dfrac{(c\rho)^c P_0}{c!} [(N-c+1)(N-c)/2], & \rho = 1 \end{cases} \quad (3.31)$$

The expected number of customers in the system is equal to the expected number of customers waiting for service to begin, plus the expected number of customers being served. Therefore,

$$L = L_q + \sum_{n=0}^{c-1} n P_n + c \sum_{n=c}^{N} P_n \qquad (3.32)$$

which reduces to

$$L = \begin{cases} L_q + \sum_{n=1}^{c-1} n P_n + \dfrac{(c\rho)^c (1 - \rho^{N-c+1}) P_0}{(c-1)!(1-\rho)}, & \rho \neq 1 \\ L_q + \sum_{n=1}^{c-1} n P_n + \dfrac{c^c(N-c+1) P_0}{(c-1)!}, & \rho = 1 \end{cases} \quad (3.33)$$

Since the effective arrival rate is given by $\tilde{\lambda} = \lambda(1 - P_N)$ then the values of W_q and W can be obtained using the relations

$$W_q = \frac{L_q}{\tilde{\lambda}} \qquad (3.34)$$

$$W = W_q + \frac{1}{\mu} \qquad (3.35)$$

To compute the probability of delay for a customer who enters the system, we again wish to develop first the probability mass function for the number of customers in the system ahead of an entering customer. We can imagine that a customer walks up to the "door" of the system, looks in, counts the customers, and decides whether or not to enter the system. In this

Infinite Population Models

case, he will enter if the number ahead of him is *less than* N. Thus, he enters with probability $1 - P_N$ and the probability mass function $\hat{P}_n$ is given as

$$\hat{P}_n = \begin{cases} \dfrac{(c\rho)^n}{n!}\hat{P}_0, & n = 0, 1, \ldots, c \\ \dfrac{c^c \rho^n}{c!}\hat{P}_0, & n = c+1, \ldots, N-1 \\ 0, & n \geq N \end{cases} \quad (3.36)$$

where $\hat{P}_0$ equals $P_0/(1 - P_N)$.

A customer who enters the system will be delayed if all servers are busy immediately before he enters the system. From (3.36), the probability of delay for an entering customer is

$$D = \sum_{n=c}^{N-1} \hat{P}_n, \quad \text{or} \quad D = \begin{cases} \dfrac{(c\rho)^c}{c!}\hat{P}_0\left[\dfrac{1 - \rho^{N-c}}{1 - \rho}\right], & \rho \neq 1 \\ \dfrac{(c\rho)^c}{c!}(N - c)\hat{P}_0, & \rho = 1 \end{cases} \quad (3.37)$$

Results for the $(M\,|\,M\,|\,c):(GD\,|\,N\,|\,\infty)$ queue are summarized in Table 3.2.

TABLE 3.2 Summary of operating characteristics for variations of the $(M\,|\,M\,|\,c)$ queue

	$(M\,|\,M\,|\,c):(GD\,|\,N\,|\,\infty)^a$	$(M\,|\,M\,|\,c):(GD\,|\,\infty\,|\,\infty)^b$
L_q	$\dfrac{\rho(c\rho)^c}{c!(1-\rho)^2}P_0[1 - \rho^{N-c+1} - (N - c + 1)\rho^{N-c}(1 - \rho)]$	$\dfrac{\rho(c\rho)^c P_0}{c!(1-\rho)^2}$
L	$L_q + \displaystyle\sum_{n=0}^{c-1} nP_n + \dfrac{(c\rho)^c(1 - \rho^{N-c+1})P_0}{(c-1)!(1-\rho)}$	$L_q + \dfrac{\lambda}{\mu}$
W_q	$\dfrac{(c\rho)^c[1 - \rho^{N-c+1} - (1 - \rho)(N - c + 1)\rho^{N-c}]P_0}{c!c\mu(1-\rho)^2(1 - P_N)}$	$\dfrac{(c\rho)^c P_0}{c!c\mu(1-\rho)^2}$
W	$W_q + \dfrac{1}{\mu}$	$W_q + \dfrac{1}{\mu}$
D	$\dfrac{(c\rho)^c(1 - \rho^{N-c})P_0}{c!(1-\rho)(1 - P_N)}$	$\dfrac{(c\rho)^c P_0}{c!(1-\rho)}$
P_0	$\left[\dfrac{(c\rho)^c(1 - \rho^{N-c+1})}{c!(1-\rho)} + \displaystyle\sum_{n=0}^{c-1}\dfrac{(c\rho)^n}{n!}\right]^{-1}$	$\left[\dfrac{(c\rho)^c}{c!(1-\rho)} + \displaystyle\sum_{n=0}^{c-1}\dfrac{(c\rho)^n}{n!}\right]^{-1}$

[a] For $\rho \neq 1$.
[b] For $\rho < 1$.

TABLE 3.3 Summary of operating characteristics for variations of the $(M \mid M \mid 1)$ queue

	$(M \mid M \mid 1):(GD \mid N \mid \infty)^a$	$(M \mid M \mid 1):(GD \mid \infty \mid \infty)^b$
L_q	$\dfrac{\rho^2[1 - \rho^N - N\rho^{N-1}(1 - \rho)]}{(1 - \rho)(1 - \rho^{N+1})}$	$\dfrac{\rho^2}{1 - \rho}$
L	$\dfrac{\rho[1 - \rho^N - N\rho^N(1 - \rho)]}{(1 - \rho)(1 - \rho^{N+1})}$	$\dfrac{\rho}{1 - \rho}$
W_q	$\dfrac{\rho[1 - \rho^N - N\rho^{N-1}(1 - \rho)]}{\mu(1 - \rho)(1 - \rho^N)}$	$\dfrac{\rho}{\mu(1 - \rho)}$
W	$\dfrac{1 - \rho^N - N\rho^N(1 - \rho)}{\mu(1 - \rho)(1 - \rho^N)}$	$\dfrac{1}{\mu(1 - \rho)}$
D	$\dfrac{\rho(1 - \rho^{N-1})}{1 - \rho^N}$	ρ
P_0	$\dfrac{1 - \rho}{1 - \rho^{N+1}}$	$1 - \rho$

[a] For $\rho \neq 1$.
[b] For $\rho < 1$.

Since the single-server queue occurs frequently in practice, results for this special case have been tabulated in Table 3.3.

Example 3.2 As an illustration of the previous development, recall the barber shop example, and let μ, λ, c, and N be 6, 12, 2, and 6, respectively. The resulting probability mass function is given in Table 3.4. (Note that $\rho = 1$ for this example.) An additional column, $\hat{P}_n$, has been added to Table 3.4. You might want to verify the values given for P_n and $\hat{P}_n$.

TABLE 3.4 Computation of values for P_n and $\hat{P}_n$ in Example 3.2

n	λ_n	μ_n	P_n	P_n	$\hat{P}_n$
0	12	0	P_0	1/13	1/11
1	12	6	$2P_0$	2/13	2/11
2	12	12	$2P_0$	2/13	2/11
3	12	12	$2P_0$	2/13	2/11
4	12	12	$2P_0$	2/13	2/11
5	12	12	$2P_0$	2/13	2/11
6	0	12	$2P_0$	2/13	0

On applying (3.31)–(3.35) and (3.37), we obtain

$$L_q = 20/13 = 1.54 \text{ customers}$$
$$L = 42/13 = 3.23 \text{ customers}$$
$$\tilde{\lambda} = 132/13 = 10.15 \text{ customers/hr}$$
$$W_q = 5/33 = 0.1515 \text{ hr/customer}$$
$$W = 7/22 = 0.3182 \text{ hr/customer}$$
$$D = 8/11 = 0.7272$$

Comparing the above results with those obtained previously for the barber shop, we see that the values of the operating characteristics for the system are increased by assuming that all arriving customers enter the shop until there are six customers in the system.

$(M|M|c):(GD|\infty|\infty)$ **results.** Given the results for the $(M|M|c):(GD|N|\infty)$ queue, it is an easy matter to obtain comparable results for the $(M|M|c):(GD|\infty|\infty)$ queue by noting that

$$(M|M|c):(GD|\infty|\infty) = \lim_{N \to \infty} (M|M|c):(GD|N|\infty)$$

In order for the limit to be defined, the infinite series in the denominator of (3.20) must converge to a finite value. Since we can be assured of convergence if $\lambda_{n-1}/\mu_n < 1$ for all values of $n > c$, we will require that $\rho =$

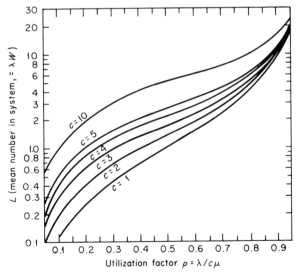

Fig. 3.4 L (mean number in system, $= \lambda W$) for different values of c versus the utilization factor ρ.

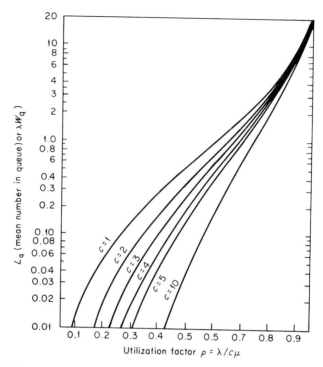

Fig. 3.5 L_q (mean number in queue, $= \lambda W_q$) for different values of c versus the utilization factor ρ.

$\lambda/c\mu < 1$. Formulas for the operating characteristics of the $(M\,|\,M\,|\,c):(GD\,|\,\infty\,|\,\infty)$ queue are given in Table 3.2, with values of L and L_q provided in Figs. 3.4 and 3.5, respectively. Also, results for the $(M\,|\,M\,|\,1):(GD\,|\,\infty\,|\,\infty)$ queue are summarized in Table 3.3.

For the $(M\,|\,M\,|\,c):(GD\,|\,\infty\,|\,\infty)$ queue, the probability distribution for the number of customers in the system can be given as follows:

$$P_n = \begin{cases} \dfrac{(c\rho)^n}{n!} P_0, & n = 0, 1, \ldots, c-1 \\[2mm] \dfrac{c^c \rho^n}{c!} P_0, & n = c, c+1, \ldots, \infty \end{cases} \tag{3.38}$$

where

$$P_0 = \left[\frac{(c\rho)^c}{c!\,(1-\rho)} + \sum_{n=0}^{c-1} \frac{(c\rho)^n}{n!} \right]^{-1} \tag{3.39}$$

Infinite Population Models

Example 3.3 Consider again the barber shop example in which customers arrive at a rate of 12/hr and are served at a rate of 6/hr. Suppose we wish to determine the effect of adding another barber in the shop. We will assume that with a third barber the $(M|M|c):(GD|\infty|\infty)$ queueing model is a reasonable approximation of the physical system.

Since $c = 3$ and $\rho = 2/3$, from (3.39), $P_0 = 1/9$. Therefore, from Table 3.2,

$$L_q = 8/9 \text{ customers}$$
$$L = 26/9 \text{ customers}$$
$$W_q = 2/27 \text{ hr/customer}$$
$$W = 13/54 \text{ hr/customer}$$
$$D = 4/9$$

Based on the above predictions of the behavior of the system with three barbers, we must now decide if the cost of the additional barber is justified. At this point, we must rely on subjective judgment in reaching a decision. In Chapter 5 we will consider the economic factors involved and develop a normative model for aiding in the choice of alternative system designs.

Waiting-Time and Total-Time Distributions[†]

At this point we have available the probability mass functions for the number of customers in the system and the number of customers in the system ahead of an arriving customer. Although we can obtain values for the expected amount of time spent waiting and total time spent in the system for a customer who enters the system, it is sometimes useful to know the probability density functions for waiting time and total time. This is particularly true when we wish to design a queueing system with a constraint on, say, the amount of time a customer has to wait for service.

$(M|M|c):(FCFS|N|\infty)$ **results.** Let us first consider how we might develop the density function for the amount of time a customer spends wating for service when service is first come–first served, i.e., $(M|M|c):(FCFS|N|\infty)$. Notice that a customer entering the system will not have to wait for service if there is an idle server available. Therefore, there is zero waiting time if there are $c - 1$ or less customers in the system ahead of an entering customer. If there are n ($n > c$) customers ahead of an entering customer, he will have to wait until $n - c + 1$ customers have been

---

[†] This section can be omitted on first reading without loss of continuity.

served. Due to the forgetfulness property of the exponential distribution, we know that the time until service completion for a customer in service is exponentially distributed with parameter μ. Furthermore, we know that all c servers are busy (or else the customer would not have to wait!). Therefore, the departure rate is $c\mu$. Letting $f(t_q)$ denote the density function for waiting time and $f(t_q \mid n)$ be the conditional density function for waiting time, we see that

$$f(t_q) = \sum_{n=0}^{N-1} f(t_q \mid n)\hat{P}_n \qquad (3.40)$$

or, on taking the Laplace transform of both sides of (3.40),

$$\mathscr{L}[f(t_q)] = \sum_{n=0}^{N-1} \mathscr{L}[f(t_q \mid n)]\hat{P}_n$$

which can be expressed as

$$\mathscr{L}[f(t_q)] = \sum_{n=0}^{c-1} \mathscr{L}[\delta(t_q - 0)]\hat{P}_n + \sum_{n=c}^{N-1} \left(\frac{c\mu}{c\mu + s}\right)^{n-c+1} \hat{P}_n \qquad (3.41)$$

The first term on the right-hand side of (3.41) represents the zero waiting time condition. As discussed in Chapter 2, the delta function is a convenient device for denoting this condition. Recall that $\mathscr{L}[\delta(t_q - 0)] = 1$. The second term on the right-hand side of (3.41) gives the transform for waiting time when the customer must wait. From Chapter 2, we recall that the transform for the distribution of the sum of independently distributed random variables is the product of the transforms of the distributions for each random variable involved in the sum. Here we have $n - c + 1$ identical exponentially distributed random variables with parameter $c\mu$.

Substituting (3.36) in (3.41) and reducing gives

$$\mathscr{L}[f(t_q)] = \frac{(c\rho)^c \hat{P}_0}{c!\,\rho} \sum_{n=1}^{N-c} \left(\frac{\lambda}{c\mu + s}\right)^n + \sum_{n=0}^{c-1} \hat{P}_n \qquad (3.42)$$

On inverting (3.42) term by term, we obtain

$$f(t_q) = \begin{cases} 1 - \dfrac{(c\rho)^c(1 - \rho^{N-c})\hat{P}_0}{c!\,(1 - \rho)}, & t_q = 0, \quad \rho \neq 1 \\[2mm] 1 - \dfrac{c^c(N - c)\hat{P}_0}{c!}, & t_q = 0, \quad \rho = 1 \\[2mm] \dfrac{(c\rho)^c \hat{P}_0}{c!\,\rho} \sum_{n=1}^{N-c} \rho^n \dfrac{(c\mu)^n}{(n-1)!} t_q^{n-1} \exp(-c\mu t_q), & t_q > 0 \end{cases} \qquad (3.43)$$

Infinite Population Models

The density function for waiting time is not particularly inviting. However, things are not quite as bad as they appear. In particular, the cumulative distribution function for waiting time can be given as

$$F_{T_q}(k) = f(t_q = 0) + \frac{(c\rho)^c \hat{P}_0}{c!\,\rho} \sum_{n=1}^{N-c} \rho^n \int_0^k g(t_q, n, c\mu)\, dt_q, \qquad k > 0 \quad (3.44)$$

where $g(t_q, n, c\mu)$ is the gamma density function, with parameters n and $c\mu$. Letting $G_n(k)$ be the cumulative distribution function for the gamma density, (3.44) becomes

$$F_{T_q}(k) = f(t_q = 0) + \frac{(c\rho)^c \hat{P}_0}{c!\,\rho} \sum_{n=1}^{N-c} \rho^n G_n(k) \quad (3.45)$$

Values for $G_n(k)$ are obtained from the appropriate tables of the incomplete gamma function.

We previously gave an indirect method for determining the expected waiting time. Specifically, we suggested the calculation of L_q and $\tilde{\lambda}$, and then the use of the relation $W_q = L_q/\tilde{\lambda}$. Since we know the density function for waiting time T_q, we can obtain the value of W_q directly:

$$W_q = E(T_q) = \sum_{n=0}^{c-1} E(T_q \mid n)\hat{P}_n + \sum_{n=c}^{N-1} E(T_q \mid n)\hat{P}_n$$

Since $E(T_q \mid n \leq c-1) = 0$ and the mean of the gamma density $E(T_q \mid n \geq c) = n/c\mu$, it follows that

$$W_q = \frac{(c\rho)^c \hat{P}_0}{c!\,\lambda} \sum_{n=1}^{N-c} n\rho^n$$

or

$$W_q = \begin{cases} \dfrac{(c\rho)^c [1 - \rho^{N-c+1} - (1-\rho)(N-c+1)\rho^{N-c}]\hat{P}_0}{c!\,c\mu(1-\rho)^2}, & \rho \neq 1 \\[2mm] \dfrac{c^c(N-c+1)(N-c)\hat{P}_0}{2\lambda c!}, & \rho = 1 \end{cases} \quad (3.46)$$

In order to obtain $f(t_T)$, the density function for total time spent in the system, notice that the random variable T_T, total time in the system, is equal to the sum of the random variables T_q, waiting time, and T_s, service time. We also note that T_q and T_s are independent. Therefore, the Laplace transform of $f(t_T)$ is equal to the product of the transforms of $f(t_q)$ and $f(t_s)$. Thus, from (3.41),

$$\mathscr{L}[f(t_T)] = \sum_{n=0}^{c-1} \left(\frac{\mu}{\mu+s}\right)\hat{P}_n + \sum_{n=c}^{N-1} \left(\frac{\mu}{\mu+s}\right)\left(\frac{c\mu}{c\mu+s}\right)^{n-c+1} \hat{P}_n$$

or

$$\mathscr{L}[f(t_T)] = \sum_{n=0}^{c-1}\left(\frac{\mu}{\mu+s}\right)\hat{P}_n + \frac{(c\rho)^c \hat{P}_0}{c!\rho}\sum_{n=1}^{N-c}\left(\frac{\lambda}{c\mu+s}\right)^n\left(\frac{\mu}{\mu+s}\right) \quad (3.47)$$

To obtain the inverse transform of (3.47) requires the use of a partial fraction expansion. The resulting expression for $f(t_T)$ (given $c > 1$) is

$$f(t_T) = \mu\exp(-\mu t_T)\sum_{n=0}^{c-1}\hat{P}_n + \frac{(c\rho)^c \hat{P}_0}{c!\rho}\sum_{n=1}^{N-c}\mu\lambda^n$$

$$\times \left[\frac{\exp(-\mu t_T)}{(c\mu-\mu)^n} - \sum_{k=1}^{n}\frac{t_T^{n-k}\exp(-c\mu t_T)}{(n-k)!(c\mu-\mu)^k}\right] \quad (3.48)$$

Unfortunately, (3.48) is not in a form that is useful to us in analyzing queueing systems. [You may wish to verify that (3.48) is indeed the inverse transform of (3.47). To do this, take the transform of (3.48). After much patient algebraic reduction, (3.47) is obtained.]

In order to obtain the expected time in the system, we recall

$$W = E(T_T) = (-1)\frac{d\mathscr{L}[f(t_T)]}{ds}\bigg|_{s=0}$$

On taking the derivative of (3.47) with respect to s, evaluating the derivative at $s = 0$, and multiplying the result by -1, we get

$$W = \frac{1}{\mu} + \sum_{n=c}^{N-1}\frac{(n-c+1)}{c\mu}\hat{P}_n, \quad \text{or} \quad W = \frac{1}{\mu} + W_q \quad (3.49)$$

as expected.

$(M\,|\,M\,|\,c)\!:\!(FCFS\,|\,\infty\,|\,\infty)$ **results.** In order to develop the waiting-time and total-time distributions for the $(M\,|\,M\,|\,c)\!:\!(FCFS\,|\,\infty\,|\,\infty)$ queue, we again require that $\rho < 1$. Taking the limit of (3.43) as $N \to \infty$, we get

$$f(t_q) = \begin{cases} 1 - \dfrac{(c\rho)^c P_0}{c!(1-\rho)}, & t_q = 0 \\ \dfrac{(c\rho)^c \mu P_0}{(c-1)!}\exp[-(c\mu-\lambda)t_q], & t_q > 0 \end{cases} \quad (3.50)$$

In order to obtain a comparable expression for $f(t_T)$, we will first take the limit of $\mathscr{L}[f(t_T)]$ as $N \to \infty$. The resulting Laplace transform is

$$\mathscr{L}[f(t_T)] = \frac{\mu}{\mu+s}\left[1 - \frac{(c\rho)^c P_0}{c!(1-\rho)}\right] + \frac{\mu^2(c\rho)^c P_0}{(c-1)!(\mu+s)(c\mu-\lambda+s)} \quad (3.51)$$

Using partial fraction expansion and obtaining the inverse Laplace transform of (3.51) gives

$$f(t_T) = \mu \exp(-\mu t_T)\left[1 - \frac{(c\rho)^c P_0}{c!(1-\rho)} + \frac{(c\rho)^c P_0}{(c-1)![c-1-(\lambda/\mu)]}\right]$$

$$- \frac{\mu(c\rho)^c P_0}{(c-1)![c-1-(\lambda/\mu)]} \exp[-(c\mu - \lambda)t_T] \qquad (3.52)$$

Notice that when there exists a single server, (3.52) reduces to

$$f(t_T) = (\mu - \lambda)\exp[-(\mu - \lambda)t_T], \qquad t_T > 0$$

which is an exponential distribution with parameter $(\mu - \lambda)$. Therefore, the probability of a customer being in the system for at least γ more time units is independent of how long he has already been in the system.

It is also interesting to note for the single-server case that the waiting-time distribution for those customers who must wait,

$$f(t_q \mid t_q > 0) = (\mu - \lambda)\exp[-(\mu - \lambda)t_q], \qquad t_q > 0 \qquad (3.53)$$

is the same as the distribution for total time in the system.

$(M\mid M\mid c):(GD\mid c\mid \infty)$ Queue: Erlang's Loss Formula

Another special case of the $(M\mid M\mid c):(GD\mid N\mid \infty)$ queueing problem occurs when $N = c$. In such a case, a waiting line does not exist. Thus, if a customer arrives and cannot be served immediately, he departs. In this case, the probability mass function for the number of customers in the system is given as

$$P_n = \frac{(c\rho)^n/n!}{\sum_{k=0}^{c}(c\rho)^k/k!}, \qquad n = 0, 1, \ldots, c \qquad (3.54)$$

If we multiply both the numerator and the denominator of (3.54) by $e^{-c\rho}$, then we obtain

$$P_n = \frac{e^{-c\rho}(c\rho)^n/n!}{\sum_{k=0}^{c} e^{-c\rho}(c\rho)^k/k!}, \qquad n = 0, 1, \ldots, c \qquad (3.55)$$

which is a truncated Poisson distribution with parameter $c\rho$. The numerator expresses the probability of a Poisson distributed random variable having value n and the denominator is the cumulative distribution function for the Poisson mass function. Thus, values of P_n are easily obtained using standard tables for the Poisson distribution.

The quantity P_c gives the proportion of the time the system is fully occupied. In the case of telephone exchanges with c lines, this is the proportion of incoming calls that will encounter a "busy" signal. The formula is called *Erlang's loss formula* and is written

$$P_c = \frac{(c\rho)^c/c!}{\sum_{k=0}^{c}(c\rho)^k/k!} \tag{3.56}$$

In Europe, (3.56) is referred to as Erlang's first formula and is denoted by $E_{1,c}(\lambda/\mu)$. Also of interest is the reference to the formula for the delay probability for an $(M \mid M \mid c):(GD \mid \infty \mid \infty)$ queue as Erlang's second formula or Erlang's delay formula, denoted $E_{2,c}(\lambda/\mu)$.

Interestingly, Erlang's loss formula holds when service times are not exponentially distributed. For a discussion of Erlang's loss formula and the conditions under which it holds, see Syski (1960), Takács (1969), and Cooper (1972), among others. Values of P_c are provided in Fig. 3.6 for selected values of c and λ/μ.

Example 3.4 As an illustration of the use of the Erlang loss formula, consider an airline reservation desk that needs to have a number of telephone lines installed such that the probability of a person being unable to complete his call will be less than 0.05. Suppose it is estimated that the demand for the telephone is Poisson distributed with a mean of 15 calls/hr. The duration of a telephone conversation is believed to be exponentially distributed with a mean of 4 min. Therefore, $\lambda/\mu = 1$. From (3.56), we wish to determine the smallest value of c such that

$$0.05 \geq \frac{e^{-1}/c!}{\sum_{n=0}^{c} e^{-1}/n!}$$

From Fig. 3.6, we see that

$$P_{c=3} > 0.05, \quad P_{c=4} < 0.05$$

Therefore, four telephone lines should be installed. As an exercise at the end of the chapter, you are asked to show for the $(M \mid M \mid c):(GD \mid c \mid \infty)$ queue that

$$L = (c\rho)(1 - P_c), \quad L_q = 0, \quad \tilde{\lambda} = L\mu, \quad W_q = 0, \quad W = \frac{1}{\mu}$$

For our example problem, with four telephone lines, only 0.985 lines are busy on the average. Therefore, we see that we are paying a price of low telephone utilization in order to provide the protection of no more than 5% lost calls.

Finite Population Models

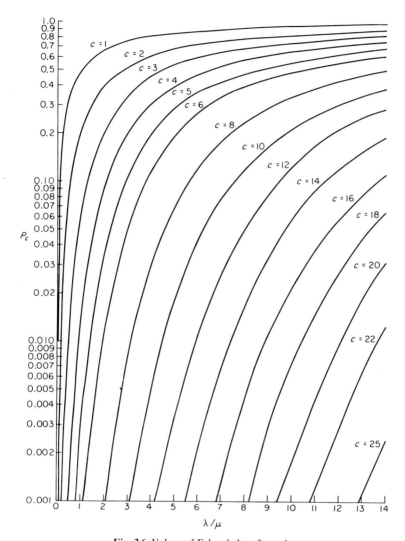

Fig. 3.6 Values of Erlang's loss formula.

FINITE POPULATION MODELS: $(M\,|\,M\,|\,c):(GD\,|\,K\,|\,K)$ QUEUE

Our previous discussion has concentrated on the set of queueing problems in which the customer population is infinitely large. The assumption of an infinite population is obviously a mathematical convenience. We know the population of customers for the barber shop example cannot really be infinite. However, we anticipate that any errors produced in our predictions of the behavior of the system will be negligible. Naturally, this assumption is not valid in some cases. For example, consider the maintenance of five textile machines by a single mechanic. We would expect, in this case, that the probability of some machine requiring maintenance during time Δt is dependent on the number in service.

In this section we will be concerned with the development of results for the $(M \mid M \mid c):(GD \mid K \mid K)$ queue. For obvious reasons we assume $K \geq c$; otherwise, we have an excess of servers. (Note that we still assume that one server performs service for one customer.) We will further assume that the system capacity is not less than the population size. To do otherwise substantially complicates the problem. Fortunately, most finite population problems encountered in practice satisfy this assumption.

As discussed in Chapter 1, when a finite population exists, we measure the arrival rate in a different way. For convenience, think of a customer as a machine that runs until service is required. If service is not immediately available, the machine waits. When a server is available, the machine is serviced and returned to a running state. Pictorially, this sequence is shown in Fig. 3.7. Three random variables are defined: T_r, running time; T_q, waiting time; and T_s, service time.

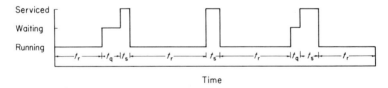

Fig. 3.7 Representation of running, waiting, and serviced states for an individual customer over time.

Suppose running time is exponentially distributed with parameter λ. Due to the memoryless property of the exponential distribution, the probability that the machine will run for at least Δt time units is

$$\int_{\Delta t}^{\infty} \lambda e^{-\lambda t} \, dt = e^{-\lambda \Delta t}$$

but since

$$e^{-\lambda \Delta t} = \sum_{k=0}^{\infty} (-\lambda \Delta t)^k$$

then for Δt sufficiently small,

$$e^{-\lambda \Delta t} \approx 1 - \lambda \Delta t$$

Finite Population Models

Now the probability that the machine will stop during Δt is one minus the probability that it will run for at least Δt. Therefore,

$$P(\text{machine stops during } \Delta t) = \lambda \, \Delta t$$

From our earlier discussion concerning the Poisson process, and the fact that the sum of Poisson distributed random variables is Poisson, we see that if there are m machines running at time t, then the probability of one of the m machines stopping before time $t + \Delta t$ is $m\lambda \, \Delta t$. Thus, if we have K machines and n are down (not running), then the arrival rate at the system is referred to as quasi-random input and is given by

$$\lambda_n = (K - n)\lambda, \qquad n = 0, 1, \ldots, K \tag{3.57}$$

Furthermore, since we have assumed exponentially distributed service time,

$$\mu_n = \begin{cases} n\mu, & n = 0, 1, \ldots, c - 1 \\ c\mu, & n = c, c + 1, \ldots, K \end{cases} \tag{3.58}$$

We can substitute λ_n and μ_n in (3.18) to obtain the probability mass function for the number of customers in the system,

$$P_n = \begin{cases} \binom{K}{n}(c\rho)^n P_0, & n = 0, 1, \ldots, c - 1 \\ \binom{K}{n}\dfrac{n!\, c^c \rho^n}{c!} P_0, & n = c, c+1, \ldots, K \\ 0, & n > K \end{cases} \tag{3.59}$$

where $\rho = \lambda/c\mu$. Solving for P_0 gives

$$P_0 = \left[\sum_{n=0}^{c-1} \binom{K}{n}(c\rho)^n + \sum_{n=c}^{K} \binom{K}{n} \dfrac{n!\, c^c \rho^n}{c!} \right]^{-1} \tag{3.60}$$

Unfortunately, (3.60) cannot be further reduced to a compact expression. In the case of a single server, (3.59) can be expressed as

$$P_n = \dfrac{[K!/(K - n)!]\rho^n}{\sum_{n=0}^{K} [K!/(K - n)!]\rho^n} \tag{3.61}$$

Cancelling $K!$ in both the numerator and the denominator of (3.61), letting γ equal $1/\rho$, multiplying the numerator and denominator by $e^{-\gamma}\gamma^K$, and substituting x for $K - n$ gives

$$P_{K-x} = \dfrac{e^{-\gamma}\gamma^x/x!}{\sum_{x=0}^{K} e^{-\gamma}\gamma^x/x!}, \qquad x = 0, 1, \ldots, K \tag{3.62}$$

Notice that the numerator of (3.62) is the probability of a Poisson distributed random variable with parameter γ having a value of $x = K - n$. The denominator is the cumulative distribution function for the Poisson distribution evaluated at K. Thus, the values of the numerator and denominator can be evaluated easily using tables of the Poisson distribution.

Operating Characteristics

In the case of multiple servers (as well as single servers) we can rely on the extensive tables of computational results provided by Peck and Hazelwood (1958) for the finite population model. A sample of the type of data tabulated by Peck and Hazelwood is given in the Appendix. To interpret these data, let the service factor X be defined as follows:

$$X = \frac{1/\mu}{(1/\lambda) + (1/\mu)} = \frac{W - W_q}{W - W_q + \lambda^{-1}} = \frac{\lambda}{\lambda + \mu} \tag{3.63}$$

and

$$F = \frac{(1/\mu) + (1/\lambda)}{(1/\mu) + (1/\lambda) + W_q} = \frac{W - W_q + \lambda^{-1}}{W + \lambda^{-1}} \tag{3.64}$$

Values for F are provided for various combinations of X, K, and c. Based on the value of F, the following values can be obtained:

$$L_q = K(1 - F) \tag{3.65}$$

$$L = K[1 - F(1 - X)] \tag{3.66}$$

$$W_q = \frac{1 - F}{\mu F X} \tag{3.67}$$

$$W = \frac{1 - F(1 - X)}{\mu F X} \tag{3.68}$$

The above expressions are derived using the following proportionalities:

$$L_q : L : K = W_q : W : (W + \lambda^{-1}) \tag{3.69}$$

An additional quantity provided by Peck and Hazelwood is the value for delay probability. However, they define the delay probability D as the probability that an arriving customer is delayed. To see how this quantity is obtained, notice that it is no longer true that an arrival is equally likely to

Finite Population Models

occur in any time increment Δt. The probability of an arrival during Δt is dependent on the number in the system. Specifically, notice that

$P(\text{customer calls for service during } \Delta t)$

$$= \sum_{n=0}^{K} P(\text{customer calls for service during } \Delta t \mid n \text{ in the system}) P_n$$

$$= \sum_{n=0}^{K} (K - n) P_n \lambda \, \Delta t \qquad (3.70)$$

Now let the delay probability be defined as follows:

$D = P(\text{customer calls for service during } \Delta t$ and has to wait)

$\div P(\text{customer calls for service during } \Delta t)$

Thus,

$$D = \frac{\sum_{n=c}^{K} (K - n) P_n \lambda \, \Delta t}{\sum_{n=0}^{K} (K - n) P_n \lambda \, \Delta t} \qquad (3.71)$$

which, on cancelling $\lambda \, \Delta t$, can be expressed as

$$D = \frac{K \sum_{n=c}^{K} P_n - \sum_{n=c}^{K} n P_n}{K - L} \qquad (3.72)$$

We can obtain the probability mass function for the number of customers in the system ahead of an arriving customer by noting that

$$P(n \text{ in system} \mid \text{arriving customer}) = \frac{P(\text{arrival} \mid n \text{ in system}) P_n}{P(\text{arrival})}$$

or, from (3.70) and (3.71),

$$\hat{P}_n = \frac{(K - n) P_n}{\sum_{n=0}^{K} (K - n) P_n} \qquad (3.73)$$

Example 3.5 Twelve automatic production machines are operated by three operators. The machines randomly require the attention of an operator. Currently, Operator A is assigned to machines O, P, Q, and R, Operator B is assigned to machines S, T, U, and V, and Operator C is assigned to machines W, X, Y, and Z. Machine running time is exponentially distributed with a mean of 1 hr. Service time is exponentially distributed with a mean of (1/9) hr. It has been observed that the operators are idle a large portion of the time. Two alternative modifications have been proposed: Proposal I involves the assignment of six machines per operator and requires only two operators; Proposal II involves the assignment of twelve machines to two operators using a service-pool concept. With a service pool, an operator can service any of the twelve machines. It is anticipated that the service-time distribution will remain the same, since all machines are centrally located.

We wish to evaluate the proposals by analyzing the operating characteristics of the system. Since λ equals 1 demand/hr and μ equals 9 services/hr, then X equals 0.10. From the Appendix, we obtain appropriate values for D and F. These values, as well as those for L, L_q, W, and W_q, are given in Table 3.5 for the present system and the two proposed systems.

TABLE 3.5 Operating characteristic values for Example 3.5

	Present system 3 $(M\mid M\mid 1):(GD\mid 4\mid 4)$ queues	Proposal I 2 $(M\mid M\mid 1):(GD\mid 6\mid 6)$ queues	Proposal II 1 $(M\mid M\mid 2):(GD\mid 12\mid 12)$ queue
D	0.294	0.475	0.361
F	0.965	0.932	0.970
L_q	0.1400 machines/system	0.408 machines/system	0.360 machines/system
L	0.5260 machines/system	0.967 machines/system	1.524 machines/system
W_q	0.0404 hr/machine	0.081 hr/machine	0.034 hr/machine
W	0.1515 hr/machine	0.192 hr/machine	0.145 hr/machine

Notice that the service pool (Proposal II) yields considerable improvement over the separation of servers (Proposal I). There are 2 × 0.408, or 0.816, machines waiting for service under Proposal I and only 0.360 under Proposal II. The average waiting time under Proposal II is less than one-half that under Proposal I. Therefore, unless there are some overriding qualitative considerations, we would prefer Proposal II. Note that with a service pool and only two operators we operate more efficiently than with three operators using separation of servers. As an example, there are 3 × 0.140, or 0.42, machines waiting under the present system and only 0.36 under Proposal II.

Waiting-Time and Total-Time Distributions $(M\mid M\mid c):(FCFS\mid K\mid K)$[†]

We can use (3.73) in the development of probability distributions for waiting time and total time spent in the system. To do this, we assume a first come–first served service discipline.

As in our previous analysis of the waiting-time distribution, we see that an arriving customer will not have to wait if all servers are not busy. Other-

---

[†] This section can be omitted on first reading without loss of continuity.

Finite Population Models

wise, the customer will have to wait until a server is available. Therefore, the density function for waiting time can be expressed as

$$f(t_q) = \sum_{n=0}^{K-1} f(t_q \mid n)\hat{P}_n$$

where $f(t_q \mid n) = \delta(t_q - 0)$ for $n = 0, 1, \ldots, c - 1$, and $f(t_q \mid n)$ is the probability distribution for the sum of $n - c + 1$ identical and independent exponentially distributed random variables with parameter $c\mu$, where $n = c, \ldots, K - 1$. Thus, $f(t_q \mid n)$ is a gamma distribution with parameters $n - c + 1$ and $c\mu$ if $n \geq c$, or

$$f(t_q) = \begin{cases} \sum_{n=0}^{c-1} \hat{P}_n, & t_q = 0 \\ \sum_{n=c}^{K-1} \frac{(c\mu)^{n-c+1}}{(n-c)!} t_q^{n-c} \exp(-c\mu t_q)\hat{P}_n, & t_q > 0 \end{cases} \quad (3.74)$$

Now, since

$$E(t_q) = \sum_{n=c}^{K-1} E(t_q \mid n)\hat{P}_n$$

then we see that

$$W_q = \sum_{n=c}^{K-1} \left(\frac{n - c + 1}{c\mu}\right)\hat{P}_n \quad (3.75)$$

To obtain the density function for time spent in the system, we observe that since $T_T = T_s + T_q$, then

$$\mathscr{L}[f(t_T)] = \mathscr{L}[f(t_s)]\mathscr{L}[f(t_q)]$$

where

$$\mathscr{L}[f(t_s)] = \frac{\mu}{\mu + s}$$

and

$$\mathscr{L}[f(t_q)] = \sum_{n=0}^{c-1} \hat{P}_n + \sum_{n=c}^{K-1} \left(\frac{c\mu}{c\mu + s}\right)^{n-c+1} \hat{P}_n$$

Therefore,

$$\mathscr{L}[f(t_T)] = \sum_{n=0}^{c-1} \left(\frac{\mu}{\mu + s}\right)\hat{P}_n + \sum_{n=c}^{K-1} \left(\frac{\mu}{\mu + s}\right)\left(\frac{c\mu}{c\mu + s}\right)^{n-c+1} \hat{P}_n \quad (3.76)$$

To obtain the inverse transform of (3.76), recall the inverse transform obtained for (3.47). Due to the similarities between (3.76) and (3.47), we see that

$$f(t_T) = \exp(-\mu t_T) \sum_{n=0}^{c-1} \hat{P}_n + \sum_{n=1}^{K-c} \mu(c\mu)^n$$
$$\times \left[\frac{\exp(-\mu t_T)}{(c\mu - \mu)^n} - \sum_{j=1}^{n} \frac{t_T^{n-j} \exp(-c\mu t_T)}{(n-j)!(c\mu - \mu)^j} \right] \hat{P}_{n+c-1} \quad (3.77)$$

The expected time in the system is

$$W = \frac{1}{\mu} + \sum_{n=c}^{K-1} \frac{n-c+1}{c\mu} \hat{P}_n \quad (3.78)$$

or, from (3.75),

$$W = \frac{1}{\mu} + W_q$$

BULK ARRIVALS: $(M^{(b)} \mid M \mid c):(GD \mid \infty \mid \infty)$ **QUEUE**

Our discussion of the $(M \mid M \mid c)$ queue began with the development of a rather general procedure for analyzing $(M \mid M \mid c)$ queueing systems. The procedure is valid as long as the steady-state, general balance equations for the $(M \mid M \mid c)$ queue are satisfied. Recall that these equations can be given as

$$P_{n+1} = \frac{P_n(\lambda_n + \mu_n)}{\mu_{n+1}} - \frac{P_{n-1}\lambda_{n-1}}{\mu_{n+1}}, \quad n \geq 1 \quad (3.79)$$

and

$$P_1 = \frac{P_0 \lambda_0}{\mu_1} \quad (3.80)$$

There exist some variations of the $(M \mid M \mid c)$ queue that do not satisfy (3.79) and (3.80). One case that comes to mind is that in which more than one customer arrives at an arrival instant. Such a situation is referred to as a batch-arrival or bulk-arrival system and designated as an $(M^{(b)} \mid M \mid c)$ queue, where the superscript b denotes the number of customers per arrival.

A number of practical situations involve multiple customers arriving at the same time. As an example of bulk arrivals, consider the arrival of a load of b parts at a machine for further processing. Bulk-arrival examples also arise in a number of transportation problems. An airplane arriving with a load of passengers is a common example.

Bulk Arrivals

We will consider a very simple bulk-arrival problem in which the number of customers per arrival is a constant and there is a single server. Specifically, we will consider an $(M^{(b)}|M|1):(GD|\infty|\infty)$ queueing system.

We begin by developing the steady-state, general balance equations for the case of b customers per arrival. First, the following equations are obtained:

$$P_0(t + \Delta t) = P_0(t)(1 - \lambda \Delta t) + P_1(t)\mu \Delta t(1 - \lambda \Delta t)$$

$$P_n(t + \Delta t) = P_n(t)(1 - \lambda \Delta t)(1 - \mu \Delta t) + P_{n+1}(t)\mu \Delta t(1 - \lambda \Delta t),$$
$$n = 1, \ldots, b - 1$$

$$P_n(t + \Delta t) = P_{n-b}(t)\lambda \Delta t(1 - \mu \Delta t) + P_n(t)(1 - \lambda \Delta t)(1 - \mu \Delta t)$$
$$+ P_{n+1}(t)\mu \Delta t(1 - \lambda \Delta t), \quad n \geq b$$

Now, by transposing $P_n(t)$ to the left-hand side, dividing by Δt, and taking the limit as Δt approaches zero, we obtain

$$\frac{dP_0(t)}{dt} = -\lambda P_0 + \mu P_1 \tag{3.81}$$

$$\frac{dP_n(t)}{dt} = -(\lambda + \mu)P_n + \mu P_{n+1}, \quad n = 1, \ldots, b - 1 \tag{3.82}$$

$$\frac{dP_n(t)}{dt} = \lambda P_{n-b} - (\lambda + \mu)P_n + \mu P_{n+1}, \quad n \geq b \tag{3.83}$$

We will be interested in steady-state results. So we set (3.81)–(3.83) to zero by taking the limit of each as $t \to \infty$. The resulting set of difference equations will be solved using generating functions. To begin, multiply both sides of (3.82) by z^n and sum from $n = 1$ to $b - 1$. Next, multiply both sides of (3.83) by z^n and sum from $n = b$ to ∞. Combining the results obtained along with (3.81) gives

$$\sum_{n=b}^{\infty} \mu z^n P_{n+1} - \sum_{n=b}^{\infty} (\lambda + \mu) z^n P_n + \sum_{n=b}^{\infty} \lambda z^n P_{n-b}$$
$$+ \sum_{n=1}^{b-1} \mu z^n P_{n+1} - \sum_{n=1}^{b-1} (\lambda + \mu) z^n P_n + \mu P_1 - \lambda P_0 = 0 \tag{3.84}$$

Therefore, on collecting terms and factoring appropriate powers of z, we obtain

$$z^{-1} \sum_{n=1}^{\infty} \mu z^{n+1} P_{n+1} - \sum_{n=1}^{\infty} (\lambda + \mu) z^n P_n + z^b \sum_{n=b}^{\infty} \lambda z^{n-b} P_{n-b} + \mu P_1 - \lambda P_0 = 0$$

$$\tag{3.85}$$

Letting $P(z)$ be the generating function for n, where $P(z) = \sum_{n=0}^{\infty} z^n P_n$, gives

$$\mu z^{-1}[P(z) - P_0 - zP_1] - (\lambda + \mu)[P(z) - P_0] + \lambda z^b P(z) + \mu P_1 - \lambda P_0 = 0 \tag{3.86}$$

Solving for $P(z)$ yields the following expression:

$$P(z) = \frac{\mu(1-z)P_0}{\mu - (\lambda + \mu)z + \lambda z^{b+1}} \tag{3.87}$$

Dividing both numerator and denominator of (3.87) by $(1 - z)$ gives

$$P(z) = \frac{\mu P_0}{\mu - [\lambda z(1-z^b)/(1-z)]} \quad \text{or} \quad P(z) = \frac{\mu P_0}{\mu - \lambda z \sum_{j=0}^{b-1} z^j}$$

Now, since $P(z)|_{z=1} = 1$, then

$$1 = \frac{\mu P_0}{\mu - \lambda \sum_{j=0}^{b-1} 1^j} \quad \text{or} \quad 1 = \frac{\mu P_0}{\mu - b\lambda}$$

Letting $\rho = b\lambda/\mu$ gives

$$P_0 = 1 - \rho$$

Therefore,

$$P(z) = \frac{\mu(1-\rho)}{\mu - \lambda z \sum_{j=0}^{b-1} z^j} \tag{3.88}$$

At this point, we have obtained the generating function for the number of customers in the system. We can recover the corresponding probability mass function by noting that

$$P_n = \frac{1}{n!} \frac{d^n P(z)}{dz^n} \bigg|_{z=0} \tag{3.89}$$

If we want to know the expected number of customers in the system, we recall that

$$L = E(n) = \frac{dP(z)}{dz}\bigg|_{z=1} \quad \text{or} \quad L = \frac{\rho(1+b)}{2(1-\rho)} \tag{3.90}$$

To obtain the variance for the number in the system, recall from Chapter 2 that

$$V(n) = \frac{d^2 P(z)}{dz^2}\bigg|_{z=1} + L - L^2$$

Bulk Arrivals

or

$$V(n) = \frac{b^2\lambda^2 - b^4\lambda^2 + 2\mu b\lambda + 4b^3\mu\lambda + 6b^2\mu\lambda}{12(\mu - b\lambda)^2} \quad (3.91)$$

In order to determine the expected number of customers in the waiting line, we recall that

$$L_q = \sum_{n=c}^{\infty}(n-c)P_n$$

or, since $c = 1$,

$$L_q = \sum_{n=1}^{\infty}(n-1)P_n = \sum_{n=1}^{\infty}nP_n - \sum_{n=1}^{\infty}P_n = L - (1 - P_0)$$

Since $P_0 = 1 - \rho$, it follows that

$$L_q = L - \frac{b\lambda}{\mu} \quad (3.92)$$

Now, from (3.27),

$$\tilde{\lambda} = (L - L_q)\mu, \quad \text{or} \quad \tilde{\lambda} = b\lambda \quad (3.93)$$

Consequently,

$$W = \frac{1 + b}{2\mu(1 - \rho)} \quad (3.94)$$

and

$$W_q = W - \frac{1}{\mu} \quad (3.95)$$

Furthermore, since there is a single server and $P_0 = 1 - \rho$, the probability that the server is idle is

$$P(\text{idle server}) = 1 - \rho$$

Since b customers arrive at an arrival instant and arrival instants are equally likely to occur in time Δt, the probability that all b customers will be delayed is ρ. With only one server, it must be true that at least $(b - 1)$ of the arriving customers will be delayed.

Example 3.6 At an inspection station in a manufacturing facility, boxes of parts arrive at random following a Poisson distribution. A box arrives, on the average, every five minutes. Each box contains 1000 parts. Ten parts are drawn at random from the box and each part is inspected separately. Inspection time is exponentially distributed with a mean of 20 sec. Determine the operating characteristics of the system.

Since $\lambda = 12$ arrivals/hr, $b = 10$ parts, and $\mu = 180$ parts/hr, then $\rho = 0.667$. From (3.90), (3.92), (3.94), and (3.95), we find that

$L = 11$ parts, $\qquad L_q = 10.333$ parts

$W = 5.5$ min/part, $\qquad W_q = 5.167$ min/part

NETWORK OF POISSON QUEUES

Thus far our analysis has been concerned with a single service facility. Although many interesting queueing problems involve a single facility, a number of queueing systems involve series and parallel combinations of queues. In this section we will show how to extend the analysis of the previous sections to situations involving a network of queues. We will continue to restrict our analysis to those problems in which arrivals and services are generated by a Poisson process.

Queues in Series with Unlimited Storage

Our analysis begins by considering the series system. In particular, we will suppose that two $(M \mid M \mid 1) : (GD \mid \infty \mid \infty)$ stations are located in series. We assume that arrivals to the first station follow a Poisson distribution and that service times are exponentially distributed with parameters μ_1 and μ_2 at each station, respectively. Negligible travel time is assumed between stations. Letting

$P_{n_1, n_2}(t)$ = probability of n_1 at station 1 and n_2 at station 2 at time t

the following balance equations are obtained:

$$P_{0,0}(t + \Delta t) = P_{0,0}(t)(1 - \lambda \, \Delta t) + P_{0,1}(t)\mu_2 \, \Delta t$$

$$P_{0,n_2}(t + \Delta t) = P_{0,n_2}(t)(1 - \lambda \, \Delta t)(1 - \mu_2 \, \Delta t)$$
$$+ P_{1, n_2 - 1}(t)(1 - \lambda \, \Delta t)\mu_1 \, \Delta t$$
$$+ P_{0, n_2 + 1}(t)(1 - \lambda \, \Delta t)\mu_2 \, \Delta t$$

$$P_{n_1, 0}(t + \Delta t) = P_{n_1, 0}(t)(1 - \lambda \, \Delta t)(1 - \mu_1 \, \Delta t)$$
$$+ P_{n_1, 1}(t)\mu_2 \, \Delta t(1 - \lambda \, \Delta t)$$
$$+ P_{n_1 - 1, 0}(t)\lambda \, \Delta t(1 - \mu_1 \, \Delta t)$$

$$P_{n_1, n_2}(t + \Delta t) = P_{n_1 n_2}(t)(1 - \lambda \, \Delta t)(1 - \mu_1 \, \Delta t)(1 - \mu_2 \, \Delta t)$$
$$+ P_{n_1 + 1, n_2 - 1}(t)\mu_1 \, \Delta t(1 - \mu_2 \, \Delta t)(1 - \lambda \, \Delta t)$$
$$+ P_{n_1, n_2 + 1}(t)\mu_2 \, \Delta t(1 - \lambda \, \Delta t)(1 - \mu_1 \, \Delta t)$$
$$+ P_{n_1 - 1, n_2}(t)\lambda \, \Delta t(1 - \mu_1 \, \Delta t)(1 - \mu_2 \, \Delta t)$$

Network of Poisson Queues

Under steady-state conditions, the general balance equations can be shown to reduce to

$$P_{n_1, n_2} = \rho_1^{n_1} \rho_2^{n_2} P_{0,0} \qquad (3.96)$$

where $\rho_1 = \lambda/\mu_1$ and $\rho_2 = \lambda/\mu_2$. The value of $P_{0,0}$ is evaluated as follows:

$$\sum_{n_1=0}^{\infty} \sum_{n_2=0}^{\infty} \rho_1^{n_1} \rho_2^{n_2} P_{0,0} = 1, \quad \text{or} \quad P_{0,0} = \left[\sum_{n_1=0}^{\infty} \rho_1^{n_1} \sum_{n_2=0}^{\infty} \rho_2^{n_2} \right]^{-1}$$

which gives

$$P_{0,0} = (1 - \rho_1)(1 - \rho_2)$$

In order to determine the marginal probability mass function for n_1, observe that

$$P_{n_1} = \sum_{n_2=0}^{\infty} P_{n_1, n_2}, \quad \text{or} \quad P_{n_1} = \rho_1^{n_1}(1 - \rho_1)$$

Similarly, the marginal probability mass function for n_2 is

$$P_{n_2} = \rho_2^{n_2}(1 - \rho_2)$$

Since $P_{n_1, n_2} = P_{n_1} P_{n_2}$, we conclude that the two stations act independently, each with a *completely random arrival rate of* λ. By induction, we can establish, for a series of Q such stations,

$$P_{n_1, n_2, \ldots, n_Q} = \prod_{i=1}^{Q} \rho_i^{n_i}(1 - \rho_i), \quad \text{and} \quad P_{n_i} = \rho_i^{n_i}(1 - \rho_i), \quad i = 1, \ldots, Q$$

(3.97)

This is a very interesting and important result. Specifically, we see that if arrivals to an $(M|M|1):(GD|\infty|\infty)$ queue occur with a mean rate of λ, the departures from the queue follow a Poisson distribution with mean rate λ. We can prove this result by noting that with probability ρ a departing customer will look back and see at least one customer present. In such a situation the time until the next departure is one service time. Likewise, with probability $(1 - \rho)$, he will see no customers present in the system. Thus, the time until the next departure is the sum of an interarrival time and a service time. The resulting Laplace transform for interdeparture time is

$$\mathscr{L}[f(t_d)] = \rho \left(\frac{\mu}{\mu + s} \right) + (1 - \rho) \left(\frac{\lambda}{\lambda + s} \right) \left(\frac{\mu}{\mu + s} \right)$$

or

$$\mathscr{L}[f(t_d)] = \frac{\lambda}{\lambda + s}$$

Therefore, the density function for interdeparture time is exponential with parameter λ.

Although we will omit the required proof, Burke (1968) has shown that the density function for interdeparture time from the $(M\,|\,M\,|\,c)$: $(GD\,|\,\infty\,|\,\infty)$ queue is also exponential with parameter λ.

We established in Chapter 2 that the sum of Poisson distributed random variables is Poisson distributed. Therefore, if arrivals from source k are Poisson distributed with mean λ_k, then the sum of the arrivals over all sources is Poisson distributed with mean $\lambda = \sum \lambda_k$.

Suppose that the destination of a customer who departs from an $(M\,|\,M\,|\,c)$: $(GD\,|\,\infty\,|\,\infty)$ queue is randomly determined. In this case, customers will arrive at their destination in a Poisson fashion. To see this, let α be the probability that a customer goes to destination A and $(1 - \alpha)$ be the probability that the customer does not go to destination A. The number of customers going to destination A per unit time is

$$Y = \sum_{j=1}^{N} X_j$$

where N is the number of customers departing the $(M\,|\,M\,|\,c)$: $(GD\,|\,\infty\,|\,\infty)$ queue per unit time and

$$X_j = \begin{cases} 1, & \text{if customer } j \text{ goes to destination } A \\ 0, & \text{otherwise} \end{cases}$$

From Chapter 2 we know that the generating function for Y is given by the nested combination of the generating functions for N and X. Thus

$$G_Y(z) = G_N[G_X(z)] \tag{3.98}$$

Since

$$G_N(z) = e^{-\lambda(1-z)}, \quad \text{and} \quad G_X(z) = (1 - \alpha) + \alpha z$$

then

$$G_Y(z) = e^{-\lambda[1-(1-\alpha)-\alpha z]} = e^{-\lambda\alpha(1-z)} \tag{3.99}$$

Therefore, Y is Poisson distributed with mean $\alpha\lambda$.

Thus, one can study a network consisting of a series of stations by treating the behavior of each system as an independent unit. However, it must be true that $c\mu > \lambda$ for each station in the network and no restriction can exist on queue length in order for the result to hold. Of course, if $c\mu \leq \lambda$ for some station, then the situation will develop in which there is always a unit available for service and the output rate will be $c\mu$.

Network of Poisson Queues

The preceding discussion can be summarized as follows:

If the probability of a customer going from station i to station j upon completion of service at station i is given by p_{ij}, if customers arrive at station i at a Poisson rate of λ_i, and if the service time at station i is exponentially distributed with parameter $c\mu$, such that $c\mu > \lambda$, then

$$\lambda_j = \lambda_{0,j} + \sum_i p_{ij}\lambda_i$$

where $\lambda_{0,j}$ is the Poisson arrival rate at station j from outside the system (or network) of service stations.

In the event that customers are dispatched from one service system to the next in some regular, but nonrandom, fashion, the interarrival-time distribution will not be exponential. To illustrate, suppose the output from station 1 is split so that units go alternately to stations 2 and 3. The interarrival-time distribution at station 2 will be the distribution for the sum of two identical and independent exponentially distributed random variables with parameter λ, the output rate from station 1. Thus, the interarrival distribution at station 2 is

$$f(t) = \lambda(\lambda t)e^{-\lambda t}$$

and the problem at station 2 can be formulated as an $(E_2 \mid M \mid 1)$ queue, or a gamma (or Erlang) arrival, exponential service queue. For additional discussion of this problem, see Chapter 4.

In general, the analysis of networks of queues is quite complex. Jackson (1957, 1963) has done considerable research on queueing networks. His decomposition theorem gives sufficient conditions under which a general network of queues may be treated as an aggregation of independent queues. Jackson shows that each service facility within the network behaves probabilistically as an independent service facility if:

(1) arrivals are Poisson,
(2) customers are routed randomly,
(3) service times are exponential, and
(4) service discipline is not a function of a customer's service time or his routing through the system.

Since a number of practical problems violate the sufficient conditions for decomposition and the necessary conditions are not known, analytical analysis is not feasible for a number of interesting queueing systems involving networks. For this reason, we recommend the use of simulation techniques in the analysis of the more complex queueing systems. The use of simulation is discussed in Chapter 9.

Example 3.7 As an example of a network of queues, consider the acquisition process in the VIP University Library. Requisitions for new books (monographs) and periodicals (serials) are received at random. Each requisition is received by Ms. Aaron, who verifies certain information and, if necessary, completes the requisition form. Requisitions for monographs are sent to Ms. Brown and requisitions for serials are sent to Mr. Craig. Ms. Brown (Mr. Craig) checks to see if the monograph (serial) is either in the library collection or on order. All requisitions are sent to Ms. Davis, who enters the necessary data on a purchase order form. The cost for the monograph (serial) is determined and the appropriate account charged.

Requisitions arrive at an average rate of 8/hr. Approximately 70% of the requisitions are for monographs. Also, 60% of the requisitions are returned to the person who initiated the request, notifying him that the monograph (serial) is either on hand or on order. Service time at each stage of the acquisition process is exponentially distributed with a mean of 6 min for Ms. Aaron, 7.5 min for Ms. Brown, 15 min for Mr. Craig, and 15 min for Ms. Davis. A schematic model of the process is given in Fig. 3.8;

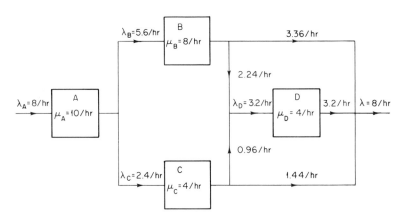

Fig. 3.8 Network representation of Example 3.7.

Values of the operating characteristics for each stage of the acquisition process are given in Table 3.6. The probability that the system is empty is

$$P_0 = P_{A0} P_{B0} P_{C0} P_{D0} = 0.0048$$

since each service facility is statistically independent of all other stages in the system. The probability mass function for the number of requisitions in the system can be obtained from the generating function for the system,

$$G(z) = G_A(z) G_B(z) G_C(z) G_D(z)$$

Network of Poisson Queues

TABLE 3.6 Operating characteristics for Example 3.7

	A	B	C	D
L	4.0	2.33	1.5	4.0
L_q	3.2	1.6333	0.9	3.2
W	0.5	0.4166	0.625	1.25
W_q	0.4	0.2916	0.375	1.00
P_0	0.2	0.3	0.4	0.2
D	0.8	0.7	0.6	0.8

where

$$G_A(z) = \frac{0.2}{1 - 0.8z}, \quad G_B(z) = \frac{0.3}{1 - 0.7z}$$

$$G_C(z) = \frac{0.4}{1 - 0.6z}, \quad G_D(z) = \frac{0.2}{1 - 0.8z}$$

Therefore,

$$G(z) = \frac{0.0048}{(1 - 0.8z)^2(1 - 0.7z)(1 - 0.6z)}$$

Using partial fraction expansion, we express $G(z)$ as

$$G(z) = 0.0048\left[\frac{32}{(1 - 0.8z)^2} - \frac{320}{1 - 0.8z} - \frac{343}{1 - 0.7z} - \frac{54}{1 - 0.6z}\right]$$

From the Appendix, we can obtain the inverse transform of $G(z)$. The resulting probability mass function for the number in the system is

$$P_n = 0.0048[32(n + 1)0.8^n - 320(0.8)^n + 343(0.7)^n - 54(0.6)^n]$$

The cumulative distribution function for the number in the system can be given as

$$P(n \leq N) = 0.0048\left[640(1 - 0.8^{N+1}) - 0.2(N + 1)0.8^N\right.$$
$$\left. - 1440(1 - 0.8^{N+1}) + \frac{343}{0.3}(1 - 0.7^{N+1})\frac{54}{0.4}(1 - 0.6^{N+1})\right]$$

The expected number in the system is the sum of the expected number at each facility,

$$L = L_A + L_B + L_C + L_D = 11.833$$

The expected time for a requisition to be processed from A through D is computed as follows:

E(time for monographs) = $0.5 + 0.4166 + 1.25 = 2.167$ hr/monograph

E(time for serials) = $0.5 + 0.625 + 1.25 = 2.375$ hr/serial

Since 70% of the requisitions are for monographs,

$W_{\text{system}} = 2.167(0.7) + 2.375(0.3) = 2.229$ hr/requisition

Queues in Series with Limited Storage

One of the restrictions placed on our discussion of analytical solutions to networks of queues was that the capacity of each system was infinitely large. In the case of assembly-line production, this is seldom the case. Many times a unit cannot leave one production station until the next station along the line becomes free. Since queueing systems located in series often have limitations on available storage space, we now modify the results of the previous section for the case of a finite queue at each station. A number of rules can be given for governing the movement of customers who have completed service. One typical rule is that if a customer has completed service at station k and is ready to go to station $k + 1$ (which is full), one of the service channels at station k will be blocked and no additional customer can enter service at station k until a departure occurs at station $k + 1$. To illustrate, consider two sequential stations each with a single server and exponential service-time distributions. Arrivals at the first station are Poisson distributed. For simplicity, let $N_1 = N_2 = 1$.

For the first station we distinguish three states:

0 (empty),   1 (busy),   b (blocked)

For the second station, we distinguish two states:

0 (empty),   1 (busy)

Letting P_{n_1, n_2} denote the steady-state probability of station j being in state n_j, steady-state balance equations will be obtained using a *rate out = rate in* approach.

The rate out = rate in method of developing the set of steady-state balance equations is based on the notion of conservation of flow employed in network analysis. Basically, the method consists of equating the rate at which the process leaves a given state and the rate at which the state is entered. In particular, for the present example a conservation of flow equation is required for each of the possible states for the system.

Network of Poisson Queues

Applying the rate out = rate in approach, the following steady-state balance equations are obtained directly:

$$\lambda P_{0,0} = \mu_2 P_{0,1}, \qquad (\lambda + \mu_2)P_{0,1} = \mu_1 P_{1,0} + \mu_2 P_{b,1}$$
$$\mu_1 P_{1,0} = \lambda P_{0,0} + \mu_2 P_{1,1}, \qquad \mu_2 P_{b,1} = \mu_1 P_{1,1},$$
$$(\mu_1 + \mu_2)P_{1,1} = \lambda P_{0,1} \tag{3.100}$$

Notice from the first balance equation that the only way to leave state $(0, 0)$ is to experience an arrival; the expected rate at which this occurs is $\lambda P_{0,0}$. Additionally, the only way to enter state $(0, 0)$ is to be in state $(0, 1)$ and experience a service at station 2; the expected rate at which this occurs is $\mu_2 P_{0,1}$. Similar arguments apply for the remaining equations in (3.100).

Solving for the steady-state probabilities recursively in terms of $P_{0,0}$ and letting $\rho_1 = \lambda/\mu_1$, $\rho_2 = \lambda/\mu_2$, gives

$$P_{0,1} = \rho_2 P_{0,0} \tag{3.101}$$

$$P_{1,1} = \frac{\lambda \rho_2}{\mu_1 + \mu_2} P_{0,0} \tag{3.102}$$

$$P_{b,1} = \frac{\mu_1 \rho_2^2}{\mu_1 + \mu_2} P_{0,0} \tag{3.103}$$

$$P_{1,0} = \rho_1 \left[1 + \frac{\lambda}{\mu_1 + \mu_2} \right] P_{0,0} \tag{3.104}$$

Since

$$P_{0,0} + P_{0,1} + P_{1,0} + P_{1,1} + P_{b,1} = 1$$

the following expression is obtained for $P_{0,0}$:

$$P_{0,0} = \left[1 + \rho_1 + \rho_2 + \rho_1 \rho_2 + \frac{\mu_1 \rho_2^2}{\mu_1 + \mu_2} \right]^{-1} \tag{3.105}$$

Considering the special case where $\mu_1 = \mu_2$, such that $\rho_1 = \rho_2 = \rho$, yields

$$P_{0,0} = \frac{2}{3\rho^2 + 4\rho + 2} \tag{3.106}$$

$$P_{0,1} = \frac{2\rho}{3\rho^2 + 4\rho + 2} \tag{3.107}$$

$$P_{1,0} = \frac{2\rho + \rho^2}{3\rho^2 + 4\rho + 2} \tag{3.108}$$

$$P_{1,1} = \frac{\rho^2}{3\rho^2 + 4\rho + 2} \tag{3.109}$$

$$P_{b,1} = \frac{\rho^2}{3\rho^2 + 4\rho + 2} \tag{3.110}$$

The expected number of units in the system is given by

$$L = \frac{4\rho + 5\rho^2}{2 + 4\rho + 3\rho^2} \quad (3.111)$$

The probability of blocking $P_{b,1}$ indicates the maximum possible utilization of the system. Notice, as the value of ρ increases, that the value of $P_{b,1}$ approaches a limiting value of 1/3. Therefore, even if the first station is kept busy all of the time, the utilization of the system will be less than 2/3 when μ_1 equals μ_2.

As the queue length in front of either or both channels is increased, the solution for P_{n_1, n_2} becomes more and more complex. For example, suppose we allow no queue before the first channel and a maximum queue length of one before the second channel, and assume $\mu_1 = \mu_2 = \mu$. The resulting steady-state equations, obtained using the rate out = rate in approach, are given by

$$\lambda P_{0,0} = \mu P_{0,1}, \qquad (\lambda + \mu) P_{0,1} = \mu P_{0,2} + \mu P_{1,0}$$

$$\mu P_{1,0} = \lambda P_{0,0} + \mu P_{1,1}, \qquad 2\mu P_{1,1} = \lambda P_{0,1} + \mu P_{1,2}$$

$$(\lambda + \mu) P_{0,2} = \mu P_{1,1} + \mu P_{b,2}, \qquad \mu P_{b,2} = \mu P_{1,2}$$

(3.112)

where

$$P_{0,0} + P_{0,1} + P_{0,2} + P_{1,0} + P_{1,1} + P_{1,2} + P_{b,2} = 1$$

The solution to the system of equations given in (3.112) is:

$$P_{0,0} = \frac{4 + \rho}{S} \quad (3.113)$$

$$P_{0,1} = \frac{\rho(4 + \rho)}{S} \quad (3.114)$$

$$P_{0,2} = \frac{2\rho^2}{S} \quad (3.115)$$

$$P_{1,0} = \frac{\rho(4 + 3\rho + \rho^2)}{S} \quad (3.116)$$

$$P_{1,1} = \frac{\rho^2(2 + \rho)}{S} \quad (3.117)$$

$$P_{1,2} = P_{b,2} = \frac{\rho^3}{S} \quad (3.118)$$

where
$$S = 4 + 9\rho + 8\rho^2 + 4\rho^3 \quad (3.119)$$

The expected number of units in the system is given by

$$L = \frac{\rho(8 + 12\rho + 9\rho^2)}{S} \quad (3.120)$$

Notice that, as the value of ρ increases the value of $P_{b,2}$ approaches a limiting value of 0.25. Thus, the maximum utilization of the system is increased from 0.67 to 0.75 by providing waiting space for one customer at the second station.

Saaty (1961) shows that in the case of infinite *waiting* space at the first station and finite *waiting* space for $N - 1$ customers at the second station, the maximum utilization of the system is given by

$$\rho_{\max} = \frac{\mu_2(\mu_1^{N+1} - \mu_2^{N+1})}{\mu_1^{N+2} - \mu_2^{N+2}} \quad (3.121)$$

which for $\mu_1 = \mu_2$ becomes

$$\rho_{\max} = \frac{N+1}{N+2}$$

From (3.121) we see that if the values of μ_1 and μ_2 are interchanged the maximum utilization is greatest when the system having the larger mean service *rate* is second in the sequence. Even though all of the assumptions of the model may not be physically realized, the results obtained from the model might enhance our understanding of the behavior of complex systems. Since the placement of the operation having the largest mean service time first in a sequence of operations supports our intuition concerning the effect on the utilization of the overall system, we have greater confidence in the results from the model.

Although we do not normally encounter queueing systems having excessively large values of ρ, (3.121) provides us with a measure of how the system behaves. It serves as a benchmark against which alternative systems can be compared.

SUMMARY

In this chapter we have developed an approach that can be used to analyze a number of Poisson queues; namely, we have formulated differential–difference equations to describe the behavior of the system. We solved the resulting system of equations under steady-state assumptions and proceeded to develop descriptive models of the operating characteristics of the system. For the most part, we have been concerned with the development of expected value measures of system performance. For those interested in developing waiting-time and total-time distributions, we have provided an approach based on the use of generating functions and Laplace transforms. In the case of multiple-server queues, the resulting distributions are formidable in their appearance. However, the distributions can be used by the analyst without extreme difficulty for specific applications, as long as he has available tables for the gamma distribution.

REFERENCES

Burke, P. J., The output process of a stationary $M \mid M \mid s$ queueing system, *Ann. Math. Statist.* **39**, 1144–1152, 1968.
Cooper, R. B., *Introduction to Queueing Theory*. New York: Macmillan, 1972.
Hillier, F. S., and Lieberman, G. J., *Introduction to Operations Research*. San Francisco: Holden-Day, 1967.
Jackson, J. R., Networks of waiting lines, *OR* **5**, 518–521, 1957.
Jackson, J. R., Jobshop-like queueing systems, *Management Sci.* **10**, 131–142, 1969.
Lee, A. M., *Applied Queueing Theory*. Toronto: Macmillan, 1966.
Peck, L. G., and Hazelwood, R. N., *Finite Queueing Tables*. New York: Wiley, 1958.
Saaty, T. L., *Elements of Queueing Theory*. New York: McGraw-Hill, 1961.
Shelton, J. R., Solution methods for waiting line problems, *J. Indust. Eng.* **11**, 293–303, 1960.
Stidham, S., A last word on $L = \lambda W$, *OR* **22**, 417–421, 1974.
Syski, R., *Introduction to Congestion Theory in Telephone Systems*. London: Oliver & Boyd, 1960.
Taha, H. A., *Operations Research: An Introduction*. New York: Macmillan, 1971.
Takács, L., On Erlang's formula, *Ann. Math. Statist.* **40**, 71–78, 1969.
Votaw, D. F., and Peck, L. G., Remarks on finite queueing tables, *OR* **16**, 1084–1086, 1968.

PROBLEMS

1. Develop closed-form expressions for P_0, L_q, L, W_q, W, and D for the $(M \mid M \mid 1) : (GD \mid N \mid \infty)$ as well as the $(M \mid M \mid 1) : (GD \mid \infty \mid \infty)$ queue, when

$$\lambda_n = \begin{cases} \lambda, & n = 0, 1, \ldots, N-1 \\ 0, & n \geq N \end{cases}$$

$$\mu_n = \begin{cases} 0, & n = 0 \\ \mu, & n = 1, 2, \ldots, N \end{cases}$$

Simplify all results.

2. Show that for the $(M \mid M \mid c) : (FCFS \mid \infty \mid \infty)$ queue the waiting-time distribution, given that one has to wait, is given as

$$f(t_q \mid t_q > 0) = (c\mu - \lambda) \exp[-(c\mu - \lambda)t_q], \quad t_q > 0$$

Problems

3. For the $(M\,|\,M\,|\,c):(GD\,|\,\infty\,|\,\infty)$ queue, show that the probability mass function for the number in the waiting line is given as

$$P_m = \begin{cases} 1 - \dfrac{\rho(c\rho)^c P_0}{c!\,(1-\rho)}, & m = 0 \\[6pt] \dfrac{c^c \rho^{m+c} P_0}{c!}, & m = 1, 2, \ldots \end{cases}$$

Also, show that the expected number in the waiting line, given that a waiting line exists, is given by $1/(1-\rho)$.

4. Show that the probability mass function for the number in the system in an $(M\,|\,M\,|\,\infty):(GD\,|\,\infty\,|\,\infty)$ queue (self-service system) is a Poisson distribution with parameter λ/μ.

5. For the $(M\,|\,M\,|\,c):(FCFS\,|\,N\,|\,\infty)$ queue, show that the waiting-time distribution for those customers who have to wait is, for $\rho \neq 1$,

$$f(t_q \,|\, t_q > 0) = \frac{1-\rho}{\rho(1-\rho^{N-c})} \sum_{n=1}^{N-c} \frac{\lambda^n}{(n-1)!}\, t_q^{n-1} \exp(-c\mu t_q), \qquad t_q > 0$$

with an expected value of

$$E(T_q \,|\, t_q > 0) = \frac{1 - \rho^{N-c+1} - (1-\rho)(N-c+1)\rho^{N-c}}{(c\mu - \lambda)(1 - \rho^{N-c})}$$

6. For the $(M\,|\,M\,|\,c):(GD\,|\,N\,|\,\infty)$ queue, show that the probability mass function for the number in the waiting line is

$$P_m = \begin{cases} 1 - \left[\dfrac{c^c \rho^{c+1} P_0}{c!}\right]\left[\dfrac{1-\rho^{N-c}}{1-\rho}\right], & m = 0 \\[6pt] \dfrac{c^c \rho^{m+c}}{c!} P_0, & m = 1, 2, \ldots, N-c \\[6pt] 0, & m > N - c \end{cases}$$

Also show that the expected number in the waiting line, given that a waiting line exists, is

$$\frac{1}{1-\rho} - \frac{(N-c)\rho^{N-c}}{1-\rho^{N-c}}$$

7. Can we obtain results for the $(M\,|\,M\,|\,c):(GD\,|\,\infty\,|\,\infty)$ queue by using the results for the $(M\,|\,M\,|\,c):(GD\,|\,K\,|\,K)$ queue, letting K approach infinity? Support your answer.

8. Show that the probability that an arriving customer is delayed in a system $(M\,|\,M\,|\,c):(GD\,|\,K\,|\,K)$ with parameter X is given by the probability that all service channels are busy in a system $(M\,|\,M\,|\,c):(GD\,|\,K-1\,|\,K-1)$ with parameter X.

9. Develop the probability mass function for the number of customers in the waiting line for the $(M\,|\,M\,|\,c):(GD\,|\,K\,|\,K)$ queue. Also determine the expected number of persons in the waiting line, given that a waiting line exists.

10. For the $(M\,|\,M\,|\,c):(GD\,|\,\infty\,|\,\infty)$ queue, show that (a) the probability that someone is waiting is $\lambda P_c/(c\mu - \lambda)$; (b) the expected waiting time, given that one must wait, is $1/(c\mu - \lambda)$.

11. For the $(M\,|\,M\,|\,1):(GD\,|\,\infty\,|\,\infty)$ queue, show that the probability of there being at most N customers in the system is $(P_0 - P_{N+1})/P_0$.

12. For the $(M \mid M \mid c) : (GD \mid c \mid \infty)$ queue, show that

$$L = c\rho(1 - P_c), \quad \tilde{\lambda} = L\mu, \quad W_q = 0, \quad \text{and} \quad W = \frac{1}{\mu}$$

13. The McDougal Sandwich Shop has two windows available for serving customers, who arrive at a Poisson rate of 40/hr. Service time is exponentially distributed with a mean of 2 min. Only one window is open as long as there are three or less customers in the shop. The second window opens when there are four or more customers in the shop. The manager of the shop helps the two attendants when there are more than six customers in the shop. When the manager is helping, the mean service time is reduced to 1.5 min. Determine the expected number of customers in the shop and the probability that the manager will be helping serve customers.

14. A dentist's office has available six parking spaces for patients' cars. Whenever all spaces are taken, arriving patients must park in a pay lot across the street. Recent data indicate that 54% of the arriving patients must park in the pay lot. Assuming that a Poisson process generates arrivals and departures, how many parking spaces should be added at the dentist's office in order to accommodate at least 80% of the arriving patients?

15. A time study is made of a tool crib and it is found that the probability of it taking more than T minutes to fill an order is given by the expression $e^{-0.2T}$. Customers are found to arrive at the tool crib at a Poisson rate of 10/hr. There is a single attendant at the tool crib. What is the probability of there being three customers in the system, given that the attendant is busy?

16. It is found that the arrival of parts at station Y on an assembly line follows the Poisson distribution. Also, the time to perform operation Y is exponentially distributed. If the average arrival rate is 5/min, what must be the average service rate in order that there will be a probability of 0.90 that the time a part spends at station Y (waiting and being serviced) is not greater than 12 min?

17. Customers arrive at a neighborhood barber shop at a Poisson rate of 4/hr. The time required to cut hair is exponentially distributed with an average time of 10 min. Experience shows that the maximum number of customers waiting for service is never greater than three. With only one barber in the shop, what is the probability that a customer will arrive at the shop and not stay to be serviced? If the barber does not wish this probability to be greater than 0.05, what must be his average service time?

18. Without using Laplace transform methods, derive the probability distribution for T_T and T_q for the $(M \mid M \mid c) : (FCFS \mid N \mid \infty)$ queue and the $(M \mid M \mid c) : (FCFS \mid K \mid K)$ queue.

19. A local carry-out sandwich shop currently has two attendants. A study of the arrival and service rates indicates that Poisson arrivals and exponential service assumptions hold. The mean service rate is found to be a function of the number in the system. It is also found that customers do not enter the shop if there are five people in the system. The accompanying tabulation relates arrival and service rates to the number in the system, for the present system:

n	λ_n	μ_n
0	2	0
1	2	1
2	2	1
3	2	2
4	2	2
5	0	3

(a) What is the probability that at least one attendant will be idle if the system is observed at some random point in time during steady-state operation?

(b) With two attendants, what is the expected number of customers waiting for service during steady-state operation?

20. A work-sampling study is performed for a storeroom employing 2 servers. After taking many observations it is found that the average number of idle servers equals 0.40. Assuming Poisson arrivals and exponential service, what is the probability of there being at least one customer in the storeroom?

21. An operator services 20 identical textile machines. Machine running time is exponentially distributed with a mean of 45 min; service time is exponentially distributed with a mean of 5 min. Determine the values of L, L_q, W, W_q, and D for this situation.

22. A maintenance man is assigned to four identical machines that randomly demand his attention. Machine running time is exponentially distributed with a mean running time of 60 min. Expected maintenance time equals $60/(1 + 0.5k)$ min, where k equals the number of machines waiting for service. Maintenance time is also exponentially distributed.

(a) Determine the expected number of machines running during steady-state operations.

(b) What is the expected number of machines being serviced by the maintenance man during steady-state operations?

23. In a barber shop there are 3 barbers. Customers arrive randomly at the shop at an average rate of 20/hr. The time required to provide service to a customer is exponentially distributed with a mean of 12 min. Past experience indicates that the probability that a customer will enter the waiting line is a function of the number of persons waiting for service. The probability mass function is given as:

x = no waiting, $p(x)$ = probability a customer enters the shop if there are x customers waiting

x:	0	1	2	3	4	≥ 5
$p(x)$:	1.00	1.00	0.80	0.50	0.00	0.00

Determine the expected number of customers in the waiting line during steady-state operation.

24. In a grocery store, the store manager assigns cashiers to registers based on the number of people in line. Unfortunately, the manager is not consistent in his decisions. Thus, the number of cashiers operating for a given number in line is a random variable. The probability distribution for the number of cashiers operating is given in the following table:

	Number of customers in system (n)			
Number of cashiers (c)	1–3	4–6	7–9	≥ 10
1	0.8	0.1	0	0
2	0.2	0.6	0.2	0
3	0	0.3	0.6	0.4
4	0	0	0.2	0.6

Customers arrive at the cashiers in a random fashion with a mean arrival rate of 30/hr. The service time per customer is exponentially distributed with a mean of 5 min. Give the differential–difference equations describing the operation of the cashiers.

25. Derive the variance for the number of people in the waiting line in an $(M|M|1)$: $(GD|\infty|\infty)$ queue.

26. At a check-out counter, customers arrive in a Poisson fashion at a mean rate of 8/hr. Service time is exponentially distributed with a mean of 4 min. A single check-out counter is used.

(a) What is the expected time a customer spends in the system given that the customer has to wait?

(b) What is the probability that a customer has to spend at least 4 min in the system?

(c) What is the probability that a customer has to spend at least 2 min waiting in line before being served?

27. Customers arrive at a self-service cafeteria at a Poisson rate λ. The time spent at the counter is exponentially distributed with a mean of 5 min. Determine the value of λ such that there is a 0.90 probability of there being no more than 50 customers in the system.

28. Develop an expression for $\tilde{\lambda}$ for the $(M|M|c):(GD|K|K)$ queue.

29. Consider the $(M^{(b)}|M|1):(GD|\infty|\infty)$ queue in which b customers arrive at an arrival instant with probability q_b, $b = 1, 2, \ldots, B$, where B is the largest number arriving at one time. Develop the steady-state generating function for the number of customers in the system.

30. Consider the $(M|M|1):(GD|N|\infty)$ queue in which a customer does not enter if there are N customers in the system. Modify the problem such that a customer enters with probability p if there are N or more customers present. Develop the steady-state probability mass function for the number of customers in the system.

31. Consider the $(M|M|1):(GD|\infty|\infty)$ queue. Suppose a customer reneges (leaves the system after entering, but before being served) during Δt with probability $p \Delta t$ if the number in the system is greater than one. Develop the steady-state probability mass function for the number of customers in the system.

32. Consider the $(M|M|2):(GD|\infty|\infty)$ queue. Suppose the first server has service rate μ_1 and the second server has service rate μ_2, where $\mu_1 \neq \mu_2$. Furthermore, suppose customers choose servers with equal probability when both servers are free. Develop the steady-state probability mass function for the number of customers in the system.

33. In Problem 32 assume that $\mu_1 > \mu_2$ and customers always choose the faster server when both servers are idle at the time the customer arrives; otherwise the customer is served by the available server.

34. In Problem 32 assume that $\mu_1 > \mu_2$ and customers choose the faster server with probability p when both servers are available; otherwise the customer is served by the available server.

35. Arrivals at a local house of questionable reputation are Poisson distributed with a mean of 10/hr. Service time is exponentially distributed with a mean of 15 min/customer. There are two servers available. Since there is a competitor located across the street, only 80% of the arriving customers stay for service if they find one person waiting in the queue for service; only 60% enter if there are two people waiting for service; only 40% enter if there are three people waiting for service; none enter if there are four people waiting for service.

(a) Determine the probability that an arriving customer will be served immediately upon arrival.

(b) Determine the expected number of customers who will be caught if the house is raided.

(c) Determine the average amount of time a person spends in the house.

36. Sales of basketball tickets at a single ticket window at VIP University Arena follow a Poisson process. A work-sampling study of the individual selling tickets shows the person to be busy selling tickets 80% of the time. What is the probability that there are less than three persons *waiting* to be serviced at the window?

Problems

37. A new automatic car wash is being opened. It is anticipated that arrivals will be Poisson distributed at a rate of 15/hr, unless there are X cars waiting, in which case the customer will go elsewhere for a car wash. Service time is exponentially distributed with a mean of 4 min. Compare the values of the effective arrival rate if X equals (a) 0, (b) 2, and (c) 4 spaces.

38. Suppose one repairman has been assigned the responsibility of maintaining four machines. For each machine, the probability distribution of the running time before a breakdown is exponential with a mean of 6 hr. The repair time also has an exponential distribution with a mean of 1.5 hr. Determine the expected number of machines that are (a) running and (b) being repaired.

39. In Problem 38, assume an infinite population and an arrival rate of $\frac{2}{3}$/hr and determine the expected number of machines that are (a) running and (b) being repaired.

40. The plating department in a plant receives material in a Poisson manner with a mean rate of 24 batches/hr. The time required to plate a batch of parts is exponentially distributed with a mean of 2 min/batch.
    (a) What are the values of L, L_q, W, W_q, and D?
    (b) What is the probability that a batch will wait longer than 18 min for plating to begin?
    (c) What is the probability of there being more than 6 batches waiting to be plated?

41. A service station has three gasoline pumps and can accommodate six cars at a given time. Cars arrive at a Poisson rate of 24/hr. The time required to service a car is exponentially distributed with a mean of 5 min. If all six spaces are full, cars go elsewhere for service. The station manager is considering adding another pump, such that eight cars could be accommodated. What are the values of L, L_q, W, W_q, and D for the present and proposed system? On the average, how many more cars are served per hour under the proposed system than under the present system?

42. A bank has four tellers working on savings accounts. The first two tellers handle withdrawals only. The third and fourth tellers handle deposits only. It has been found that the service-time distributions for both deposits and withdrawals are exponential with mean service time of three min/customer. Depositors are found to arrive in a Poisson fashion throughout the day with a mean arrival rate of 30/hr. Withdrawers also arrive in a Poisson fashion with mean arrival rate of 24/hr. What would be the effect on the average waiting time for depositors and withdrawers if each teller could handle both withdrawals and deposits and there were (a) 4, (b) 3 tellers available?

43. Arrivals at a tool crib are Poisson at a rate of 20/hr. Service time by the tool crib attendant is exponentially distributed with a mean of 2.4 min. Compare the values of L, L_q, W, W_q, and D under the present system and those that would exist by doubling the arrival rate and the number of attendants.

44. Consider an $(M \mid M \mid c):(GD \mid \infty \mid \infty)$ queue in which

$$\mu_n = \begin{cases} n\mu, & n \leq c \\ \left(\dfrac{n}{c}\right)^{\alpha} c\mu, & n \geq c \end{cases}$$

and $\lambda_n = \lambda$ for $n = 0, 1, 2, \ldots$. Show that

$$P_n = \begin{cases} \dfrac{(\lambda/\mu)^n P_0}{n!}, & n < c \\ \dfrac{(\lambda/\mu)^n P_0}{c!\,(n!/c!)^{\alpha} c^{(1-\alpha)(n-c)}}, & n > c \end{cases}$$

140   3. Poisson Queues

45. In Problem 44, let

$$\lambda_n = \begin{cases} \lambda, & n \leq c - 1 \\ \left(\dfrac{c}{n+1}\right)^{\beta} \lambda, & n \geq c - 1 \end{cases}$$

and show that

$$P_n = \begin{cases} \dfrac{(\lambda/\mu)^n P_0}{n!}, & n \leq c \\[2mm] \dfrac{(\lambda/\mu)^n P_0}{c!\,(n!/c!)^{\gamma+\beta} c^{(1-\alpha-\beta)(n-c)}}, & n > c \end{cases}$$

46. Consider an $(M \mid M \mid 1):(GD \mid \infty \mid \infty)$ queue in which

$$\mu_n = \begin{cases} 0, & n = 0 \\ \mu, & n > 0 \end{cases}$$

$$\lambda_n = \lambda \exp \dfrac{-\alpha n}{\mu}, \qquad n = 0, 1, 2, \ldots$$

and show that

$$P_n = P_0 \left(\dfrac{\lambda}{\mu}\right)^n \exp \dfrac{-n(n-1)\alpha}{2\mu}$$

47. Customers arrive at a public telephone booth with a mean frequency of 30/hr, with exponential interarrival times. Conversation durations are exponentially distributed with mean duration of 3 min. Customers balk if they find people waiting to use the telephone. Suppose the arrival rate (in arrivals per hour), due to balking, is

(a) $\lambda_n = 30 \dfrac{1}{n+1}, \qquad n = 0, 1, 2, \ldots$

(b) $\lambda_n = 30 e^{-n/20}, \qquad n = 0, 1, 2, \ldots$

(c) $\lambda_n = \begin{cases} 30 & n = 0, 1, 2, 3 \\ 0 & n = 4 \end{cases}$

(d) $\lambda_0 = 30, \quad \lambda_1 = 30, \quad \lambda_2 = 20, \quad \lambda_3 = 10, \quad \lambda_n = 0, \quad n \geq 4.$

(e) $\lambda_n = \begin{cases} 30, & n = 0, 1, 2 \\ 10, & n \geq 3 \end{cases}$

Determine the value for the expected number in the system. What is the probability that the telephone booth is empty?

48. Arrivals at a tollgate on a freeway are Poisson distributed with a mean of 1.5 vehicles/min. Service time is exponentially distributed with a mean of 30 sec. Determine the values of the

operating characteristics for the system. What is the probability that a vehicle will have to wait more than 6 sec before being admitted to service?

49. Consider the $(M^{(b)} \mid M \mid c):(GD \mid \infty \mid \infty)$ queue in which b customers arrive per arrival instant. Determine the steady-state generating function for the number of customers in the system when (a) $b \leq c$, (b) $b > c$.

50. Three exponential channels are arranged in parallel. The three channels have mean service rates μ_1, μ_2, and μ_3, respectively. Input to the system is Poisson with mean rate λ. Assume that if more than one channel is open the entering unit selects the channel it will enter in a random fashion, with each channel having an equal probability of being chosen. Develop the steady-state balance equations for this case, assuming a system capacity of N.

51. Consider the $(M^{(b)} \mid M \mid 1):(GD \mid \infty \mid \infty)$ queue, where b is a random variable with generating function $Q(z)$ and mean $E(b)$. Show that

$$P(z) = [\mu - \lambda E(b)](1 - z)/[\mu(1 - z) - \lambda z(1 - Q(z))]$$

52. Boxes of parts are delivered to inspection station Y on an assembly line at a Poisson rate of 4/min. Five parts are drawn from each box at random and each is inspected for defects; inspection time is exponentially distributed with a mean of 2 sec. Determine the operating characteristics for the system.

53. At a local airport, loads of baggage are taken to an inspection station for a security check. Loads arrive at a Poisson rate of 20/hr, with 10 pieces of luggage contained in a load 50% of the time and 5 pieces of luggage contained in a load 50% of the time. Two inspectors are available to inspect the luggage, with inspection time per inspector exponentially distributed with a mean of 20 sec. Determine the operating characteristics for the system.

54. Three production stations are located in series along a conveyor line, with no waiting space available between stations. Arrivals at the first station are Poisson distributed at an average rate of λ. All service times are exponentially distributed with a mean of $1/\mu_1$ at the first station, $1/\mu_2$ at the second station, and $1/\mu_3$ at the third station. Waiting space in front of the first station is unlimited. Determine the steady-state balance equations for the system.

55. Consider the two station production process depicted below. New parts arrive at station 1 at a Poisson rate of 10/hr; production time at stage 1 is exponentially distributed with a mean of 4 min, including inspection. If a unit is defective and can be reworked, it is placed at the end of the waiting line and is reprocessed at station 1. Experience shows that 20% of the items inspected are reworked, 10% of the items rejected and scrapped, and 70% of the inspected items sent to production station 2. Processing time at station 2 is exponentially distributed with a mean of 6 min. All items processed at station 2 leave the system. Determine the operating characteristics for each station.

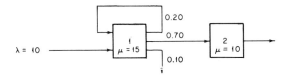

56. In a production line there are 4 stations in series. Interarrival times at station 1 are exponentially distributed with a mean of 15 min; service times are exponentially distributed at all four stations with means of 8, 10, 12, and 10 min, respectively. Waiting space for 1 unit exists between stations; an unlimited waiting space exists for station 1. Determine the operating characteristics for the system.

142   3. Poisson Queues

57. Consider the network of $(M\,|\,M\,|\,1):(GD\,|\,\infty\,|\,\infty)$ queues shown below. Determine the operating characteristics for each station in the network and for the total system.

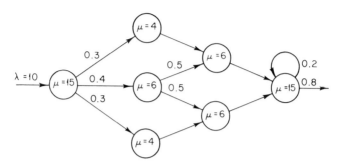

58. Consider the network of $(M\,|\,M\,|\,c):(GD\,|\,\infty\,|\,\infty)$ queues shown below. Determine the operating characteristics for each station in the network and for the total system.

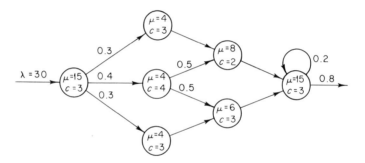

59. Consider the network of $(M\,|\,M\,|\,c):(GD\,|\,\infty\,|\,\infty)$ queues shown below. Determine the operating characteristics for each station in the network and for the total system.

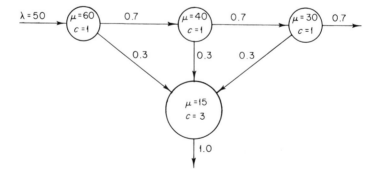

Chapter 4

NON-POISSON QUEUES

INTRODUCTION

In the preceding discussion we employed differential–difference equations in analyzing Poisson queues. Obviously, in a number of practical applications Poisson assumptions are not appropriate. Therefore, we need to consider how we will analyze non-Poisson queues. In this chapter we will consider some non-Poisson queues that can be modeled rather simply using the methods employed in analyzing Poisson queues. Specifically, we will concentrate on gamma and hyperexponential queues. Additionally, we will treat the $(M|G|1):(GD|\infty|\infty)$ queue and develop the very powerful Pollaczek–Khintchine formula for such queues. As in the previous chapter, our concern will be with steady-state results.

We postpone a treatment of the more complicated non-Poisson queues until Chapters 6 and 9. If the queueing system under study is non-Poisson and cannot be reasonably represented as a gamma or hyperexponential queue of the variety treated in this chapter, then a decision must be made concerning the approach to be taken. Either an analytic or a simulation approach can be used. If an analytic approach is taken, the discussion of Chapter 6 is appropriate; if a simulation approach is to be followed, the material presented in Chapter 9 will be beneficial.

Our discussion of non-Poisson queues in this chapter begins with a treatment of the $(M|G|1):(GD|\infty|\infty)$ queue and the Pollaczek–Khintchine formula. Next, we treat a variety of Erlang and hyperexponential queues using the "method of stages" approach. The method of stages involves the development of an alternate (but equivalent) queueing system consisting of several stages, each of which is a Poisson queue. As seen in Table 4.1, we analyze a variety of queues using the method of stages approach. Furthermore, in a number of cases we provide general expressions for the operating characteristics of the system; in the remaining cases, steady-state balance equations are presented and solutions provided for specific values of the system parameters.

TABLE 4.1 Summary of non-Poisson queues analyzed via method of stages

Case	Classification	Solution
1	$(M \mid E_k \mid 1):(GD \mid \infty \mid \infty)$	general
2	$(M \mid E_2 \mid 2):(GD \mid 2 \mid \infty)$	general
3	$(E_2 \mid M \mid 1):(GD \mid N \mid \infty)$	specific
4	$(E_2 \mid M \mid 2):(GD \mid N \mid \infty)$	specific
5	$(M \mid HE_2 \mid 1):(GD \mid \infty \mid \infty)$	general
6	$(M \mid HE_2 \mid 2):(GD \mid 2 \mid \infty)$	specific
7	$(HE_2 \mid M \mid 1):(GD \mid N \mid \infty)$	specific
8	$(HE_2 \mid M \mid 2):(GD \mid N \mid \infty)$	specific
9	$(HE_2 \mid E_2 \mid 1):(GD \mid 1 \mid \infty)$	specific

In analyzing non-Poisson queues using the method of stages our emphasis is on the development of a model of the system in the form of the steady-state balance equations for the system under study. Once the appropriate equations are obtained and the values of the system parameters are known, it is normally a simple matter to obtain a solution to the set of simultaneous linear equations for a specific problem.

POLLACZEK–KHINTCHINE FORMULA

Even though arrivals to a service facility often follow a Poisson distribution, it is not as common to find that service times are exponentially distributed. For this reason, the $(M \mid M \mid c)$ results presented in Chapter 3 might not be appropriate for the queueing problem under study. Often, results are desired for the $(M \mid G \mid c)$ queue. Unfortunately, the analysis required for such a situation is a bit advanced for the novice. However, it is possible to gain some insight into the effects of nonexponential service times on the operating characteristics of the system by considering the $(M \mid G \mid 1):(GD \mid \infty \mid \infty)$ queue. We will only consider the development of expressions for L, L_q, W, W_q, and D. The expression for W_q for the $(M \mid G \mid 1):(GD \mid \infty \mid \infty)$ queue is referred to as the Pollaczek–Khintchine formula. A more detailed treatment of the $(M \mid G \mid c)$ queue is reserved for Chapter 6.

Pollaczek–Khintchine Formula

In this analysis the system will be "glimpsed" immediately after service is completed and the associated customer has departed. Just as customer n departs from the system, he looks back and counts the number of customers in the system. This quantity is represented by q_n. Let ξ_n represent the number of arrivals during service of customer n. Therefore, we see that

$$q_{n+1} = \begin{cases} q_n - 1 + \xi_{n+1}, & q_n > 0 \\ \xi_{n+1}, & q_n = 0 \end{cases} \quad (4.1)$$

The quantities q_n and ξ_n are random variables. Since service times are identically and independently distributed, then $\xi_1, \xi_2, \ldots$ are identically and independently distributed random variables. Furthermore, ξ_n is independent of $q_0, q_1, \ldots, q_{n-1}$.

If we let

$$U(x) = \begin{cases} 1, & \text{if } x > 0 \\ 0, & \text{if } x \leq 0 \end{cases} \quad (4.2)$$

then (4.1) can be written compactly as

$$q_{n+1} = q_n - U(q_n) + \xi_{n+1} \quad (4.3)$$

Squaring (4.3) gives

$$q_{n+1}^2 = q_n^2 + U^2(q_n) + \xi_{n+1}^2 + 2q_n\xi_{n+1} - 2\xi_{n+1}U(q_n) - 2q_n U(q_n) \quad (4.4)$$

Observe that $U^2(q_n) = U(q_n)$ and $q_n U(q_n) = q_n$. Taking expected values of (4.4) gives

$$E(q_{n+1}^2) = E(q_n^2) + E[U(q_n)] + E(\xi_{n+1}^2) + 2E(q_n)E(\xi_{n+1}) \\ - 2E(\xi_{n+1})E[U(q_n)] - 2E(q_n) \quad (4.5)$$

since ξ_{n+1} and q_n are independent. However, in the steady state $E(q_{n+1}^2) = E(q_n^2)$. Furthermore, it is established in Chapter 6 that $E[U(q_n)] = \lambda/\mu = \rho$. Therefore, (4.5) reduces to

$$E(q_n) = \frac{\rho + E(\xi_{n+1}^2) - 2\rho E(\xi_{n+1})}{2[1 - E(\xi_{n+1})]}$$

Now, $E(\xi_{n+1}^2) = E(\xi_n^2) = V(\xi_n) + E(\xi_n)^2$. From Problem 2.43, we see that $E(\xi_n) = \rho$ and $V(\xi_n) = \lambda^2 V(t) + \rho$, where $V(t)$ is the variance of the service-time distribution. Therefore,

$$E(q_n) = \rho + \frac{\lambda^2 V(t) + \rho^2}{2(1 - \rho)} \quad (4.6)$$

Even though $E(q_n)$ is the expected number of customers in the system immediately following the departure of customer n, a result discussed by

Cooper (1972) establishes that the probability distribution for q_n is identical to the probability distribution for the number of customers in the system, when observed at any instant in time. Thus, the expected number of customers in the system can be given by

$$L = \rho + \frac{\lambda^2 V(t) + \rho^2}{2(1 - \rho)} \tag{4.7}$$

Furthermore, since $W = L/\lambda$, then the expected time a customer spends in the system is given by

$$W = \frac{1}{\mu} + \frac{\lambda[V(t) + 1/\mu^2]}{2(1 - \rho)} \tag{4.8}$$

Additionally, since $W_q = W - (1/\mu)$, then the Pollaczek–Khintchine formula is

$$W_q = \frac{\lambda[V(t) + 1/\mu^2]}{2(1 - \rho)} \tag{4.9}$$

Finally, by employing the result, $L_q = \lambda W_q$, it follows that

$$L_q = \frac{\lambda^2[V(t) + 1/\mu^2]}{2(1 - \rho)} \tag{4.10}$$

Also, since $E[U(q_n)] = \rho$ and $P_0 = 1 - E[U(q_n)]$, we see that the delay probability can be expressed as

$$D = \rho \tag{4.11}$$

Interestingly, the operating characteristics for the $(M|G|1):(GD|\infty|\infty)$ queue, L, W, W_q, and L_q, are linear functions of the variance of the service-time distribution. Furthermore, knowing the first two moments of the service-time distribution provides all the information necessary to develop the operating characteristics of the system.

Expressing (4.10) in terms of α, the coefficient of variation of service time, gives

$$L_q = \frac{\rho^2(\alpha^2 + 1)}{2(1 - \rho)}$$

where $\alpha^2 = \mu^2 V(t)$. Values of L and L_q are provided in Figs. 4.1 and 4.2, respectively, for various values of ρ and α^2.

If service time is exponentially distributed, $\alpha = 1$, whereas a constant service time yields a value of $\alpha = 0$. Thus, if service time can be standardized such that the variance of service time is approximately equal to zero, then the expected number of customers waiting for service would be reduced to

Pollaczek-Khintchine Formula

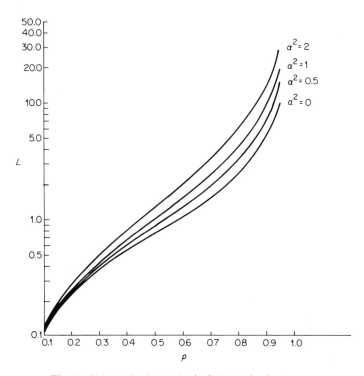

Fig. 4.1 Values of L for the $(M|G|1):(GD|\infty|\infty)$ queue.

one-half of what would result with exponentially distributed service times. Consequently, in some situations it might be advantageous to standardize service even if such standardization increases the expected value of service time.

To illustrate the effect of standardization, let ρ_e and ρ_c denote the values of ρ with exponentially distributed and constant service times, respectively. In order to achieve the same expected number of customers in the system, $L_e = L_c$, the following relation must hold:

$$\frac{\rho_e}{1-\rho_e} = \rho_c + \frac{\rho_c^2}{2(1-\rho_c)}$$

Letting L_e equal $\rho_e/(1-\rho_e)$ and reducing gives

$$\rho_c^2 - 2(1+L_e)\rho_c + 2L_e = 0$$

Solving for ρ_c, we obtain the following result:

$$\rho_c = (1+L_e) \pm (1+L_e)^{1/2}$$

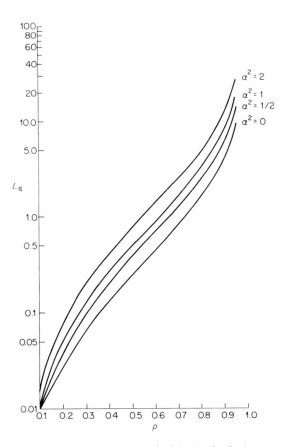

Fig. 4.2 Values of L_q for the $(M\,|\,G\,|\,1):(GD\,|\,\infty\,|\,\infty)$ queue.

However, since $\rho_c < 1$, then

$$\rho_c = (1 + L_e) - (1 + L_e^2)^{1/2}$$

To illustrate the effect of standardization on ρ_c, consider the results provided in Table 4.2. For a given value of ρ_e, the value of L_e is computed; given the value of L_e, the value of ρ_c that yields $L_e = L_c$ is determined; and the ratio ρ_c/ρ_e is formed. If λ remains the same when the service time becomes standardized (i.e., constant), then the ratio ρ_c/ρ_e is the same as the ratio of the constant service time to the expected value of the exponentially distributed service time. Consequently, if the constant service time resulting from standardization is less than ρ_c/ρ_e times the expected value of the exponentially distributed service time, then $L_c < L_e$. Observe that the ratio ρ_c/ρ_e achieves a maximum value at $\rho_e = 1/2$.

Pollaczek–Khintchine Formula

TABLE 4.2 Comparison of constant and exponential service times

ρ_e	L_e	ρ_c	ρ_c/ρ_e
0.10	0.111	0.105	1.050
0.20	0.250	0.219	1.096
0.30	0.429	0.341	1.135
0.40	0.667	0.465	1.162
0.50	1.000	0.586	1.171
0.60	1.500	0.697	1.162
0.70	2.333	0.795	1.135
0.80	4.000	0.877	1.096
0.90	9.000	0.945	1.050

Example 4.1 As an example of the use of the operating characteristics for the $(M\,|\,G\,|\,1):(GD\,|\,\infty\,|\,\infty)$ queue, consider a local newsstand that sells Playman magazine. It has been found that arrivals on the day of the new Playman issue follow a Poisson distribution with an average time between arrivals of 30 sec. There is one attendant at the newsstand to handle demands for Playman. Time-study results indicate she has the following service-time distribution:

Service time (min)	Occurrence (%)
0.1	10
0.2	30
0.3	30
0.4	20
0.5	10

Based on the observed distribution for service time, what is the average time a customer spends in the system? What is the average length of the waiting line?

The expected value and variance for service time are:

$$E(T_s) = 0.1(0.1) + 0.2(0.3) + 0.3(0.3) + 0.4(0.2) + 0.5(0.1)$$

$$= 0.29 \text{ min} = \frac{1}{\mu}$$

$$V(T_s) = E(T_s^2) - E(T_s)^2 = 0.01(0.1) + 0.04(0.3) + 0.09(0.3)$$
$$+ 0.16(0.2) + 0.25(0.1) - (0.29)^2$$

$$= 0.0129 \text{ min}^2$$

Since $\lambda = 2$ customers/min, from (4.10) and (4.12),

E(time in system) $= 0.521$ min/customer

E(number waiting) $= 0.462$ customers

Notice that the attendant will be idle 42% of the time.

Now suppose Playman is sold through a coin-operated machine with a constant service time of 0.20 min. In this case, $V(t) - 0$ and

E(time in system) $= 0.266$ min/customer

E(number waiting) $= 0.133$ customers

The newsstand dealer must decide if the increase in customer service justifies the use of a coin-operated machine.

For our example, suppose we had blindly applied the results of Chapter 3 using a value of 0.29 min for the average service time. In this case, we would have concluded incorrectly that

$W = 0.690$ min/customer, and $L_q = 0.60$ customers

Furthermore, in order for $L_c = L_e$, with $\mu_c = 5$/min, then $\mu_e = 5.75$/min.

The Pollaczek–Khintchine results allow us to analyze a number of interesting non-Poisson queues. However, there still remain many non-Poisson queues that cannot be classified as $(M \mid G \mid 1) : (GD \mid \infty \mid \infty)$ queues. In the subsequent discussion we will examine a class of non-Poisson queues that can be analyzed in a straightforward manner using the balance equation approach of Chapter 3.

METHOD OF STAGES

In this section we will describe a technique that can be used to simplify the analysis of a special class of non-Poisson queues. The technique is variously referred to as the simulation technique (Cox and Smith, 1961) and the method of stages (Cooper, 1972). We choose the latter to avoid any confusion between the present discussion and that given in Chapter 9 on Monte Carlo simulation.

To motivate the discussion, suppose the random variable Y is defined as the sum of k identical and independent exponentially distributed random variables having parameter β, i.e.,

$$Y = \sum_{j=1}^{k} X_j \qquad (4.12)$$

Method of Stages

where
$$f(x_j) = \beta \exp(-\beta x_j), \quad x_j > 0 \qquad (4.13)$$

From Chapter 2, we know that

$$f(y) = \frac{\beta(\beta y)^{k-1}}{(k-1)!} e^{-\beta y}, \quad y > 0 \qquad (4.14)$$

since Y is gamma distributed with integer parameter k. Therefore,

$$E(Y) = \frac{k}{\beta}, \quad \text{and} \quad V(Y) = \frac{k}{\beta^2} \qquad (4.15)$$

If we replace β by $k\mu$ in (4.15), we obtain the usual representation of the Erlang density:

$$f(y) = \frac{(k\mu)^k}{(k-1)!} y^{k-1} e^{-k\mu y}, \quad y > 0 \qquad (4.16)$$

with

$$E(Y) = \frac{1}{\mu}, \quad \text{and} \quad V(Y) = \frac{1}{k\mu^2} \qquad (4.17)$$

Suppose we observe that service times are gamma distributed with integer parameter k. From the above discussion, we see that it is equivalent to considering that each customer must be processed through a series of k identical and independent exponentially distributed service *stages*, each having parameter β. Our equivalent representation requires that at most one customer be allowed in the series of service stages at a given time. The equivalence for a single-service system is represented graphically in Fig. 4.3.

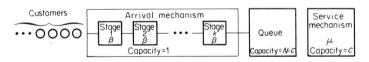

Fig. 4.3 Method of stages representation of a gamma arrival process and an exponential service process.

Customers arrive at the queue and eventually enter service, where service time is given by the sum of k identical and independent exponentially distributed random variables each having parameter β. At most one customer is in the service mechanism at any time.

In the case where interarrival time is gamma distributed with integer parameter k, we can equivalently represent the interarrival time by a series of

k identical and independent exponentially distributed service *stages*, each having parameter β. Thus, we see that our discussion of series of queues in Chapter 3 will be useful in analyzing queueing systems involving interarrival times and service times that are gamma distributed with integer parameters.

From (4.17) we see that if the Erlang distribution has the same mean value as the exponential distribution, then the variance of the Erlang is less than that of the exponential. Another characteristic of the Erlang distribution concerns the value of the mode of the distribution. Specifically, the mode can be obtained by differentiating (4.16) with respect to y, setting the result equal to zero, and solving for y, to obtain a value of $(k-1)/k\mu$ for the mode. Thus, as the value of k increases, the mode of the distribution increases and the value of the variance decreases. Hence, when we consider the analysis of arrival data and service-time data in Chapter 7, we will be interested in testing the fit of an integer parameter gamma distribution with the data. Such a fit would be attempted when the variance of the data is less than the square of the mean obtained from the data.

Of course, the variance might be greater than the square of the mean. In this case, our series arrangement of stages would not be appropriate—a different approach is suggested for this situation. Namely, suppose that with probability α_j a customer has, say, a service time that is exponentially distributed with parameter μ_j. The unconditional service-time distribution would be given as

$$f(t_s) = \sum_{j=1}^{k} f(t_s \mid \mu_j)\alpha_j \tag{4.18}$$

or

$$f(t_s) = \sum_{j=1}^{k} \alpha_j \mu_j \exp(-\mu_j t_s) \tag{4.19}$$

where

$$\sum_{j=1}^{k} \alpha_j = 1$$

The mean and variance of service time will be

$$E(t_s) = \sum_{j=1}^{k} \frac{\alpha_j}{\mu_j}, \quad \text{and} \quad V(t_s) = \sum_{j=1}^{k} \frac{2\alpha_j}{\mu_j^2} - E^2(t_s)$$

Suppose we let $k = 2$; then (4.19) reduces to

$$f(t_s) = \alpha \mu_1 \exp(-\mu_1 t_s) + (1-\alpha)\mu_2 \exp(-\mu_2 t_s) \tag{4.20}$$

Method of Stages

Now, let $\mu_1 = 2\alpha\mu$ and $\mu_2 = 2(1 - \alpha)\mu$. Therefore,

$$E(t_s) = \frac{\alpha}{\mu_1} + \frac{1-\alpha}{\mu_2} = \frac{1}{\mu} \tag{4.21}$$

and

$$V(t_s) = \frac{2\alpha}{\mu_1^2} + \frac{2(1-\alpha)}{\mu_2^2} - \frac{1}{\mu^2} = \frac{1}{\mu^2}\left[1 + \frac{(1-2\alpha)^2}{2(1-\alpha)\alpha}\right] \tag{4.22}$$

From (4.22), if $\alpha < 0.5$, then $V(t_s) > (1/\mu^2)$. Thus, the variance would be greater than that for the exponential with a mean of $1/\mu$. The density function given by (4.20) is referred to as a hyperexponential distribution. The hyperexponential distribution is sometimes referred to as a member of the family of Erlangian distributions (Cooper, 1972). However, it is not a form of the Erlang distribution.

Now, how can we model a queueing system having hyperexponentially distributed interarrival and/or service times? One approach is to employ the method of stages again, except we now consider parallel stages, rather than a series of stages.

As an illustration, suppose service time is found to have the distribution given by (4.20). An equivalent representation of such a system would consist of each server having two service-time distributions. Each service-time distribution is an exponential, but with parameters $2\alpha\mu$ and $2(1 - \alpha)\mu$. The service time for the customer will be drawn from the exponential distribution having parameter $2\alpha\mu$, a proportion of the time equal to α, and from the other exponential distribution $(1 - \alpha)$ percent of the time.

In order to illustrate the use of the method of stages in analyzing queueing systems having either gamma or hyperexponential interarrival and service times, we will present a number of special cases. In all cases the procedure will be the same, namely, we will develop the steady-state balance equations for the equivalent system having series or parallel stages. Next, we will determine the operating characteristics for the system under steady-state operation.

$(M \mid E_k \mid 1) : (GD \mid \infty \mid \infty)$ Queue

We will begin with an analysis of the $(M \mid E_k \mid 1) : (GD \mid \infty \mid \infty)$ queue. Since service time is k-Erlang distributed, we will employ the method of stages to represent equivalently the service time by a series of service stages. Specifically, we will consider that a customer who enters the service channel must pass successively through k stages of service. The time required to pass through stage j is exponentially distributed with parameter $k\mu$. Only one customer can be in service at a single time. Therefore, a customer cannot enter stage 1 until all stages are empty. Moreover, only one event can occur in time Δt.

The state of the system is defined by two parameters: n, the number in the system (including the customer in service), and j, the service stage occupied by a customer in service. Let

$P_{n,j}(t)$ = probability of finding the system in state n, j at time t

and

$P_0(t)$ = probability of an empty system at time t

Therefore, recalling that stage k is the exit stage from service, we can write the following state equations for the system:

$$P_0(t + \Delta t) = P_0(t)(1 - \lambda \, \Delta t) + P_{1,k}(t) k\mu \, \Delta t(1 - \lambda \, \Delta t)$$

$$P_{1,j}(t + \Delta t) = P_{1,j}(t)(1 - \lambda \, \Delta t)(1 - k\mu \, \Delta t)$$
$$+ P_{1,j-1}(t) k\mu \, \Delta t(1 - \lambda \, \Delta t), \quad 1 < j \le k$$

$$P_{1,1}(t + \Delta t) = P_{1,1}(t)(1 - \lambda \, \Delta t)(1 - k\mu \, \Delta t)$$
$$+ P_0(t)\lambda \, \Delta t + P_{2,k}(t) k\mu \, \Delta t(1 - \lambda \, \Delta t)$$

$$P_{n,j}(t + \Delta t) = P_{n,j}(t)(1 - \lambda \, \Delta t)(1 - k\mu \, \Delta t)$$
$$+ P_{n-1,j}(t)\lambda \, \Delta t(1 - k\mu \, \Delta t)$$
$$+ P_{n,j-1}(t) k\mu \, \Delta t(1 - \lambda \, \Delta t), \quad n > 1, \quad 1 < j \le k$$

$$P_{n,1}(t + \Delta t) = P_{n,1}(t)(1 - \lambda \, \Delta t)(1 - k\mu \, \Delta t)$$
$$+ P_{n-1,1}(t)\lambda \, \Delta t(1 - k\mu \, \Delta t)$$
$$+ P_{n+1,k}(t) k\mu \, \Delta t(1 - \lambda \, \Delta t), \quad n > 1$$

In words, our balance equations represent the following mutually exclusive and collectively exhaustive situations:

1. Zero customers will be in the system at time $(t + \Delta t)$ if either (a) there were zero customers in the system at time t and no arrivals occurred during Δt; or (b) there was only one customer in the system at time t, that customer was in the kth service stage, a service was performed during Δt, and no arrival occurred during Δt.

2. One customer will be in the system at time $(t + \Delta t)$ and he will be in the jth service stage ($1 < j \le k$) if either (a) there was only one customer in the system at time t, he was in service stage j, and neither an arrival nor a service occurred during Δt; or (b) there was only one customer in the system at time t, he was in service stage $j - 1$, and a service and no arrival occurred during Δt.

Method of Stages

3. One customer will be in the system at time $(t + \Delta t)$ and he will be in the first service stage if either (a) the customer was in the first service stage at time t and there were no services or arrivals in Δt; (b) there were no customers in the system at time t, and one arrival, but no service, occurred in Δt; or (c) there were two customers in the system at time t, the customer in service was in service stage k, and there was one service, but no arrival, in Δt.

4. n customers $(n > 1)$ will be in the system at time $(t + \Delta t)$ and the customer in service will be in service stage j $(1 < j \leq k)$ if either (a) n customers were in the system at time t, the customer in service was in service stage j, and neither an arrival nor a service occurred in Δt; (b) $n - 1$ customers were in the system at time t, the customer in service was in service stage j, and one arrival and no service occurred in Δt; or (c) n customers were in the system at time t, the customer in service was in service stage $j - 1$, and one service and no arrival occurred in Δt.

5. n customers $(n > 1)$ will be in the system at time $(t + \Delta t)$ and the customer in service will be in service stage 1 if either (a) n customers were in the system at time t, the customer in service was in service stage 1, and there were no arrivals or services in Δt; (b) $n - 1$ customers were in the system at time t, the customer in service was in service stage 1, and there were one arrival and no services in Δt; or (c) $n + 1$ customers were in the system at time t, the customer in service was in service stage k, and there were one service and no arrivals in Δt.

On taking the appropriate term to the left-hand side of each equation, dividing by Δt, taking the limit as Δt approaches zero, and setting the resulting derivative equal to zero, we obtain the following steady-state equations:

$$\lambda P_0 = k\mu P_{1,k} \tag{4.23}$$

$$(\lambda + k\mu)P_{1,j} = k\mu P_{1,j-1}, \quad 1 < j \leq k \tag{4.24}$$

$$(\lambda + k\mu)P_{1,1} = \lambda P_0 + k\mu P_{2,k} \tag{4.25}$$

$$(\lambda + k\mu)P_{n,j} = \lambda P_{n-1,j} + k\mu P_{n,j-1}, \quad n > 1, \quad 1 < j \leq k \tag{4.26}$$

$$(\lambda + k\mu)P_{n,1} = \lambda P_{n-1,1} + k\mu P_{n+1,k}, \quad n > 1 \tag{4.27}$$

The probability distribution for the number in the system is obtained by

$$P_n = \sum_{j=1}^{k} P_{n,j}$$

The steady-state equations can be expressed compactly by defining a new variable m, which equals the number of stages awaiting service. Since each arrival undergoes k stages of service, m can be written as $m = nk - j + 1$. The stage occupied by a customer is counted due to the forgetfulness property of the exponential distribution. With the change in parameters, the steady-state equations become

$$\lambda P_0 = k\mu P_1 \qquad (4.28)$$

$$(\lambda + k\mu)P_m = k\mu P_{m+1} + \lambda P_{m-k}, \qquad m \geq 1 \qquad (4.29)$$

with the understanding that any probability with a negative subscript is zero. Multiplying (4.29) by z^m and summing over $m \geq 1$ gives

$$\sum_{m=1}^{\infty} (\lambda + k\mu)z^m P_m = \sum_{m=1}^{\infty} (k\mu z^m P_{m+1} + \lambda z^m P_{m-k})$$

Defining the generating function as

$$G(z) = \sum_{m=0}^{\infty} z^m P_m$$

gives

$$(\lambda + k\mu)[G(z) - P_0] = \frac{k\mu}{z}[G(z) - P_0 - zP_1] + \lambda z^k G(z)$$

From (4.28), $P_1 = (\lambda/k\mu)P_0$. Therefore, solving for $G(z)$ gives

$$G(z) = \frac{k\mu(1-z)P_0}{k\mu(1-z) - z\lambda(1-z^k)}$$

Letting $\mu_1 = k\mu$ and dividing both numerator and denominator of $G(z)$ by $(1-z)$ gives

$$G(z) = \frac{\mu_1 P_0}{\mu_1 - z\lambda[(1-z^k)/(1-z)]} = \frac{\mu_1 P_0}{\mu_1 - z\lambda \sum_{j=0}^{k-1} z^j}$$

Now, since

$$G(z)\bigg|_{z=1} = 1$$

then

$$\frac{\mu_1 P_0}{\mu_1 - \lambda \sum_{j=0}^{k-1} 1^j} = \frac{\mu_1 P_0}{\mu_1 - k\lambda} = 1, \quad \text{and} \quad P_0 = 1 - \frac{k\lambda}{\mu_1}$$

But, $\mu_1 = k\mu$. Therefore,

$$P_0 = 1 - \frac{\lambda}{\mu}, \quad \text{or} \quad P_0 = 1 - \rho \qquad (4.30)$$

Method of Stages

Thus,

$$G(z) = \frac{\mu_1(1-\rho)}{\mu_1 - z\lambda \sum_{j=0}^{k-1} z^j} \quad (4.31)$$

which is similar to (3.88) for bulk arrivals. This should not be too surprising, since an arrival requires k services by the method of stages and an arrival requires b services in the case of bulk arrivals.

Unfortunately, we can not obtain a general solution for P_m by inverting $G(z)$. However, for a specific application where the values of the system parameters are known, recall from (3.89) that a power series expansion of $G(z)$ can be employed to obtain the desired value of P_m.

Taking the first derivative of $G(z)$ with respect to z, letting $z \to 1$, and reducing gives the following expression for the expected value of m:

$$E(m) = \frac{k+1}{2} \frac{\lambda}{\mu - \lambda} \quad (4.32)$$

Since, on the average, an arriving customer will see $E(m)$ stages to be served ahead of him in the system and the average service time per stage is $(k\mu)^{-1}$, the expected waiting time is $E(m)/k\mu$, or

$$W_q = \frac{k+1}{2k\mu} \frac{\lambda}{\mu - \lambda} \quad (4.33)$$

Therefore, since $W = W_q + (1/\mu)$ and $L = \lambda W$,

$$W = \frac{k+1}{2k\mu} \frac{\lambda}{\mu - \lambda} + \frac{1}{\mu} \quad (4.34)$$

$$L_q = \frac{k+1}{2k\mu} \frac{\lambda^2}{\mu - \lambda} \quad (4.35)$$

$$L = \frac{k+1}{2k\mu} \frac{\lambda^2}{\mu - \lambda} + \frac{\lambda}{\mu} \quad (4.36)$$

Note that we could have obtained (4.33) through (4.36) by appropriately substituting (4.17) in (4.7)–(4.10).

Example 4.2 As an example of an $(M \mid E_k \mid 1) : (GD \mid \infty \mid \infty)$ queue, consider the drive-in window at a local bank. Let us assume we have observed the arrival and service pattern of customers, have analyzed the data using the techniques of Chapters 7 and 8, and have found that arrivals are Poisson distributed with a mean of 10 arrivals/hr and that service time is Erlang distributed with a mean of 3 min and a variance of 3 min².

From (4.17) we see that $\mu = 20$ customers/hr and $k = 3$. Since $\lambda = 10$ customers/hr, from (4.33)–(4.36), we find that

$W_q = 2$ min/customer, $W = 5$ min/customer

$L_q = 0.333$ customers, $L = 0.833$ customers

The probability that an arriving customer will be delayed, $1 - P_0$, is 0.50 for this example.

$(M \mid E_2 \mid 2) : (GD \mid 2 \mid \infty)$ Queue

The method of stages can also be used to analyze multiserver systems. However, the analysis can become tedious. As an illustration of an approach that can be taken, we will consider the $(M \mid E_2 \mid 2) : (GD \mid 2 \mid \infty)$ queue.

Since each server is equivalently represented by pseudoseries service channels, it is necessary to keep track of customers being served by each of the servers. To accomplish this, we let

P_{nij} = probability of n customers in the system when the customer being served by server 1 is in the ith stage of service and the customer being served by server 2 is in the jth stage of service, where $i = 0, j = 0$, denotes idle servers.

We will assume that an arriving customer will choose with equal probability either server if both are available. To distinguish between the servers, let μ_h denote the mean service rate for server h. Therefore, using the rate out = rate in approach, the steady-state balance equations can be written as follows:

$\lambda P_{000} = \mu_1 P_{120} + \mu_2 P_{102}$, $(\mu_1 + \mu_2)P_{211} = \lambda P_{110} + \lambda P_{101}$

$(\lambda + \mu_2)P_{101} = 0.5\lambda P_{000} + \mu_1 P_{221}$, $(\mu_1 + \mu_2)P_{212} = \lambda P_{102} + \mu_2 P_{211}$

$(\lambda + \mu_1)P_{110} = 0.5\lambda P_{000} + \mu_2 P_{212}$, $(\mu_1 + \mu_2)P_{221} = \lambda P_{120} + \mu_1 P_{211}$

$(\lambda + \mu_2)P_{102} = \mu_2 P_{101} + \mu_1 P_{222}$, $(\mu_1 + \mu_2)P_{222} = \mu_2 P_{221} + \mu_1 P_{212}$

$(\lambda + \mu_1)P_{120} = \mu_1 P_{110} + \mu_2 P_{222}$

Since we have assumed Erlang service with two phases, then $\mu_1 = \mu_2 = 2\mu$. Moreover, since each server is equally likely to be chosen for service when an arriving customer finds both servers idle, then

$P_{101} = P_{110}$, $P_{102} = P_{120}$,   and $P_{212} = P_{221}$

Method of Stages

Therefore, our set of balance equations reduces to

$$\lambda P_{000} = 4\mu P_{102}, \qquad\qquad 2\mu P_{211} = \lambda P_{101}$$
$$(\lambda + 2\mu)P_{101} = 0.5\lambda P_{000} + 2\mu P_{212}, \qquad 4\mu P_{212} = \lambda P_{102} + 2\mu P_{211}$$
$$(\lambda + 2\mu)P_{102} = 2\mu P_{101} + 2\mu P_{222}, \qquad P_{222} = P_{212}$$

Solving in terms of P_{000} gives

$$P_{101} = P_{110} = P_{102} = P_{120} = \frac{\lambda}{4\mu} P_{000} \qquad (4.37)$$

$$P_{211} = P_{212} = P_{221} = P_{222} = \frac{\lambda^2}{8\mu^2} P_{000} \qquad (4.38)$$

Since the sum of the state probabilities must equal one, it follows that

$$P_{000} = \frac{2}{2 + 2\rho + \rho^2} \qquad (4.39)$$

where $\rho = \lambda/\mu$. The expected number in the system becomes

$$L = \frac{2(\rho + \rho^2)}{2 + 2\rho + \rho^2} \qquad (4.40)$$

Furthermore, since no waiting line exists,

$$W = \frac{1}{\mu} \qquad (4.41)$$

and the effective arrival rate is given by

$$\tilde{\lambda} = \frac{2\mu(\rho + \rho^2)}{2 + 2\rho + \rho^2} \qquad (4.42)$$

Two aspects of our development are of particular interest. First, we provided steady-state balance equations for the general case in which $\mu_1 \neq \mu_2$. Thus, we can handle the situation in which our servers have different mean service rates. Second, we can also account for unequal preferability of servers by changing the probability from 0.5 to, say, α. Combining these two features of our formulation we can easily handle the case in which, say, one server is faster than the other and customers always choose the faster server if both are available; otherwise, our customer is served by whichever server is available. An exploration of these possibilities is suggested in the exercises at the end of the chapter.

Example 4.3 As an illustration of the $(M \mid E_2 \mid 2):(GD \mid 2 \mid \infty)$ queueing results, consider a telephone switchboard consisting of two operators. Calls arrive in a Poisson fashion at an average rate of 10/min. The time required to service the customer is Erlang distributed with a mean of 12 sec and a variance of 72. Therefore, $\lambda = 10$, $\mu = 5$, and $k = 2$. From (4.37) and (4.38),

$$P_{101} = P_{110} = P_{102} = P_{120} = 0.5 P_{000}$$

$$P_{211} = P_{212} = P_{221} = P_{222} = 0.5 P_{000}$$

and, from either (4.39)–(4.42) or Erlang's loss formula,

$P_{000} = 0.2$, $\qquad L = 1.2$ customers

$W = 0.2$ min/customer, $\qquad \tilde{\lambda} = 6$ customers/min

Notice that $\tilde{\lambda}$ is also equal to $\lambda(1 - P_2)$, where

$$P_2 = P_{211} + P_{212} + P_{221} + P_{222} = \frac{\rho^2}{2 + 2\rho + \rho^2}$$

$(E_2 \mid M \mid 1):(GD \mid N \mid \infty)$ **Queue**

We now turn our attention to those queueing problems in which the interarrival time is not exponentially distributed, but is instead Erlang distributed with two phases. In this case, the interarrival distribution is

$$f(t_a) = 4\lambda^2 t_a \exp(-2\lambda t_a), \qquad t_a > 0 \qquad (4.43)$$

with $E(t_a) = 1/\lambda$ and $V(t_a) = 1/2\lambda^2$. According to Morse (1958, p. 88), the 2-Erlang distribution is the type of arrival distribution more often encountered than any other except the Poisson distribution.

As shown in Fig. 4.3, we assume there always exists a supply of customers waiting to enter the arrival mechanism. We will identify the state of the system by using two subscripts: n, the number in the system, and j, the arrival stage occupied by the next arriving customer. Thus, in steady state we let $P_{n,j}$ denote the probability of finding the system in state n, j. Applying the rate out = rate in approach, the resulting steady-state equations are

$$2\lambda P_{01} = \mu P_{11}, \qquad 2\lambda P_{02} = 2\lambda P_{01} + \mu P_{12}$$

$$(2\lambda + \mu)P_{n1} = 2\lambda P_{n-1,2} + \mu P_{n+1,1},$$

$$(2\lambda + \mu)P_{n2} = 2\lambda P_{n1} + \mu P_{n+1,2} \qquad (n > 0)$$

$$(2\lambda + \mu)P_{N1} = 2\lambda P_{N-1,2} + 2\lambda P_{N2}, \qquad (2\lambda + \mu)P_{N2} = 2\lambda P_{N1}$$

For given values of λ, μ, and N we can determine the values for P_{nj}.

Example 4.4 As an illustration of the approach employed when analyzing an $(E_2 \mid M \mid 1):(GD \mid N \mid \infty)$ queue, consider a manufacturing situation involving two production stations. Raw material is always available at the first

Method of Stages

station, where two operations are performed consecutively, each exponentially distributed with a mean of 0.5 min. Once the two operations are performed, the semifinished part is placed on a belt conveyor, which transports the part to the second production station. At the second station there is waiting space for only one part. Therefore, if a part arrives at the second station and two parts are ahead of it at the station, the part is automatically removed from the production line. The time required to perform the operation required at the second production station is exponentially distributed with a mean of 0.5 min.

The interarrival distribution for parts at the second production station follows the E_2 distribution. Consequently, the second production station can be analyzed as an $(E_2 \mid M \mid 1) : (GD \mid 2 \mid \infty)$ queue, with $\lambda = 1/\text{min}$ and $\mu = 2/\text{min}$. Letting $\rho = \lambda/\mu = 0.5$, we obtain the following equations of balance for the system:

$$P_{01} = \frac{1}{2\rho} P_{11}, \qquad P_{02} = P_{01} + \frac{1}{2\rho} P_{12}$$

$$P_{11} = \left(\frac{2\rho}{2\rho + 1}\right) P_{02} + \left(\frac{1}{2\rho + 1}\right) P_{21},$$

$$P_{12} = \left(\frac{2\rho}{2\rho + 1}\right) P'_{11} + \left(\frac{1}{2\rho + 1}\right) P_{22}$$

$$P_{21} = \left(\frac{2\rho}{2\rho + 1}\right) P_{12} + \left(\frac{2\rho}{2\rho + 1}\right) P_{22}$$

Solving in terms of P_{01} gives

$$P_{02} = (8/5) P_{01}, \qquad P_{11} = P_{01}$$
$$P_{12} = (3/5) P_{01}, \qquad P_{21} = (2/5) P_{01}, \qquad P_{22} = (1/5) P_{01}$$

Since $P_{01} + P_{02} + P_{11} + P_{12} + P_{21} + P_{22} = 1$, it follows that

$$P_{01} = 5/24, \qquad P_{12} = 3/24$$
$$P_{02} = 8/24, \qquad P_{21} = 2/24$$
$$P_{11} = 5/24, \qquad P_{22} = 1/24$$

Therefore,

$$L = 14/24 \text{ parts}, \quad \text{and} \quad L_q = 3/24 \text{ parts}$$

Consequently, the effective arrival rate is

$$\tilde{\lambda} = \mu(L - L_q) = 22/24 \text{ parts/min}$$

Thus, it follows that

$$W = \frac{L}{\tilde{\lambda}} = 14/22 \text{ min/part}, \quad \text{and} \quad W_q = \frac{L_q}{\tilde{\lambda}} = 3/22 \text{ min/part}$$

Example 4.5 It has been proposed that no waiting space be provided at the second production station. In order to analyze the resulting $(E_2|M|1)$: $(GD|1|\infty)$ queue, we first obtain the steady-state balance equations for the system:

$$P_{01} = \frac{\mu}{2\lambda} P_{11}, \qquad\qquad P_{02} = P_{01} + \frac{\mu}{2\lambda} P_{12}$$

$$P_{11} = \frac{2\lambda}{2\lambda + \mu} P_{02} + \frac{2\lambda}{2\lambda + \mu} P_{12}, \qquad P_{12} = \frac{2\lambda}{2\lambda + \mu} P_{11}$$

The solution to this system of four equations of state gives:

$$P_{01} = \frac{\mu}{2(2\lambda + \mu)}, \qquad P_{02} = \frac{4\lambda\mu + \mu^2}{2(2\lambda + \mu)^2}$$

$$P_{11} = \frac{\lambda}{2\lambda + \mu}, \qquad P_{12} = \frac{2\lambda^2}{(2\lambda + \mu)^2}$$

Since the expected time spent in the system is the expected service time and $L = \tilde{\lambda}W$, it follows that the effective arrival rate for the system is $\tilde{\lambda} = \mu L$. For the case of $\lambda = 1$ and $\mu = 2$, we find that

$$P_0 = 10/16, \qquad P_1 = 6/16, \qquad L = 6/16 \text{ parts}, \qquad \tilde{\lambda} = 12/16 \text{ parts/min}$$

$(E_2 \,|\, M \,|\, 2) : (GD \,|\, N \,|\, \infty)$ Queue

In this section we generalize our preceding discussion to include multiple servers. We will consider a system having 2 servers and a capacity of N customers. The following steady-state equations hold for this system:

$$2\lambda P_{01} = \mu P_{11},$$
$$2\lambda P_{02} = 2\lambda P_{01} + \mu P_{12},$$
$$(2\lambda + \mu)P_{11} = 2\lambda P_{02} + 2\mu P_{21},$$
$$(2\lambda + \mu)P_{12} = 2\lambda P_{11} + 2\mu P_{22},$$
$$(2\lambda + 2\mu)P_{n1} = 2\lambda P_{n-1,2} + 2\mu P_{n+1,1} \qquad (1 < n < N)$$
$$(2\lambda + 2\mu)P_{n2} = 2\lambda P_{n1} + 2\mu P_{n+1,2} \qquad (1 < n < N)$$
$$(2\lambda + 2\mu)P_{N1} = 2\lambda P_{N-1,2} + 2\lambda P_{N2}$$
$$(2\lambda + 2\mu)P_{N2} = 2\lambda P_{N1}$$

Method of Stages

As before, for given values of λ, μ, and N, we can solve the balance equations to obtain the values for $P_{n,j}$. As usual, we make use of the fact that

$$\sum_{n=0}^{N} \sum_{j=1}^{2} P_{n,j} = 1$$

Example 4.6 An illustration of an $(E_2 \mid M \mid 2):(GD \mid N \mid \infty)$ queue consider the previous manufacturing example. It has been proposed that two operators be placed at the second production station. However, as in Example 4.5, no waiting space will be provided. With $N = 2$, $\lambda = 1$, and $\mu = 1$, the state equations become

$$P_{01} = (1/2)P_{11}, \qquad P_{02} = P_{01} + (1/2)P_{12}$$
$$P_{11} = (2/3)P_{02} + (2/3)P_{21}, \qquad P_{12} = (2/3)P_{11} + (2/3)P_{22}$$
$$P_{21} = (1/2)P_{12} + (1/2)P_{22}, \qquad P_{22} = (1/2)P_{21}$$

We can write the set of homogeneous state equations in matrix form as follows:

$$\begin{pmatrix} -1 & 0 & 1/2 & 0 & 0 & 0 \\ 1 & -1 & 0 & 1/2 & 0 & 0 \\ 0 & 2/3 & -1 & 0 & 2/3 & 0 \\ 0 & 0 & 2/3 & -1 & 0 & 2/3 \\ 0 & 0 & 0 & 1/2 & -1 & 1/2 \\ 0 & 0 & 0 & 0 & 1/2 & -1 \end{pmatrix} \begin{pmatrix} P_{01} \\ P_{02} \\ P_{11} \\ P_{12} \\ P_{21} \\ P_{22} \end{pmatrix} = \begin{pmatrix} 0 \\ 0 \\ 0 \\ 0 \\ 0 \\ 0 \end{pmatrix}$$

Since we have homogeneous equations, there is an infinite number of solutions available. However, we require a particular solution; namely, we desire a solution such that

$$P_{01} + P_{02} + P_{11} + P_{12} + P_{21} + P_{22} = 1$$

Consequently, we replace one of the state equations with $\sum_{n=0}^{N} \sum_{j=1}^{2} P_{n,j} = 1$. Replacing the last state equation results in the following matrix solution:

$$\begin{pmatrix} P_{01} \\ P_{02} \\ P_{11} \\ P_{12} \\ P_{21} \\ P_{22} \end{pmatrix} = \begin{pmatrix} -1 & 0 & 1/2 & 0 & 0 & 0 \\ 1 & -1 & 0 & 1/2 & 0 & 0 \\ 0 & 2/3 & -1 & 0 & 2/3 & 0 \\ 0 & 0 & 2/3 & -1 & 0 & 2/3 \\ 0 & 0 & 0 & 1/2 & -1 & 1/2 \\ 1 & 1 & 1 & 1 & -1 & 1 \end{pmatrix}^{-1} \begin{pmatrix} 0 \\ 0 \\ 0 \\ 0 \\ 0 \\ 1 \end{pmatrix} = \begin{pmatrix} 7/58 \\ 13/58 \\ 14/58 \\ 12/58 \\ 8/58 \\ 4/58 \end{pmatrix}$$

Therefore,

$$L = 50/58 \text{ parts}, \quad \text{and} \quad L_q = 12/58 \text{ parts}$$

Consequently, the effective arrival rate is

$$\tilde{\lambda} = \mu(L - L_q) = 38/58 \text{ parts/min}$$

the expected time in the system is

$$W = \frac{L}{\tilde{\lambda}} = 50/38 \text{ min/part}$$

and the expected time spent waiting for service is

$$W_q = \frac{L_q}{\tilde{\lambda}} = 12/38 \text{ min/part}$$

$(M \mid HE_2 \mid 1) : (GD \mid \infty \mid \infty)$ **Queue**

In this and subsequent sections we will examine cases involving an hyperexponential distribution of interarrival or service times. As discussed earlier we will employ parallel stages to represent the equivalent system involving hyperexponential distributions. We begin our discussion with a treatment of the $(M \mid HE_2 \mid 1) : (GD \mid \infty \mid \infty)$ queue.

Recall, for the case of hyperexponential service times (HE_2), that we assume customers enter the waiting line and when chosen for service choose service branch 1 with probability α and service branch 2 with probability $(1 - \alpha)$. Service in branch 1 is performed according to an exponential service-time distribution with parameter $\mu_1 = 2\alpha\mu$; service in branch 2 follows an exponential distribution with parameter $\mu_2 = 2(1 - \alpha)\mu$. Naturally, service must be performed in either branch 1 or branch 2; customers are served one at a time.

Let

$P_{n,j}$ = probability of n customers in the system with the customer in service being in service branch j, $\quad n > 0, \quad j = 1, 2$

P_0 = probability of no customers in the system

Therefore, the steady-state equations for the system can be given as:

$$\lambda P_0 = \mu_1 P_{11} + \mu_2 P_{12} \tag{4.44}$$

$$(\lambda + \mu_1)P_{11} = \alpha\lambda P_0 + \alpha\mu_1 P_{21} + \alpha\mu_2 P_{22} \tag{4.45}$$

$$(\lambda + \mu_2)P_{12} = (1 - \alpha)\lambda P_0 + (1 - \alpha)\mu_1 P_{21} + (1 - \alpha)\mu_2 P_{22} \tag{4.46}$$

$$(\lambda + \mu_1)P_{n1} = \lambda P_{n-1,1} + \alpha\mu_1 P_{n+1,1} + \alpha\mu_2 P_{n+1,2} \tag{4.47}$$

$$(\lambda + \mu_2)P_{n2} = \lambda P_{n-1,2} + (1 - \alpha)\mu_1 P_{n+1,1} + (1 - \alpha)\mu_2 P_{n+1,2} \tag{4.48}$$

Method of Stages

As in the discussion of the $(M\,|\,E_k\,|\,1):(GD\,|\,\infty\,|\,\infty)$ queue, we will develop the generating function for the number in the system. To achieve this, we will let

$$G_1(z) = \sum_{n=1}^{\infty} z^n P_{n1} \qquad (4.49)$$

$$G_2(z) = \sum_{n=1}^{\infty} z^n P_{n2} \qquad (4.50)$$

$$G(z) = \sum_{n=0}^{\infty} z^n P_n$$

Notice that

$$G(z) = \sum_{n=1}^{\infty} z^n P_n + P_0$$

which can be expressed as

$$G(z) = \sum_{n=1}^{\infty} z^n (P_{n1} + P_{n2}) + P_0, \quad \text{or} \quad G(z) = G_1(z) + G_2(z) + P_0 \qquad (4.51)$$

Multiplying (4.47) by z^n and summing over n from 2 to infinity gives

$$\sum_{n=2}^{\infty} z^n P_{n1} - \frac{\lambda}{\lambda+\mu_1} \sum_{n=2}^{\infty} z^n P_{n-1,1} - \frac{\alpha\mu_1}{\lambda+\mu_1} \sum_{n=2}^{\infty} z^n P_{n+1,1}$$
$$- \frac{\alpha\mu_2}{\lambda+\mu_1} \sum_{n=2}^{\infty} z^n P_{n+1,2} = 0 \qquad (4.52)$$

Multiplying (4.45) by z, adding the resulting expression to (4.52), and reducing gives

$$\left[1 - \frac{\lambda z}{\lambda+\mu_1} - \frac{\alpha\mu_1 z^{-1}}{\lambda+\mu_1}\right] G_1(z) - \frac{\alpha\mu_2 z^{-1}}{\lambda+\mu_1} G_2(z)$$
$$= \frac{\alpha\lambda z}{\lambda+\mu_1} P_0 - \frac{\alpha\mu_1}{\lambda+\mu_1} P_{11} - \frac{\alpha\mu_2}{\lambda+\mu_1} P_{12} \qquad (4.53)$$

Multiplying both sides of (4.53) by $(\lambda+\mu_1)$ gives

$$(\lambda+\mu_1 - \lambda z - \alpha\mu_1 z^{-1}) G_1(z) - \alpha\mu_2 z^{-1} G_2(z)$$
$$= \alpha\lambda z P_0 - \alpha\mu_1 P_{11} - \alpha\mu_2 P_{12} \qquad (4.54)$$

Multiplying (4.48) by z^n, summing over n from 2 to infinity, adding the resulting expression to the product of z and (4.46), and reducing gives

$$-(1-\alpha)\mu_1 z^{-1} G_1(z) + [\lambda + \mu_2 - \lambda z - (1-\alpha)\mu_2 z^{-1}] G_2(z)$$
$$= (1-\alpha)\lambda z P_0 - (1-\alpha)\mu_1 P_{11} - (1-\alpha)\mu_2 P_{12} \qquad (4.55)$$

From (4.44),

$$\alpha\mu_1 P_{11} + \alpha\mu_2 P_{12} = \alpha\lambda P_0, \quad \text{and} \quad (1-\alpha)\mu_1 P_{11} + (1-\alpha)\mu_2 P_{12}$$
$$= (1-\alpha)\lambda P_0$$

Therefore, (4.54) and (4.55) can be written as

$$(\lambda + \mu_1 - \lambda z - \alpha\mu_1 z^{-1}) G_1(z) - \alpha\mu_2 z^{-1} G_2(z) = \alpha\lambda P_0(z-1) \quad (4.56)$$

$$-(1-\alpha)\mu_1 z^{-1} G_1(z) + [\lambda + \mu_2 - \lambda z - (1-\alpha)\mu_2 z^{-1}] G_2(z)$$
$$= (1-\alpha)\lambda P_0(z-1) \qquad (4.57)$$

Solving (4.56) and (4.57) simultaneously for $G_1(z)$ and $G_2(z)$ gives

$$G_1(z) = \frac{\alpha[\lambda(1-z) + \mu_2]}{k(z)} P_0 \qquad (4.58)$$

$$G_2(z) = \frac{(1-\alpha)[\lambda(1-z) + \mu_1]}{k(z)} P_0 \qquad (4.59)$$

where

$$k(z) = \frac{\mu_1 \mu_2}{\lambda z} + \frac{1}{z}[\alpha\mu_1 + (1-\alpha)\mu_2] - \lambda(1-z) - (\mu_1 + \mu_2)$$

Now, since

$$G(z) = G_1(z) + G_2(z) + P_0 \qquad (4.60)$$

and $G(1)$ must equal one, then

$$\left[\frac{\alpha\mu_2}{k(1)} + \frac{(1-\alpha)\mu_1}{k(1)} + 1\right] P_0 = 1$$

Since

$$k(1) = \frac{\mu_1 \mu_2}{\lambda} - (1-\alpha)\mu_1 - \alpha\mu_2$$

then

$$P_0 = \frac{\mu_1\mu_2 - (1-\alpha)\lambda\mu_1 - \alpha\lambda\mu_2}{\mu_1\mu_2} \qquad (4.61)$$

Method of Stages

Substituting (4.58)–(4.60) in (4.61) gives

$$G(z) = \left[\frac{\lambda(1-z) + \alpha\mu_2 + (1-\alpha)\mu_1}{k(z)}\right]\left[\frac{\mu_2\mu_1 - (1-\alpha)\lambda\mu_1 - \alpha\lambda\mu_2}{\mu_1\mu_2}\right] \quad (4.62)$$

From (4.62) we can obtain the following expressions for the operating characteristics for the system:

$$L = G'_z(z)\bigg|_{z=1} \quad (4.63)$$

$$W = L/\lambda \quad (4.64)$$

$$W_q = W - \frac{\alpha}{\mu_1} - \frac{(1-\alpha)}{\mu_2} \quad (4.65)$$

$$L_q = L - \frac{\lambda\alpha}{\mu_1} - \frac{\lambda(1-\alpha)}{\mu_2} \quad (4.66)$$

Recalling that we let $\mu_1 = 2\alpha\mu$ and $\mu_2 = 2(1-\alpha)\mu$ and letting $\rho = \lambda/\mu$, (4.62) becomes

$$G(z) = \frac{[\rho(1-z) + 4\alpha(1-\alpha)]\rho(1-\rho)z}{4\alpha(1-\alpha)(1-\rho) + (2-\rho z)(1-z)\rho} \quad (4.67)$$

Solving for the expected number in the system gives

$$L = \rho + \frac{\rho^2}{4\alpha(1-\alpha)(1-\rho)} \quad (4.68)$$

Corresponding values for W, W_q, and L_q can be obtained from (4.64), (4.65), or (4.66), or the results derived from the Pollaczek–Khintchine formula, (4.8)–(4.10).

Example 4.7 Consider a check-out counter at a tool crib having a single attendant. Customers arrive at a Poisson rate of 10/hr. Service time is hyperexponentially distributed with a mean of 4 min and a variance of 19.6 min. From (4.21) and (4.22), we find that $\mu = 15$/hr and $\alpha = 0.2$. Since $\lambda = 10$/hr, then $\rho = 2/3$ and, from (4.68),

$$L = 2.75 \text{ customers}$$

From (4.64)–(4.66), we find that

$W = 16.5$ min/customer, $W_q = 12.5$ min/customer

$L_q = 2.08$ customers

$(M \mid HE_2 \mid 2) : (GD \mid 2 \mid \infty)$ Queue

In this section we will consider a simple multiserver problem involving hyperexponentially distributed service times. Specifically, we will consider a 2-server problem in which no queue is allowed. As in an earlier discussion we let

P_{nij} = probability of n customers in the system, with the customer being served by server 1 being in service branch i and the customer being served by server 2 being in service branch j. If $i = 0$ or $j = 0$, the respective server is idle

We will also let

μ_{jk} = exponential service rate in service branch j for server k, $j = 1, 2$, $k = 1, 2$

α = probability of customer choosing branch 1

$1 - \alpha$ = probability of customer choosing branch 2

β = probability of customer choosing server 1

$1 - \beta$ = probability of customer choosing server 2

P_0 = probability of no customers in the system

The steady-state equations for our system can be written as:

$$\lambda P_0 = \mu_{11} P_{110} + \mu_{21} P_{120} + \mu_{12} P_{101} + \mu_{22} P_{102}$$
$$(\lambda + \mu_{11}) P_{110} = \alpha\beta\lambda P_0 + \mu_{12} P_{211} + \mu_{22} P_{212}$$
$$(\lambda + \mu_{21}) P_{120} = (1 - \alpha)\beta\lambda P_0 + \mu_{12} P_{221} + \mu_{22} P_{222}$$
$$(\lambda + \mu_{12}) P_{101} = \alpha(1 - \beta)\lambda P_0 + \mu_{11} P_{211} + \mu_{21} P_{221}$$
$$(\lambda + \mu_{22}) P_{102} = (1 - \alpha)(1 - \beta)\lambda P_0 + \mu_{11} P_{212} + \mu_{21} P_{222}$$
$$(\mu_{11} + \mu_{12}) P_{211} = \alpha\lambda P_{101} + \alpha\lambda P_{110}$$
$$(\mu_{11} + \mu_{21}) P_{212} = \alpha\lambda P_{102} + (1 - \alpha)\lambda P_{110}$$
$$(\mu_{21} + \mu_{12}) P_{221} = (1 - \alpha)\lambda P_{101} + \alpha\lambda P_{120}$$
$$(\mu_{21} + \mu_{22}) P_{222} = (1 - \alpha)\lambda P_{102} + (1 - \alpha)\lambda P_{120}$$

Note that we have assumed an arriving customer will be served by the idle server if only one server is idle upon arrival of the customer. If both servers are idle, server 1 is chosen with probability β and server 2 is chosen with

Method of Stages

probability $(1 - \beta)$. Given values of μ_{jk}, λ, α, and β, we can solve for P_{nij} by making use of the fact that

$$P_0 + P_{110} + P_{120} + P_{101} + P_{102} + P_{211} + P_{212} + P_{221} + P_{222} = 1$$

Example 4.8 As an illustration of the computational approach to be taken in analyzing an $(M\,|\,HE_2\,|\,2):(GD\,|\,2\,|\,\infty)$ queue, consider an airport that provides two parking spaces at the terminal entrance for persons "dropping off" a rental car. Attendants transfer the cars to the appropriate parking lot. If persons arrive with their cars and find both spaces occupied they take the cars to the appropriate lots and park them. Persons arrive at a Poisson rate of 2/min to drop off cars. The time a car spends in one of the parking spaces is exponentially distributed with a mean of 2.5 min if it is a Hurts car and 0.625 min if it is an Avex car. The probability that an arriving car is a Hurts car is 0.20.

Based on the data for this example, we see that $\alpha = 0.2$, $\beta = 0.5$, $\mu_{11} = \mu_{12} = 2\alpha\mu$, $\mu_{21} = \mu_{22} = 2(1 - \alpha)\mu$, $\mu = 1$, and $\lambda = 2$. The steady-state equations become

$$2P_0 = 0.4P_{110} + 1.6P_{120} + 0.4P_{101} + 1.6P_{102}$$

$$2.4P_{110} = 0.2P_0 + 0.4P_{211} + 1.6P_{212}$$

$$3.6P_{120} = 0.8P_0 + 0.4P_{221} + 1.6P_{222}$$

$$2.4P_{101} = 0.2P_0 + 0.4P_{211} + 1.6P_{221}$$

$$3.6P_{102} = 0.8P_0 + 0.4P_{212} + 1.6P_{222}$$

$$0.8P_{211} = 0.4P_{101} + 0.4P_{110}$$

$$2P_{212} = 0.4P_{102} + 1.6P_{110}$$

$$2P_{221} = 1.6P_{101} + 0.4P_{120}$$

$$3.2P_{222} = 1.6P_{102} + 1.6P_{120}$$

Since $P_{110} = P_{101}$, $P_{120} = P_{102}$, and $P_{212} = P_{221}$, the number of equations can be reduced to

$$2P_0 = 0.8P_{110} + 3.2P_{120},$$

$$2.4P_{110} = 0.2P_0 + 0.4P_{211} + 1.6P_{221},$$

$$3.6P_{120} = 0.8P_0 + 0.4P_{221} + 1.6P_{222},$$

$$0.8P_{211} = 0.8P_{110}$$

$$2P_{221} = 1.6P_{110} + 0.4P_{120}$$

$$3.2P_{222} = 3.2P_{120}$$

Solving for P_{nij} in terms of P_0 gives
$$P_{110} = P_{120} = P_{211} = P_{221} = P_{222} = 0.5P_0$$
Therefore, since
$$P_0 + P_{110} + P_{101} + P_{120} + P_{102} + P_{211} + P_{221} + P_{212} + P_{222} = 1$$
then $P_0 = 0.2$ and $P_{nij} = 0.1$. Consequently,
$$L = 1.2 \text{ cars}, \qquad L_q = 0$$
$$W = 1 \text{ min/car}, \qquad W_q = 0$$
and
$$\tilde{\lambda} = \mu(L - L_q) = 1.2 \text{ cars/min}$$
or
$$\tilde{\lambda} = \lambda(1 - P_2) = 2(1 - 0.4) = 1.2 \text{ cars/min}$$

$(HE_2 \mid M \mid 1):(GD \mid N \mid \infty)$ Queue

We now turn our attention to hyperexponential interarrival times. Using the method of stages, we assume there always exist customers available to enter our arrival mechanism, which consists of two arrival branches. Arrival branch 1 is chosen with probability α and has parameter λ_1; arrival branch 2 is chosen with probability $(1 - \alpha)$ and has parameter λ_2. Exactly one customer is in the arrival mechanism at all times. Our analysis begins with a treatment of a single-server system having a capacity of N customers.

Let

P_{nj} = probability of n customers in the system with a customer in branch j of the arrival mechanism, $\quad j = 1, 2$

The steady-state equations for the system are
$$\lambda_1 P_{01} = \mu P_{11}, \qquad \lambda_2 P_{02} = \mu P_{12}$$
$$(\lambda_1 + \mu)P_{n1} = \alpha\lambda_1 P_{n-1,1} + \alpha\lambda_2 P_{n-1,2} + \mu P_{n+1,1} \qquad (0 < n < N)$$
$$(\lambda_2 + \mu)P_{n2} = (1-\alpha)\lambda_1 P_{n-1,1} + (1-\alpha)\lambda_2 P_{n-1,2} + \mu P_{n+1,2}$$
$$(\lambda_1 + \mu)P_{N1} = \alpha\lambda_1 P_{N-1,1} + \alpha\lambda_2 P_{N-1,2} + \alpha\lambda_1 P_{N1} + \alpha\lambda_2 P_{N2}$$
$$(\lambda_2 + \mu)P_{N2} = (1-\alpha)\lambda_1 P_{N-1,1} + (1-\alpha)\lambda_2 P_{N-1,2} + (1-\alpha)\lambda_1 P_{N1}$$
$$+ (1-\alpha)\lambda_2 P_{N2}$$

As in our previous analyses, we can solve for the values of P_{nj} once values are assigned to $\lambda_1, \lambda_2, \mu, \alpha$, and N.

Example 4.9 As an illustration of the computational approach for the $(HE_2 \mid M \mid 1):(GD \mid N \mid \infty)$ queue, consider a business phone in a massage

Method of Stages

parlor. Interarrival data were gathered and it was found that demands for the phone arrived at an average rate of 5/hr. The variance for the interarrival time was found to be 0.04133. Since the variance was greater than the square of the mean, a goodness of fit test was run on the data using an HE_2 distribution with $\alpha = 0.4$ and $\lambda = 5$ customers/hr. Based on the test results, it was concluded that the HE_2 distribution reasonably approximated the interarrival distribution. The length of a telephone call was observed to be exponentially distributed with a mean of 20 min. Furthermore, no customer waiting could take place. Consequently, for this example the data are: $\alpha = 0.4$, $\lambda_1 = 2\alpha\lambda$, $\lambda_2 = 2(1-\alpha)\lambda$, $\lambda = 5$, $\mu = 3$, and $N = 1$. Therefore,

$$4P_{01} = 3P_{11}, \qquad 7P_{11} = 1.6P_{01} + 2.4P_{02} + 1.6P_{11} + 2.4P_{12}$$

$$6P_{02} = 3P_{12}, \qquad 9P_{12} = 2.4P_{01} + 3.6P_{02} + 2.4P_{11} + 3.6P_{12}$$

Solving for the values of the steady-state probabilities gives

$$P_{01} = 9/42, \qquad P_{02} = 7/42, \qquad P_{11} = 12/42, \qquad P_{12} = 14/42$$

Therefore,

$$L = 26/42 \text{ customers}, \qquad L_q = 0$$

$$W = 1/3 \text{ hr}, \qquad W_q = 0$$

and

$$\tilde{\lambda} = \mu(L - L_q) = 13/7 \text{ customers/hr}$$

Observe that since arrivals are not Poisson distributed

$$\tilde{\lambda} \neq \lambda(1 - P_1)$$

$(HE_2 \mid M \mid 2) : (GD \mid N \mid \infty)$ Queue

Let us now consider a two-server generalization of our previous discussion of hyperexponential interarrival times. The steady-state balance equations are

$$\lambda_1 P_{01} = \mu P_{11}, \qquad \lambda_2 P_{02} = \mu P_{12}$$

$$(\lambda_1 + \mu)P_{11} = \alpha\lambda_1 P_{01} + \alpha\lambda_2 P_{02} + 2\mu P_{21}$$

$$(\lambda_2 + \mu)P_{12} = (1-\alpha)\lambda_1 P_{01} + (1-\alpha)\lambda_2 P_{02} + 2\mu P_{22}$$

$$(\lambda_1 + 2\mu)P_{n1} = \alpha\lambda_1 P_{n-1,1} + \alpha\lambda_2 P_{n-1,2} + 2\mu P_{n+1,1} \qquad (1 < n < N)$$

$$(\lambda_2 + 2\mu)P_{n2} = (1-\alpha)\lambda_1 P_{n-1,1} + (1-\alpha)\lambda_2 P_{n-1,2} + 2\mu P_{n+1,2}$$

$$(\lambda_1 + 2\mu)P_{N1} = \alpha\lambda_1 P_{N-1,1} + \alpha\lambda_2 P_{N-1,2} + \alpha\lambda_1 P_{N1} + \alpha\lambda_2 P_{N2}$$

$$(\lambda_2 + 2\mu)P_{N2} = (1-\alpha)\lambda_1 P_{N-1,1} + (1-\alpha)\lambda_2 P_{N-1,2} + (1-\alpha)\lambda_1 P_{N1}$$
$$+ (1-\alpha)\lambda_2 P_{N2}$$

Example 4.10 In order to illustrate the $(HE_2 \mid M \mid 2):(GD \mid N \mid \infty)$ queue, suppose a second phone is installed in the massage parlor described in Example 4.9. Therefore, the data are $\alpha = 0.4$, $\lambda_1 = 2\alpha\lambda$, $\lambda_2 = 2(1-\alpha)\lambda$, $\lambda = 5$, $\mu = 3$, and $N = 2$. The state equations reduce to

$$4P_{01} = 3P_{11}, \quad 6P_{02} = 3P_{12}$$

$$7P_{11} = 1.6P_{01} + 2.4P_{02} + 6P_{21}$$

$$9P_{12} = 2.4P_{01} + 3.6P_{02} + 6P_{22}$$

$$10P_{21} = 1.6P_{11} + 2.4P_{12} + 1.6P_{21} + 2.4P_{22}$$

$$12P_{22} = 2.4P_{11} + 3.6P_{12} + 2.4P_{21} + 3.6P_{22}$$

Solving for the values of the steady-state probabilities, we have

$$P_{01} = 261/1746, \quad P_{12} = 362/1746$$

$$P_{02} = 181/1746, \quad P_{21} = 264/1746$$

$$P_{11} = 348/1746, \quad P_{22} = 330/1746$$

Therefore,

$$P_0 = 442/1746, \quad P_1 = 710/1746, \quad P_2 = 594/1746$$

Consequently,

$$L = 1898/1746 \text{ customers}, \quad L_q = 0$$

and

$$\tilde{\lambda} = \mu(L - L_q), \quad \text{or} \quad \tilde{\lambda} = 5694/1746 \text{ customers/hr}$$

Of course, for this example,

$$W = 1/3 \text{ hr/customer}, \quad \text{and} \quad W_q = 0$$

$(HE_2 \mid E_2 \mid 1):(GD \mid 1 \mid \infty)$ Queue

We conclude our discussion of the method of stages technique for analyzing non-Poisson queues with an analysis of the $(HE_2 \mid E_2 \mid 1):(GD \mid 1 \mid \infty)$ queue. The system is represented graphically in Fig. 4.4. To facilitate the analysis, let

P_{njk} = probability of n customers in the system when the arriving customer is in arrival branch j and the customer in service is in service stage k, $n = 0, 1, j = 1, 2, k = 1, 2$

P_{0j} = probability of no customers in the system when the arriving customer is in arrival branch j, $j = 1, 2$

Method of Stages

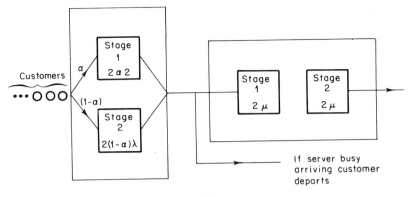

Fig. 4.4 Method of stages representation of the $(HE_2 \mid E_2 \mid 1):(GD \mid 1 \mid \infty)$ queue.

The steady-state equations for this queueing system are written as

$$\lambda_1 P_{01} = 2\mu P_{112}, \qquad \lambda_2 P_{02} = 2\mu P_{122}$$

$$(\lambda_1 + 2\mu)P_{111} = \alpha\lambda_1 P_{01} + \alpha\lambda_2 P_{02} + \alpha\lambda_1 P_{111} + \alpha\lambda_2 P_{121}$$

$$(\lambda_2 + 2\mu)P_{121} = (1-\alpha)\lambda_1 P_{01} + (1-\alpha)\lambda_2 P_{02} + (1-\alpha)\lambda_1 P_{111}$$
$$+ (1-\alpha)\lambda_2 P_{121}$$

$$(\lambda_1 + 2\mu)P_{112} = 2\mu P_{111} + \alpha\lambda_1 P_{112} + \alpha\lambda_2 P_{122}$$

$$(\lambda_2 + 2\mu)P_{122} = 2\mu P_{121} + (1-\alpha)\lambda_1 P_{112} + (1-\alpha)\lambda_2 P_{122}$$

Example 4.11 As an illustration of the computations involved in the analysis of the $(HE_2 \mid E_2 \mid 1):(GD \mid 1 \mid \infty)$ queue, recall the massage parlor described in Example 4.9. We now consider that situation, except the duration of a phone call is now E_2 distributed with a mean of 20 min. Thus, in this case the system parameters have the following values: $\lambda = 5$, $\alpha = 0.4$, $\lambda_1 = 2\alpha\lambda$, $\lambda_2 = 2(1-\alpha)\lambda$, and $\mu = 3$. Therefore,

$$4P_{01} = 6P_{112}, \qquad 6P_{02} = 6P_{122}$$

$$10P_{111} = 1.6P_{01} + 2.4P_{02} + 1.6P_{111} + 2.4P_{121}$$

$$12P_{121} = 2.4P_{01} + 3.6P_{02} + 2.4P_{111} + 3.6P_{121}$$

$$10P_{112} = 6P_{111} + 1.6P_{112} + 2.4P_{122}$$

$$12P_{122} = 6P_{121} + 2.4P_{112} + 3.6P_{122}$$

Replacing the last state equation with the equation

$$P_{01} + P_{02} + P_{111} + P_{121} + P_{112} + P_{122} = 1$$

and writing the state equations in matrix form results in the following formulation:

$$\begin{bmatrix} -4 & 0 & 0 & 0 & 6 & 0 \\ 0 & -6 & 0 & 0 & 0 & 6 \\ 1.6 & 2.4 & -8.4 & 2.4 & 0 & 0 \\ 2.4 & 3.6 & 2.4 & -8.4 & 0 & 0 \\ 0 & 0 & 6 & 0 & -8.4 & 2.4 \\ 1 & 1 & 1 & 1 & 1 & 1 \end{bmatrix} \begin{bmatrix} P_{01} \\ P_{02} \\ P_{111} \\ P_{121} \\ P_{112} \\ P_{122} \end{bmatrix} = \begin{bmatrix} 0 \\ 0 \\ 0 \\ 0 \\ 0 \\ 1 \end{bmatrix}$$

Using an appropriate matrix inversion routine, the state probabilities are obtained:

$$\begin{bmatrix} P_{01} \\ P_{02} \\ P_{111} \\ P_{121} \\ P_{112} \\ P_{122} \end{bmatrix} = \begin{bmatrix} 57/262 \\ 43/262 \\ 36/262 \\ 45/262 \\ 38/262 \\ 43/262 \end{bmatrix}$$

Therefore, the probability that the phone is idle is

$$P_0 = 100/262$$

and the probability that the phone is busy is

$$P_1 = 162/262$$

Consequently,

$$L = 162/262 \text{ customers}, \quad L_q = 0$$

and

$$\tilde{\lambda} = \mu(L - L_q) \text{ customers/hr}, \quad \text{or} \quad \tilde{\lambda} = 486/262$$

NUMERICAL SOLUTION OF STEADY-STATE BALANCE EQUATIONS

An obvious disadvantage of the method of stages approach to analyzing certain non-Poisson queues is the large number of steady-state balance equations that might need to be solved. As we pointed out in the previous section, the set of linear state equations can be solved using matrix inversion routines. One such procedure that we have found to be extremely efficient in solving a variety of sets of linear state equations is a linear iteration technique called the *method of successive displacements* or, alternately, the *Gauss–Seidel Method* (Cooper, 1972).

Numerical Solution of Steady-State Balance Equations

To motivate the discussion of the iteration technique, we will use the data from Example 4.5. Recall that the steady-state balance equations were given as follows:

$$P_{01} = \frac{\mu}{2\lambda} P_{11}, \qquad\qquad P_{02} = P_{01} + \frac{\mu}{2\lambda} P_{12}$$

$$P_{11} = \frac{2\lambda}{2\lambda + \mu} P_{02} + \frac{2\lambda}{2\lambda + \mu} P_{12}, \qquad P_{12} = \frac{2\lambda}{2\lambda + \mu} P_{11}$$

These equations were solved by using the *normalizing equation*

$$1 = P_{01} + P_{02} + P_{11} + P_{12}$$

In matrix notation, the balance equations can be expressed as

$$\mathbf{p} = \mathbf{pA}$$

where

$$\mathbf{p} = (P_{01}, P_{02}, P_{11}, P_{12}),$$

and

$$\mathbf{A} = \begin{pmatrix} 0 & 1 & 0 & 0 \\ 0 & 0 & \frac{2\lambda}{2\lambda + \mu} & 0 \\ \frac{\mu}{2\lambda} & 0 & 0 & \frac{2\lambda}{2\lambda + \mu} \\ 0 & \frac{\mu}{2\lambda} & \frac{2\lambda}{2\lambda + \mu} & 0 \end{pmatrix}$$

Note that $\mathbf{A}$ can be expressed as the sum of two matrices $\mathbf{C}$ and $\mathbf{D}$, where

$$\mathbf{C} = \begin{pmatrix} 0 & 0 & 0 & 0 \\ 0 & 0 & 0 & 0 \\ \frac{\mu}{2\lambda} & 0 & 0 & 0 \\ 0 & \frac{2\mu}{\lambda} & \frac{2\lambda}{2\lambda + \mu} & 0 \end{pmatrix} \qquad \mathbf{D} = \begin{pmatrix} 0 & 1 & 0 & 0 \\ 0 & 0 & \frac{2\lambda}{2\lambda + \mu} & 0 \\ 0 & 0 & 0 & \frac{2\lambda}{2\lambda + \mu} \\ 0 & 0 & 0 & 0 \end{pmatrix}$$

Thus, $\mathbf{p} = \mathbf{pC} + \mathbf{pD}$, which suggests the use of an iterative solution procedure called the method of successive displacements, $\mathbf{p}^{(k+1)} = \mathbf{p}^{(k)}\mathbf{C} + \mathbf{p}^{(k+1)}\mathbf{D}$. The vector $\mathbf{p}^{(k)}$ contains the estimates of the probabilities obtained on the kth iteration. Thus, the method of successive displacements consists of the following steps:

1. Choose an arbitrary initial-approximation vector:

$$\mathbf{p}^{(0)} = (p_1^{(0)}, p_2^{(0)}, \ldots, p_n^{(0)})$$

2. Generate successive approximations $p_i^{(k)}$ by the iteration

$$p_i^{(k+1)} = \sum_{j=1}^{i-1} c_{ij} p_j^{(k)} + \sum_{j=i}^{n} d_{ij} p_j^{(k+1)}$$

for $i = 1, 2, \ldots, n$ and $k = 0, 1, \ldots$.

3. Apply the normalizing equation, $\sum_{i=1}^{n} p_i = 1$, after each complete round of iterations by dividing each "probability" value by the sum of the "probability" values obtained during the iteration round.

4. Continue until an appropriate convergence criterion is satisfied. Typical convergence criteria include

(a) $\max\limits_{1 \le i \le n} \dfrac{|p_i^{(k+1)} - p_i^{(k)}|}{|p_i^{(k+1)}|} < \varepsilon$, for some prescribed ε

(b) $\sum\limits_{i=1}^{n} |p_i^{(k+1)} - p_i^{(k)}| < \alpha$, for some prescribed α

(c) $k = M$ for a prescribed integer M

For the example problem, the kth iteration can be expressed as follows:

$$P_{01}^{(k)} = 0 P_{01}^{(k-1)} + 0 P_{02}^{(k-1)} + \frac{\mu}{2\lambda} P_{11}^{(k-1)} + 0 P_{12}^{(k-1)} \tag{4.69}$$

$$P_{02}^{(k)} = 1 P_{01}^{(k)} + 0 P_{02}^{(k-1)} + 0 P_{11}^{(k-1)} + \frac{\mu}{2\lambda} P_{12}^{(k-1)} \tag{4.70}$$

$$P_{11}^{(k)} = 0 P_{01}^{(k)} + \frac{2\lambda}{2\lambda + \mu} P_{02}^{(k)} + 0 P_{11}^{(k-1)} + \frac{2\lambda}{2\lambda + \mu} P_{12}^{(k-1)} \tag{4.71}$$

$$P_{12}^{(k)} = 0 P_{01}^{(k)} + 0 P_{02}^{(k)} + \frac{2\lambda}{2\lambda + \mu} P_{11}^{(k)} + 0 P_{12}^{(k-1)} \tag{4.72}$$

To illustrate the method of successive displacements, suppose $\lambda = 1$ and $\mu = 2$ and let the following initial probability estimates be employed: $P_{01}^{(0)} = P_{02}^{(0)} = P_{11}^{(0)} = P_{12}^{(0)} = 1/4$. From (4.69) the value of $P_{01}^{(1)}$ is obtained as follows:

$$P_{01}^{(1)} = P_{11}^{(0)} = 1/4$$

Thus, from (4.70)–(4.72), respectively,

$$P_{02}^{(1)} = P_{01}^{(1)} + P_{12}^{(0)} = 1/2$$
$$P_{11}^{(1)} = 0.5P_{02}^{(1)} + 0.5P_{12}^{(0)} = 3/8$$
$$P_{12}^{(1)} = 0.5P_{11}^{(1)} = 3/16$$

Applying the normalizing equation yields the following revised estimates of the probabilities:

$$P_{01}^{(1)} = 4/21, \quad P_{02}^{(1)} = 8/21, \quad P_{11}^{(1)} = 6/21, \quad P_{12}^{(1)} = 3/21$$

On the second iteration, the estimates of the "probabilities" are revised as follows:

$$P_{01}^{(2)} = 6/21, \quad P_{02}^{(2)} = 9/21, \quad P_{11}^{(2)} = 12/42, \quad P_{12}^{(2)} = 6/42$$

Applying the normalizing equation yields the following values of the probabilities:

$$P_{01}^{(2)} = 1/4, \quad P_{02}^{(2)} = 3/8, \quad P_{11}^{(2)} = 1/4, \quad P_{12}^{(2)} = 1/8$$

On the third iteration the following estimates of the probabilities are obtained:

$$P_{01}^{(3)} = 1/4, \quad P_{02}^{(3)} = 3/8, \quad P_{11}^{(3)} = 1/4, \quad P_{12}^{(2)} = 1/8$$

Since no changes occurred in the values from the second to the third iteration, the desired probability values have been obtained, namely,

$$P_{01} = 1/4, \quad P_{02} = 3/8, \quad P_{11} = 1/4, \quad \text{and} \quad P_{12} = 1/8$$

Variations of the method of successive displacements are discussed by Cooper (1972) as well as convergence conditions for the iteration method. In particular, the normalizing equation can be applied after the convergence criterion has been satisfied. Another variation is explored in the exercises at the end of the chapter.

SUMMARY

In this chapter, we have presented a number of examples of the method of stages technique for analyzing non-Poisson queues. Additionally, the Pollaczek–Khintchine formula for the $(M \mid G \mid 1):(GD \mid \infty \mid \infty)$ queue was presented. By no means have we treated the subject of non-Poisson queues exhaustively. However, we have examined the majority of non-Poisson queues that can be treated simply. Whenever you wish to analyze queueing systems that are more complex than those we have examined, either more complex analytical tools are required or you should employ the Monte Carlo simulation techniques discussed in Chapter 9.

It should be emphasized that the class of non-Poisson queues we addressed in this chapter is quite "rich," in that a wide range of distribution shapes can be accurately represented by a gamma or a hyperexponential distribution. Once the appropriate distribution is identified and the corresponding steady-state balance equations are written, say, in matrix form, then a matrix inversion routine can be employed to determine the state probabilities. One such approach, the method of successive displacements, was presented for obtaining the values of the state probabilities. Knowing the values of the state probabilities the values of the system's operating characteristics are easily obtained.

REFERENCES

Cooper, R. B., *Introduction to Queueing Theory*. New York: Macmillan, 1972.
Cox, D. R., and Smith, W. L., *Queues*. London: Methuen, 1961.
Gue, R. L., and Thomas, M. E., *Mathematical Methods in Operations Research*. New York: Macmillan, 1968.
Kendall, D. G., Some problems in the theory of queues, *J. Roy. Statist. Soc., Ser. B* **13**, 151–185, 1951.
Morse, P., *Queues, Inventories, and Maintenance*. New York: Wiley, 1958.
Ruiz-Pala, E., Avila-Beloso, C., and Hines, W. W., *Waiting-Line Models: An Introduction to Their Theory and Application*. New York: Reinhold, 1967.
Saaty, T. L., *Elements of Queueing Theory*. New York: McGraw-Hill, 1961.
Schmidt, J. W., and Taylor, R. E., *Simulation and Analysis of Industrial Systems*. Homewood, Illinois: Irwin, 1970.

PROBLEMS

1. Patients arrive in a Poisson fashion at the emergency room of a local hospital. The average arrival rate is 4/hr. The time required to admit an emergency patient is uniformly distributed over the interval from 5 through 15 min. There is only one admission clerk at the hospital during the late evening shift. During the shift, what is the expected amount of time a patient spends in the hospital?

2. In the University Computer Center, it has been found that students arrive to submit or pick up jobs in a manner that may be regarded as Poisson. The average arrival rate is 4/min. There is one person available to handle the student requests. This person is found to spend T minutes in servicing a student, where time-study results have yielded the following distribution for T:

T (min)	%
0.1	20
0.2	40
0.3	30
0.4	10

Using the observed distribution for service time, what is the average time a student spends in the system? What is the average length of the waiting line?

3. Arrivals at a cashier's window in a local bank meet the conditions of a Poisson process. The mean interarrival time is 15 min. Service times are gamma distributed with a mean of 6 min and a standard deviation of 2 min. During steady state, determine (a) the expected waiting time per customer, and (b) the variance for the number of customers at the window, including the one being served

4. Customers arrive at an information desk in a department store in a Poisson fashion at a rate of 15/hr. Service time, measured in minutes, is uniformly distributed over the interval (0, 3). Assuming steady-state operations, what are the values for the mean and variance of the number of customers at the desk, including the one being served?

5. Service times per customer by a local M.D. are believed to be uniformly distributed in the interval (0, 5) min. Arrivals are Poisson distributed. The mean time between arrivals is five times the mean service time. Find (a) the mean and variance for the number of patients in the doctor's office, and (b) the expected waiting time per customer.

6. Arrivals at a truck-weighing station are Poisson distributed. The mean arrival rate is 5/hr. The service time is a constant of 6 min/truck. Find (a) the mean and variance for the number of trucks at the weighing station, and (b) the probability that a truck will have to wait for service.

7. Arrivals at a tollgate on a freeway are Poisson distributed with a mean of 1.5 vehicles/min. Service time in the exact-change line is constant and equals 30 sec. Determine the values of the operating characteristics for the system. Compare your answers with those obtained from Problem 3.48.

8. Customers arrive at the cashier's station in a restaurant at a Poisson rate of 10/hr. Service time is normally distributed with a mean of 2 min and a standard deviation of 0.25 min. Determine the values of the operating characteristics for the system.

9. Parts arrive at an inspection station at a Poisson rate of 20/hr. Currently inspection is performed manually, with inspection time normally distributed with a mean of 2 min and a standard deviation of 0.50 min. It has been proposed that the inspection be performed automatically. With automatic inspection, the inspection time will be constant and equal to X min. Compare the operating characteristics for the manual inspection system with those for the automatic inspection system when X equals (a) 1.5, (b) 2, (c) 2.5, (d) 30 min.

10. Solve Problem 4.1 assuming admission time is Erlang distributed with a mode of 10 min and a mean of 15 min.

11. Solve Problem 4.2 assuming service time is Erlang distributed with a mean of 0.25 min and a standard deviation of 0.125 min.

12. In Problem 4.11, what is the probability that a departing customer can look back and see someone waiting for service? What is the probability that an arriving customer will be delayed?

13. Write the steady-state balance equations for the $(M|E_2|2):(GD|2|\infty)$ queue assuming server 1 always provides service to an arriving customer when both servers are idle at the time a customer arrives. Will the values of L, L_q, W, and W_q be affected by this change in the internal operation of the system?

14. In Example 4.3 let $\mu_1 = 4$, $k_1 = 2$, $\mu_2 = 6$, $k_2 = 2$, and $\lambda = 10$. If both servers are idle when a call arrives, server 2 handles the call. Determine the values of the operating characteristics for the system. What is the probability that server 1 will be idle? that server 2 will be idle? that both servers will be idle simultaneously?

15. In Example 4.4, let $N = 3$ and determine the values of the operating characteristics for the system.

16. Write out the state equations for the $(E_K|M|1):(GD|N|\infty)$ queue.

17. Modify the state equations for the $(E_2|M|2):(GD|N|\infty)$ queue to allow for nonidentical servers. Let server 1 be chosen for service with probability p when both servers are available for service; otherwise, assume the customer is served by the available server.

18. Compare the results from Example 4.6 with those obtained by letting $N = 2$, $\lambda = 2$, and $\mu = 2$.

19. Solve Example 4.6 for the case of $N = 4$, $\lambda = 1$, and $\mu = 1$.

20. Plot the expected number of customers waiting for service versus α and β based on Eq. (4.14).

21. Determine the mode of the distribution given in Eq. (4.20).

22. Plot Eq. (4.36) showing the effect of changes in k and $\rho = \lambda/\mu$ on L.

23. Plot Eq. (4.40) showing the effect of ρ on L.

24. Solve Example 4.7 for the case of $\mu = 15/\text{hr}$, $\lambda = 10/\text{hr}$, and (a) $\alpha = 0.10$, (b) $\alpha = 0.25$, (c) $\alpha = 0.40$.

25. Solve Example 4.8 for the case of $\alpha = 0.2$, $\beta = 1.0$, $\mu_{11} = 0.8$, $\mu_{12} = 0.4$, $\mu_{21} = 3.2$, $\mu_{22} = 1.6$, and $\lambda = 2$. What physical interpretation can be given to the values of $\beta, \mu_{11}, \mu_{12}, \mu_{21}$, and μ_{22} in this case?

26. Consider a queueing system in which arrivals are Poisson distributed and service times are gamma distributed. In particular, consider the $(M|\Gamma|2):(GD|N|\infty)$ queue with service time having the following probability density function:

$$f(t_s) = \frac{\mu^n}{(n-1)!} t_s^{n-1} \exp(-\mu t_s), \qquad t_s > 0$$

where n is integer valued. Write out the state equations for this system.

27. Consider the $(M|E_2|1):(GD|2|\infty)$ queue in which $\lambda = \mu = 5$ and $k = 2$. Determine the steady-state values for L, L_q, W, W_q, and D.

28. A repairman is responsible for maintaining two identical machines. Service time is exponentially distributed with a mean of 20 min. Machine running time (in hours) has the following probability density function:

$$f(t_r) = 16 t_r \exp(-4t_r), \qquad t_r > 0$$

Under steady-state conditions, determine the values for L, L_q, W, W_q, and D.

29. Employ the method of stages in analyzing a single-server queue in which interarrival time is gamma distributed with a mean of 10 min and a variance of 50. Service time is exponentially distributed with a mean of 6 min. Assume no waiting space exists. Develop the steady-state difference equations describing the system. Determine the probability that the server is idle.

30. Consider the queueing network shown below. Station A always has available raw material for processing. Service times are exponentially distributed for all stations except station G; its service time is normally distributed with a mean of 10 min and a standard deviation of 2 min. All stations are manned by a single operator. Travel time between stations is negligible. The average service times are shown on the network, along with the probabilities that a unit of product moves along a particular arc. Assume steady-state operations.

(a) What is the expected amount of time required for a unit to arrive at station F? Measure time from the beginning of service at station A until the arrival at station F, given that a unit arrives at station F.

(b) What problems will develop in the network? Why?
(c) What is the expected number of units waiting to be processed at station G?

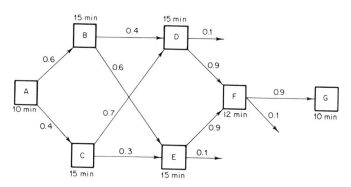

31. Cooper (1972) suggests the following variation of the method of successive displacements:

$$p^{(k+1)} = \omega[p^{(k)}C + p^{(k+1)}D] + (1 - \omega)p(k)$$

The resulting iteration is referred to as the method of successive *overrelaxation* when $\omega \geq 1$ and the method of successive *underrelaxation* when $\omega < 1$. Cooper suggests that $\omega \approx 1.3$ often yields better results than $\omega = 1$. Apply the method of successive overrelaxation in obtaining the values of the probabilities for Example 4.5; let $\omega = 1.2, 1.3, 1.5$.

Chapter 5

DECISION MODELS

INTRODUCTION

In the two preceding chapters we have discussed and presented the development of mathematical models that can be used to *describe* the behavior of a variety of queueing systems. The descriptive models were developed to assist us in analyzing various operating characteristics of the system under study.

It should be evident that the model required for a particular analysis will depend on both the characteristics of the system and the objectives to be achieved. For example, suppose that the probability mass function of the total number of units in a system is to be determined. The model representing this distribution will depend on such system characteristics as the number of service channels in parallel and/or series, the density functions of service time per service channel, and the distribution of customer interarrival times, to mention but a few. Thus, although the objective of the analysis may be the same for several different systems, the expression for the model will be different for systems having different characteristics. In a similar manner, different objectives for the same system will dictate different models.

In large measure, the models presented thus far have had as their objective the description of some characteristic of the queueing system investigated; that is, we have been concerned with describing mathematically the behavior of certain characteristics of the system. Frequently, however the objectives of the analysis call for more than describing the behavior o some characteristic of a system. For example, suppose that the purpose i

Introduction

analyzing a particular queueing system was to prescribe the mean service rate for each service channel in the system that would minimize the expected total annual cost resulting from the operation. The first task would be to develop a model representing expected total annual cost as a function of the service rates for the service channels included in the system. Once a model that estimates the characteristic expected total annual cost is developed, we are faced with the problem of operating on the model in a manner that will prescribe those service rates that will yield the minimum expected total annual cost.

The example just cited raises two important points. Every commercial enterprise must concern itself with the cost of operations and the profit realized as a result of those operations. Hence, the queueing analyst should have a fundamental understanding of and exposure to the development of cost and profit models for waiting-line systems. The second point relates to the manipulation of the model once it is developed; that is, the analyst should be able to operate on the model in a manner that will allow him to prescribe a strategy of operation that will lead to minimum cost or maximum profit depending on the objective of the analysis.

The purpose of this chapter is twofold: first, to present the fundamental concepts and methods of classical and exploratory optimization techniques; and second, to introduce the reader to decision modeling in the queueing environment. The first section of the chapter will be devoted to optimization theory. Here we will discuss classical optimization of unconstrained systems and optimization of systems through the application of search techniques. Classical methods using the derivative are probably the most efficient means of finding the optimum values of the decision variables for a given model where they apply. However, quite frequently the model to be optimized is sufficiently complex that the computational difficulties arising from the use of classical optimization techniques render the application of classical methods inefficient or ineffective. In such cases, the analyst may wish to seek an optimum solution through one of many available search techniques.

The second part of the chapter deals with the development of decision models to assist the decision maker in designing a queueing system in a manner consistent with his value system. The decision models we consider are prescriptive models; thus, an objective function is required that is consistent with the objectives of the manager. Initially, we assume the manager is interested in minimizing the expected total cost for the queueing system. As an alternative, we then consider the use of an aspiration level approach in which the system is designed to provide a specified level of performance. Finally, we examine the process of obtaining values for the cost parameters used in the decision models.

CLASSICAL OPTIMIZATION

Continuous Variables

Consider the problem of finding the value of the decision variable x that minimizes or maximizes the continuous function $f(x)$, where $-\infty < x < \infty$. If x^* is a value of x that either minimizes or maximizes $f(x)$, then x^* is called an *extreme point* for $f(x)$. For the present, let us assume that there is one and only one finite or *interior extreme point* x^* for $f(x)$. If x^* is a minimum for $f(x)$, then

$$f(x^*) < f(x) \tag{5.1}$$

for all $x \neq x^*$, and if x^* is a maximum, then

$$f(x^*) > f(x) \tag{5.2}$$

for all $x \neq x^*$. Functions having a single interior maximum and minimum are shown in Figs. 5.1 and 5.2.

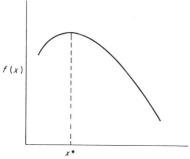

Fig. 5.1 Function having a single interior maximum at x^*.

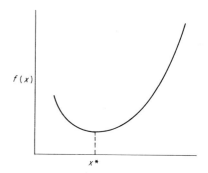

Fig. 5.2 Function having a single interior minimum at x^*.

Our problem is to determine a method through which the extreme point or points of a given function can be identified. Suppose that $f(x)$ is continuous with first derivative $f'(x)$ defined for all x. Then $f'(x)$ measures the slope of the function at x. Now consider the function illustrated in Fig. 5.3. As shown, $f(x)$ has a minimum at x^*. To the left of x^*, $f'(x)$ is negative, and to the right, positive. Hence, $f'(x)$ must pass through 0 at x^*. By a similar argument, if x^* is an interior maximum for $f(x)$, then $f'(x)$ is positive to the left of x^*, negative to the right of x^*, and zero at x^*. Hence, if x^* is an interior extreme point for $f(x)$, then

$$f'(x^*) = 0 \tag{5.3}$$

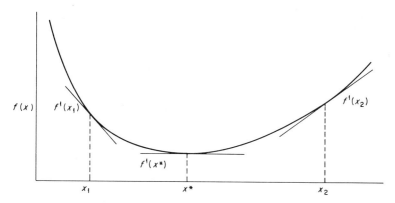

Fig. 5.3 Function having negative, zero, and positive slopes.

The condition given in (5.3) is said to be *necessary* for the existence of an extreme point at x^*. However, not every point x^* satisfying (5.3) is necessarily an extreme point. To illustrate, consider the functions shown in Fig. 5.4. For both of the functions shown there exist points x_1^* and x_2^* satisfying (5.3). However, neither point is an extreme point. Observe that in both cases in Fig. 5.4 $f'(x)$ does not change sign as x passes through the points x_1^* and x_2^*. If x^* is a point such that $f'(x^*) = 0$ but $f'(x)$ does not change sign at x^*, then x^* is called a *point of inflection*. A point satisfying (5.3) is called a *stationary point*, whether it is a maximum, minimum, or point of inflection.

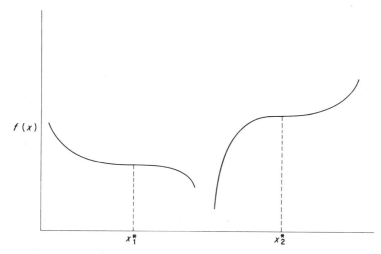

Fig. 5.4 Functions having points of inflection at x_1^* and x_2^*, respectively.

As we have already noted, a point satisfying the necessary condition for an extreme point given in (5.3) may or may not in fact be an extreme point. In addition, if the point x^* is an extreme point, the necessary condition for an extreme point tells us nothing about the nature of the point, that is, whether it is a maximum or a minimum. To resolve these problems we use the second derivative of $f(x)$ evaluated at x^*. Recall that $f'(x)$ changes from a positive to a negative at x^* if x^* is an interior maximum for $f(x)$. Hence, $f'(x)$ is a decreasing function near x^*, or the slope of $f'(x)$ is negative at all points sufficiently close to x^* and zero at x^* itself. In other words, if $f''(x^*) < 0$, x^* is an interior maximum. Alternatively, if $f''(x^*) > 0$, x^* is an interior minimum for $f(x)$ by a similar argument. If $f''(x^*) = 0$, x^* may be an interior maximum, minimum, or a point of inflection.

TABLE 5.1 Necessary and sufficient conditions for identifying and determining the nature of a stationary point x^*

Condition	Analysis	Conclusion
Necessary	$f'(x^*) = 0$	Stationary point
Sufficient	$f^i(x^*) = 0, i = 2, 3, \ldots, m - 1$	Minimum if m even and $f^m(x^*) > 0$
	$f^m(x^*) \neq 0$	Maximum if m even and $f^m(x^*) < 0$
		Point of inflection if m odd

If $f''(x^*) = 0$, we continue to take higher-order derivatives to determine the nature of x^*. Assume that $f^i(x^*) = 0$, $i = 2, 3, \ldots, m - 1$, and $f^m(x^*) \neq 0$. If m is odd, then x^* is a point of inflection. If m is even and $f^m(x^*) < 0$, then x^* is an interior maximum, and if m is even and $f^m(x^*) > 0$, it is an interior minimum for $f(x)$. Thus, if x^* is a point such that $f'(x^*) = 0$, we find the first higher-order derivative of $f(x)$ at x^* that is nonzero. If the first higher-order derivative evaluated at x^* is of odd order, x^* is a point of inflection, and if the first higher-order derivative at x^* is of even order, it is an extreme point. These conditions are said to be *sufficient* for determining the nature of a point satisfying (5.3). The necessary and sufficient conditions for identifying and determining the nature of a stationary point are summarized in Table 5.1.

Example 5.1 Find and determine the nature of the stationary point for the following function:

$$f(x) = \exp[-(x - K)^2], \quad -\infty < x < \infty$$

Classical Optimization

From Table 5.1, x^* is a stationary point for $f(x)$ if $f'(x^*) = 0$. Taking the first derivative of $f(x)$, we have

$$f'(x) = -2(x - K)\exp[-(x - K)^2]$$

Now $f'(x) = 0$ for $x = K$. Hence, the stationary point for $f(x)$ is $x^* = K$. To determine the nature of x^*, we evaluate the second derivative of $f(x)$ at x^*:

$$f''(x) = -2\exp[-(x - K)^2] + 4(x - K)^2 \exp[-(x - K)^2]$$

and

$$f''(x^*) = -2$$

Since $f''(x^*) < 0$ and the order of this derivative is even, x^* is an interior maximum for $f(x)$.

Thus far our discussion has been limited to the case where the function under consideration has only one stationary point. It should be obvious that a function may contain several stationary points. An illustration of such a function is given in Fig. 5.5, where maxima are located at x_1^* and x_4^*, minima at x_2^* and x_6^*, and points of inflection at x_3^* and x_5^*.

Since the analyst may have no way of knowing a priori how many stationary points a given function has, it is generally necessary to identify all those points satisfying (5.3), stationary points, and to determine the nature of each using the sufficient conditions summarized in Table 5.1. Suppose

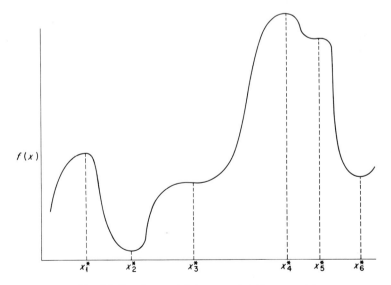

Fig. 5.5 Function containing several stationary points.

that a function has m minima, called *local minima*, and n maxima, called *local maxima*, and we wish to find the point x_1^* that yields the least value of $f(x)$, the *global minimum*, and the point x_2^* that yields the maximum value of $f(x)$, the *global maximum*. To resolve this problem we can evaluate $f(x)$ at each local minimum and each local maximum to identify the global minimum and global maximum.

Example 5.2 Find and identify the nature of each stationary point for the function

$$f(x) = 3x^4 - 4x^3$$

The stationary points for $f(x)$ are given by

$$f'(x) = 12x^3 - 12x^2 = 0, \quad \text{and} \quad x^* = 0, 1$$

The second derivative of $f(x)$ is given by

$$f''(x) = 36x^2 - 24x$$

At $x^* = 0$,

$$f''(x^*) = 0$$

and at $x^* = 1$,

$$f''(x^*) = 12$$

Since $f''(1) > 0$, $x^* = 1$ is a local minimum. However, the nature of $x^* = 0$ cannot be determined at this point, since $f''(0) = 0$. Taking the third derivative, we have

$$f^3(x) = 72x - 24, \quad \text{and} \quad f^3(0) = -24$$

Since $f^3(0) \neq 0$ and the first higher-order derivative is of odd order, $x^* = 0$ is a point of inflection.

More often than not, the system to be optimized is a function of several decision variables rather than a single decision variable. Let $f(x_1, x_2, \ldots, x_n)$ be a function of the n decision variables $x_1, x_2, \ldots, x_n$. The procedure for finding and identifying the extreme points for $f(x_1, x_2, \ldots, x_n)$ is similar to that employed for functions of a single variable. We first identify the stationary points for $f(x_1, x_2, \ldots, x_n)$ using first partial derivatives and then identify the nature of each stationary point using second partial derivatives.

Let $x_1^*, x_2^*, \ldots, x_n^*$ be a stationary point for $f(x_1, x_2, \ldots, x_n)$. Then

$$\frac{\partial}{\partial x_i} f(x_1^*, x_2^*, \ldots, x_n^*) = 0, \quad i = 1, 2, \ldots, n \quad (5.4)$$

Classical Optimization

As in the case of functions of a single decision variable, a point satisfying (5.4) may or may not be an extreme point. In this sense (5.4) is a necessary condition for the existence of an extreme point at $x_1^*, x_2^*, \ldots, x_n^*$; that is, if $x_1^*, x_2^*, \ldots, x_n^*$ is an extreme point, then it must satisfy (5.4). If the stationary point satisfying (5.4) is not an extreme point then it is referred to as a *saddle point*. An illustration of a saddle point is shown in Fig. 5.6.

To determine whether a stationary point is a local minimum, a local maximum, or a saddle point, we evaluate all of the second partial derivatives of $f(x_1, x_2, \ldots, x_n)$ at the stationary point. Let

$$f_{ij}^* = \frac{\partial^2}{\partial x_i \, \partial x_j} f(x_1^*, x_2^*, \ldots, x_n^*), \qquad i, j = 1, 2, \ldots, n \qquad (5.5)$$

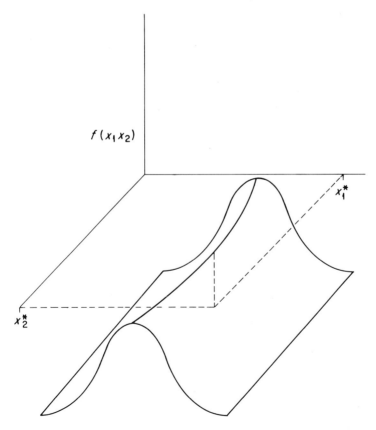

Fig. 5.6 Function $f(x_1, x_2)$ having a saddle point at x_1^*, x_2^*.

and let H be the matrix of all second partial derivatives evaluated at the stationary point, defined by

$$H = \begin{bmatrix} f_{11}^* & f_{12}^* & \cdots & f_{1n}^* \\ f_{21}^* & f_{22}^* & \cdots & f_{2n}^* \\ \vdots & & & \\ f_{n1}^* & f_{n2}^* & & f_{nn}^* \end{bmatrix} \quad (5.6)$$

The matrix H is called the *Hessian*. The ith leading principal minor of H, denoted H_i, is defined as the determinant of the first i rows and columns of H:

$$H_i = \begin{vmatrix} f_{11}^* & f_{12}^* & \cdots & f_{1i}^* \\ f_{21}^* & f_{22}^* & \cdots & f_{2i}^* \\ \vdots & & & \\ f_{i1}^* & f_{i2}^* & \cdots & f_{ii}^* \end{vmatrix} \quad (5.7)$$

If $H_i > 0$, $i = 1, 2, \ldots, n$, then the stationary point $x_1^*, x_2^*, \ldots, x_n^*$ is a local minimum for $f(x_1, x_2, \ldots, x_n)$. If $(-1)^i H_i > 0$, $i = 1, 2, \ldots, n$, then the stationary point is a local maximum. If $H_i \not> 0$, $i = 1, 2, \ldots, n$, $(-1)^i H_i \not> 0$, $i = 1, 2, \ldots, n$, and $H_i \neq 0$, $i = 1, 2, \ldots, n$, then $x_1^*, x_2^*, \ldots, x_n^*$ is a saddle point. If $H_i = 0$ for some i, then the nature of the stationary point is undetermined and may be a saddle point or an extreme point. The necessary and sufficient conditions for identifying and determining the nature of a stationary point are summarized in Table 5.2.

TABLE 5.2 Necessary and sufficient conditions for identifying and determining the nature of a stationary point $x_1^*, x_2^*, \ldots, x_n^*$

Condition	Analysis	Conclusion
Necessary	$\dfrac{\partial}{\partial_i} f(x_1^*, x_2^*, \ldots, x_n^*) = 0$	Stationary point
Sufficient	$H_i > 0, \quad i = 1, 2, \ldots, n$	Minimum
	$(-1)^i H_i > 0, \quad i = 1, 2, \ldots, n$	Maximum
	$\left. \begin{array}{l} H_i \not> 0, \quad i = 1, 2, \ldots, n \\ (-1)^i H_i \not> 0, \quad i = 1, 2, \ldots, n \\ H_i \neq 0, \quad i = 1, 2, \ldots, n \end{array} \right\}$	Saddle point
	$H_i = 0, \quad \text{some } i$	None

Example 5.3 Find and identify the stationary points for the following function:

$$f(x_1, x_2) = x_1^3 + x_2^3 + 2x_1^2 + 4x_2^2$$

Classical Optimization

From the necessary conditions, we have

$$\frac{\partial f(x_1, x_2)}{\partial x_1} = 3x_1^2 + 4x_1, \qquad \frac{\partial f(x_1, x_2)}{\partial x_2} = 3x_2^2 + 8x_2$$

Setting both partial derivatives equal to zero and solving for x_1^* and x_2^* yields

$$x_1^* = 0, -\tfrac{4}{3}, \qquad x_2^* = 0, -\tfrac{8}{3}$$

Hence, the stationary points for this function are given by the vectors

$$\begin{bmatrix} 0 \\ 0 \end{bmatrix}, \begin{bmatrix} 0 \\ -8/3 \end{bmatrix}, \begin{bmatrix} -4/3 \\ 0 \end{bmatrix}, \begin{bmatrix} -4/3 \\ -8/3 \end{bmatrix}$$

which will be denoted by X_1^*, X_2^*, X_3^*, and X_4^*, respectively. The second partial derivatives of $f(x_1, x_2)$ are given by

$$\frac{\partial^2 f(x_1, x_2)}{\partial x_1^2} = 6x_1 + 4, \qquad \frac{\partial^2 f(x_1, x_2)}{\partial x_2^2} = 6x_2 + 8,$$

$$\frac{\partial^2 f(x_1, x_2)}{\partial x_1 \, \partial x_2} = 0$$

The Hessian matrix is then given by

$$H = \begin{bmatrix} 6x_1^* + 4 & 0 \\ 0 & 6x_2^* + 8 \end{bmatrix}$$

The values of the leading principal minors for each stationary point are summarized in Table 5.3.

TABLE 5.3

i	X_i	H_1	H_2
1	X_1^*	4	32
2	X_2^*	4	−32
3	X_3^*	−4	−32
4	X_4^*	−4	32

For X_1^*, $H_1 > 0$ and $H_2 > 0$, and X_1^* is therefore a local minimum. Since $H_1 > 0$, $H_2 < 0$ for X_2^*, X_2^* is a saddle point, as is X_3^* since $H_1 < 0$ and $H_2 < 0$. X_4^* is a local maximum, since $H_1 < 0$ and $H_2 > 0$.

Constrained Optimization

The optimization techniques presented thus far are appropriate where the values that the decision variables may assume are unlimited. However, one often finds that the decision variables with which he must deal are constrained. For example, suppose that we were to determine the optimum service rate for a particular single-service channel. It is obvious that the service rate cannot be negative. In addition, there may be a technological upper limit b on the service rate. Therefore, if μ^* is the optimum service rate, then $0 \leq \mu^* \leq b$.

In many cases one can treat a constrained problem as though it were unconstrained; that is, we apply the techniques already presented and, after determining the optimum unconstrained values of the decision variables, determine whether any of the constraints are violated. If the resulting unconstrained optimum does not violate any of the constraints, then this solution is also the optimum for the constrained problem. However, if the unconstrained optimum violates any of the constraints, the analyst must resort to more sophisticated methods of analysis, such as the method of Lagrange multipliers or one of several mathematical programming techniques. For a discussion of these techniques, the reader should see Schmidt (1974), Gue and Thomas (1968), Wilde and Beightler (1967), and Beveridge and Schechter (1970), among others.

Discrete Variables

In this section we will develop methods for identifying interior extreme points for functions of a single discrete variable. We will deal only with functions of integer-valued variables, although the methods presented can be generalized to include functions of variables that are discrete but not necessarily integer valued. Suppose that $f(x)$ is a function of the integer-valued variable x, with a local minimum at x^*. Then

$$f(x^* - 1) \geq f(x^*) \leq f(x^* + 1) \tag{5.8}$$

Now let $\Delta f(x)$ be the *first forward difference*, defined by

$$\Delta f(x) = f(x + 1) - f(x) \tag{5.9}$$

If x^* is a local minimum

$$\Delta f(x^*) = f(x^* + 1) - f(x^*) \geq 0 \tag{5.10}$$

and

$$\Delta f(x^* - 1) = f(x^*) - f(x^* - 1) \leq 0 \tag{5.11}$$

Thus, at a local minimum x^*,

$$\Delta f(x^* - 1) \leq 0 \leq \Delta f(x^*) \tag{5.12}$$

Now let us consider the case where x^* is a local maximum. In this case
$$f(x^* - 1) \leq f(x^*) \geq f(x^* + 1) \tag{5.13}$$
or
$$\Delta f(x^*) \leq 0 \tag{5.14}$$
and
$$\Delta f(x^* - 1) \geq 0 \tag{5.15}$$
Hence, for a local maximum x^*,
$$\Delta f(x^* - 1) \geq 0 \geq \Delta f(x^*) \tag{5.16}$$

Example 5.4 Let
$$f(x) = \frac{\lambda^x}{x!} e^{-\lambda}, \qquad x = 0, 1, 2, \ldots$$
Find the value of x that maximizes $f(x)$.

For this function, $\Delta f(x)$ is given by
$$\Delta f(x) = \frac{\lambda^{x+1}}{(x+1)!} e^{-\lambda} - \frac{\lambda^x}{x!} e^{-\lambda} = \frac{\lambda^x}{x!} e^{-\lambda} \left(\frac{\lambda}{x+1} - 1 \right)$$
If x^* is a maximum for $f(x)$, then
$$\frac{\lambda^{x^*-1}}{(x^*-1)!} e^{-\lambda} \left(\frac{\lambda}{x^*} - 1 \right) \geq 0 \geq \frac{\lambda^{x^*}}{x^*!} e^{-\lambda} \left(\frac{\lambda}{x^*+1} - 1 \right)$$
by (5.16), and
$$\left(\frac{\lambda}{x^*} - 1 \right) \geq 0 \geq \frac{\lambda}{x^*} \left(\frac{\lambda}{x^*+1} - 1 \right)$$
or
$$(\lambda - x^*) \geq 0 \geq \lambda \left(\frac{\lambda}{x^*+1} - 1 \right)$$

Therefore x^* is the largest integer value of x such that $x \leq \lambda$.

SEARCH TECHNIQUES

In the preceding sections of this chapter we have explored analytic methods for optimizing a function of one or more decision variables. However, these methods are sometimes difficult to apply, particularly when the function to be optimized is complex. To illustrate, consider the following simple function:

$$f(x) = x^2 + e^{-x} \tag{5.17}$$

Taking the first derivative yields

$$f'(x) = 2x - e^{-x} \tag{5.18}$$

Setting $f'(x)$ equal to zero, we have

$$2x - e^{-x} = 0 \tag{5.19}$$

It is readily seen that we cannot solve for the stationary value of x directly. In this case, we could apply rather simple numerical techniques to (5.19) to find the stationary value of x. However, as the complexity of the function to be optimized and the number of decision variables included increase, the application of numerical techniques to the resolution of problems of the type posed by the function in (5.17) is not so straightforward.

To resolve the dilemma posed by functions of the type just discussed, the analyst can resort to the application of a search technique. To illustrate what is meant by a search technique, consider the problem of a man attempting to reach the peak of a hill in a dense fog. Since he cannot identify a direction of increasing altitude visually he must take a step in an arbitrary direction. If the initial step does not lead to an increase in elevation, he must explore by taking a step in another direction, continuing to explore in this manner until a favorable direction is ascertained. He would then move off in that direction until he finds that no further improvement in elevation is achieved. At this point he would again explore to determine a new direction that will increase his elevation. This process is repeated until he reaches a point where exploration tells him that he cannot further improve his position. This point would be accepted as the peak that he sought to achieve.

Search techniques operate, in large measure, in a manner similar to that described for the man climbing a hill in a fog. The function to be optimized is evaluated at a starting point, defined by values of the decision variables. The values of one or more decision variables are then changed, yielding a new point, and the objective function is evaluated at this point. If this point does not yield an improvement over the initial point, the search technique returns to the initial point and explores in an effort to find a more favorable set of values for the decision variables. Once a set of values of the decision variables is identified that yields an improvement in the objective function, the search technique moves to this point and begins exploration again. This procedure is then repeated until a point is reached where exploration indicates that further improvement of the objective function is not possible. This point is taken as the desired optimum.

The reader will notice that a search technique may lead to a false optimum. For example, the objective function may have several local minima and maxima. A search technique may lead to a point that is near one of these local optima; however the indicated optimum may be far from the global optimum. Hence, a search technique may be expected to perform best when there is reason to believe that the objective function has only one local maximum or minimum. When the analyst has no idea whether the objective function has multiple local optima or not, he can apply the search technique several times, choosing a different starting point in each instance. If, in each case, the search terminates at or near the same point, he may be justified in concluding that the indicated optimum is a global optimum, although this conclusion is not necessarily valid. However, multiple application of a search technique can be quite time consuming and expensive, particularly when executed on a digital computer. In addition, the procedure is by no means foolproof. Fortunately, functions that possess several local maxima and minima are encountered only infrequently in practice. For this reason, the analyst often assumes (or, more correctly, hopes) that the function he is dealing with is unimodal, and he proceeds to apply a search technique on the basis of this assumption.

Even when the objective function is, in fact, unimodal one cannot be sure that the point indicated as optimum by the search technique is the true optimum. In most cases, the indicated optimum for a unimodal function will lie near enough to the true optimum to be accepted as a satisfactory approximation to the optimum solution. However, there are cases where certain search techniques will fail to identify approximate optimum solutions for certain classes of unimodal objective functions. Problems of this type frequently arise when the objective function contains a ridge. Ridge problems will be discussed further later in this chapter.

As we shall see, search techniques offer the analyst a reasonable alternative to functional optimization when classical methods fail. The problems that have been mentioned regarding the application of search techniques are not intended to discourage the analyst from their use, but are pointed out only as a precaution. These techniques have been applied with considerable success to a wide variety of problems and one can expect their increased utilization in the future.

The Sequential One-Factor-at-a-Time Method

One of the simplest search techniques available is the *sectioning* or *sequential one-factor-at-a-time method* of Friedman and Savage (1947). Let $f(x_1, x_2, \ldots, x_n)$ be the function to be minimized, where x_i, $i = 1, 2, \ldots, n$, are the decision variables. To apply this technique, the analyst fixes the values of

$n - 1$ decision variables and varies the remaining variable, say x_i, incrementally seeking reduction in the value of $f(x_1, x_2, \ldots, x_n)$. When further reduction in $f(x_1, x_2, \ldots, x_n)$ cannot be achieved through further variation in x_i, the value of that variable is fixed at x_i^o, the value of x_i that leads to the least observed value of $f(x_1, x_2, \ldots, x_n)$. At this point the process is repeated, varying another variable x_j in the same manner as described for x_i. The process terminates when a further increase or decrease in any single variable fails to reduce the objective function.

To formalize the algorithm for this search technique, let $(x_1, x_2, \ldots, x_n)$ be the starting point for the search and $f(x_1, x_2, \ldots, x_n)$ the value of the objective function at this point. In addition, let $(x_1^o, x_2^o, \ldots, x_n^o)$ be the point yielding the minimum observed value of $f(x_1, x_2, \ldots, x_n)$ up to a given point in the search. Hence, at the initial point in the search, $(x_1, x_2, \ldots, x_n)$ is the optimum point achieved and

$$(x_1^o, x_2^o, \ldots, x_n^o) = (x_1, x_2, \ldots, x_n) \tag{5.20}$$

Now we increase the value of x_1 to $x_1^o + \delta_1$ and evaluate $f(x_1^o + \delta_1, x_2^o, \ldots, x_n^o)$. If $f(x_1^o + \delta_1, x_2^o, \ldots, x_n^o) < f(x_1^o, x_2^o, \ldots, x_n^o)$, then

$$(x_1^o, x_2^o, \ldots, x_n^o) = (x_1^o + \delta_1, x_2^o, \ldots, x_n^o) \tag{5.21}$$

and x_1 is again incremented from x_1^o to $x_1^o + \delta_1$. This procedure is repeated until a point $(x_1^o + \delta_1, x_2^o, \ldots, x_n^o)$ is found such that

$$f(x_1^o, x_2^o, \ldots, x_n^o) \leq f(x_1^o + \delta_1, x_2^o, \ldots, x_n^o) \tag{5.22}$$

At this point further variation of x_1 is terminated.

If the initial increment of x_1 is such that $f(x_1^o, x_2^o, \ldots, x_n^o) < f(x_1^o + \delta_1, x_2, \ldots, x_n)$, then x_1 is decremented from x_1^o to $x_1^o - \delta_1$. If $f(x_1^o - \delta_1, x_2, \ldots, x_n) < f(x_1^o, x_2^o, \ldots, x_n^o)$, then $(x_1^o, x_2^o, \ldots, x_n^o) = (x_1^o - \delta_1, x_2^o, \ldots, x_n^o)$. The value of x_1 is successively decreased in increments δ_1 until a point $(x_1^o - \delta_1, x_2^o, \ldots, x_n^o)$ is reached such that $f(x_1^o, x_2^o, \ldots, x_n^o) \leq f(x_1^o - \delta_1, x_2^o, \ldots, x_n^o)$. At this point the search over x_1 is terminated.

Once variation of x_1 in increments δ_1 fails to reduce the objective function further, variation of x_2 in increments δ_2 commences. The search over x_2 proceeds in a manner identical to that for x_1, with the process repeated for $x_3, x_4, \ldots, x_n$. Once the search over x_n is completed, the procedure returns to x_1, initiating a second cycle where the increments $\delta_1, \delta_2, \ldots, \delta_n$ are the same as those used in the first cycle. The search continues through successive cycles until a point $(x_1^o, x_2^o, \ldots, x_n^o)$ is found such that

$$f(x_1^o, x_2^o, \ldots, x_n^o) < f(x_1^o, x_2^o, \ldots, x_i^o \pm \delta_i, \ldots, x_n^o), \qquad i = 1, 2, \ldots, n \tag{5.23}$$

Search Techniques

At this point, further variation in any x_i using the initial increments will not yield a value of the objective function less than $f(x_1^0, x_2^0, \ldots, x_n^0)$ and a complete iteration of the search is completed. When the condition in (5.23) is achieved, the search can be terminated. However, it may prove useful to refine the search by reducing the increments $\delta_1, \delta_2, \ldots, \delta_n$ and reinitiating the search.

The algorithm for the sequential-one-factor-at-a-time method is summarized in Fig. 5.7, where all variables and the objective function are as defined above. In addition, M is the number of complete iterations to be allowed in the search, j the iteration number, δ_{ji} the magnitude of the change in the ith variable at the jth iteration, N the number of consecutive changes

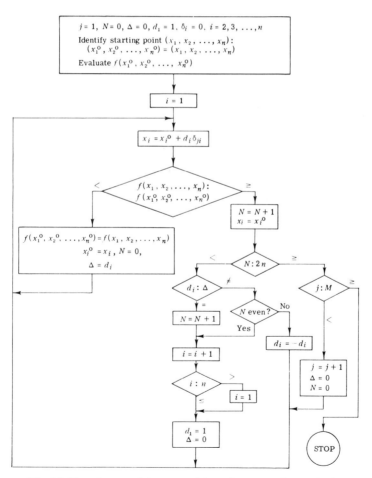

Fig. 5.7 Flow diagram of the sequential-one-factor-at-a-time method.

```
C     ************************************************************
C     *                                                          *
C     *   SEQUENTIAL ONE-FACTOR-AT-A-TIME METHOD                 *
C     *                                                          *
C     *   THIS PROGRAM IS DESIGNED TO MINIMIZE THE OBJECTIVE FUNCTION GIVEN IN *
C     *   SUBROUTINE SUBSYS. FOR MAXIMIZATION OF AN OBJECTIVE FUNCTION MULTIPLY *
C     *   THE OBJECTIVE FUNCTION BY -1 AND MINIMIZE THE RESULTING FUNCTION.    *
C     *                                                          *
C     *   DEFINITION OF VARIABLES                                *
C     *     NVAR = NUMBER OF DECISION VARIABLES                  *
C     *     ITER = NUMBER OF COMPLETE ITERATIONS OF THE SEARCH   *
C     *     IPRINT = 1, PRINTOUT AT EVERY POINT IN THE SEARCH AS WELL AS AT FINISH *
C     *            = 0, PRINTOUT AT END OF SEARCH ONLY           *
C     *     X(I) = VALUE OF THE ITH DECISION VARIABLE AT ANY POINT IN THE SEARCH *
C     *     STEP(I,J) = INCREMENTAL CHANGE IN VARIABLE I AT THE JTH ITERATION OF *
C     *                 THE SEARCH                               *
C     *     IND = 0, SUBROUTINE SUBSYS READS IN DATA PERTINENT TO THE OBJECTIVE *
C     *              FUNCTION TO BE OPTIMIZED                    *
C     *         = 1, SUBROUTINE SUBSYS EVALUATES THE OBJECTIVE FUNCTION *
C     *     SUBSYS = SUBROUTINE USED TO EVALUATE THE OBJECTIVE FUNCTION *
C     *     YY = OBJECTIVE FUNCTION EVALUATION AT ANY POINT IN THE SEARCH *
C     *     ICOUNT = CUMULATIVE NUMBER OF CONSECUTIVE CHANGES IN VARIABLE VALUES *
C     *              WITHOUT IMPROVEMENT OF THE OBJECTIVE FUNCTION *
C     *     JTER = ITERATION NUMBER                              *
C     *     XIMP = 0, NO IMPROVEMENT IN THE OBJECTIVE FUNCTION AT A GIVEN POINT *
C     *          = 1, IMPROVEMENT IN THE OBJECTIVE FUNCTION AT A GIVEN POINT *
C     *     DIR(I) = DIRECTION OF CHANGE IN THE ITH DECISION VARIABLE *
C     *     XOPT(I) = CURRENT INDICATED OPTIMUM VALUE OF THE ITH DECISION VARIABLE *
C     *     INCR = INDEX OF THE DECISION VARIABLE WHOSE VALUE IS CURRENTLY CHANGING *
C     *     YMIN = CURRENT INDICATED OPTIMUM VALUE OF THE OBJECTIVE FUNCTION *
C     ************************************************************
      DIMENSION DIR(10),STEP(10,10),XOPT(10),X(10)
      COMMON X
      READ(5,100) NVAR,ITER,IPRINT
      WRITE(6,100) NVAR,ITER,IPRINT
      READ(5,200) (X(I),I=1,NVAR)
      WRITE(6,200) (X(I),I=1,NVAR)
      DO 1 I=1,ITER
      READ(5,200) (STEP(I,J),J=1,NVAR)
    1 WRITE(6,200) (STEP(I,J),J=1,NVAR)
      IND=0
      CALL SUBSYS(IND,YY)
      IND=1
      ICOUNT=0
      JTER=1
      XIMP=0
      DO 2 I=1,NVAR
      DIR(I)=0
    2 XOPT(I)=X(I)
      INCR=1
      DIR(1)=1
      CALL SUBSYS(IND,YY)
      YMIN=YY
      IF(IPRINT.EQ.1) WRITE(6,330) YY,(I,X(I),I=1,NVAR)
    3 X(INCR)=XOPT(INCR)+DIR(INCR)*STEP(JTER,INCR)
      CALL SUBSYS(IND,YY)
      IF(IPRINT.EQ.1) WRITE(6,330) YY,(I,X(I),I=1,NVAR)
      IF(YY.GT.YMIN) GO TO 4
      ICOUNT=0
      XOPT(INCR)=X(INCR)
      YMIN=YY
      XIMP=DIR(INCR)
      GO TO 3
    4 ICOUNT=ICOUNT+1
      X(INCR)=XOPT(INCR)
      IF(ICOUNT.GE.2*NVAR) GO TO 7
      IF(DIR(INCR).NE.XIMP) GO TO 5
      ICOUNT=ICOUNT+1
      GO TO 6
    5 JCOUNT=ICOUNT/2
      JCOUNT=2*JCOUNT
      IF(JCOUNT.EQ.ICOUNT) GO TO 6
      DIR(INCR)=-DIR(INCR)
      GO TO 3
    6 INCR=INCR+1
      IF(INCR.GT.NVAR) INCR=1
      DIR(INCR)=1
      XIMP=0
      GO TO 3
    7 IF(JTER+1.GT.ITER) GO TO 8
      JTER=JTER+1
      XIMP=0
      ICOUNT=0
      GO TO 3
    8 WRITE(6,400)
      WRITE(6,300) YMIN,(I,XOPT(I),I=1,NVAR)
      STOP
  100 FORMAT(1X,3I5)
  200 FORMAT(1X,F10.4)
  300 FORMAT(1X,'YY=',E14.7/(1X,'X(',I3,')=',E14.7))
  400 FORMAT(////1X,'INDICATED OPTIMUM SOLUTION'//)
      END
```

Fig. 5.8 FORTRAN IV program for the sequential-one-factor-at-a-time method.

Search Techniques

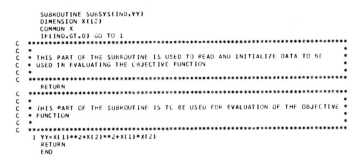

Fig. 5.8 (*continued*)

in variable values without improving the objective function, and Δ the direction of the variable change at the last improvement. A FORTRAN IV program for this search technique is given in Fig. 5.8.

Example 5.5 Using the sequential-one-factor-at-a-time method, attempt to determine the minimum for the function

$$f(x_1, x_2) = x_1^2 + x_1 x_2 + x_2^2$$

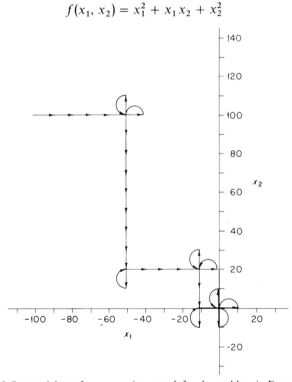

Fig. 5.9 Sequential-one-factor-at-a-time search for the problem in Example 5.5.

Let the starting point for the search be $(-101, 101)$. Carry out one complete iteration using $\delta_{1i} = 10$, $i = 1, 2$.
Each step of the search is given in Table 5.4 and shown graphically in Fig. 5.9.

TABLE 5.4 Sequential-one-factor-at-a-time search for the Problem given in Example 5.5

Point number	(x_1, x_2)	$f(x_1, x_2)$	(x_1^o, x_2^o)	Remarks
1	$(-101, 101)$	10201	$(-101, 101)$	optimum
2	$(-91, 101)$	9291	$(-91, 101)$	optimum
3	$(-81, 101)$	8581	$(-81, 101)$	optimum
4	$(-71, 101)$	8072	$(-71, 101)$	optimum
5	$(-61, 101)$	7761	$(-61, 101)$	optimum
6	$(-51, 101)$	7651	$(-51, 101)$	optimum
7	$(-41, 101)$	7741	$(-51, 101)$	
8	$(-51, 111)$	9261	$(-51, 101)$	
9	$(-51, 91)$	6241	$(-51, 91)$	optimum
10	$(-51, 81)$	5031	$(-51, 81)$	optimum
11	$(-51, 71)$	4021	$(-51, 71)$	optimum
12	$(-51, 61)$	3211	$(-51, 61)$	optimum
13	$(-51, 51)$	2601	$(-51, 51)$	optimum
14	$(-51, 41)$	2191	$(-51, 41)$	optimum
15	$(-51, 31)$	1981	$(-51, 31)$	optimum
16	$(-51, 21)$	1971	$(-51, 21)$	optimum
17	$(-51, 11)$	2161	$(-51, 21)$	
18	$(-41, 21)$	1261	$(-41, 21)$	optimum
19	$(-31, 21)$	751	$(-31, 21)$	optimum
20	$(-21, 21)$	441	$(-21, 21)$	optimum
21	$(-11, 21)$	331	$(-11, 21)$	optimum
22	$(-1, 21)$	421	$(-11, 21)$	
23	$(-11, 31)$	741	$(-11, 21)$	
24	$(-11, 11)$	121	$(-11, 11)$	optimum
25	$(-11, 1)$	111	$(-11, 1)$	optimum
26	$(-11, -9)$	301	$(-11, 1)$	
27	$(-1, 1)$	1	$(-1, 1)$	optimum
28	$(9, 1)$	91		
29	$(-1, 11)$	111		
30	$(-1, -9)$	91		terminate

The sequential-one-factor-at-a-time method is quite effective for functions of several variables where there is little or no interaction among the decision variables. For such functions, the response contours are often circular or nearly so. This method is also efficient when the response contours are elliptical but with major and minor axes parallel to the coordinate axes.

Search Techniques 201

However, when the contours are elliptical, or roughly so, with the major and minor axes tilted, the sequential-one-factor-at-a-time method may fail completely. Examples illustrating the performance of the sequential-one-factor-at-a-time method applied to functions with contours of the type just described are shown in Fig. 5.10. As suggested in Fig. 5.10, where strong interaction among the decision variables exists the sequential-one-factor-at-a-time method may terminate at a point that is far from the optimum. This is particularly true when a sharp ridge exists of the type shown in Fig. 5.10(c).

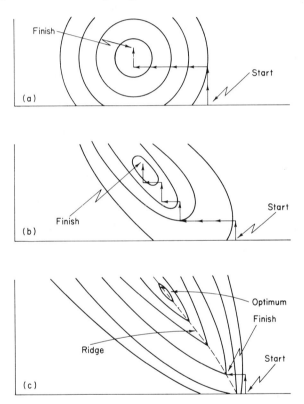

Fig. 5.10 Effect of variable interaction on the sequential-one-factor-at-a-time method. (a) No interaction, (b) mild interaction, (c) strong interaction.

Pattern Search

As we have indicated, the sequential-one-factor-at-a-time method may prove ineffective when applied to functions containing a ridge. Even when it is successful in following a ridge to the vicinity of the optimum, it is often inefficient in doing so. The *pattern search* is generally more efficient in following a ridge than the sequential-one-factor-at-a-time method. Because of its efficiency in this regard, the pattern search is often referred to as a ridge-following technique.

The philosophy underlying the pattern search is that if a series of adjustments in the decision variables is successful it is worth trying again. The technique is initiated at a starting point specified by the analyst and adjustments in the decision variables are executed in small steps. However, with repeated success the steps grow; that is, the technique accelerates in a direction that seems to be promising.

Let the starting point be defined as $(x_1, x_2, \ldots, x_n)$ and define the *temporary head* T as

$$T = (x_1, x_2, \ldots, x_n) \tag{5.24}$$

We will also define this point as the *base point* B_1. Let

$$D_i = (0, 0, \ldots, \delta_i, \ldots, 0), \quad i = 1, 2, \ldots, n \tag{5.25}$$

The first step in the search is to increase T by D_1 and evaluate $f(T + D_1)$. If $f(T) > f(T + D_1)$,

$$T = T + D_1 \tag{5.26}$$

On the other hand, if $f(T) < f(T + D_1)$, we decrement T by D_1 and evaluate $f(T - D_1)$. If $f(T) > f(T - D_1)$,

$$T = T - D_1 \tag{5.27}$$

We now increase T by D_2 and evaluate $f(T + D_2)$. Again if $f(T) > f(T + D_2)$,

$$T = T + D_2 \tag{5.28}$$

Otherwise, T is decreased by D_2. Again if $f(T) > f(T - D_2)$,

$$T = T - D_2 \tag{5.29}$$

This process is repeated n times; that is, each variable x_i, $i = 1, 2, \ldots, n$, is increased and if necessary decreased from the best point achieved in increments δ_i before this portion of the search terminates. If $f(T) < f(B_1)$, then

$$B_2 = T \tag{5.30}$$

If $f(T) = f(B_1)$, the increment vectors D_i, $i = 1, 2, \ldots, n$, are divided by two and the entire process is repeated until a point T is found such that $f(T) < f(B_1)$. Once such a point is found, B_2 is defined as in (5.30).

Search Techniques

Once B_2 is defined, the new temporary head T is given by the acceleration

$$T = 2B_2 - B_1 \tag{5.31}$$

and

$$B_1 = B_2 \tag{5.32}$$

At this point in the search $f(T)$ is *not* evaluated. Rather, we first evaluate $f(T + D_1)$, and if $f(T + D_1) < f(B_1)$,

$$T = T + D_1$$

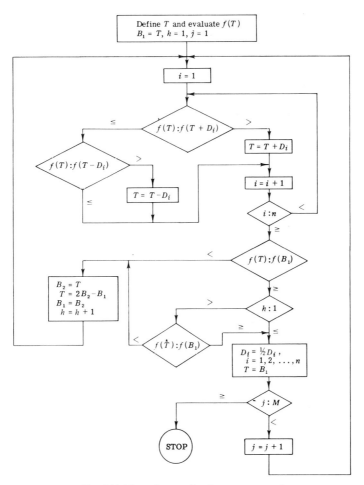

Fig. 5.11 Flow diagram for the pattern search.

```
C     ****************************************************************
C     *                                                              *
C     * PATTERN SEARCH                                               *
C     *                                                              *
C     * THIS PROGRAM IS DESIGNED TO MINIMIZE THE OBJECTIVE FUNCTION GIVEN IN *
C     * SUBROUTINE SUBSYS. FOR MAXIMIZATION OF AN OBJECTIVE FUNCTION MULTIPLY*
C     * THE OBJECTIVE FUNCTION BY -1 AND MINIMIZE THE RESULTING FUNCTION.   *
C     *                                                              *
C     * DEFINITION OF VARIABLES                                      *
C     *    NPROB = NUMBER OF SETS OF DATA FOR WHICH THE MODEL IN SUBROUTINE *
C     *            SUBSYS IS TO BE OPTIMIZED                         *
C     *    NVAR = NUMBER OF DECISION VARIABLES                       *
C     *    NSTEP = NUMBER OF TIMES THE INCREMENTAL CHANGE IN THE DECISION VARIABLES*
C     *            MAY BE CHANGED BEFORE TERMINATION OF THE SEARCH   *
C     *    IPRINT = 1, PRINTOUT AT EVERY POINT IN THE SEARCH AS WELL AS AT FINISH*
C     *           = 0, PRINTOUT AT END OF SEARCH ONLY                *
C     *    THEAD(I) = ITH ELEMENT OF THE TEMPORARY HEAD VECTOR       *
C     *    DELT(I) = INITIAL INCREMENT FOR THE ITH DECISION VARIABLE *
C     *    IND = 0, SUBROUTINE SUBSYS READS IN DATA PERTINENT TO THE OBJECTIVE*
C     *            FUNCTION TO BE OPTIMIZED                          *
C     *        = 1, SUBROUTINE SUBSYS EVALUATES THE OBJECTIVE FUNCTION*
C     *    SUBSYS = SUBROUTINE USED TO EVALUATE THE OBJECTIVE FUNCTION*
C     *    YY = OBJECTIVE FUNCTION EVALUATION AT ANY POINT IN THE SEARCH*
C     *    BASE(J,I) = VALUE OF THE ITH ELEMENT OF THE JTH BASE VECTOR, J=1,2*
C     *    X(I) = VALUE OF THE ITH DECISION VARIABLE AT ANY POINT IN THE SEARCH*
C     *    YMIN = CURRENT INDICATED OPTIMUM VALE OF THE OBJECTIVE FUNCTION*
C     *    HEAD = NUMBER OF TEMPORARY HEADS YIELDING IMPROVEMENTS IN THE OBJECTIVE*
C     *           FUNCTION                                           *
C     *    IBASE = BASE POINT NUMBER                                 *
C     *    IMP = 0, NO IMPROVEMENT IN THE OBJECTIVE FUNCTION         *
C     *        = 1, AN IMPROVEMENT IN THE OBJECTIVE FUNCTION HAS BEEN ACHIEVED*
C     *    DIR(I) = DIRECTION OF CHANGE IN THE ITH DECISION VARIABLE *
C     *                                                              *
C     ****************************************************************
      DIMENSION THEAD(10),BASE(2,10),X(10),DIR(10),DELT(10)
      COMMON X
      READ(5,100) NPROB
      WRITE(6,100) NPROB
      DO 15 IPROB=1,NPROB
      WRITE(6,500)
      READ(5,100) NVAR,NSTEP,IPRINT
      WRITE(6,100) NVAR,NSTEP,IPRINT
      READ(5,200) (THEAD(I),DELT(I),I=1,NVAR)
      WRITE(6,200) (THEAD(I),DELT(I),I=1,NVAR)
      IND=0
      CALL SUBSYS(IND,YY)
      IND=1
      DO 1 I=1,NVAR
      BASE(1,I)=THEAD(I)
    1 X(I)=BASE(1,I)
      CALL SUBSYS(IND,YY)
      IF(IPRINT.EQ.1) WRITE(6,300) YY,(I,X(I),I=1,NVAR)
      YMIN=YY
      HEAD=1
      IBASE=1
      DO 14 J=1,NSTEP
    2 IMP=0
      DO 5 I=1,NVAR
      DIR(I)=1
    3 X(I)=THEAD(I)+DELT(I)*DIR(I)
      CALL SUBSYS(IND,YY)
      IF(IPRINT.EQ.1) WRITE(6,300) YY,(K,X(K),K=1,NVAR)
      IF(YY.LT.YMIN) GO TO 4
      X(I)=THEAD(I)
      IF(DIR(I).LT.0.) GO TO 5
      DIR(I)=-1.
      GO TO 3
    4 THEAD(I)=X(I)
      IMP=1
      YMIN=YY
    5 CONTINUE
      IF(HEAD.EQ.1.AND.IMP.EQ.0) GO TO 12
      IF(IMP.NE.0) GO TO 9
      DO 6 I=1,NVAR
    6 X(I)=THEAD(I)
      CALL SUBSYS(IND,YY)
      IF(IPRINT.EQ.1) WRITE(6,300) YY,(I,X(I),I=1,NVAR)
      IF(YY.LT.YMIN) GO TO 7
      GO TO 12
    7 YMIN=YY
      DO 8 I=1,NVAR
    8 THEAD(I)=X(I)
    9 IBASE=IBASE+1
      DO 10 I=1,NVAR
   10 BASE(IBASE,I)=THEAD(I)
      HEAD=HEAD+1
      DO 11 I=1,NVAR
      THEAD(I)=2.*BASE(IBASE,I)-BASE(IBASE-1,I)
      BASE(1,I)=BASE(2,I)
   11 X(I)=THEAD(I)
      IBASE=1
      GO TO 2
```

Fig. 5.12 FORTRAN IV program for the pattern search.

```
     12 DO 13 I=1,NVAR
        DELT(I)=DELT(I)/2.
        THEAD(I)=BASE(1,I)
     13 X(I)=BASE(1,I)
        HEAD=1
        IBASE=1
     14 CONTINUE
        WRITE(6,400)
        WRITE(6,300) YMIN,(I,THEAD(I),I=1,NVAR)
     15 CONTINUE
        STOP
    100 FORMAT(1X,3I5)
    200 FORMAT(1X,2F10.4)
    300 FORMAT(1X,'YY=',E14.7/(1X,'X(',I5,')=',E14.7))
    400 FORMAT(////1X,'INDICATED OPTIMUM SOLUTION')
    500 FORMAT(1H1)
        END
```

Fig. 5.12 (*continued*)

If $f(T + D_1) \geq f(B_1)$, we evaluate $f(T - D_1)$, and if $f(T - D_1) < f(B_1)$,

$$T = T - D_1$$

As before, we repeat this sequence of evaluations n times. If at the end of this cycle $f(T) < f(B_1)$, then B_2 is given by (5.30), the new temporary head by (5.31), and B_1 by (5.32). On the other hand, if $f(B_1)$ is the least value of the function thus far, we evaluate $f(T)$, where T is the original temporary head for this cycle, and if $f(T) < f(B_1)$, B_2 is given by (5.30), T by (5.31), B_1 by (5.32), and the third cycle is initiated. If $f(T) = f(B_1)$, the pattern is broken and we return to B_1 by letting

$$T = B_1 \qquad (5.33)$$

and

$$D_i = \tfrac{1}{2} D_i, \qquad i = 1, 2, \ldots, n \qquad (5.34)$$

Once a pattern is broken, the search is reinitiated at the last base point B_1, using increments that are one-half their previous value.

The algorithm for the pattern search is summarized in the flow diagram in Fig. 5.11, where the notation used is identical to that used above, and M is the number of times the increment D_i, $i = 1, 2, \ldots, n$, may be halved before termination of the search, and h the index that counts the number of temporary heads identified. A FORTRAN IV program for the pattern search is given in Fig. 5.12.

Example 5.6 Using the pattern search, attempt to find the minimum of the following function:

$$f(x_1, x_2) = 0.1(x_2 - x_1^2)^2 + 0.001(1 - x_1)^2$$

Let the starting point be

$$(x_1, x_2) = (-20, 30)$$

and the initial increments

$$(\delta_1, \delta_2) = (4, 4)$$

The search is to terminate when $\delta_i < 2$, $i = 1, 2$.

The results of the pattern search are shown graphically in Fig. 5.13, where temporary heads and base points have been subscripted for ease in following the development of the pattern.

At $(-20, 30)$, the value of the objective function is 13,690.440. As indicated in Fig. 5.13, at the termination of the search the indicated optimum is $(-6, 38)$, where the value of the objective function is 0.049. The true optimum is at $(1, 1)$, where the value of the objective function is 0.000. Although the distance between the indicated optimum and the true optimum may appear large, the difference in the respective values of the objective function is quite small. Hence, for practical purposes the indicated optimum could be taken as an acceptable solution to the problem.

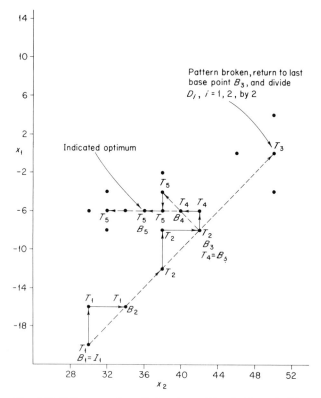

Fig. 5.13 Pattern search applied to the problem in Example 5.6.

COST MODELS

In this section we concentrate on the development of mathematical models of the cost of operating the queueing system. We will assume, for the present, the availability of estimates for the relevant cost coefficients. Subsequently, we explore the process of obtaining these estimates.

Our concern will be with modeling a particular queueing system. To achieve this objective, we will present a wide variety of examples and develop the corresponding cost models. Additionally, a number of exercises are presented at the end of the chapter to provide an opportunity for you to test your modeling ability. To facilitate the presentation, we will restrict our coverage of queueing models to include those presented in Chapter 3. However, the incorporation of results from non-Poisson models in the expected cost models will normally be quite straightforward. Thus, one would obtain appropriate values of the operating characteristics of the system using either the models in Chapter 4 or Monte Carlo simulation, as described in Chapter 9.

Example 5.7 As an illustration of a queueing problem that occurs quite commonly, consider the operation of a storeroom that can be classified as a $(M \mid M \mid c):(GD \mid \infty \mid \infty)$ queueing system. Management is faced with the decision of the number of attendants to be assigned to the storeroom. As servers are added to the storeroom, the cost of customers waiting for service decreases; however, the cost of providing service increases. The proper balance of these two costs is desired. A rather general formulation of the optimization problem can be given as follows:

minimize

$$TC(c) = C_1 L_q + C_2(L - L_q) + C_3 \sum_{n=0}^{c} (c-n)P_n$$
$$+ C_4 \left[c - \sum_{n=0}^{c} (c-n)P_n \right] \qquad (5.35)$$

where C_1 is the cost per unit of time spent waiting by a customer, C_2 the cost per unit of time spent in service by a customer, C_3 the cost per unit of time a server is idle, and C_4 the cost per unit of time a server spends serving a customer. The parameters P_n, L, and L_q are as previously defined in Chapter 3 for the $(M \mid M \mid c):(GD \mid \infty \mid \infty)$ queue and $TC(c)$ is the expected total cost per unit time as a function of the decision variable c, the number of servers.

Since the expected number of customers being served equals the expected number of busy servers, (5.35) reduces to

minimize

$$TC(c) = C_1 L_q + (C_2 + C_4)(L - L_q) + C_3(c - L + L_q) \qquad (5.36)$$

or, equivalently,

minimize

$$TC(c) = C_1 L_q + (C_2 + C_4)\frac{\lambda}{\mu} + C_3\left(c - \frac{\lambda}{\mu}\right) \qquad (5.37)$$

Discarding those cost terms not a function of the decision variable c and expressing L_q as a function of c yields

$$VC(c) = C_1 L_q(c) + C_3 c \qquad (5.38)$$

where it is assumed that C_2, C_4, λ, and μ are not functions of c, and $VC(c)$ denotes the expected variable cost per unit time as a function of the decision variable c.

To minimize $VC(c)$, we wish to determine the value c^* such that

$$VC(c^*) - VC(c^* - 1) \leq 0, \quad \text{and} \quad VC(c^*) - VC(c^* + 1) \leq 0$$

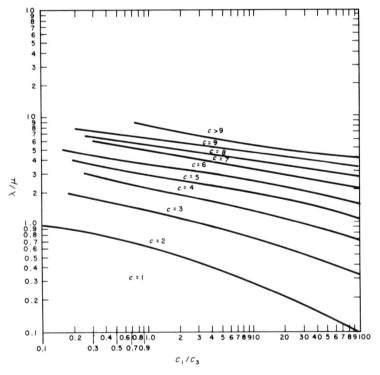

Fig. 5.14 Optimum number of service channels with service times exponentially distributed (after Mangelsdorf, 1959).

Cost Models

From (5.38), the necessary conditions for c^* to be the optimum value of c reduce to

$$\Delta L_q(c^*) \geq -\frac{C_3}{C_1} \geq \Delta L_q(c^* - 1) \qquad (5.39)$$

where Δ is the first forward difference operator, defined such that

$$\Delta L_q(x) = L_q(x + 1) - L_q(x)$$

Although a mathematical proof of the convexity of $VC(c)$ is not available, considerable computational experience is provided by Manglesdorf (1959) to support the claim. Figure 5.14 provides a number of curves developed by Mangelsdorf to assist in determining the value of c^*.

To complete the example problem, suppose $\lambda = 30$ arrivals/hr and $\mu = 20$ services/hr. Moreover, let $C_1 = \$15/\text{hr}$ and $C_3 = \$5/\text{hr}$. From Fig. 5.14, with $\lambda/\mu = 1.5$ and $C_1/C_3 = 3.0$, it is apparent that $c^* = 3$. To verify this result, a one-at-a-time search is used, with the results shown in Table 5.5. Since it is required that $\lambda/c\mu < 1$, then $c^* \geq 2$. Thus, the search begins with $c = 2$. Values of $L_q(c)$ are obtained indirectly from Fig. 5.15.

TABLE 5.5 Solution for Example 5.7

c	$L_q(c)$	$L_q(c) - L_q(c-1)$
2	1.930	$-\infty$
3	0.237	$-1.693 < -0.333$
4	0.046	$-0.191 > -0.333$

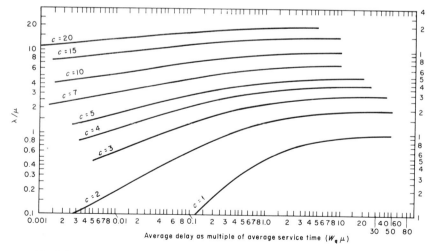

Fig. 5.15 Average delay for the $(M|M|c)$: $(GD|\infty|\infty)$ queue (after Mangelsdorf, 1959).

An indication of the sensitivity of the number of servers to errors in estimating the cost coefficients can be gained from Fig. 5.14. Since the ratio $\lambda/\mu = 1.5$ for the example, notice that for $0.6 \leq C_1/C_3 \leq 6.0$ the optimum number of servers remains 3. Consequently, it is not required that the values of C_1 and C_3 be known exactly. Similarly, for $C_1/C_3 = 3.0$ the ratio λ/μ can take on values from 1.00 to approximately 1.70 without changing the optimum number of servers.

Example 5.8 The next illustration of a queueing system involves the design of a parking lot. A decision is to be made concerning the number of parking spaces to provide in the lot. If a parking space is not available when a customer arrives, he goes elsewhere for service. Thus, the queueing system is viewed as an $(M|G|c):(GD|c|\infty)$ queue.

A cost model for this example might be formulated as follows:

minimize

$$TC(c) = cC_5 + C_6 \lambda P_c \qquad (5.40)$$

where C_5 is the cost per unit capacity of the system per unit time, C_6 the cost per customer who balks and does not enter the system, and the value of P_c is obtained from Erlang's loss formula (3.56). For the parking lot example, C_5 is the daily cost of providing a parking space in the lot and C_6 the lost profit plus goodwill cost resulting from a customer finding the parking lot full and going elsewhere for service.

To complete the analysis, suppose $\lambda = 50$ cars/hr, $\mu = 2$ cars/hr, C_5 has a value of \$0.50/hr per parking space, and C_6 has a value of \$2/customer.

TABLE 5.6 Solution for Example 5.8

c	P_c	P_{c-1}	$P_{c-1} - P_c$	$TC(c)$ (\$/hr)
2	0.9232	0.9615	0.0383	93.32
⋮	⋮	⋮	⋮	⋮
10	0.6224	0.6592	0.0368	67.24
⋮	⋮	⋮	⋮	⋮
20	0.2799	0.3109	0.0310	37.99
⋮	⋮	⋮	⋮	⋮
30	0.0526	0.0666	0.0140	20.26
31	0.0407	0.0526	0.0119	19.57
32	0.0308	0.0407	0.0099	19.08
33	0.0228	0.0308	0.0080	18.70
34	0.0165	0.0228	0.0063	18.65
35	0.0116	0.0165	0.0049	18.66
36	0.0080	0.0116	0.0036	18.80
37	0.0054	0.0080	0.0026	19.04

Cost Models

Necessary conditions for an optimum solution to (5.40) reduce to

$$P_{c^*} - P_{c^*+1} \le \frac{C_5}{\lambda C_6} \le P_{c^*-1} - P_{c^*}$$

Thus, from Table 5.6 it is seen that $c^* = 34$. Notice that for the example $c^* = 34$ so long as the value of C_5/C_6 is contained in the closed interval [0.245, 0.315]. Thus, in this example the optimum solution is relatively sensitive to errors in estimating the values of the cost coefficients. However, the total cost function is very flat in the region around the optimum solution. Consequently, it is not critical that accurate values of C_5 and C_6 be provided for this example.

Example 5.9 As an example of the design of a queueing system involving two decision variables, consider the design of a production department. Decisions are to be made concerning the number of machine centers that must be provided to process production orders, as well as the amount of space to provide for in-process inventory. If a production order arrives and no additional waiting space is available, the order is taken to a special waiting area and is processed by a separate department at a higher cost. Thus, the problem is viewed as an $(M\,|\,M\,|\,c) : (GD\,|\,N\,|\,\infty)$ queue with decision variables c and N.

Combining the cost terms in the two previous examples, the following cost model is obtained:

$$\begin{aligned} TC(c, N) = &\, C_1 L_q + (C_2 + C_4)(L - L_q) \\ &+ C_3(c - L + L_q) \\ &+ C_5 N + C_6 \lambda P_N \end{aligned} \quad (5.41)$$

To determine the optimum values of c and N, a search procedure is used.

For the production department, orders arrive at a Poisson rate of 8/hr and service time is exponentially distributed with a mean of 30 min. The cost of a machine center is estimated to be \$8/hr when the machine center is idle and \$20/hr when the machine center is busy. The cost of a production order passing through the production department is estimated to be \$10/hr, whether it is waiting or being processed through a machine center. The cost of in-process inventory space required for production orders waiting to be processed is estimated to be \$1/hr per space. When production orders are transported to another department for processing due to a lack of space in the production department an incremental cost of \$25/order results.

Based on the cost data available, the following cost function is to be minimized:

$$TC(c, N) = 10L_q + (10 + 20)(L - L_q) + 8(c - L + L_q) + N + 200P_N$$

or

$$TC(c, N) = 22L - 12L_q + 8c + N + 120P_N$$

To solve the example problem, we enumerate over N and c; for each value of N, we determine the optimum value of c. To illustrate the shape of the total cost function, values of $TC(c, N)$ are given in Table 5.7 for various values of c and N. As indicated in Table 5.7, the optimum values of c and N are $c^* = 5$ and $N^* = 7$, respectively. Notice how flat the cost function is in the region surrounding the optimum solution. In this case, it appears that having too many servers $(c > c^*)$ is less costly than having too few servers $(c - c^*)$; also, having too much space $(N > N^*)$ appears preferable to having too little space $(N < N^*)$.

TABLE 5.7 Values of $TC(c, N)$ for Example 5.9

				N					
c	3	4	5	6	7	8	9	10	N^*	$TC(c, N^*)$
3	165.48	161.79	163.53	168.06	174.28	181.63	189.79	198.56	4	161.79
4	—	158.80	153.92	153.21	154.66	157.37	160.88	164.93	6	153.21
5	—	—	155.30	150.76	149.31	149.40	150.33	151.70	7	149.31
6	—	—	—	155.12	151.84	150.77	150.80	151.37	8	150.77
7	—	—	—	—	158.03	156.22	155.90	156.28	9	155.90

Example 5.10 A number of optimization problems arise in the design of queueing systems involving finite populations. As an illustration, consider the problem of determining the size of a pool of repairmen who maintain a number of identical production machines. Specifically, the number of servers c is to be determined for an $(M \mid M \mid c) : (GD \mid K \mid K)$ queueing system.

Based on the arguments given for Example 5.7, the expected variable cost model for this example becomes

$$VC(c) = C_1 L_q + (C_2 - C_3 + C_4)(L - L_q) + C_3 c$$

where, as before, C_1 is the unit cost of a customer waiting for service, C_2 the unit cost of a customer being served, C_3 the idle server cost, and C_4 the busy server cost. Since, from (3.65) and (3.66),

$$L_q = K(1 - F), \quad \text{and} \quad L - L_q = KFX$$

Cost Models

then $VC(c)$ can be expressed as

$$VC(c) = C_1 K(1 - F) + (C_2 - C_3 + C_4)KFX + C_3 c \qquad (5.42)$$

The necessary conditions for c^* to be the optimum number of servers can be written as follows:

$$F_{c*+1} - F_{c*} \leq \frac{C_3}{K} \bigg/ [C_1 - (C_2 - C_3 + C_4)X] \leq F_{c*} - F_{c*-1}$$

where F_c denotes the value of F from the Appendix, given c servers. Recall from Chapter 3 that the quantity X is defined as $X = \lambda/(\lambda + \mu)$.

To complete the example, suppose there are 20 machines to be maintained. Machines requiring service continue to run and produce defective product, which must be scrapped. Thus, $C_1 = \$2000/\text{shift}$ for each machine, whereas the normal cost of machine downtime $C_2 = \$1000/\text{shift}$ for each machine. The repairmen are assumed to cost the same, whether busy or idle. Thus, $C_3 = C_4 = \$200/\text{shift}$ for each repairman. The machines are expected to require service at a rate of 2.22 times/shift; a service rate of 20/shift is anticipated.

With $X = 0.10$, values of F are obtained from the Appendix for various values of c. The necessary conditions for an optimum solution reduce to

$$F_{c*+1} - F_{c*} \leq 0.00526 \leq F_{c*} - F_{c*-1}$$

From Table 5.8, the optimum number of repairmen is seen to be four.

TABLE 5.8 Values of F_c for Example 5.10[a]

c	F_c	$F_{c+1} - F_c$	$F_c - F_{c-1}$
5	0.999	0.001	0.004
4	0.995	0.004	0.020
3	0.975	0.020	0.097
2	0.878	0.097	0.378
1	0.500	0.378	∞

[a] $X = 0.10$, $K = 20$.

Example 5.11 In the previous example the number of repairmen was determined to service a given number of machines. Another decision problem that occurs frequently in determining economic machine assignments concerns the number of machines to assign to an operator. Examples of the latter decision problem occur in both the chemical and textile processing industries, among others.

Since the number of machines to assign an operator is to be determined, a slightly different objective function is called for. Observe that it is not appropriate to base the decision on the number of machines that will yield a minimum total cost for an $(M \mid M \mid 1) : (GD \mid K \mid K)$ queueing system, since one machine would be preferred. However, if a given total number of machines is to be allocated, then the total cost over all single-server facilities should be minimized.

Letting K_j be the number of machines to assign to server j, the optimization problem becomes:

minimize

$$TC(K, n) = \sum_{j=1}^{n} [C_{1j} L_q^{(j)} + (C_{2j} - C_{3j} + C_{4j})(L^{(j)} - L_q^{(j)}) + C_{3j}] \quad (5.43)$$

subject to

$$\sum_{j=1}^{n} K_j = \mathcal{K}, \quad K_j = 1, 2, \ldots$$

where subscripts are employed on the cost parameters and the decision variable and superscripts are used with the operating characteristics to designate the jth single-server service facility. $\mathcal{K}$ represents the total number of machines to be assigned.

Dynamic programming can be used to solve (5.43). However, for those unfamiliar with the dynamic-programming solution procedure, an alternate approach can be used in a special case for either large values of $\mathcal{K}$ or those instances where there is some flexibility in choosing the value of $\mathcal{K}$. We will consider the alternate approach here and leave as an exercise further consideration of the use of dynamic programming.

As a special case of (5.43) suppose the values of the cost parameters are the same for all values of j. In this case, each server will be assigned an equal number of machines, if possible. Consider the situation in which an equal number of machines is assigned to each server. For the resulting total cost to be a minimum, it must be true that

$$\frac{\mathcal{K}}{K} [C_1 L_q + (C_2 - C_3 + C_4)(L - L_q) + C_3] \quad (5.44)$$

is a minimum, where K is the number of machines assigned to an operator.

Since minimizing (5.44) is equivalent to minimizing the average total cost per machine, the following optimization problem is considered:

minimize

$$VC(K) = \frac{C_1 L_q + (C_2 - C_3 + C_4)(L - L_q) + C_3}{K} \quad (5.45)$$

Cost Models

subject to

$$K = 1, 2, \ldots$$

where $VC(K)$ is the expected variable cost per machine. For the $(M|M|1):(GD|K|K)$ queue, $VC(K)$ becomes

$$VC(K) = C_1(1 - F_K) + (C_2 - C_3 + C_4)F_K X + \frac{C_3}{K} \quad (5.46)$$

The necessary conditions for K^* to be the optimum population size can be written as

$$K^*(K^* - 1)(F_{K^*-1} - F_{K^*}) \leq \frac{C_3}{C_1 - (C_2 - C_3 + C_4)X}$$

$$\leq K^*(K^* + 1)(F_{K^*} - F_{K^*+1})$$

where F_K is the value of F for population size K provided in the Appendix.

To illustrate the solution procedure, suppose $C_1 = \$20$/machine-hour, $C_2 = \$8$/machine-hour, $C_3 = C_4 = \$8$/operator-hour, and $X = 0.10$. Therefore, given $c = 1.0$ and $X = 0.10$, values of F are obtained from the Appendix for the value of K satisfying the necessary conditions,

$$K^*(K^* - 1)(F_{K^*-1} - F_{K^*}) \leq 0.41667 \leq K^*(K^* + 1)(F_{K^*} - F_{K^*-1})$$

For $K = 5$, it is found that the necessary conditions are satisfied. Therefore, 5 machines should be assigned to one operator to minimize the expected cost per machine.

If $\mathcal{K}$ is not an integer multiple of K^*, then adjustments will need to be made in the assignments among operators. For example, suppose 12 machines are available for assignment in the situation just considered.. A number of assignment combinations present themselves. Specifically, 3 servers could be used, with each assigned 4 machines; another possibility is to assign 6 machines to each of 2 operators. When the combinatorial problem becomes extremely complex, a dynamic-programming approach might be justified. However, for this particular problem an enumeration procedure can be employed. Letting

$$C(K) = \frac{C_1 L_q + (C_2 - C_3 + C_4)(L - L_q) + C_3}{K}$$

expected costs per facility are computed for various values of K, with the results given in Table 5.9. Based on the data in Table 5.9, the total cost per unit time for a (4, 4, 4) assignment is 41.664; for a (5, 7) assignment, 42.227; and for a (6, 6) assignment, 41.268. Thus, a (6, 6) assignment is preferred.

TABLE 5.9
Solution for Example 5.11

K	C(K)
4	13.888
5	16.800
6	20.634
7	25.427

Example 5.12 Having given a reasonable argument for the objective of minimizing the expected total cost per machine in the previous example, we now pose yet another approach, which we believe is also reasonable. In this case, we are not faced with a total number of machines to be assigned. Rather, we are designing a new production system and wish to assign the number of machines to an operator that will minimize the cost per unit of product produced by the machines.

During the period of time that a machine is running, it is producing units of product. We assume production does not take place either during the waiting or the servicing time for the machine. Thus, in order to minimize the expected cost per unit of product produced, we will minimize the expected total cost per *running* machine. It is assumed that the cost of the server is the same whether busy or idle.

The objective function to be minimized can be written as follows:

$$TC(K) = \frac{C_0(K - L) + C_1 L_q + C_2(L - L_q) + C_3}{K - L} \tag{5.47}$$

where C_0 is the cost per unit time of a running machine and $K - L$ is the expected number of machines running.

For the $(M \mid M \mid 1) : (GD \mid K \mid K)$ queue, (5.47) reduces to

$$TC(K) = C_0 + \frac{C_2 X}{1 - X} + \frac{C_1(1 - F) + C_3/K}{F(1 - X)} \tag{5.48}$$

Eliminating the nonvariable costs, the objective function becomes

$$VC(K) = \frac{C_1(1 - F) + C_3/K}{F(1 - X)} \tag{5.49}$$

For the data given in the previous example, the following values of $VC(K)$ were obtained:

$$VC(4) = 3.1088, \quad VC(5) = 3.0409, \quad VC(6) = 3.2070$$

Cost Models

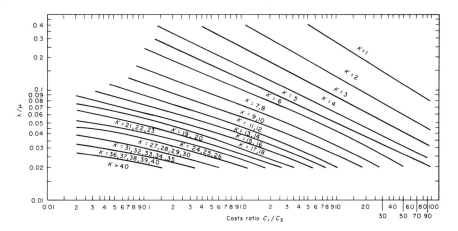

Fig. 5.16 Optimum machine assignment for the $(M|M|1):(GD|K|K)$ queue (after Palm, 1958).

Thus, the optimum number of machines to assign an operator remains 5 under the alternate formulation. In fact, it is not uncommon to find that identical solutions are provided by the two approaches.

To facilitate the minimization of (5.49), Fig. 5.16, which is based on the work of Palm (1958), provides the optimum value of K^* when service times are exponentially distributed; based on the work of Ashcroft (1950), Fig. 5.17 portrays the optimum solution when service times are constant.

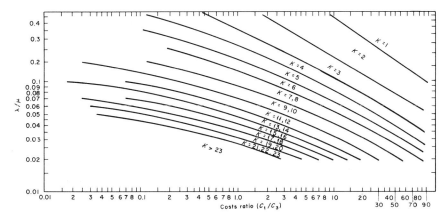

Fig. 5.17 Optimum machine assignment for the $(M|D|1):(GD|K|K)$ queue (after Ashcroft, 1950).

Example 5.13 In some cases, one of the design parameters for a queueing system is the service rate. This is especially true when service is provided by a machine and a variety of machine speeds is available. To illustrate the consideration of the service rate as a decision variable, consider an $(M\,|\,M\,|\,1):(GD\,|\,\infty\,|\,\infty)$ queue. To be determined is the service rate that balances the waiting cost of customers and the cost of providing the particular service rate.

Letting $f(\mu)$ be the cost per unit time of providing a service rate of μ, the following expected cost model results:

$$TC(\mu) = C_1 L_q + C_2(L - L_q) + f(\mu) \tag{5.50}$$

For the $(M\,|\,M\,|\,1):(GD\,|\,\infty\,|\,\infty)$ queue, (5.50) reduces to

$$TC(\mu) = \frac{C_1 \lambda^2}{\mu(\mu - \lambda)} + \frac{C_2 \lambda}{\mu} + f(\mu) \tag{5.51}$$

Differentiating (5.51) with respect to μ and setting the result equal to zero yields the following necessary condition for an optimum service rate:

$$\frac{df(\mu^*)}{d\mu^*} = \frac{C_2 \lambda \mu^{*2} + (C_1 - C_2)\lambda^2(2\mu^* - \lambda)}{\mu^{*2}(\mu^* - \lambda)^2} \tag{5.52}$$

When $C_1 = C_2$ and $f(\mu) = C_7 \mu$, then (5.52) reduces to

$$\mu^* = \lambda + \left(\frac{C_1 \lambda}{C_7}\right)^{1/2} \tag{5.53}$$

As an application of (5.53), suppose arrivals occur at a rate of 10/hr, $C_1 = C_2 = \$10/\text{hr}$, and $C_7 = \$4/\text{customer}$. Thus,

$$\mu^* = 10 + \left(\frac{10(10)}{4}\right)^{1/2} = 15 \text{ customers/hr}$$

Example 5.14 In determining the service rate to be provided in a queueing system, two decision variables should be considered: μ, the service rate per server, and c, the number of servers. Examples of such a decision problem include the determination of the number and sizes of maintenance crews, as well as the number and type of material-handling equipment to be purchased (Fetter and Galliher, 1958). The feasible values of μ would normally be finite, with μ_k the service rate for, say, a maintenance crew of size k or material-handling equipment of type k. In the case of a finite number of values, the optimization problem reduces to:

minimize

$$TC(\mu, c) = C_1 L_q + (C_2 + C_4)(L - L_q) + C_3(c - L + L_q) + cf(\mu) \tag{5.54}$$

Cost Models

subject to

$$c = 1, 2, \ldots, \quad \mu \in S$$

where S is the set of feasible service rates per server and $f(\mu)$ is the cost per server of providing the service rate μ. The cost parameters C_1, C_2, C_3, and C_4 are defined as in Example 5.7.

To solve the optimization problem, note that for a given value of μ, the problem reduces to the one considered in Example 5.7. Thus, a search procedure can be employed over the range of values for μ and the corresponding "optimum" value of c determined; the latter determination can be made by using Fig. 5.14 when $C_1 = C_2$ and $C_3 = C_4$. For small numbers of feasible values for μ, complete enumeration is reasonable.

As an illustration of the recommended solution procedure, suppose 4 types of material handling equipment are under consideration for use in a warehouse. The first two types of equipment are manually operated, electrically powered industrial trucks; the second two equipment types are more expensive due to the incorporation of computer control systems on the powered industrial trucks. The service rates and related costs for the equipment types are given in Table 5.10. These include only those costs associated with the equipment. The additional server related costs C_3 and C_4 are both equal to \$10/hr for each unit of material handling equipment. Requests for material handling occur at a Poisson rate of 20/hr. The cost of a request being processed (waiting plus service) is estimated to be \$40/hr.

TABLE 5.10 Data for Example 5.14

k	μ_k (per hour)	$f(\mu_k)$(\$/hr)
1	10	5
2	15	10
3	25	40
4	40	90

The cost model for the example problem can be expressed as

$$TC(\mu, c) = C_1 L + [C_3 + f(\mu)]c \qquad (5.55)$$

Thus, for specific values of $C_1/[C_3 + f(\mu)]$ and λ/μ, the "optimum" value of c can be obtained from Fig. 5.14. Given the values of λ/μ and c, the appropriate value of $W_q \mu$ is obtained from Fig. 5.15 and the corresponding value of L computed. The total cost value is then determined for each value of μ and the minimum cost solution obtained.

For $\mu = 10$, $\lambda/\mu = 2.00$ and $C_1/[C_3 + f(\mu)] = 2.67$. Thus, from Fig. 5.14, $c = 4$. From Fig. 5.15, $W_q \mu = 0.09$. Therefore, a calculation establishes that $L = 2.18$. From (5.55), $TC(10, 4) = 125.40/\text{hr}$. Similar calculations for $\mu = 15$, 25, and 40 yield the following total cost values:

$$TC(15, 3) = \$103.98, \qquad TC(25, 2) = \$108.32, \qquad TC(40, 1) = \$130.00$$

Therefore, the second type of material-handling equipment is the indicated choice on the basis of the costs involved.

Example 5.15 The previous examples involved one customer per arrival; we now consider a problem involving bulk arrivals. Specifically, consider a material-handling problem requiring the determination of the number of parts to transport by industrial truck to a work station for processing.

Each trip by the truck is assumed to cost C_8. Parts are to be supplied to the work station at a rate of P units/hr. If b units are transported per trip to the work station, then the arrival rate λ will be P/b and the hourly cost of transporting parts to the work station will be $C_8 P/b$. The hourly cost of parts waiting to be processed at the work station will be $C_1 L_q$; the hourly cost of parts being processed at the work station will be $C_2(L - L_q)$.

The expected cost per unit time for this example can be written as

$$TC(b) = C_1 L_q + C_2(L - L_q) + \frac{C_8 P}{b}$$

where, from Chapter 3,

$$L = \frac{b\lambda(1 + b)}{2(\mu - b\lambda)}, \qquad L_q = L - \frac{b\lambda}{\mu}$$

Since $P = b\lambda$, $TC(b)$ can be given as

$$TC(b) = C_1 \left[\frac{P(1 + b)}{2(\mu - P)} - \frac{P}{\mu} \right] + \frac{C_2 P}{\mu} + \frac{C_8 P}{b} \qquad (5.56)$$

Omitting the fixed costs, the expected variable cost per unit time becomes

$$VC(b) = K_1 b + \frac{K_2}{b} \qquad (5.57)$$

where

$$K_1 = \frac{C_1 P}{2(\mu - P)}, \qquad K_2 = C_8 P$$

Necessary and sufficient conditions for b^* to be the optimum bulk arrival quantity reduce to

$$b^*(b^* - 1) \leq \frac{K_2}{K_1} \leq b^*(b^* + 1) \qquad (5.58)$$

Cost Models

Thus, if $P = 100$ parts/hr, $\mu = 150$ parts/hr, $C_1 = \$20$/hr, and $C_8 = \$10$/trip, then $K_1 = 20$ and $K_2 = 1000$. Thus, a calculation establishes that $b^* = 7$ parts/delivery.

Example 5.16 As the next illustration of the development of cost models for queueing systems, we will consider a situation involving the determination of the values of λ and c. Our discussion will be based on that of Hillier (1963). To motivate the treatment of queueing systems involving the choice of λ and c, consider the problem of determining the number and size of branch banks in a metropolitan area. The number of service facilities provided will clearly affect the arrival rate of customers at the branch banks. Thus, indirectly the arrival rate becomes a decision variable when the number of facilities is to be decided.

As a first approach at modeling the decision problem, consider the following expected cost model:

$$TC(n, c) = n[C_1 L_q + (C_2 + C_4)(L - L_q) + C_3(c - L + L_q) + C_9] \quad (5.59)$$

where C_9 is the fixed cost per unit time of providing a service facility and n the number of service facilities. If the arrival rate for the overall population is λ and n identical service facilities are to be provided, then the arrival rate λ at each service facility will be λ/n. Of course, depending on the locations of the service facilities relative to the customers' locations, a uniform distribution of customers among service facilities might not be realistic, but for now we may accept that assumption.

Substituting λ/λ for n and letting the decision variables be λ and c, (5.59) is expressed as

$$TC(\lambda, c) = \frac{\lambda}{\lambda}[C_1 L_q + (C_2 + C_4)(L - L_q) + C_3(c - L + L_q) + C_9] \quad (5.60)$$

where L and L_q are functions of both λ and c. Recalling that $W = L/\lambda$ and $W_q = L_q/\lambda$, the optimization problem reduces to

minimize
λ, c

$$\frac{TC(\lambda, c)}{\lambda} = C_1 W_q + \frac{(C_2 - C_3 + C_4)}{\mu} + \frac{(cC_3 + C_9)}{\lambda} \quad (5.61)$$

for the $(M|M|c):(GD|\infty|\infty)$ queue. By omitting the nonvariable cost $(C_2 - C_3 + C_4)/\mu$, the following variable cost is to be minimized:

$$VC(\lambda, c) = C_1 W_q + \frac{cC_3 + C_9}{\lambda} \quad (5.62)$$

subject to

$$\lambda = \frac{\lambda}{n}, \quad n = 1, 2, \ldots, \quad \frac{\lambda}{c} < \mu$$

The optimum value of λ in (5.62) can be shown to be $\lambda^* = \lambda$. Thus, only one service facility would be used to service the entire population. The optimum value of c is then obtained from Fig. 5.14, given the optimum value of λ.

The use of one large service facility rather than several small service facilities can be justified intuitively by considering a finite population problem involving four machines and two operators. If each operator is assigned to two specific machines, then both machines assigned to operator A might be running at the same time both machines assigned to operator B are down, requiring the attention of an operator. Thus, operator A is idle while a machine assigned to operator B is waiting for service. Such a situation would not occur if the two operators constituted a service pool that provided service to any of the four machines requiring service. The benefits of *server collaboration* over *separation of service* were also examined in Chapter 3.

Of course, the policy of server collaboration is not always an optimum policy. Otherwise, there would be no shopping centers, branch banks, and multiple service stations selling the same gasoline product. If the customer population is not captive, many customers will tend to go to the service facility that is the closest. Thus, in order to compete for a customer's business, a grocery chain will build stores throughout a metropolitan area. In the case of captive populations, a server collaboration policy might not be optimum, depending on the cost of travel time for the customers. Such a situation is considered in the next example.

Example 5.17 In this example we treat the server collaboration problem in which the customer's travel time is considered. Specifically, it is assumed that the cost C_{10} of a customer traveling to the service facility is expressed as cost per unit distance traveled. Typically C_{10} is expressed as the ratio of C_1, the waiting time cost for the customer, and v, the average velocity at which customers travel to and from the service facility.

If the total area to be served is A, and n equal-sized service regions are to be defined, then A/n will be the area to be served by each facility. To determine the expected distance traveled by a customer within a given service region, it will be assumed that customers are distributed uniformly over the service region. Also, we will employ service region configurations that minimize the expected distance traveled.

Two distance measures commonly used in facilities-location analysis are the rectilinear and Euclidean distances (Francis and White, 1974). The recti-

Cost Models

linear distance between the points $X_1 = (x_1, y_1)$ and $X_2 = (x_2, y_2)$ is defined as

$$d_1(X_1, X_2) = |x_1 - x_2| + |y_1 - y_2|$$

The Euclidean, or straight line, distance is

$$d_2(X_1, X_2) = [(x_1 - x_2)^2 + (y_1 - y_2)^2]^{1/2}$$

It has been shown previously that the service region configurations that will minimize the expected distance traveled by customers going to and from the service facility are the diamond and circle shown in Fig. 5.18. The

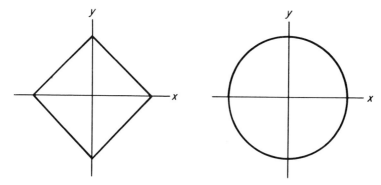

Fig. 5.18 Service configuration for (a) rectilinear and (b) Euclidean distances.

diamond-shaped service region minimizes the expected rectilinear distance traveled, with an expected distance of $\frac{2}{3}(2A/n)^{1/2}$; the circular-shaped service region minimizes the expected Euclidean distance traveled, with an expected distance of $\frac{4}{3}(A/\pi n)^{1/2}$. These expected distances are twice the expected distance from the customer's location to the service facility, since the customer returns to his original location. For a detailed discussion of the derivation of the expected distance traveled see Francis and White (1974).

From (5.60), the expected total cost model can be given as

$$TC(\lambda, c) = \frac{\lambda}{\lambda}[C_1 L_q + (C_2 - C_3 + C_4)(L - L_q) + C_3 c + C_9 + C_{10} D] \tag{5.63}$$

where D, the expected distance traveled by a customer in going to and from the service facility, is defined as follows:

$$D = \begin{cases} \dfrac{2}{3}\left(\dfrac{2A\lambda}{\lambda}\right)^{1/2}, & \text{for rectilinear distances} \\ \dfrac{4}{3}\left(\dfrac{A\lambda}{\pi\lambda}\right)^{1/2}, & \text{for Euclidean distances} \end{cases}$$

When an $(M\,|\,M\,|\,c):(GD\,|\,\infty\,|\,\infty)$ queueing system exists at each service facility, the optimization problem reduces to:

minimize

$$VC(\lambda, c) = C_1 W_q + \frac{cC_3 + C_9}{\lambda}\frac{\alpha}{\sqrt{\lambda}} \qquad (5.64)$$

subject to

$$\lambda = \frac{\lambda}{n}, \quad n = 1, 2, \ldots, \quad \frac{\lambda}{c} < \mu$$

where

$$\alpha = \begin{cases} \dfrac{2C_{10}}{3}\left(\dfrac{2A}{\lambda}\right)^{1/2}, & \text{for rectilinear distances} \\ \dfrac{4C_{10}}{3}\left(\dfrac{2A}{\lambda}\right)^{1/2}, & \text{for Euclidean distances} \end{cases}$$

The solution procedure suggested by Hillier (1963) for solving (5.64) is to search over λ, with the optimum value of c (given λ) obtained by minimizing the expression $C_1 L_q + cC_3$, using Fig. 5.14. However, for large values of λ/μ, Fig. 5.14 cannot be used. In such cases, a pattern search can be used to determine the values of λ^* and c^*.

As an illustration of the pattern-search procedure, suppose a service area of 128 mi^2 has been defined in a metropolitan region, $\lambda = 10{,}000$ customers/day, and $\mu = 100$ customers/server-day. The values of the cost coefficients have been assigned as follows:

$C_1 = C_2 = \$60/\text{customer-day}, \qquad C_9 = \$300/\text{facility-day}$

$C_3 = C_4 = \$40/\text{server-day}, \qquad C_{10} = \$6/\text{mi}$

Assuming rectilinear travel and $(M\,|\,M\,|\,c):(GD\,|\,\infty\,|\,\infty)$ queueing systems at each service facility, (5.64) reduces to:

minimize

$$VC(\lambda, c) = 60W_q + \left(\frac{40c + 300}{\lambda}\right)\left(\frac{0.64}{\sqrt{\lambda}}\right)$$

subject to

$$\lambda = \frac{10{,}000}{n}, \quad n = 1, 2, \ldots, \quad \frac{\lambda}{c} < 100$$

Cost Models

Using a pattern search, the optimum values obtained were $\lambda^* = 1,666.67$ and $c^* = 21$. Therefore, six service facilities should be constructed to service the 128-mi^2 area.

Example 5.18 As a special case of Example 5.17, suppose only one server is to be located at each service facility. Examples would include an automatic banking machine at branch bank locations, automobile inspection stations, single-station car washes, and rescue squad stations. Also, suppose each service facility can be represented as an $(M \mid G \mid 1):(GD \mid \infty \mid \infty)$ queue, such that the Pollaczek–Khintchine formulas are used. In this case, the expected variable cost model reduces to

$$VC(\lambda) = \frac{C_1 \lambda(\mu^2 \sigma^2 + 1)}{2\mu(\mu - \lambda)} + \frac{C_9' + C_{10}D}{\lambda} \tag{5.65}$$

where $C_9' = C_3 + C_9$ and σ^2 is the variance of the service time.

To illustrate the use of (5.65) consider the previous example problem, but let $\sigma^2 = 0.001$. The optimization problem becomes:

minimize

$$VC(\lambda) = \frac{6.6\lambda}{100 - \lambda} + \frac{280}{\lambda} + \frac{0.64}{\sqrt{\lambda}}$$

subject to

$$\lambda = \frac{10{,}000}{n}, \quad n = 101, 102, \ldots$$

A search over λ yields a value of 39.526 for λ^*. Therefore, 253 single-server facilities should be provided. The effect of the value of C_{10} on the optimum solution is indicated in Table 5.11. Clearly, the optimum solution is not

Table 5.11 Solution for Example 5.18

C_{10}	n^*	λ^*	$VC(\lambda^*)$
1	253	39.526	11.414
⋮			
6	253	39.526	11.500
⋮			
10	253	39.526	11.567
15	252	39.683	11.652
20	252	39.683	11.737
40	250	40.000	12.075
60	248	40.323	12.411
80	247	40.486	12.747
100	245	40.816	13.081

sensitive to errors in estimating the value of C_{10}. Note that as C_{10} increases in value from $1 to $100, the number of facilities decreases by approximately 3% and the minimum variable cost per unit time increases by less than 15%.

ASPIRATION LEVEL MODELS

Depending on the objectives of the decision maker, the design of a queueing system might be influenced more by the level of service to be provided than by the costs of providing the service. As an illustration, a store manager may wish to ensure that, say, no more than 10% of his customers have to wait for service upon entering his store. The manager's objectives may at first appear to be unreasonable when compared with the "minimum expected cost solution." However, lessons from decision theory remind us that managers have unique value systems.

If the manager's utility function for money is nonlinear, then it is unlikely that he will choose the system design that minimizes expected cost. Furthermore, the manager will often have multiple objectives, rather than the single objective of cost minimization. Thus, for example, the manager's utility for having customers served without delay must be compared with his utility for money.

Due to the number of instances in which decision makers have "service-level" objectives, as opposed to "cost-minimization" objectives, aspiration level models are often used to assist the manager in designing the queueing system. We have previously introduced the notion of aspiration levels in designing queueing systems. Specifically, in Example 3.4 we wished to determine the minimum number of telephone lines to be installed in order to insure that no more than 5% of those placing calls would encounter a busy signal.

Basically, the approach employed in using aspiration level models is to minimize an appropriate objective function subject to one or more service level constraints. The decision problem in Example 3.4, for instance, could be stated formally as:

minimize c

subject to

$$\frac{e^{-cp}(cp)^c/c!}{\sum_{n=0}^{c} e^{-cp}(cp)^n/n!} \leq 0.05$$

In general, the optimization problem can be formulated as:

minimize $f(\lambda, \mu, c, N, K)$ (5.66)

subject to

$$g_i(\lambda, \mu, c, N, K) \leq \alpha_i, \quad i = 1, \ldots, m$$

where $f(\ )$ can be either an expected cost objective function or a service level objective function and $g_i(\ )$ either an expected cost constraint or a service level constraint. In the following discussion of the use of aspiration level models, a number of different applications of (5.66) will be examined.

Example 5.19 As an illustration of the use of aspiration level models in designing queueing systems, consider the situation posed in Example 5.7. The number of servers to provide in a storeroom is to be determined. Customers arrive at a Poisson rate of 30/hr; each server can service an average of 20 customers/hr, and service time is exponentially distributed. The storeroom manager wishes to provide the minimum number of servers necessary to achieve the following levels of service:

$$\text{delay probability} \leq 0.15$$
$$\text{average waiting time} \leq 1.00 \text{ min}$$

From Table 3.3, for the $(M\,|\,M\,|\,c):(GD\,|\,\infty\,|\,\infty)$ queue it is observed that $D = (c\mu - \lambda)W_q$. Therefore, the design problem reduces to determining the minimum value of c that satisfies the following inequalities:

$$(20c - 30)W_q \leq 0.15, \qquad W_q \leq 0.0167 \text{ hr}, \qquad c \geq 2$$

The latter inequality stems from the requirement that $\rho < 1$.

Figure 5.15 provides values of $W_q\mu$ for combinations of λ/μ and c. Shown in Table 5.12 are results obtained from Fig. 5.15. Notice that the binding constraint is the one involving the delay probability. To insure that $D \leq 0.15$ requires four servers; whereas to have $W_q \leq 0.0167$ requires only three servers. Also, notice that a delay probability of 0.2250 is produced by three servers.

TABLE 5.12 Solution for Example 5.19

c	$W_q\mu$	W_q	$(20c - 30)W_q$
2	1.30	0.06500	0.6500
3	0.15	0.00750	0.2250
4	0.007	0.00035	0.0175

It is interesting to consider the imputed value for customers waiting for service, given the objective of maintaining a delay probability of no more than 0.15. From Fig. 5.14, for $c^* = 4$, $4 \leq C_1/C_3 \leq 20$. Thus, stipulating that $D \leq 0.15$ is equivalent to stating that $C_1 \geq 4C_3$, or the cost of customers waiting for service is at least four times the cost of the time servers spend in providing service. Also, notice that with four servers the average server utilization will be only 37.5%.

One additional (and obvious) point should be made concerning the use of aspiration level models; namely, the feasible solution space for (5.66) must be nonempty. Put another way, the service level objectives, which are stated as constraints in (5.66), must not be contradictory. As an illustration, for the storeroom example, suppose it is also required that the servers be utilized at least 50% of the time; thus, $\lambda/c\mu \geq 0.50$, or $c \leq 3$. In this case, it would be impossible to satisfy simultaneously both the delay probability objective and the server utilization objective. Thus, care must be taken to ensure that the service level objectives are not mutually exclusive.

Example 5.20 Next, we consider the situation posed by Example 5.9 in which a production department is to be designed. Specifically, the values of c and N are to be determined for an $(M \mid M \mid c):(GD \mid N \mid \infty)$ queue. Management wishes to minimize the investment cost in production machines and in-process inventory space subject to the following service level constraints. First, it is desired that no more than 20% of the arriving orders be dispatched to a special processing area, due to lack of space at the production department. Second, it is desired that the production machines be utilized at least 70% of the time. Third, on the average, no more than 4 orders are to be waiting to be processed.

The optimization problem reduces to:

minimize

$$TC(c, N) = C_3(c - L + L_q) + C_4(L - L_q) + C_5 N \quad (5.67)$$

subject to

$$\frac{\tilde{\lambda}}{\lambda} \geq 0.80, \qquad \frac{\tilde{\lambda}}{c\mu} \geq 0.70, \qquad L_q \leq 4$$

where $\tilde{\lambda}$ is the effective arrival rate and is represented mathematically by $\tilde{\lambda} = \mu(L - L_q)$.

Depending on the definition of "investment cost in production equipment," C_3 and C_4 might be equal. If the investment cost is to include only the cost of capital recovery (Thuesen et al., 1971) then $C_3 = C_4$, since capital recovery cost includes only depreciation cost plus the interest on the original investment. However, if investment cost is to include the costs of operating and maintaining the equipment, then C_3 and C_4 will be different. For our purposes, we will assume they are not the same and let $C_3 = \$8/\text{hr}$, $C_4 = \$20/\text{hr}$, and $C_5 = \$1/\text{hr}$. Additionally, we let $\lambda = 8/\text{hr}$ and $\mu = 2/\text{hr}$.

Shown in Table 5.13 are the values of $TC(c, N)$ for various values of c and N. Also shown are the values of $\tilde{\lambda}/\lambda$, $\tilde{\lambda}/c\mu$, and L_q. Observe that the optimum values of c and N are 4 and 6, respectively.

TABLE 5.13 Results for Example 5.20

c	N	$TC(c, N)$	$\dfrac{\tilde{\lambda}}{\lambda}$	$\dfrac{\lambda}{c\mu}$	L_q
2	4	42.82	0.475[a]	0.951	1.311
2	5	44.42	0.488[a]	0.976	2.176
2	6	45.72	0.494[a]	0.988	3.099
2	7	46.86	0.497[a]	0.994	4.055[a]
2	8	47.93	0.499[a]	0.997	5.030[a]
3	4	57.98	0.625[a]	0.833	0.375
3	5	60.99	0.666[a]	0.889	0.917
3	6	63.22	0.692[a]	0.923	1.558
3	7	65.03	0.709[a]	0.945	2.269
3	8	66.58	0.720[a]	0.961	3.032
4	4	69.09	0.689[a]	0.689[a]	0.000
4	5	73.62	0.763[a]	0.763	0.237
4	6	76.80	0.808	0.808	0.575
4	7	79.28	0.839	0.839	0.965
4	8	81.35	0.861	0.861	1.385
5	5	83.44	0.801	0.641[a]	0.000
5	6	87.41	0.863	0.690[a]	0.137
5	7	90.25	0.901	0.721	0.322
5	8	92.48	0.927	0.741	0.518
6	6	96.38	0.883	0.589[a]	0.000
6	7	99.52	0.928	0.618[a]	0.072
6	8	101.79	0.954	0.636[a]	0.161

[a] Infeasible combination due to constraint violation.

Example 5.21 As another illustration of the use of aspiration levels in designing queueing systems, consider the problem of determining the number of service facilities and the number of servers per service facility, as depicted previously in Example 5.17. Suppose it is desired that the location of service facilities be such that the expected distance traveled by customers in going to the nearest service facility will be no greater than d distance units. Furthermore, suppose it is desired that the delay probability at each service facility be no greater than q. Also, servers at each facility must be utilized, on the average, at least $p\%$ of the time. Finally, the cost of service facilities plus servers is to be minimized subject to the service level constraints.

A mathematical statement of the problem can be given for the rectilinear distance case as follows:

minimize

$$TC(\lambda, c) = \frac{\lambda}{\tilde{\lambda}}[C_3 c + C_9] \qquad (5.68)$$

subject to

$$\frac{2}{3}\left(\frac{2A\lambda}{\lambda}\right)^{1/2} \leq d, \quad D \leq q, \quad p\mu \leq \frac{\lambda}{c} < \mu, \quad \lambda = \frac{\lambda}{n}, \quad n = 1, 2, \ldots$$

Using the data from Example 5.17 and letting $d = 2$ mi, $q = 0.40$, and $p = 0.50$, (5.68) becomes:

minimize

$$TC(\lambda, c) = \frac{10{,}000}{\lambda}[40c + 300]$$

subject to

$$\lambda \leq 351.562, \quad D \leq 0.40, \quad 50 \leq \frac{\lambda}{c} < 100, \quad \lambda = \frac{10{,}000}{n}, \quad n = 29, 30, \ldots$$

The last condition establishing a lower limit of 29 on n is based on the constraint that $\lambda \leq 351.562$. The value for the delay probability D is obtained from the appropriate expression in Table 3.3.

Using a pattern search solution procedure it is found that $\lambda^* = 270.27$ and $c^* = 4$. Therefore, 37 service facilities are to be provided, with four servers at each facility. The optimum solution produces an expected distance traveled of 1.75 mi, a delay probability of 0.392, and a server utilization of 67.57%. To determine the effects of the delay probability constraint on the optimum solution, values of λ^* and c^* are also given in Table 5.14 for various values of q. Observe that for the values of q considered, c^* remains equal to 4 and λ^* increases with increasing values of q. Furthermore, substantial reductions in total cost result from increases in the allowable delay probability.

Table 5.14 Sensitivity analysis for Example 5.21

q	λ^*	c^*	n^*	$TC(\lambda^*, c^*)$
0.40	270.27	4	37	17,020
0.50	294.12	4	34	15,640
0.60	312.50	4	32	14,720
0.80	344.83	4	29	13,340

Example 5.22 As a final illustration of the use of an aspiration level model, suppose it is desired that the average waiting time per customer be minimized subject to constraints on the number of servers and other service levels. Specifically, suppose the number of servers is to be determined for an

$(M\,|\,M\,|\,c):(GD\,|\,\infty\,|\,\infty)$ queue subject to the average server utilization being at least 70%, the delay probability being no greater than 0.20, and the number of servers being no more than 4. To complete the example, let $\lambda = 20$ customers/hr and $\mu = 6$ customers/hr.

The decision problem can be formulated as follows:

minimize c

subject to

$$0.70 \leq \frac{20}{6c} \leq 1.00, \qquad D \leq 0.20, \qquad c = 1, 2, 3, 4$$

Comparing the first, second, and fourth constraints indicates that the number of serves must be equal to 4. Furthermore, on recalling the relationship between D and W_q, the second constraint can be expressed as

$$(6c - 20)W_q \leq 0.20$$

From Fig. 5.15 with $\lambda/\mu = 3.33$ and $c = 4$, it is observed that $6W_q \doteq 0.90$, or $W_q \doteq 0.15$. Therefore, $D \doteq 0.60$, which violates the delay probability constraint. Thus, there does not exist a feasible solution to the example.

The data for this example were chosen purposely to illustrate the situation that can develop if the analyst does not exercise extreme caution in formulating the aspiration level model. It often occurs that the manager, when asked for his service level objectives, will wish for the impossible, a system with no customers waiting and no idle servers. If minimum customers waiting is wanted, a price of low server utilization must be paid and vice versa. The design of a queueing system ultimately involves a trade-off between the conflicting objectives of providing a high level of service to the customers and achieving a high utilization of the servers.

COST DETERMINATION

In this section we consider the determination of the values of the cost coefficients employed in the models developed previously. Our discussion of the determination of costs for economic queueing models will be heavily influenced by the writings of Hillier (1964, 1965, 1966), whose research on the subject provides considerable insight into the operational aspects of measuring costs associated with queueing systems.

In estimating the values of the cost coefficients for a specific decision model, a number of points should be considered: namely, precise estimates of the values of the cost coefficients are seldom required due to the insensitivity of the optimum solution to errors in estimating cost coefficients. Additionally, incremental, rather than accounting, costs are desired, since the latter often include such nonincremental elements as overhead. Another aspect of cost estimation to remember is that *future* costs are needed, as opposed to *past* or *present* costs, since the queueing system is to be designed to meet the needs of the future.

For the most part, the cost coefficients presented previously can be categorized as either customer-related, server-related, or facility-related costs. Customer-related cost coefficients include: C_1, the waiting-time cost coefficient; C_2, the servicing-time cost coefficient; C_6, the balking cost coefficient; C_8, the batch-arrival cost coefficient; C_{10}, the travel-time cost coefficient; and for the finite population case C_0, the running-time cost coefficient. Server-related cost coefficients include: C_3, the idle-time cost coefficient; C_4, the busy-time cost coefficient; and C_7, the service-rate cost coefficient. Facility-related cost coefficients include C_5, the facility capacity cost coefficient, and C_9, the coefficient for the fixed cost of a facility.

Hillier (1965) has observed that the greatest obstacle to the use of cost models in designing queueing systems appears to be due to the difficulty in assigning values to the customer-related cost coefficients, with special emphasis being given to the measurement of C_1, the cost of waiting. It is difficult to understand why proportionately more discussion has centered around the determination of the waiting cost for customers in a queueing system rather than the determination of, say, the carrying costs or stock-out costs in an inventory system. However, such appears to be the case! Certainly the determination of the values of the costs for a queueing model is not a trivial matter: but, in our experience, it is not a trivial matter to determine the values of the costs required for a linear programming model, an inventory model, a facilities location model, or a sequencing model, among others. Yet, the process of measuring cost values for the latter classes of models is a subject seldom discussed in textbooks.

Considering the customer-related costs, the costs of a customer waiting for service and being served have many common elements. Indeed, in many cases the values of the costs C_1 and C_2 will be the same. However, the two costs need not have the same value. As an illustration, when the customer is an automatic production machine, such as those found in paper, chemical, and textile processing plants, the machine might be producing defective product throughout the waiting time until the operator is able to shut off the machine and make the necessary repairs. During the servicing of the machine, no defective product is being produced. Thus, the two costs C_1 and C_2 would not have the same value in this application.

Hillier provides the following lucid discussion of the process of estimating the value of C_1:[†]

[†]The following discussion is taken from Hillier (1965) by permission of the publisher. It should be noted that the arguments given apply as well to the determination of the value of C_2.

Before plunging into the detailed discussion, a summary of the primary components of C_1 is given below.

C_1 = cost of a customer waiting one unit of time
   = consequent net reduction in long run earnings of firm
   = long run net incremental income foregone
   + long run net incremental expenses incurred.

Net incremental income foregone includes that due to

1. any lost productive output,
2. any deterioration in customer relations.

Net incremental expenses incurred include those due to

1. any idle in-process inventories,
2. any increased expediting, supervision, and administrative costs.

Any other variable incomes or expenses that would occur in the particular situation of concern should, of course, also be included.

Granted that these four items seem to be the primary candidates for inclusion in C_1, the next question is how to estimate these component costs. These costs will now be considered in turn, beginning with income foregone due to lost productive output.

The cost of lost productive output may be incurred whenever the members of the waiting line would otherwise be engaged in some productive activity for the firm. Examples of such members are employees of the firm and machines waiting to be repaired. It should be emphasized that this cost would not be incurred in cases where the members of the waiting line are customers or jobs rather than employees or other normally productive components of the firm.

The first step in evaluating the cost of lost productive output is to ascertain just how much, if any, output would be lost in the long run due to a delay in the waiting line of one unit of time. Perhaps this time can and would be made up if the work pace is flexible and a fixed work load is rigidly adhered to, in which case there is no net loss in output in the long run. This would tend to be the case, for example, in a job shop situation where the work load is controlled basically by the customers rather than by the shop. On the other hand, if the work pace is quite constant while the work load is not tightly controlled, the net loss in productive output may be just the production that would have taken place during the lost time; this would seem to be an appropriate and convenient assumption in many situations such as when the work pace is machine controlled and a standard product is being produced for an essentially unlimited market. Or perhaps the situation lies somewhere in between these extremes, so that one would need to estimate the portion of the output foregone in the lost time that will not be made up in the long run by a faster work pace. For the situation where both the work pace and the work load are rigid, so that lost time is made up by extra overtime work, the net income foregone by (temporary) lost production would be the net increase in expenses caused by the extra overtime work. Finally, it should be stressed that one is seeking total lost productive output, and not just output lost at the operation involved in the waiting. In other words, if the waiting operation is not independent of other operations, for example, as in an assembly line, so that this waiting creates a chain reaction of waiting on down the line, the total amount of waiting and consequent lost output needs to be ascertained.

After determining the total net long run loss in productive output due to one member of the population waiting one unit of time, the next step is to ascertain the net loss in income attributable to this lost output. This is quite simple in the kind of situation where the lost output consists of whole units of salable items, so that the net loss in income is the sales price minus the expenses saved by not producing the items. Here, as elsewhere, the expenses saved should include the taxes saved on the lost net income. More commonly, the lost productive output involves only a portion of the total manufacturing work on a certain product. In other words, one or perhaps several work stations had to postpone their work on the product, thereby decreasing the time available for subsequent productive output by that amount, whereas the other work stations were unaffected. For this kind of situation, it becomes necessary to assign a monetary value to the lost productive output, where this monetary value may be interpreted as the amount by which this output would have increased the value of the goods-in-process. It usually would be difficult to estimate this value directly. However, an indirect approach can be used if it is agreed that the value to the firm of the productive output of economic resources equals the compensation for those economic resources (where "economic resources" is used in the broad, all-inclusive sense of the economists, and therefore includes managerial skill, capital, and so forth). Thus, the value of productive output equals the compensation to the labor (wages), equipment (capital recovery costs), material, indirectly contributing resources (overhead), and capital (interest payments and dividends) responsible for that output. Thus, in using this analysis to evaluate the value of lost productive output, in effect, one is estimating the long run loss in sales income. This analysis should not be too difficult since the total direct and indirect expenses attributable to each operation should have been estimated during the product analysis, to which one need only add a contribution to profit (in an accounting sense). The resulting estimate of the value of the lost productive output is not yet equal to the desired estimate of the net income foregone because of the lost output. The one step remaining is to subtract the incremental expenses foregone (in a cash flow sense) because of the lost output, such as, for example, taxes and the cost of any material saved.

One common approach to determining the cost of an employee waiting is to assign the wage of that employee. The reader is referred to Hillier (1964), where it was argued that this is usually not a valid approach. One exception might be where both the work load and work pace are fixed so that additional employees must be hired to compensate for employees delayed in the waiting line, and where no new machinery or other expenses would be required. Otherwise, the outlined procedure should be followed.

The second component of C_1 mentioned, income foregone due to any deterioration in customer relations, may not be relevant. For example, under conditions of pure competition, when the firm can always sell at the market price, this

Cost Determination

factor would not apply. However, when there is an intimate vendor–customer relationship, perhaps even a contractual agreement, the promptness of deliveries and the firm's competitive position for quick deliveries may become rather crucial. Beyond any contractual penalty costs involved, this is a difficult cost, or income foregone, to evaluate. It should probably be an average imputed value based on the judgment regarding the worth of promptness in deliveries. The question to be answered is, how much is it worth (that is, how much expenditure would be justified) from the standpoint of customer relations and meeting competition for speedy deliveries to save one unit of time in the delivery. Keep in mind, also, that a delay on one operation does not necessarily mean that delivery will be delayed by the same amount; subsequent operations may have sufficient buffer stock and may be sufficiently slower to avoid being delayed.

The incremental net expenses incurred due to idle in-process inventories should be relatively simple to determine. It is merely the interest costs on the capital tied up by those in-process inventories which are idled by one member of the population waiting one unit of time. Fetter and Galliher (1958) include some discussion regarding this particular cost.

Finally, incremental net expenses incurred due to any increased expediting, supervision, and administrative costs should be included in C_1. While this cost is difficult to pin down, it would usually be a rather minor item. It would probably be treated as an imputed penalty cost for the nuisance caused by a member of the population waiting one unit of time.

In concluding this discussion of the determination of costs, the point of view presented in Hillier (1963) and Hillier (1964) should be reiterated. While a reliable value of C_1 cannot be ascertained with only a minimum of effort, and while a precise estimate is practically unobtainable, a usable approximation should be obtainable after a reasonable amount of competent effort. Furthermore, this penetrating analysis of the consequences of waiting is essential if a sound decision is to be made regarding the proper balance between the amount of service and the amount of waiting for that service. Using superficial intuitive criteria in order to avoid this analysis may lead to a far-from-optimal decision that costs much more in the long run than the analysis would have. Since almost any decision regarding the proper balance between service and waiting is based ultimately on a comparison of the cost of service and the prevailing conception of the consequences of waiting for that service, this decision should be based on the fundamental measures of the consequences of waiting, C_1, if possible.

However, recognizing the difficulty of obtaining estimates of C_1 which warrant some confidence, are there any reasonable alternatives to obtaining and blindly using a single estimate of each C_1? One alternative is to develop a range of values for each C_1, obtain the solutions as described in the preceding sections, and then study these answers and the consequent characteristics of the waiting line. By intuitively evaluating the desirability of the alternative solutions given by the alternative values of each C_1, the final solution can be selected. Furthermore, the corresponding value of each C_1 becomes an imputed value that can be used in the future. A second alternative is to proceed simultaneously with a study using superficial criteria and with another study using an estimate for each

C_1. Then, by analyzing the imputed value for each C_1 from the first study and the waiting line characteristics from the second study, the input parameters for both studies would be revised repeatedly until the differences are completely reconciled and a final solution obtained.

In addition to the cost of a customer waiting for service and the cost of a customer being served, the values of a number of other customer-related costs must be measured. The cost C_6 of a customer balking due to a lack of waiting space should include the cost of lost profit, plus a goodwill cost, in the case of a customer who leaves the system and takes his business elsewhere. If the customer is captive and is simply routed elsewhere for service, then the incremental cost incurred by the customer being served elsewhere should be included in C_6.

The batch arrival cost coefficient C_8 is measured in terms of the incremental cost of material handling required to transport a load of parts to a production department. Fixed costs such as the investment in the material handling equipment and the operator of the equipment would not normally be included unless the addition of such equipment was required as a result of the batch-arrival quantity. A more detailed discussion of material handling costs is given by Fetter and Galliher (1958). In general, those cost elements are to be included in C_8 that actually vary in a cash flow sense with the number of material handling trips made to the service facility per unit time.

The cost per unit time C_{10} of a customer traveling to and from the service facility includes many of the cost elements found in C_1 and C_2. However, C_{10} also includes those expense items that are actually affected by the incremental distance a customer travels per unit time. Such expense items would include, for example, the cost of vehicular transportation, where appropriate.

Since C_0, the cost per unit time of a producing machine, does not enter into the optimization in Example 5.12, it is not crucial that its value be estimated. However, for completeness we should reiterate that incremental costs are required. Thus, included in C_0 are those costs whose values vary in a cash flow sense with the number of machines that are in a producing state.

Hillier also discusses the process of determining estimates of C_3, C_4, and C_7, the server-related cost coefficients. He points out that the direct cost of service per server per unit time is merely the net incremental cost attributable to either the addition of one server in the case of C_3 and C_4 or the addition of one unit of service rate in the case of C_7. This would include the total compensation directly associated with the server as well as any other expenses that vary in a cash flow sense with either the number of busy servers, the number of idle servers, or the magnitude of the service rate per server. As with the case of C_1 and C_2, the costs C_3 and C_4 will normally

Cost Determination

have the same values; however, depending on the wage payment plan (incentive or nonincentive pay) the cost of an idle server could be quite different from the cost of a busy server.

In the case of facility-related costs, it should be emphasized again that fixed administrative and other overhead expenses should not be included in the estimation of the value of either C_5 or C_9. Instead, C_5 should include only those expenses that actually vary in a cash flow sense with the number of waiting spaces in the facility; C_9 should include only those elements of cost that change in a cash flow sense due to the provision of the service facility. In the determination of the value of C_9, one expense not to be forgotten is the capital recovery cost of any investment involved.

Given that either cost models or aspiration level models are to be used, the values of the cost coefficients are determined based on the considerations stated above. However, there still remains a very basic question concerning the appropriateness of cost models and aspiration level models as models of the decision problem. Some argue that modified versions of the aspiration level models should be used; namely, it is frequently suggested that the analyst should simply determine the set of feasible solutions to the aspiration level problem (5.66), rather than specify an objective function that might not accurately reflect the decision maker's objectives. Unfortunately, this approach is sometimes used for the wrong reasons: namely, a justification often given for not developing an expected cost model and for, instead, simply determining the set of feasible solutions to the aspiration level model is the difficulty in obtaining accurate estimates of the required cost coefficients. It should be emphasized that the choice of the decision model to employ should be based more on the objectives of the decision maker, rather than on the ease of data collection for the decision model.

When an analyst is either unable or unwilling to develop an objective function for the aspiration level model, Hillier (1964) suggests the following approach:

1. Determine the cost of service per unit time for the possible levels of service, as shown in Fig. 5.19(a). Let the unit of time be the average interarrival time.
2. Determine the values of W_q for the various possible levels of service, as shown in Fig. 5.19(b).
3. Combine the two figures to obtain a plot of W_q versus cost of service per arrival, as shown in Fig. 5.19(c).

The problem now consists of determining the point on the curve in Fig. 5.19(c) that seems to give the best balance between the average delay in being serviced and the cost of providing that service.

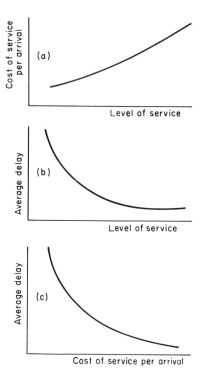

Fig. 5.19 Approach for balancing average delay and cost of service.

For additional discussion of the merits of cost models to assist the decision maker, see the cogent arguments of Hillier (1966) and Taha (1966) for cost models and a modified version of aspiration level models, respectively. Our recommended approach is to model as closely as possible the manager's decision process, based on his utility function. As mentioned previously, we do not believe that this choice should be based wholly on the ease of data collection.

SUMMARY

In this chapter we have presented a number of illustrations of queueing system design problems and have modeled them using cost models and aspiration level models. In order to emphasize the modeling aspects rather than the queueing analysis aspects we restricted our attention to models of Poisson queueing systems. Consequently, the solution to the design problems was quite straightforward in most of the illustrations.

Recognizing that non-Poisson queueing systems are more the rule than the exception and that in many cases simulation results will be used in the decision models, we provided a brief introduction to the subject of optimization. Furthermore, since the mathematical models of the decision problem normally cannot be solved analytically, we emphasized the use of search techniques as opposed to exact, analytical techniques.

Specifically, we presented two search techniques that have proven to be effective, simple to apply, and reasonably efficient for most practical situations. Both techniques can be applied to functions of either discrete or continuous variables or both. However, where some or all of the decision variables are discrete, the analyst must be careful to define the increments δ_i, $i = 1, 2, \ldots, n$, such that the discrete character of the decision variables is maintained throughout each iteration of the search.

The discussion of search techniques presented is by no means exhaustive. There are a multitude of alternatives to the two methods discussed available to the analyst. In addition, research concerned with the improvement of existing techniques and the development of new techniques continues. For a more extensive treatment of this subject, the reader should see Wilde and Beightler (1967) and Beveridge and Schechter (1970), among others.

The estimation process required to determine the values of the cost coefficients employed in the decision models was considered in detail. In that discussion we emphasized repeatedly that incremental costs be used and that the decision model employed be based on the objectives of the decision maker.

REFERENCES

Ashcroft, H., The productivity of several machines under the care of one operator, *J. Royal Statistical Soc., Series B* **12**, 145–151, 1950.

Beveridge, G. S. G., and Schechter, R. S., *Optimization: Theory and Practice*. New York: McGraw-Hill, 1970.

Denn, M. M., *Optimization by Variational Methods*. New York: McGraw-Hill, 1969.

Fetter, R. B., The assignment of operators to service automatic machines, *J. Indust. Eng.* **6** (5), 22–30, 1955.

Fetter, R. B., and Galliher, H. P., Waiting-line models in materials handling, *J. Indust. Eng.* **9** (3), 202–208, 1958.

Francis, R. L., and White, J. A., *Facility Layout and Location: An Analytical Approach*, Englewood Cliffs, New Jersey: Prentice-Hall, 1974.

Friedman, M., and Savage, L. J., Planning experiments seeking maxima, in *Techniques of Statistical Analysis*. (C. Wisenhart et al., ed.). New York: McGraw-Hill, 1947.

Gue, R. L., and Thomas, M. E., *Mathematical Methods in Operations Research*. London: Macmillan, 1968.

Hillier, F. S., Economic models for industrial waiting line problems, *Management Sci.* **10** (1), 119–130, 1963.

Hillier, F. S., The application of waiting line theory to industrial problems, *J. Indust. Eng.* **15** (1), 3-8, 1964.

Hillier, F. S., Cost models for the application of priority waiting line theory to industrial problems, *J. Indust. Eng.* **16** (3), 178-185, 1965.

Hillier, F. S., Reader comments, *J. Indust. Eng.* **17** (5), 283, 1966.

Hillier, F. S., and Lieberman, G. J., *Introduction to Operations Research*. San Francisco: Holden-Day, 1967.

Hooke, R., and Jeeves, T. A., Direct search solution of numerical and statistical problems, *J. Assn. Comp. Mach.* **8** (2), 1961.

King, J. R., On the optimum size of workforce engaged in the servicing of automatic machines, *Internat. J. Production Res.* **8** (3), 207-220, 1970.

Mangelsdorf, T. M., Waiting line theory applied to manufacturing problems, in *Analyses of Industrial Operations* (E. H. Bowman and R. B. Fetter, eds.). Homewood, Illinois: Irwin, 1959.

Morris, W. T., *Analysis for Materials Handling Management*. Homewood, Illinois: Irwin, 1962.

Morris, W. T., *The Analysis of Management Decisions*. Homewood, Illinois: Irwin, 1964.

Palm, D. C., The assignment of workers in servicing automatic machines, *J. Indust. Eng.* **9** (1), 28-42, 1958.

Peck, L. G., and Hazlewood, R. N., *Finite Queueing Tables*. New York: Wiley, 1958.

Schmidt, J. W., *Mathematical Foundations of Management Science and Systems Analysis*. New York, San Francisco, and London: Academic Press, 1974.

Shelton, J. R., Solution methods for waiting line problems, *J. Indust. Eng.* **11** (4), 293-303, 1960.

Taha, H. A., Reader comments, *J. Indust. Eng.* **17** (5), 282, 1966.

Taha, H. A., A case study comparison of independent channels versus a combined pool, *J. Indust. Eng.* **19** (3), 137-143, 1968.

Taha, H. A., *Operations Research: An Introduction*. New York: Macmillan, 1971.

Thuesen, H. G., Fabrycky, W. J., and Thuesen, G. J., *Engineering Economy*, 4th edition. Englewood Cliffs, New Jersey: Prentice-Hall, 1971.

Wilde, D. J., and Beightler, C. S., *Foundations of Optimization*. Englewood Cliffs, New Jersey: Prentice-Hall, 1967.

PROBLEMS

1. Find and identify the stationary points of

$$f(x) = 8x^3 - 30x^2 + 36x - 12$$

2. Show that the function

$$f(x) = ax^3 + b$$

has no interior extreme points.

3. If

$$f(x) = ax^3 + bx^2 + cx + d$$

find conditions for a, b, c, and d such that any value x^* satisfying

$$f'(x^*) = 0$$

is a point of inflection for $f(x)$.

Problems

4. Find and identify the stationary points for

$$f(x) = \frac{a}{4}x^4 - \frac{a}{3}x^3 + b$$

(a) for $a < 0$, (b) for $a > 0$.

5. Find the mode (maximum) of the Erlang density function.

6. Find the value of β that minimizes the function

$$f(\beta) = \int_{-\infty}^{\infty} (x - \beta)^2 \frac{1}{(2\pi)^{1/2}} \exp[-\tfrac{1}{2}(x - \mu)^2]\, dx$$

7. Find and identify the stationary points of

$$f(x_1, x_2) = x_1^3 + 2x_1^2 - 3x_1 + 4x_2^2 - 3x_1 x_2$$

8. Find and identify the stationary points of:

(a) $f(x_1, x_2) = x_1^3 + \tfrac{1}{2}x_2^2$,   (b) $f(x_1, x_2) = x_1^2 - 4x_1 + 4x_2^2$
(c) $f(x_1, x_2) = x_1 x_2 + x_1^2$,   (d) $f(x_1, x_2) = x_1^3 - 3x_k - x_2^2$

9. Find the minimum of Rosenbrock's function given by

$$f(x_1, x_2) = 100(x_2 - x_1^2)^2 + (1 - x_1)^2$$

analytically.

10. Find the mode (maximum) of the Poisson probability mass function.

11. Find the extreme points of $f(x)$, where

$$f(x) = 3x^4 - 4x^3, \quad x = 0, \pm 1, \pm 2, \ldots$$

12. Find the value of x that maximizes

$$f(x) = \frac{5!}{x!(5 - x)!}(0.5)^5, \quad x = 0, 1, 2, 3, 4, 5$$

13. Find and identify the extreme points of the following function:

$$f(x) = 2x^3 - 6x^2 + 1, \quad x = 0, \pm 1, \pm 2, \ldots$$

14. Find and identify the extreme points for

$$f(x) = \tfrac{1}{4}x^4 - 2x^2, \quad x = 0, \pm 1, \pm 2, \ldots$$

15. The manufacturer of a particular product guarantees its operation for T years. If a unit fails within the guarantee period, the unit is replaced at a cost C_F. L units are produced each year. The cost of producing units with a mean life of t years is proportional to t and is given by $C_L t$ per unit. Time x until failure is a random variable that is approximately normally distributed with mean t and variance σ^2. Therefore, the expected annual cost of producing and replacing units is given by

$$C(t) = C_L t L + C_F L \int_{-\infty}^{T} \frac{1}{\sigma(2\pi)^{1/2}} \exp\left[-\frac{(x - t)^2}{2\sigma^2}\right] dt$$

Find the mean design life t that will minimize $C(t)$, using the sequential one-factor-at-a-time method.

16. Let

$$f(x_1, x_2) = x_1^2[\exp -x_1 + 1] + x_2^2 \exp -0.1x_2$$

Find the point (x_1^*, x_2^*) that minimizes $f(x_1, x_2)$: (a) using the sequential-one-factor-at-a-time method, and (b) using the pattern search.

17. Solve Problem 9: (a) using the sequential-one-factor-at-a-time method, and (b) using the pattern search.

18. Consider the problem of determining the number of servers for the $(M \mid M \mid c) : (GD \mid \infty \mid \infty)$ queue. Suppose $\lambda = 3.50$, $\mu = 2$, $C_1 = C_2 = \$10$/customer-hr, and $C_3 = C_4 = \$6$/server-hr. Determine the value of c^*.

19. Solve Problem 18 for the $(M \mid M \mid c) : (GD \mid 4 \mid \infty)$ queue.

20. Two operators are being considered for a particular job. The first operator will be paid at a rate of \$5/hr and will have an exponentially distributed service time with a mean of 10 min. The second operator will be paid at a rate of \$8/hr and will have a normally distributed service time with a mean of 8 min and a standard deviation of 2 min. Customers arrive at a Poisson rate of 5/hr and have a cost of \$10/hr spent in the system. Which operator should be chosen on the basis of minimizing expected costs?

21. Arrivals at a storeroom occur at a Poisson rate of 15/hr; service time is normally distributed with a mean of 4 min and a standard deviation of 1 min. Each minute a customer spends in the system costs \$0.15; each server costs \$0.10/min. Determine the economic number of servers based on an objective of minimizing the expected cost of the system per unit time.

22. In a textile plant, automatic running time for machines is exponentially distributed with a mean of 45 min; service time is exponentially distributed with a mean of 5 min. One operator is to be assigned to keep a specified number of machines running. $C_1 = \$30$/machine-hr, $C_2 = \$15$/machine-hr, and $C_3 = C_4 = \$6$/operator-hr. Determine the economic number of machines to assign the operator.

23. In Problem 22, suppose a crew of repairmen is to be assigned to service 30 machines of the variety described. Determine the economic number of operators to be assigned to service the machines.

24. In Problem 22, suppose the number of machines to be assigned must minimize the expected cost per *running* machine. Determine the economic assignment.

25. In Problem 24, suppose $C_1 = C_2 = \$25$/machine-hr. Determine the economic assignment based on the expected cost per running machine assuming service time is (a) exponentially distributed, (b) constant.

26. As a service to its customers, a discount store provides free parking in its parking lot. Also, if customers arrive and find the lot full the department store pays a fee to a nearby commercial lot to cover the parking expense of the customer. Arrivals are Poisson and occur at a rate of 60/hr; parking time is uniformly distributed between 10 and 30 min. Each parking space owned by the discount store is estimated to cost \$1.80/hr, whereas the average parking fee at the nearby lot is \$0.60/customer. Determine the optimum size for the discount store's parking lot.

27. Solve the problem depicted in Example 5.9 for the case of $C_3 = \$6$/hr when the machine center is idle; $C_4 = \$18$/hr when the machine center is busy; and $C_5 = \$2.50$/hr per waiting space.

28. A population of 30 chemical mixers is to be operated by crews of operators. Crews of size 2, 3, 5, and 6 are being considered. The cost of the crews is \$10, \$14, \$24, and \$30/hr, respectively. The service rates for the various crew sizes will be 6, 8, 12, and 14, respectively. Crews are to be

Problems

assigned to specific mixers. Mixers demand service at a rate of once per hour. The hourly cost of a mixer not running is $100. Determine the economic crew size assuming a Poisson process.

29. Cargo ships arrive at a dock for unloading at a rate of 4/day; unloading time averages 2 days/ship. The cost of a cargo ship being at the dock for unloading is $1500/day; the cost to rent space for unloading a ship is $400/day. Assuming a Poisson process, determine the optimum number of unloading spaces.

30. In Problem 29, suppose a fleet of 30 ships is to be handled at the docks and each ship spends an average of 10 days at sea. Determine the optimum number of unloading spaces.

31. Product is to be delivered to the packaging department at an average rate of 250 parts/hr using lift trucks. A unit load is to be designed for the lift truck in order to minimize expected costs. Each trip by the lift truck will cost $6. Packaging time is exponentially distributed with a mean of 5 sec/part using the automatic packaging machine. The cost of parts waiting to be packaged and being packaged is $10/hr. Assuming Poisson arrivals, determine the optimum size for the unit load.

32. Solve Example 5.17 assuming $C_1 = C_2 = \$80$/customer-day, $C_3 = C_4 = \$30$/server-day, $C_9 = \$200$/facility-day, and $C_{10} = \$3$/mi.

33. Solve Example 5.18 using the data provided in Problem 32.

34. A factory occupies an area of 800,000 ft². Rest-room facilities are to be provided in the factory. Demands for the rest-room facilities are anticipated to occur at a Poisson rate of 4000/day; service time is exponentially distributed with a mean of 6 min. The values of the cost coefficients have been estimated as follows:

$$C_1 = C_2 = \$80\text{/customer-day}, \qquad C_9 = \$100\text{/facility-day}$$

$$C_3 = C_4 = \$10\text{/server-day}, \qquad C_{10} = \$3.50\text{/mi}$$

Determine the optimum number of rest rooms for the factory, as well as the optimum number of "servers" per facility.

35. Arrivals at a group of toll booths occur at a Poisson rate of 600/hr; service time is exponentially distributed with a mean of 15 sec. Determine the minimum number of booths required to ensure that no more than four cars will be waiting with probability of 0.90 and expected waiting time will be no greater than (a) 0.10 min, (b) 0.25 min, (c) 0.50 min, (d) 1 min.

36. Twenty textile machines demand service at random. Running time is exponentially distributed with a mean of 1 hr; service time is exponentially distributed with a mean of 15 min. Determine the minimum number of servers required to service the machines if the delay probability must be less than 0.20 and the expected machine waiting time must be no greater than 10 min/machine.

37. Given the solution to Problem 36, if in Example 5.10 $C_1 = C_2$ and $C_3 = C_4$, what range of values is imputed to the ratio C_3/C_1 in order for the solution to remain optimum?

38. Customers arrive at a storeroom at a Poisson rate of 18/hr. Service time is exponentially distributed with a mean of 6 min. It is desired to determine the minimum number of servers in order to provide a delay probability no greater than 0.15 and to ensure that the average number of customers waiting for service is no greater than four. How many servers are required?

39. In Problem 38, suppose there exists a system capacity of 6 customers and the minimum number of servers is to be determined in order to ensure that no more than 20% of the arriving customers do not enter the system and to guarantee that the utilization of the servers is at least 60%. Can these conditions be satisfied? If so, how many servers should be used? If not, why not?

40. In Problem 38 suppose a server costs $6/hr. What range of values for customer waiting cost would yield the same number of servers when minimizing expected cost?

41. A gasoline bulk terminal is being built by Get-Rich Oil. The terminal will service the 30-truck fleet, with trucks arriving at random intervals. The time required to service a truck is exponentially distributed with a mean of 20 min. A typical truck requires an average of 80 min to deliver its load and return to the terminal, with the total time away from the terminal being exponentially distributed.

(a) Determine the minimum number of loading stations required at the terminal to ensure that, on the average, no more than 3 trucks will be waiting for service.

(b) Determine the optimum number of loading stations if each station costs $100/hr and the cost of a truck waiting for service is $30/hr.

(c) Determine the optimum number of loading stations given the costs in (b) subject to the constraint in (a).

Chapter 6

TRANSIENT ANALYSIS AND SPECIAL TOPICS

INTRODUCTION

In this chapter several important topics in the study and analysis of queueing systems will be presented. The non-Poisson $(M|G|1)$: $(GD|\infty|\infty)$ queue, treated in Chapter 4, will be extensively analyzed using the concept of the imbedded Markov process as introduced by Kendall (1953). This most powerful method of analysis will also be used to establish results for the $(GI|M|1)$: $(GD|\infty|\infty)$ queue. An expression for the distribution function of the busy period for the $(M|G|1)$: $(GD|\infty|\infty)$ queue also will be derived. The busy period for a queueing system is defined as the length of time from the instant a customer enters an empty system until the next instant of time at which the system becomes empty. Another important topic treated in this chapter is that of priority service discipline. In light of the demand placed on present-day computer systems, priority service considerations are of prime importance. Both preemptive and nonpreemptive disciplines will be discussed. The remaining sections of the chapter will be devoted to obtaining transient solutions to selected queueing systems. As would be expected, transient solutions are difficult to obtain and are generally complex for even the simplest of queueing systems. This difficulty arises when one attempts to solve the resulting birth–death differential equations for the system under study. In view of this difficulty, Runge–Kutta methods will be described and illustrated. These methods provide the analyst with a means for numerically solving the birth–death equations.

$(M\,|\,G\,|\,1):(GD\,|\,\infty\,|\,\infty)$ QUEUE†

In this section we analyze the $(M\,|\,G\,|\,1):(GD\,|\infty\,|\infty)$ queue. Here arrivals are generated by a Poisson process, but we do not specify any particular service-time distribution. We only require that the service time for all customers be independently and identically distributed. To analyze this system, we can no longer use the birth–death equations discussed in Chapter 4. These equations were derived on the premise that the state of the system at any instant $t + \Delta t$ was dependent solely on the state of the system at time t and not on any particular route the system may have taken in reaching the given state at time $t + \Delta t$. This property, as we have previously shown, will only be satisfied when the time between successive changes of state is exponentially distributed. A process that has this Markov property is naturally called a Markov process. The $(M\,|\,M\,|\,c)$ queue that was extensively analyzed in Chapter 3 is an example of such a process.

Clearly, the $(M\,|\,G\,|\,1)$ queue is not a Markov process since service time is not necessarily exponentially distributed. Thus, to model this system we not only need to know what state the system is in at time t, but also how long it has been there.

Just as clearly as we recognize that the previous modes of analysis will fail when applied to this system we see the need for a new method of analysis. Such a method has been described by Kendall (1953). He suggests modeling the system by viewing it only at time points where the Markov property holds. These points are referred to as *renewal points*; that is, t is a renewal point if and only if the future states of the system depend solely on the state at time t. Certainly, in a Markov process every point t is a renewal point. Thus, by "glimpsing" the system only at renewal points, we are in essence imbedding a Markov process into a non-Markovian environment, hence the terminology imbedded Markov process.

With regard to the $(M\,|\,G\,|\,1)$ queue, those instances at which customers complete service and leave the system are the renewal points. For whenever a customer leaves the system, either a waiting customer starts service or the system becomes empty; in either case the future state of the system depends solely on the number of customers in the system at present. Thus this system will be glimpsed immediately after service is completed and the associated customer has departed.

As in Chapter 4, we let q_n represent the number of customers in the system when the nth customer departs from the system, and we let ξ_n represent the number of arrivals during service of customer n. Then

$$q_{n+1} = \begin{cases} q_n - 1 + \xi_{n+1}, & q_n > 0 \\ \xi_{n+1}, & q_n = 0 \end{cases} \quad (6.1)$$

† This discussion is based on that of Cox and Smith (1961).

$(M|G|1):(GD|\infty|\infty)$ Queue

Clearly, the quantities q_n and ξ_n are random variables. Furthermore, since service times are identically and independently distributed, $\xi_1, \xi_2, \ldots$ are identically and independently distributed. Of course, ξ_n is independent of q_0, $q_1, \ldots, q_{n-1}$.

Using the Heaviside unit function,

$$U(x) = \begin{cases} 1, & \text{if } x > 0 \\ 0, & \text{if } x \leq 0 \end{cases} \tag{6.2}$$

(6.1) can be written compactly as

$$q_{n+1} = q_n - U(q_n) + \xi_{n+1} \tag{6.3}$$

Now, if the service time for customer n is t_s, then ξ_n has a Poisson distribution with mean λt_s, since arrivals are Poisson. Therefore,

$$P[\xi_n = k \mid t_s] = \frac{\exp(-\lambda t_s)(\lambda t_s)^k}{k!}$$

The unconditional distribution of ξ_n is thus given by

$$P[\xi_n = k] = \int_0^\infty P[\xi_n = k \mid t_s] f(t_s) \, dt_s \tag{6.4}$$

where $f(t_s)$ is the service-time distribution. Taking the generating function of (6.4) we have, upon interchanging the summation and integral operations,

$$\mathscr{G}(\xi_n) = \int_0^\infty \sum_{k=0}^\infty P[\xi_n = k \mid t_s] z^k f(t_s) \, dt_s$$

or

$$\mathscr{G}(\xi_n) = \int_0^\infty \sum_{k=0}^\infty \frac{\exp(-\lambda t_s)(\lambda t_s)^k}{k!} z^k f(t_s) \, dt_s$$

which reduces to

$$\mathscr{G}(\xi_n) = \int_0^\infty \exp[-\lambda t_s(1-z)] f(t_s) \, dt_s \tag{6.5}$$

Recall that

$$\sum_{k=0}^\infty \frac{\exp(-\lambda t_s)(\lambda t_s)^k z^k}{k!} = \exp[-\lambda t_s(1-z)]$$

is the generating function for the Poisson distribution.

Let $s = \lambda(1-z)$; then (6.5) can be written as

$$\mathscr{G}(\xi_n) = \int_0^\infty \exp(-st_s) f(t_s) \, dt_s \bigg|_{s=\lambda(1-z)}$$

or

$$\mathcal{G}(\xi_n) = \mathcal{L}[f(t_s)]\Big|_{s=\lambda(1-z)} \quad (6.6)$$

Thus the generating function $\mathcal{G}(\xi_n)$ for the number of arrivals during the service of the nth customer can be expressed in terms of the Laplace transform for the service-time distribution.

Since we have the generating function for ξ_n, the first factorial moment of ξ_n is given by

$$E(\xi_n^{(1)}) = E(\xi_n) \quad (6.7)$$

where

$$E(\xi_n^{(1)}) = \frac{d\mathcal{G}(\xi_n)}{dz}\Big|_{z=1} \quad (6.8)$$

and the second *factorial* moment of ξ_n by

$$E(\xi_n^{(2)}) = E[\xi_n(\xi_n - 1)] \quad (6.9)$$

or

$$E(\xi_n^{(2)}) = \frac{d^2\mathcal{G}(\xi_n)}{dz^2}\Big|_{z=1} \quad (6.10)$$

From (6.6), $s = \lambda(1 - z)$, implying that $ds = -\lambda\, dz$, and we have

$$\frac{d\mathcal{G}(\xi_n)}{dz}\Big|_{z=1} = \frac{d\mathcal{L}[f(t_s)]}{ds}\frac{ds}{dz}\Big|_{z=1}$$

or

$$E(\xi_n) = \lambda E(t_s) = \frac{\lambda}{\mu} = \rho \quad (6.11)$$

Also,

$$E(\xi_n^{(r)}) = \lambda^r E(t_s^r) \quad (6.12)$$

Thus, the variance of ξ_n can be shown to be

$$\text{Var}(\xi_n) = \lambda^2 \text{Var}(t_s) + \rho \quad (6.13)$$

Let P_k denote the stationary probability of k customers in the system following a service completion. In order to obtain the steady-state probability distribution, we assume $\rho < 1$. In steady state, the probability distribution for q_n is identical to that for q_{n+1}. Hence, $E(q_n) = E(q_{n+1})$. Notice further that $E[U(q_n)] = P(q_n \neq 0)$. However, from (6.3),

$$E(q_{n+1}) = E(q_n) - E[U(q_n)] + E(\xi_{n+1}), \quad \text{or} \quad E[U(q_n)] = E(\xi_{n+1})$$

which from (6.11) gives

$$E[U(q_n)] = \rho$$

Therefore, $P(q_n = 0) = 1 - \rho$, the probability that an arriving customer does not have to wait before being served.

The generating function for q_{n+1} can be written

$$\mathscr{G}(q_{n+1}) = E(z^{q_{n+1}}) = E(z^{q_n - U(q_n) + \xi_{n+1}})$$

or, since ξ_{n+1} is independent of q_n, as

$$\mathscr{G}(q_{n+1}) = E(z^{\xi_{n+1}}) E(z^{q_n - U(q_n)}) \tag{6.14}$$

However,

$$E(z^{q_n - U(q_n)}) = z^{0 - U(0)} P_0 + z^{1 - U(1)} P_1 + \cdots + z^{k - U(k)} P_k + \cdots$$

or, since $U(k) = 1$ for $k > 0$,

$$E(z^{q_n - U(q_n)}) = P_0 + \sum_{k=1}^{\infty} P_k z^{k-1}$$

which can be written as

$$E(z^{q_n - U(q_n)}) = P_0 + z^{-1} \left[\sum_{k=0}^{\infty} P_k z^k - P_0 \right]$$

or

$$E(z^{q_n - U(q_n)}) = (1 - \rho) + \frac{\mathscr{G}(q_n) - (1 - \rho)}{z} \tag{6.15}$$

Recalling that $E(z^{\xi_{n+1}}) = E(z^{\xi_n}) = \mathscr{G}(\xi_n)$ and $\mathscr{G}(q_{n+1}) = \mathscr{G}(q_n)$ in steady state, we find upon substituting (6.15) into (6.14) that

$$\mathscr{G}(q_n) = \frac{(1 - \rho)(1 - z)\mathscr{G}(\xi_n)}{\mathscr{G}(\xi_n) - z} \tag{6.16}$$

where

$$\mathscr{G}(\xi_n) = \mathscr{L}[f(t_s)] \Big|_{s = \lambda(1 - z)}$$

The steady-state probability distribution for the number in the system can be obtained from (6.16) using a power series expansion or tables of transform pairs. Specifically, individual probabilities can be obtained from the relation

$$P_k = \frac{1}{k!} \frac{d^k \mathscr{G}(q_n)}{dz^k} \Big|_{z=0}$$

The factorial moments of the distribution of q_n can now be obtained from (6.16) by differentiation. In particular, $E(q_n) = \mathscr{G}'(1)$, which gives (4.5), the previously obtained result from the Pollaczek–Khintchine formula of Chapter 4.

Example 6.1 To illustrate the usefulness of (6.16), consider the $(M \mid E_k \mid 1):(GD \mid \infty \mid \infty)$ queue. For this system, the service-time density is given by

$$f(t_s) = \frac{k\mu}{(k-1)!}(k\mu t_s)^{k-1}e^{-k\mu t_s}$$

and from (6.6) we have

$$\mathscr{G}(\xi_n) = \int_0^\infty e^{\lambda t_s(z-1)} \frac{k\mu}{(k-1)!}(k\mu t_s)^{k-1}e^{-k\mu t_s}\,dt_s$$

$$= \frac{(k\mu)^k}{[\lambda(1-z)+k\mu]^k} = \left(\frac{1-\beta}{1-\beta z}\right)^k$$

where

$$\beta = \frac{\lambda}{\lambda + k\mu}$$

Hence, from (6.16), we have

$$\mathscr{G}(q_n) = \frac{(1-\rho)(1-z)(1-\beta)^k}{(1-\beta)^k - z(1-\beta z)^k}$$

For the special case of $k = 1$, this model reduces to the $(M \mid M \mid 1):(GD \mid \infty \mid \infty)$ case. Another special case can be obtained by allowing $k \to \infty$. Note that as $k \to \infty$, $\text{Var}(t_s) \to 0$, which is characteristic of a constant service time. Hence, for the $(M \mid D \mid 1):(GD \mid \infty \mid \infty)$ model, we find

$$\mathscr{G}(q_n) = \frac{(1-\rho)(1-z)e^{\rho(z-1)}}{e^{\rho(z-1)} - z}$$

To determine the probability distribution for the amount of time spent in the system, notice that q_n represents the number of arrivals that occurred after customer n entered the system. Letting $f(t_T)$ represent the probability distribution for the amount of time a customer spends in the system, it follows that

$$P(q_n = k) = \int_0^\infty \frac{\exp(-\lambda t_T)(\lambda t_T)^k}{k!} f(t_T)\,dt_T$$

or
$$\mathscr{G}(q_n) = \int_0^\infty \sum_{k=0}^\infty \frac{\exp(-\lambda t_\mathrm{T})(\lambda t_\mathrm{T})^k}{k!} z^k f(t_\mathrm{T}) \, dt_\mathrm{T}$$

which gives

$$\mathscr{G}(q_n) = \int_0^\infty \exp[-\lambda t_\mathrm{T}(1-z)] f(t_\mathrm{T}) \, dt_\mathrm{T},$$

or

$$\mathscr{G}(q_n) = \mathscr{L}[f(t_\mathrm{T})]\bigg|_{s=\lambda(1-z)} \tag{6.17}$$

However, from (6.16) it follows that

$$\mathscr{L}[f(t_\mathrm{T})]\bigg|_{s=\lambda(1-z)} = \frac{(1-\rho)(1-z)\mathscr{G}(\xi_n)}{\mathscr{G}(\xi_n) - z} \tag{6.18}$$

Since we desire the Laplace transform for time in the system, we rewrite (6.18) in terms of $\mathscr{L}[f(t_s)]$ and s. Therefore,

$$\mathscr{L}[f(t_\mathrm{T})] = \frac{(1-\rho)s\mathscr{L}[f(t_s)]}{\lambda \mathscr{L}[f(t_s)] - \lambda + s} \tag{6.19}$$

Equation (6.19) is commonly called the Pollaczek–Khintchine equation. Furthermore, from (6.17) we have

$$\frac{d\mathscr{G}(q_n)}{dz}\bigg|_{z=1} = \frac{d\mathscr{L}[f(t_\mathrm{T})]}{ds} \frac{ds}{dz}\bigg|_{z=1}, \quad \text{or} \quad E(q_n) = \lambda E(t_\mathrm{T})$$

The Laplace transform for the time in the waiting line can be obtained directly from (6.19) as

$$\mathscr{L}[f(t_q)] = \frac{(1-\rho)s}{\lambda \mathscr{L}[f(t_s)] - \lambda + s} \tag{6.20}$$

since $\mathscr{L}[f(t_\mathrm{T})] = \mathscr{L}[f(t_s)] \cdot \mathscr{L}[f(t_q)]$. Thus from (6.19) and (6.20), we find that the Laplace transforms for waiting time in the system as well as in the queue can be expressed in terms of the Laplace transform of the service-time distribution.

Interestingly, (6.20) can be easily inverted due to a result of Benes (1957), who shows that the waiting-time density function is given by

$$f(t_q) = (1-\rho) \sum_{j=0}^\infty \rho^j h^{*j}(t_q) \tag{6.21}$$

where $h^{*j}(t_q)$ is the j-fold convolution with itself of the density function

$$h(t_q) = \mu \left[1 - \int_0^{t_q} f(t_s)\, dt_s \right] \quad (6.22)$$

To establish (6.21), let us rewrite (6.20) as

$$\mathscr{L}[f(t_q)] = \frac{1 - \rho}{1 - \rho[\mu(1 - \mathscr{L}[f(t_s)])/s]} \quad (6.23)$$

Expanding, we have

$$\mathscr{L}[f(t_q)] = (1 - \rho) \sum_{j=0}^{\infty} \rho^j \left[\frac{\mu(1 - \mathscr{L}[f(t_s)])}{s} \right]^j \quad (6.24)$$

which can be written as

$$\mathscr{L}[f(t_q)] = (1 - \rho) \sum_{j=0}^{\infty} \rho^j \{\mathscr{L}[h(t_q)]\}^j \quad (6.25)$$

Term-by-term inversion yields the desired result given by (6.21).

Example 6.2 Consider the $(M|M|1)$: $(FCFS|\infty|\infty)$ queue. For this system, it follows that

$$h(t_q) = \mu \exp(-\mu t_q) \quad (6.26)$$

Hence the j-fold convolution of $h(t_q)$ with itself is the gamma density function given by

$$h^{*j}(t_q) = \frac{\mu^j}{\Gamma(j)} t_q^{j-1} \exp(-\mu t_q), \quad j \neq 0 \quad (6.27)$$

Of course, $h^{*0}(t_q) = 0$ by definition. Substituting (6.27) into (6.21), we obtain

$$\begin{aligned} f(t_q) &= (1 - \rho) \left[\sum_{j=1}^{\infty} \rho^j \frac{\mu^j}{\Gamma(j)} t_q^{j-1} \exp(-\mu t_q) \right] \\ &= (1 - \rho) \left[\lambda \exp(-\mu t_q) \sum_{j=1}^{\infty} \frac{(\lambda t_q)^{j-1}}{(j - 1)!} \right] = (1 - \rho) \lambda \exp[-(\mu - \lambda) t_q] \end{aligned} \quad (6.28)$$

∎

BUSY PERIOD

A concept that often proves useful in the analysis of queueing systems is that of the busy period. The busy period is defined to be the length of time from the instant a previously empty system is entered until the next time at which the system becomes completely empty.

Busy Period

Busy-period analysis has found increasing usage in recent years, particularly in the design and study of real-time computer systems. Of paramount importance to the analysis is the development of the probability density function of the duration of a busy period. In this section, we shall derive such an expression for the $(M \mid G \mid 1) : (GD \mid \infty \mid \infty)$ queue. The results presented are mainly due to Takacs (1962).

Let $b(t)$ denote the probability density function of the busy period and $b_j(t)$ the probability density function of the j-busy period. A j-busy period is defined as the length of time a system remains in continuous operation prior to becoming empty given that service did not start until $j \geq 1$ customers were in the system. Of course, all new arrivals occurring during this period must also be served. Obviously, $b_1(t) = b(t)$.

It should be apparent that the order in which waiting customers are served has no effect on the length of the busy period. Therefore, for modeling purposes, we can assume a first come–first served discipline and find the j-busy period as the sum of j 1-busy periods. Since these periods are independent, we conclude that the density function for the j-busy period is the j-fold convolution with itself of the density function $b(t)$ for the 1-busy period. Therefore,

$$b_j(t) = b^{*j}(t). \tag{6.29}$$

Now a 1-busy period will not exceed a time t if and only if the time to service the customer is less than t, say t_s, and the j customers who arrive during this service time generate a j-busy period that does not exceed $t - t_s$. Thus, the distribution function $B(t)$ for the 1-busy period can be written as

$$B(t) = \sum_{j=0}^{\infty} \int_0^t \frac{(\lambda t_s)^j}{j!} \exp(-\lambda t_s) B_j(t - t_s) f(t_s) \, dt_s \tag{6.30}$$

where $f(t_s)$ is the service-time probability density function. Differentiating both sides of (6.30) with respect to t and taking the Laplace transform, we obtain

$$\mathscr{L}[b(t)] = \sum_{j=0}^{\infty} \int_0^{\infty} \frac{(\lambda t_s)^j}{j!} \exp(-\lambda t_s) \exp(-s t_s)$$
$$\times \mathscr{L}[b_j(t)] f(t_s) \, dt_s \tag{6.31}$$

$$= \int_0^{\infty} \exp(-\{s + \lambda - \lambda \mathscr{L}[b(t)]\} t_s) f(t_s) \, dt_s \tag{6.32}$$

Therefore,

$$\mathscr{L}[b(t)] = \mathscr{L}[f(t_s)] \bigg|_{s' = s + \lambda - \lambda \mathscr{L}[b(t)]} \tag{6.33}$$

By direct inversion using (6.33), Takacs (1962) obtained the desired result, namely,

$$b(t) = \sum_{n=1}^{\infty} \frac{1}{n} \frac{(\lambda t)^{n-1}}{(n-1)!} e^{-\lambda t} f^{*n}(t) \qquad (6.34)$$

where $f^{*n}(t)$ is the n-fold convolution of the service time density function with itself.

Let b be the mean of the busy period. Differentiating (6.33) and using the fact that

$$-\frac{d}{ds} \mathscr{L}[f(t_s)]\bigg|_{s=0} = \mu^{-1}, \quad \text{and} \quad -\frac{d}{ds} \mathscr{L}[b(t)]\bigg|_{s=0} = b$$

we obtain

$$\frac{d}{ds} \mathscr{L}[b(t)]\bigg|_{s=0} = \left(1 - \lambda \frac{d}{ds} \mathscr{L}[b(t)]\right) \frac{d}{ds'} \mathscr{L}[f(t_s)]\bigg|_{s=0}$$

or

$$b = \frac{1}{\mu}(1 + \lambda b)$$

Solving for b, we find

$$b = \frac{1}{\mu - \lambda} \qquad (6.35)$$

It is of interest to note that Takacs (1967) also derived (6.34) through an entirely different approach, using combinatorial methods and advanced theorems of queueing theory. The reader is encouraged to consult this work for further details. Interestingly, Takacs also develops an expression for the probability density function for a j-busy period:

$$b_j(t) = \sum_{n=j}^{\infty} \frac{j}{n} \frac{(\lambda t)^{n-j}}{(n-j)!} e^{-\lambda t} f^{*n}(t) \qquad (6.36)$$

Note that for $j = 1$, (6.36) reduces to (6.34). Furthermore, Takacs showed that the factor j/n in (6.36) could be interpreted as the probability that in a j-busy period exactly $n - j$ customers arrive in such a fashion as to make this a busy period.

Before concluding our study of the busy period, we should call to the reader's attention that the busy-period distribution actually measures the rate at which the system approaches the no-queue state. With a little thought, the reader should be able to justify this fact. Thus, the probability density function of the busy period $b(t)$ can also be obtained from $dP_0(t)/dt$; that is, $b(t) = dP_0(t)/dt$ (see Problem 5).

$(GI\,|\,M\,|\,1):(GD\,|\,\infty\,|\,\infty)$ QUEUE

In the preceding section, the $(M\,|\,G\,|\,1)$ queue was modeled as an imbedded Markov process with renewal points taken at the instances of service departure. In this section, we will analyze the $(GI\,|\,M\,|\,1)$ queue by once again calling upon this very powerful mode of analysis. For this non-Poisson system, however, the renewal points occur at the epochs of customer arrival. Thus, we will model the $(GI\,|\,M\,|\,1)$ queue by glimpsing the system immediately after an arrival occurs. Specifically, let us define

q_n = number of customers in the system ahead of the newly arrived customer n

x_{n+1} = number of customers served between the arrival of customer n and customer $n+1$

Then

$$q_{n+1} = q_n + 1 - x_{n+1} \qquad (6.37)$$

where $x_{n+1} \leq q_n$. Since the service distribution is exponential with rate μ, it follows that

$$P[x_{n+1}\,|\,t_a] = \frac{(\mu t_a)^{x_{n+1}} \exp(-\mu t_a)}{x_{n+1}!} \qquad (6.38)$$

where t_a is the time between consecutive arrivals.

Let $f(t_a)$ be the interarrival time distribution; then

$$P(x_{n+1}) = \int_0^\infty \frac{(\mu t_a)^{x_{n+1}} \exp(-\mu t_a)}{x_{n+1}!} f(t_a)\, dt_a \qquad (6.39)$$

To ensure steady-state conditions, we assume $E(t_a)\mu > 1$. To obtain the steady-state distribution of q_n, notice that the random variables q_n and q_{n+1} are identically distributed in the steady state. Thus, from (6.37) it follows that

$$P(q_{n+1} = m) = \sum_{j=0}^{\infty} P(q_n = m - 1 + j) P(x_{n+1} = j), \qquad m \geq 1 \quad (6.40)$$

and

$$P(q_{n+1} = 0) = \sum_{j=0}^{\infty} P(q_n = j)\left[1 - \sum_{k=0}^{j} P(x_{n+1} = k)\right] \qquad (6.41)$$

Cox and Smith (1961) provide a solution to (6.40) of the form

$$P(q_{n+1} = m) = a\theta^m \qquad (6.42)$$

where the values of $a > 0$ and m are determined so that both (6.40) and (6.41) are satisfied.

To obtain the value of θ, we substitute (6.42) into (6.40) to give

$$a\theta^m = \sum_{j=0}^{\infty} a\theta^{m-1+j} P(x_{n+1} = j), \qquad (6.43)$$

Dividing both sides of (6.43) by $a\theta^{m-1}$, we obtain

$$\theta = \sum_{j=0}^{\infty} \theta^j P(x_{n+1} = j) \qquad (6.44)$$

Notice that the right-hand side of (6.44) is in the form of a generating function with $z = \theta$. Therefore, θ must satisfy the relation

$$\theta = G(\theta) \qquad (6.45)$$

where $G(\theta)$ is the generating function of x_{n+1} given by

$$G(\theta) = \sum_{j=0}^{\infty} \theta^j P(x_{n+1} = j) \qquad (6.46)$$

Substituting (6.39) into (6.46), we obtain

$$G(\theta) = \sum_{j=0}^{\infty} \theta^j \int_0^{\infty} \frac{(\mu t_a)^j \exp(-\mu t_a)}{j!} f(t_a) \, dt_a \qquad (6.47)$$

and upon interchanging the order of integration and summation in (6.47), we have

$$G(\theta) = \int_0^{\infty} \exp(-\mu t_a) \left[\sum_{j=0}^{\infty} \frac{(\theta \mu t_a)^j}{j!} \right] f(t_a) \, dt_a \qquad (6.48)$$

The quantity in brackets in (6.48) reduces to $\exp(\theta \mu t_a)$. Therefore,

$$G(\theta) = \int_0^{\infty} \exp[-\mu t_a (1 - \theta)] f(t_a) \, dt_a \qquad (6.49)$$

Letting $s = \mu(1 - \theta)$, then upon substituting into (6.49) we obtain

$$G(\theta) = \int_0^{\infty} \exp(-s t_a) f(t_a) \, dt_a \qquad (6.50)$$

which is the Laplace transform for the interarrival distribution. Thus, on combining (6.45) and (6.50), we find

$$\theta = \mathscr{L}[f(t_a)]\Big|_{s=\mu(1-\theta)} \qquad (6.51)$$

It can be shown that there always exists a unique solution θ_0 to (6.51) such that $0 < \theta_0 < 1$.

The value of a can be found by summing both sides of (6.42) over all values of m to obtain

$$\sum_{m=0}^{\infty} a\theta^m = 1 \qquad (6.52)$$

Substituting θ_0 for θ in (6.52) and reducing, we have

$$a = 1 - \theta_0$$

Consequently,

$$P(q_n = m) = (1 - \theta_0)\theta_0^m, \quad m = 0, 1, 2, \ldots \qquad (6.53)$$

which is a geometric distribution with mean $\theta_0/(1 - \theta_0)$.

We now obtain the distribution for time spent in the system by a newly arrived customer. Note that $f(t_T \mid q_n = m)$ is a gamma distribution, since service times are exponentially distributed. Specifically,

$$f(t_T \mid q_n = m) = \frac{\mu(\mu t_T)^m}{m!} \exp(-\mu t_T),$$

and

$$f(t_T) = \sum_{m=0}^{\infty} \frac{\mu(\mu t_T)^m}{m!} \exp(-\mu t_T)(1 - \theta_0)\theta_0^m$$

which reduces to

$$f(t_T) = \mu(1 - \theta_0) \exp[-\mu(1 - \theta_0)t_T] \qquad (6.54)$$

the exponential distribution with mean $[\mu(1 - \theta_0)]^{-1}$.

Using the above results one can now obtain useful results for the $(E_k \mid M \mid 1):(GD \mid \infty \mid \infty)$ and $(D \mid M \mid 1):(GD \mid \infty \mid \infty)$ queues (see Problems 12 and 13).

PRIORITY SERVICE DISCIPLINES

Thus far, a first come–first served service discipline has been assumed. In many cases customers are served according to a priority scheme. In this section, both nonpreemptive priorities and preemptive-resume priorities are treated. A nonpreemptive-priority scheme exists when the customers in the *waiting line* are ordered according to priority numbers, rather than time of arrival. Within a given priority class, customers are served on a first come–first served basis. A preemptive-priority scheme exists when at any time t the customers in the *system* are ordered according to priority number. Thus, a customer already being served can be removed from service by the arrival of a higher-priority customer. If service for the preempted customer must be repeated, the priority scheme is preemptive repeat, whereas a preemptive-resume priority discipline exists when service for the preempted customer commences from the point of interruption.

Only a brief treatment of priority service disciplines is given in this section. The reader interested in pursuing the topic of priorities in depth is referred to Jaiswal (1968), whose text is devoted entirely to priority queueing systems.

Nonpreemptive Priority (*NPRP*)

We will first treat a nonpreemptive priority service discipline for the $(M|G|1)$: $(NPRP|\infty|\infty)$ queue, followed by a special case of the $(M|M|c)$: $(NPRP|\infty|\infty)$ queue. For the $(M|G|1)$ queue† suppose there are k priority classes with the highest priority denoted priority 1 and the lowest denoted priority k. Thus, a j-customer (customer with priority j) has nonpreemptive priority over $j+1, \ldots, k$ customers. Customers within a given priority class are served on a first come–first served basis. Assume that j-customers arrive in a Poisson fashion with rate λ_j.

The time scale is selected such that $\sum_{j=1}^{k} \lambda_j = 1$. In this way, since customers arrive randomly and independently, λ_j is the probability of an arriving customer being a class j customer.

Let the service-time distribution for a class j customer be given by $f_j(t)$. Therefore, the combined service-time distribution is

$$f(t) = \sum_{j=1}^{k} \lambda_j f_j(t)$$

Let $\rho_j = \lambda_j/\mu_j$, where $1/\mu_j$ is the expected service time for a j-customer. Also, let $\rho = \sum_{j=1}^{k} \rho_j$.

Just as in the earlier section on the $(M|G|1)$: $(GD|\infty|\infty)$ queue, the system will be glimpsed immediately after service is completed. Such instances will be called *epochs* and a j-epoch will occur if a class j customer is to be served next. A 0-epoch occurs if the system is empty. Define an event R_j as having occurred when a j-epoch occurs. Let P_j be the stationary probability of R_j.

†This discussion follows that of Cox and Smith (1961).

Priority Service Disciplines 259

So far as a *j*-customer is concerned, all customers of higher priority are of the same priority class, say class $j - 1$. Therefore, since we are concerned with *j*-customers, we begin by fixing an integer greater than 2 and considering a modified system with customers of classes 1 through $j - 1$ combined into a single priority class H. For convenience, assume all customers in H are served on a first come–first served basis. However, H-customers have nonpreemptive priority over classes $j, j + 1, \ldots, k$. In the modified system, only events $R_H, R_j, \ldots, R_k$ occur. Observe that the consideration of the modified system in no way changes the occurrence of *j*-epochs. The *j*-epochs occur at the same instants in the original and the modified system. Since a *j*-epoch can only occur when there are no H-customers in the system and the periods during which H-customers are being served are the same for the original and modified system, the probability distribution for the number of *j*-customers in the system at a *j*-epoch is the same in both systems.

Having defined a modified system, it is convenient to let $j = 2$. We will show later that the expected number of 2-customers in the system at a 2-epoch equals

$$m_{22} = 1 + \frac{\lambda_2 E(t^2)}{2\rho(1 - \rho_1)(1 - \rho_1 - \rho_2)} \tag{6.55}$$

Now, if we let λ_1 approach zero,

$$\lim_{\lambda_1 \to 0} m_{22} = 1 + \frac{\lambda_2 E(t^2)}{2\rho(1 - \rho_2)}$$

which implies that

$$m_{11} = 1 + \frac{\lambda_1 E(t^2)}{2\rho(1 - \rho_1)} \tag{6.56}$$

Now, generalizing (6.55), we obtain

$$m_{jj} = 1 + \frac{\lambda_j E(t^2)}{2\rho(1 - \rho_H)(1 - \rho_H - \rho_j)} \tag{6.57}$$

which reduces to

$$m_{jj} = 1 + \frac{\lambda_j E(t^2)}{2\rho(1 - \sum_{i=1}^{j-1} \rho_i)(1 - \sum_{i=1}^{j} \rho_i)} \tag{6.58}$$

since

$$\rho_H = \sum_{i=1}^{j-1} \rho_i$$

From the results obtained earlier for the $(M \mid G \mid 1) : (GD \mid \infty \mid \infty)$ queue, we know that $P_0 = 1 - \rho$. Therefore, since epochs occur only after a service is completed, and since λ_j is the probability that an arriving customer is a j-customer, the probability that there will be a j-customer at the head of the line when an epoch occurs is $\lambda_j \rho$.

Consider an arbitrary epoch T and the following epoch $T + 1 = T'$. We begin by letting T be a j-epoch, with the period (T, T') the time required to serve a j-customer. Also, let

x_{j1} = number of 1-customers arriving in (T, T')

x_{j2} = number of 2-customers arriving in (T, T')

t_j = service time of the j-customer who enters service at T

Since arrivals are independent and Poisson distributed, x_{j1}, and x_{j2} are Poisson distributed with parameters $\lambda_1 t_j$ and $\lambda_2 t_j$, respectively. Now define the conditional generating function as

$$G(z_1, z_2 \mid t_j) = E(z_1^{x_{j1}} z_2^{x_{j2}} \mid t_j) = \exp[-\lambda_1 t_j(1 - z_1) - \lambda_2 t_j(1 - z_2)]$$

Removing the condition on t_j yields

$$G_j(z_1, z_2) = \int_0^\infty \exp\{-[\lambda_1(1 - z_1) + \lambda_2(1 - z_2)]t_j\} f(t_j) \, dt_j$$

Now, let $s = \lambda_1(1 - z_1) + \lambda_2(1 - z_2)$. Then

$$G_j(z_1, z_2) = \mathscr{L}[f(t_j)] \bigg|_{s = \lambda_1(1 - z_1) + \lambda_2(1 - z_2)}$$

If T is a 1-epoch, then the number of 1- and 2-customers at T' is given, respectively, by

$$n_1' = n_1 + x_{11} - 1, \quad \text{and} \quad n_2' = n_2 + x_{12}$$

If T is a 2-epoch, the number of 1- and 2-customers in the system at T' is, respectively,

$$n_1' = x_{21}, \quad \text{and} \quad n_2' = n_2 + x_{22} - 1$$

If T is a j-epoch, where $2 < j \leq k$, then

$$n_1' = x_{j1}, \quad \text{and} \quad n_2' = x_{j2}$$

Finally, if T is a 0-epoch, the server becomes free at T and with probability λ_j the next customer to arrive is a j-customer. Therefore, with probability λ_j,

$$n_1' = x_{j1}, \quad \text{and} \quad n_2' = x_{j2}$$

since the first to arrive goes immediately into service.

Priority Service Disciplines

Employing the definition of n'_1 and n'_2, the generating function for n'_1 and n'_2 can be written as

$$E(z_1^{n'_1} z_2^{n'_2}) = \sum_{j=0}^{k} E(z_1^{n'_1} z_2^{n'_2} \mid R_j) P_j$$

or, on expanding,

$$E(z_1^{n'_1} z_2^{n'_2}) = P_1 E[z_1^{n_1 + x_{11} - 1} z_2^{n_2 + x_{12}}] + P_2 E[z_1^{x_{21}} z_2^{n_2 + x_{22} - 1}]$$

$$+ \sum_{j=3}^{k} P_j E[z_1^{x_{j1}} z_2^{x_{j2}}] + P_0 \sum_{j=1}^{k} \lambda_j E[z_1^{x_{j1}} z_2^{x_{j2}}]$$

Recalling that x_{ji} and n_i are independent, we have

$$E(z_1^{n'_1} z_2^{n'_2}) = \frac{P_1}{z_1} E(z_1^{x_{11}} z_2^{x_{12}}) E(z_1^{n_1} z_2^{n_2}) + \frac{P_2}{z_2} E(z_1^{x_{21}} z_2^{x_{22}}) E(z_2^{n_2})$$

$$+ \sum_{j=3}^{k} P_j E(z_1^{x_{j1}} z_2^{x_{j2}}) + P_0 \sum_{j=1}^{k} \lambda_j E(z_1^{x_{j1}} z_2^{x_{j2}})$$

Letting

$$G_1(z_1, z_2) = E(z_1^{n_1} z_2^{n_2}), \quad \text{and} \quad G_2(z_2) = E(z_2^{n_2})$$

and recalling that

$$E(z_1^{x_{j1}} z_2^{x_{j2}}) = \mathscr{L}[f(t_j)] \bigg|_{s = \lambda_1(1 - z_1) + \lambda_2(1 - z_2)}$$

we obtain

$$E(z_1^{n'_1} z_2^{n'_2}) = \frac{P_1}{z_1} \mathscr{L}[f(t_1)] G_1(z_1, z_2) + \frac{P_2}{z_2} \mathscr{L}[f(t_2)] G_2(z_2)$$

$$+ \sum_{j=3}^{k} P_j \mathscr{L}[f(t_j)] + P_0 \sum_{j=1}^{k} \lambda_j \mathscr{L}[f(t_j)] \quad (6.59)$$

Now, for steady-state conditions, the distributions for (n'_1, n'_2) and (n_1, n_2) are the same and, thus, have the same generating functions and expected values. Notice that

$$E(z_1^{n_1} z_2^{n_2}) = \sum_{j=0}^{k} E(z_1^{n_1} z_2^{n_2} \mid R_j) P_j$$

and that

$$E(z_1^{n_1} z_2^{n_2} \mid R_2) = E(z_2^{n_2})$$

since $n_1 = 0$ if a 2-epoch occurs. Furthermore,
$$E(z_1^{n_1} z_2^{n_2} \mid R_j) = 1, \quad \text{for } j \neq 1 \text{ or } 2$$
since $n_1 = n_2 = 0$ if a j-epoch occurs, $j \neq 1$ or 2. Therefore,
$$E(z_1^{n_1} z_2^{n_2}) = P_1 G_1(z_1, z_2) + P_2 G_2(z_2) + \sum_{j=3}^{k} P_j + P_0 \sum_{j=1}^{k} \lambda_j$$
or
$$E(z_1^{n_1} z_2^{n_2}) = P_1 G_1(z_1, z_2) + P_2 G_2(z_2) + 1 - P_1 - P_2 \tag{6.60}$$
Equating (6.59) and (6.60) gives
$$P_1 G_1(z_1, z_2)[1 - \mathscr{L}[f(t_1)]/z_1] + P_2 G_2(z_2)[1 - \mathscr{L}[f(t_2)]/z_2]$$
$$+ \sum_{j=3}^{k} P_j [1 - \mathscr{L}[f(t_j)]] + P_0 \left[1 - \sum_{j=1}^{k} \lambda_j \mathscr{L}[f(t_j)]\right] = 0$$
where $s = \lambda_1(1 - z_1) + \lambda_2(1 - z_2)$ in the Laplace transform for service time.

Cox and Smith (1961) obtain, after taking second derivatives with respect to z_1 and z_2 and evaluating at $z_1 = z_2 = 1$, the following expressions:
$$(1 - \rho_1) m_{11} = 1 - \rho_1 + (2\rho)^{-1} \lambda_1 E(t_s^2)$$
$$(1 - \rho_2) m_{22} = 1 - \rho_2 + \rho_1 m_{12} + (2\rho)^{-1} \lambda_2 E(t_s^2)$$
$$(1 - \rho_1) \frac{m_{12}}{\lambda_2} = \frac{m_{11}}{\mu_1} + \frac{m_{22}}{\mu_2} + \rho^{-1} E(t_s^2) - \frac{1}{\mu_1} - \frac{1}{\mu_2}$$

Solving for m_{22} gives (6.55).

Since j-customers arrive randomly at a rate λ_j, the expected waiting time for a j-customer is given by
$$E_j(t_q) = \rho E_j(t_q \mid t_q > 0) + (1 - \rho) E_j(t_q \mid t_q = 0)$$
or
$$E_j(t_q) = \rho \left(\frac{m_{jj} - 1}{\lambda_j} \right) = \frac{\rho}{\lambda_j} (m_{jj} - 1) \tag{6.61}$$

Substituting (6.58) into (6.61) and simplifying, we obtain
$$E_j(t_q) = \frac{E(t_s^2)}{2(1 - \sum_{i=1}^{j-1} \rho_i)(1 - \sum_{i=1}^{j} \rho_i)} \tag{6.62}$$

The expected waiting time for all customers is therefore given by
$$E(t_q) = \sum_{j=1}^{k} \lambda_j E_j(t_q)$$

Cox and Smith (1961) also show that if w_j is the cost per unit time for a j-customer waiting for service, then expected cost is minimized by assigning priorities to customers based on the quantity $(\mu_j w_j)^{-1}$. The customer class having the lowest ratio of expected service time to cost of waiting is assigned the highest priority, i.e., priority 1. The customer class having the highest ratio is assigned the lowest priority, priority k. Consequently, when all customers have the same unit cost of waiting, the customers should be served using a shortest-service-time priority in order to minimize expected waiting cost for all customers.

A second nonpreemptive priority model to be considered is that for the $(M|M|c): (NPRP|\infty|\infty)$ queue. Saaty (1961) provides the following results for the case of each customer class having the same service-time distribution. The expected waiting time for a j-customer is given by

$$E_j(t_q) = \frac{E(T_0)}{[1 - (1/c\mu) \sum_{i=1}^{j-1} \lambda_i][1 - (1/c\mu) \sum_{i=1}^{j} \lambda_i]} \quad (6.63)$$

where

$$E(T_0) = \frac{(c\rho)^c}{[(c\rho)^c/c!(1 - \rho)] + c!(1 - \rho) \sum_{i=0}^{c-1} (c\rho)^i/i!} \quad (6.64)$$

with $\rho = \lambda/c\mu$ and $\lambda = \sum_{i=1}^{k} \lambda_i$. For equilibrium conditions, $\sum_{i=1}^{j} \lambda_i < c\mu$.

Preemptive Priority

We will now consider the preemptive-resume priority discipline for the $(M|M|1)$ queue. Only two priority classes will be considered and the time scale will be chosen such that $\lambda_1 + \lambda_2 = 1$. By preemptive-resume we mean that a class 1 customer will be served immediately upon arrival if there are no class 1 customers already in the system. Thus a class 1 customer can preempt a class 2 customer already in the service channel. If a class 2 customer is preempted he goes to the "head of the line" for class 2 customers and, when served, service is resumed, not repeated. Since service time is exponentially distributed, the memoryless property of the service-time distribution simplifies the preemptive-resume analysis. Moreover, it should be apparent that the following results also hold for a preemptive-repeat priority discipline.

Under equilibrium conditions $(\lambda_1 < \mu_1)$, the probability distribution for the number of 1-customers in the system is

$$P_{1,n} = \rho_1^n (1 - \rho_1)$$

where $\rho_1 = \lambda_1/\mu_1$. The expected time in the system for a 1-customer is

$$W_1 = \frac{1}{\mu_1 - \lambda_1}$$

To begin the analysis for 2-customers, let

$P_{m,n}$ = the steady-state probability of m 1-customers and n 2-customers in the system at time $t + \Delta t$

Using the fact that in the steady state rate out = rate in, we obtain the following balance equations:

$$(\lambda_1 + \lambda_2 + \mu_1)P_{m,n} = \lambda_1 P_{m-1,n} + \lambda_2 P_{m,n-1} + \mu_1 P_{m+1,n}$$
$$m > 0, \quad n > 0$$

$$(\lambda_1 + \lambda_2 + \mu_2)P_{0,n} = \mu_1 P_{1,n} + \mu_2 P_{0,n+1} + \lambda_2 P_{0,n-1}$$
$$m = 0, \quad n > 0$$

$$(\lambda_1 + \lambda_2)P_{0,0} = \mu_1 P_{1,0} + \mu_2 P_{0,1}$$
$$m = 0, \quad n = 0$$

$$(\lambda_1 + \lambda_2 + \mu_1)P_{m,0} = \lambda_1 P_{m-1,0} + \mu_1 P_{m+1,0}$$
$$m > 0, \quad n = 0$$

These steady-state balance equations can be written compactly as

$$P_{m,n}\{\lambda_1 + \lambda_2 + \mu_1 \varepsilon(m) + \mu_2 \varepsilon(n)[1 - \varepsilon(m)]\}$$
$$= \lambda_1 P_{m-1,n} + \lambda_2 P_{m,n-1} + \mu_1 P_{m+1,n} + \mu_2[1 - \varepsilon(m)]P_{m,n+1}$$

where

$$\varepsilon(m) = \begin{cases} 1, & \text{if } m \neq 0 \\ 0, & \text{if } m = 0 \end{cases} \quad \text{and} \quad \varepsilon(n) = \begin{cases} 1, & \text{if } n \neq 0 \\ 0, & \text{if } n = 0 \end{cases}$$

Also, $P_{-1,n}$ and $P_{m,-1}$ are defined to be zero. The expected number of 2-customers in the system is given by

$$\sum_{n=0}^{\infty} n P_{m,n} = \frac{\rho_2}{1 - \rho_1 - \rho_2}\left[1 + \frac{\mu_2 \rho_1}{\mu_1(1 - \rho_1)}\right]$$

The expected time in the system for customer class 1 is given by

$$W_1^{(P)} = \frac{1}{\mu_1 - \lambda_1} \qquad (6.65)$$

Transient Analysis

and for customer class 2 by

$$W_2^{(P)} = \frac{1}{\mu_2(1 - \rho_1 - \rho_2)}\left[1 + \frac{\mu_2\rho_1}{\mu_1(1 - \rho_1)}\right] \quad (6.66)$$

The expected waiting time in the queue for each customer class is given by

$$W_{q,1}^{(P)} = \frac{\rho_1}{\mu_1(1 - \rho_1)} \quad (6.67)$$

and

$$W_{q,2}^{(P)} = \frac{1}{\mu_2(1 - \rho_1 - \rho_2)}\left[\rho_1 + \rho_2 + \frac{\mu_2\rho_1}{\mu_1(1 - \rho_1)}\right] \quad (6.68)$$

Equations (6.67) and (6.68) can be compared with the analogous results for the nonpreemptive priority system:

$$W_{q,1} = \frac{\rho_1/\mu_1 + \rho_2/\mu_2}{1 - \rho_1} \quad (6.69)$$

and

$$W_{q,2} = \frac{\rho_1/\mu_1 + \rho_2/\mu_2}{(1 - \rho_1)(1 - \rho_1 - \rho_2)} \quad (6.70)$$

The expected waiting time for all customers is $\lambda_1 W_{q,1} + \lambda_2 W_{q,2}$. Without any priorities, the expected waiting time is given by

$$W_q = \frac{\rho_1/\mu_1 + \rho_2/\mu_2}{1 - \rho_1 - \rho_2} \quad (6.71)$$

TRANSIENT ANALYSIS

In this section we treat the problem of solving the birth–death differential–difference equations given by (3.13). For completeness these equations are reproduced below:

$$\frac{d}{dt}P_n(t) = -(\lambda_n + \mu_n)P_n(t) + \lambda_{n-1}P_{n-1}(t) + \mu_{n+1}P_{n+1}(t), \quad n > 0$$

$$(6.72)$$

$$\frac{d}{dt}P_0(t) = -\lambda_0 P_0 + \mu_1 P_1(t), \quad n = 0$$

$$(6.73)$$

As alluded to in Chapter 3, an exact solution of Eqs. (6.72) and (6.73) for the simplest of queueing systems, the $(M \mid M \mid 1) : (GD \mid \infty \mid \infty)$ queue, is extremely difficult to obtain and is quite complex. To verify this statement we present the solution procedure and results for this special queueing system. We also treat certain variations of (6.72) and (6.73), which allow a reasonable solution. In particular, we consider the pure-birth process, the pure-death process, and the $(M \mid M \mid \infty) : (GD \mid \infty \mid \infty)$ queueing system.

We remind the reader of the fact that time-dependent solutions to queueing systems are either unmanageable or impossible to obtain. For this reason, steady-state results were found in Chapter 3 by first taking the limit as $t \to \infty$ throughout the birth–death equations and then proceeding to solve the resulting difference equations to obtain P_n. However, by definition P_n is the limiting result of the time-dependent solution $P_n(t)$, i.e., $P_n = \lim_{t \to \infty} P_n(t)$. Since P_n was not obtained in this manner, it is reasonable to inquire into the validity of the balance-equation approach. Fortunately, for practical queueing systems the steady-state balance-equation approach is almost always valid. This is comforting, since it is by far the easier of the two approaches. Exact conditions for the validity of the approach are found below in the section describing the two-state process.

The Solution of Linear Partial Differential Equations

As we shall see in subsequent sections, the generating function

$$G(z, t) = \sum_{n=0}^{\infty} P_n(t) z^n \tag{6.74}$$

for many transient queueing situations satisfies a partial differential equation of the form

$$Q(z, t) \frac{\partial G(z, t)}{\partial z} + R(z, t) \frac{\partial G(z, t)}{\partial t} = T(z, t) \tag{6.75}$$

subject to some appropriate boundary conditions. To solve this partial differential equation, we first form the subsidiary equations given by

$$\frac{dz}{Q(z, t)} = \frac{dt}{R(z, t)} = \frac{dG(z, t)}{T(z, t)} \tag{6.76}$$

We now find two independent solutions of the subsidiary equations writing them in the form

$$u(z, t, G) = \text{const}, \qquad v(z, t, G) = \text{const} \tag{6.77}$$

The general solution of (6.75) is now given by

$$\Phi(u, v) = 0, \qquad \text{or} \qquad \Psi(v) = u \tag{6.78}$$

Transient Analysis 267

where $\Phi(u, v)$ and $\Psi(v)$ are arbitrary functions. Their precise form can be determined from the boundary conditions; hence the arbitrariness of the functions disappears.

Examples of the use of this technique can be found in the majority of the remaining sections of this chapter. If the reader is confused at this point as to the actual mechanics of applying the above method, we ask him to bear with us, for all will be cleared up shortly.

Pure-Birth Process

Consider the birth–death process in which $\lambda_n = \lambda$ and $\mu_n = 0$ for all values of n. Such a process is referred to as a pure-birth process. For this situation, the birth–death equations become

$$\frac{d}{dt} P_n(t) = \lambda P_{n-1}(t) - \lambda P_n(t), \qquad n > 0 \tag{6.79}$$

$$\frac{d}{dt} P_0(t) = -\lambda P_0(t), \qquad n = 0 \tag{6.80}$$

Let $P_0(0) = 1$ indicate the initial condition. Thus, the pure-birth process is precisely the Poisson process treated in Chapter 3. However, in light of the partial differential equation approach introduced in the last section, we will find it instructive to consider this process once again.

Multiplying (6.79) by z^n, summing from $n = 1$ to $n = \infty$, and adding to (6.80), we obtain

$$\sum_{n=0}^{\infty} \frac{d}{dt} P_n(t) z^n = \lambda \sum_{n=1}^{\infty} P_{n-1}(t) z^n - \lambda \sum_{n=0}^{\infty} P_n(t) z^n$$

or

$$\frac{d}{dt} \sum_{n=0}^{\infty} P_n(t) z^n = \lambda z \sum_{n=1}^{\infty} P_{n-1}(t) z^n - \lambda \sum_{n=0}^{\infty} P_n(t) z^n \tag{6.81}$$

Denoting the generating function of $P_n(t)$ by $G(z, t)$, we can rewrite (6.81) as

$$\frac{\partial G(z, t)}{\partial t} = -\lambda(1 - z) G(z, t) \tag{6.82}$$

The subsidiary equations are

$$\frac{dz}{0} = \frac{dt}{1} = \frac{dG(z, t)}{-\lambda(1 - z) G(z, t)}$$

which imply
$$dz = 0, \quad \text{or} \quad z = c_1$$
where c_1 is a constant. Note that $dz/0$ should not be interpreted as the undefined operation of division by zero. Also, we have
$$dt[-\lambda(1-z)] = \frac{dG(z,t)}{G(z,t)}$$
which upon solving can be written as
$$\ln G(z,t) = -\lambda t(1-z) + c_2', \quad \text{or} \quad G(z,t)e^{\lambda t(1-z)} = c_2$$
where c_2 is the constant $\exp c_2'$. Therefore, the general solution to (6.82) is
$$\Psi(z) = G(z,t)e^{\lambda t(1-z)} \tag{6.83}$$
or
$$G(z,t) = e^{-\lambda t(1-z)}\Psi(z) \tag{6.84}$$
Invoking the boundary condition, we have at time $t = 0$ that
$$G(z,0) = \sum_{n=0}^{\infty} P_n(0)z^n = 1$$
Therefore, at time $t = 0$, (6.83) becomes
$$\Psi(z) = 1$$
and from (6.84)
$$G(z,t) = e^{-\lambda t(1-z)}$$
which is the generating function for a Poisson mass function with parameter λt, i.e.,
$$P_n(t) = \frac{(\lambda t)^n e^{-\lambda t}}{n!}$$

Pure-Death Process

Another useful modification of the birth–death equations is to set $\lambda_n = 0$, $\mu_n = n\mu$ for all n and to impose the initial condition $P_m(0) = 1$. Such a queueing process would occur when arrivals were allowed to "build up" in an m-channel Poisson service system to a level of m before service began, and once begun additional arrivals would be turned away. Although this system would rarely occur in practice, its solution will provide valuable insight into the modeling techniques of transient queues.

Transient Analysis

For this system, the birth–death equations become

$$\frac{d}{dt} P_m(t) = -m\mu P_m(t), \qquad n = m \qquad (6.85)$$

$$\frac{d}{dt} P_n(t) = (n+1)\mu P_{n+1}(t) - n\mu P_n(t), \qquad 0 \le n < m \qquad (6.86)$$

Multiplying (6.86) by z^n, (6.85) by z^m, and adding for all n ($n = 0, 1, 2, \ldots, m$), we have

$$\frac{d}{dt} \sum_{n=0}^{m} P_n(t) z^n = \sum_{n=0}^{m} n\mu P_n(t) z^{n-1} - \sum_{n=0}^{m} n\mu P_n(t) z^n$$

$$= \mu \frac{d}{dz} \sum_{n=0}^{m} P_n(t) z^n - z\mu \frac{d}{dz} \sum_{n=0}^{m} P_n(t) z^n$$

which gives rise to the partial differential equation

$$\frac{\partial G(z,t)}{\partial t} - \mu(1-z) \frac{\partial G(z,t)}{\partial z} = 0 \qquad (6.87)$$

The subsidiary equations are

$$\frac{dt}{1} = \frac{dz}{-\mu(1-z)} = \frac{dG(z,t)}{0}$$

which imply that

$$G(z,t) = c_1, \qquad \text{and} \qquad \mu\, dt = -\frac{dz}{1-z} \qquad (6.88)$$

Solving (6.88), we obtain our second solution, namely,

$$c_2 = (1-z) e^{-\mu t}$$

The general solution of (6.87) can now be given by

$$\Psi[(1-z)e^{-\mu t}] = G(z,t) \qquad (6.89)$$

Invoking the initial condition $P_m(0) = 1$, we find

$$G(z, 0) = \sum_{n=0}^{m} P_n(0) z^n = z^m$$

Therefore, from (6.99) we have

$$\Psi(1-z) = z^m$$

Letting $w = 1 - z$, then $z = 1 - w$, and we can write

$$\Psi(w) = (1 - w)^m$$

Hence, it follows from (6.89) that

$$G(z, t) = [1 - (1 - z)e^{-\mu t}]^m$$

which is the generating function for a binomial mass function with parameter $e^{-\mu t}$; that is,

$$P_n(t) = \binom{m}{n}(1 - e^{-\mu t})^{m-n} e^{-n\mu t}$$

(See Problem 24.)

Two-State Process

Consider the birth–death process in which

$$\lambda_n = \begin{cases} \lambda, & n = 0 \\ 0, & n \neq 0 \end{cases} \qquad \mu_n = \begin{cases} 0, & n = 0 \\ \mu, & n \neq 0 \end{cases}$$

In this case, the system can be in only two states: empty (idle) and full (busy). The differential–difference equations corresponding to this situation can be given as

$$\frac{d}{dt} P_0(t) = -\lambda P_0(t) + \mu P_1(t) \tag{6.90}$$

$$\frac{d}{dt} P_1(t) = \lambda P_0(t) - \mu P_1(t) \tag{6.91}$$

where 0 and 1 denote the empty and full states, respectively. Adding (6.90) and (6.91), we have

$$\frac{d}{dt} [P_0(t) + P_1(t)] = 0$$

Thus

$$P_0(t) + P_1(t) = c \tag{6.92}$$

where c must be unity, since for a given value of t, $P_0(t) + P_1(t) = 1$. Substituting $P_1(t) = 1 - P_0(t)$ into (6.90) yields

$$\frac{d}{dt} P_0(t) + (\lambda + \mu) P_0(t) = \mu$$

Transient Analysis

Taking the Laplace transform, we have

$$s\mathscr{L}\{P_0(t)\} - P_0(0) + (\lambda + \mu)\mathscr{L}\{P_0(t)\} = \frac{\mu}{s}$$

or

$$\mathscr{L}\{P_0(t)\} = \frac{\mu}{(\lambda + \mu)s} - \frac{\mu}{\lambda + \mu}\left[\frac{1}{s + (\lambda + \mu)}\right] + \frac{P_0(0)}{s + (\lambda + \mu)}$$

Inverting, we obtain

$$P_0(t) = \frac{\mu}{\lambda + \mu} + \left[P_0(0) - \frac{\mu}{\lambda + \mu}\right]e^{-(\lambda+\mu)t} \tag{6.93}$$

and by (6.92)

$$P_1(t) = \frac{\lambda}{\lambda + \mu} + \left[P_1(0) - \frac{\lambda}{\lambda + \mu}\right]e^{-(\lambda+\mu)t} \tag{6.94}$$

The steady-state solution can be found by taking the limit as $t \to \infty$ of (6.93) and (6.94):

$$\lim_{t \to \infty} P_0(t) = P_0 = \frac{\mu}{\lambda + \mu} \tag{6.95}$$

and

$$\lim_{t \to \infty} P_1(t) = P_1 = \frac{\lambda}{\lambda + \mu} \tag{6.96}$$

It can be readily verified that (6.95) and (6.96) are also the solutions of the steady-state balance equations

$$\mu P_1 - \lambda P_0 = 0, \quad \text{and} \quad \lambda P_0 - \mu P_1 = 0$$

Thus it appears that the steady-state distribution can be obtained in two different ways: (1) Solve the differential–difference birth–death equations to obtain $P_n(t)$ and calculate the limits

$$\lim_{t \to \infty} P_n(t) = P_n$$

and (2) take the limit as $t \to \infty$ throughout the birth–death equations, equate to zero, and solve the resulting set of difference equations while normalizing so that $\sum_{n=0}^{\infty} P_n' = 1$. Obviously, for any queueing system method 1 is valid. But when is method 2 legal, since in practice it is a far simpler approach? By consulting any text on stochastic processes, we find that a steady-state solution exists and can be found for the birth–death equations (6.72) and (6.73) by method 2 if all $\mu_j > 0$ and the series

$$S = 1 + \frac{\lambda_0}{\mu_1} + \frac{\lambda_0 \lambda_1}{\mu_1 \mu_2} + \cdots + \frac{\lambda_0 \lambda_1 \cdots \lambda_{n-1}}{\mu_1 \mu_2 \cdots \mu_n}$$

converges. In this situation $P_0 = S^{-1}$ and

$$P_n = \frac{\lambda_0 \lambda_1 \cdots \lambda_{n-1}}{\mu_1 \mu_2 \cdots \mu_n} S^{-1}, \qquad n = 1, 2, 3, \ldots$$

$(M \mid M \mid \infty) : (GD \mid \infty \mid \infty)$ **Queue**

Consider a queueing process having exponential arrivals and exponential service with an infinite number of available service channels. Such a system is referred to as a self-service queueing system for rather apparent reasons. However, one should not conclude that this model is applicable only to self-service systems. It has been used to model, for example, bed occupancy levels for hospitals.

Specifically, for this case $\lambda_n = \lambda$ and $\mu_n = n\mu$ for all n, and the differential–difference equations for this system are

$$\frac{d}{dt} P_0(t) = -\lambda P_0(t) + \mu P_1(t) \tag{6.97}$$

$$\frac{d}{dt} P_n(t) = (n+1)\mu P_{n+1}(t) - (\lambda + n\mu) P_n(t) + \lambda P_{n-1}(t), \qquad n > 0 \tag{6.98}$$

Multiplying (6.98) by z^n, summing from 1 to ∞, and adding to (6.97), yields

$$\frac{d}{dt} \sum_{n=0}^{\infty} P_n(t) z^n = \sum_{n=0}^{\infty} (n+1)\mu P_{n+1}(t) z^n - \sum_{n=0}^{\infty} (\lambda + n\mu) P_n(t) z^n$$

$$+ \lambda \sum_{n=1}^{\infty} P_{n-1}(t) z^n$$

Letting

$$G(z, t) = \sum_{n=0}^{\infty} P_n(t) z^n$$

then we have

$$\frac{\partial G(z,t)}{\partial t} = \mu \frac{\partial}{\partial z} G(z,t) - \lambda G(z,t) - \mu z \frac{\partial G(z,t)}{\partial z} + \lambda z G(z,t)$$

or
$$\frac{\partial G(z, t)}{\partial t} - \mu(1 - z)\frac{\partial G(z, t)}{\partial z} = -\lambda(1 - z)G(z, t) \quad (6.99)$$

The solution of this partial differential equation can be shown to be

$$G(z, t) = e^{-\alpha(1-z)}[1 - (1 - z)e^{-\mu t}]^i \quad (6.100)$$

where $P_i(0) = 1$ is the initial condition and $\alpha = (\lambda/\mu)(1 - e^{-\mu t})$. Thus, when $i = 0$,

$$G(z, t) = e^{-\alpha(1-z)} \quad (6.101)$$

and

$$P_n(t) = \frac{e^{-\alpha}\alpha^n}{n!} \quad (6.102)$$

which is the Poisson distribution with parameter $(\lambda/\mu)(1 - e^{-\mu t})$. Taking the limit of (6.100) as $t \to \infty$ yields

$$\lim_{t \to \infty} G(z, t) = \exp\left[-\frac{\lambda}{\mu}(1 - z)\right]$$

which implies that

$$P_n = \frac{\exp(-\lambda/\mu)(\lambda/\mu)^n}{n!} \quad (6.103)$$

Thus, the steady-state solution for this system is Poisson with parameter λ/μ. To obtain $P_n(t)$, recall that

$$P_n(t) = \frac{d^n G(z, t)}{dz^n} \frac{1}{n!}\bigg|_{z=0}$$

$(M|M|1):(GD|\infty|\infty)$ Queue

We now consider the simplest of all the queueing models that have found widespread usage. This model, as we have seen in Chapter 3, can be extensively analyzed in the steady state. In this section we consider its time-dependent properties.

The birth–death differential equations for this system are given by

$$\frac{dP_0(t)}{dt} = -\lambda P_0(t) + \mu P_1(t) \quad (6.104)$$

$$\frac{dP_n(t)}{dt} = -(\lambda + \mu)P_n(t) + \lambda P_{n-1}(t) + \mu P_{n+1}(t), \quad n \geq 1 \quad (6.105)$$

The initial conditions are taken to be

$$P_0(0) = 1 \quad P_n(0) = 0, \quad n > 0$$

Note that (6.104) is of a different form than (6.105). This difference arises as a consequence of the nonnegativity of n and complicates the solution procedure. To avoid this difficulty, let us assume that (6.105) holds for all values of n, positive, zero, and negative. Now if we can find a solution to (6.105) that also satisfies the requirement that

$$P_0(t) = \frac{\lambda}{\mu} P_{-1}(t) \tag{6.106}$$

then (6.104) and (6.105) will be satisfied, since from (6.105)

$$\frac{dP_0(t)}{dt} = -(\lambda + \mu)P_0(t) + \lambda P_{-1}(t) + \mu P_1(t)$$

which by (6.106) can be written as

$$\frac{dP_0(t)}{dt} = -(\lambda + \mu)P_0(t) + \mu P_0(t) + \mu P_1(t) = -\lambda P_0(t) + \mu P_1(t)$$

which gives (6.104).

Multiplying (6.105) by z^n and summing over all integer values of n, we obtain

$$\frac{d}{dt} G(z, t) = \left[-(\lambda + \mu) + \left(\lambda z + \frac{\mu}{z} \right) \right] G(z, t) \tag{6.107}$$

where

$$G(z, t) = \sum_{n=-\infty}^{\infty} P_n(t) z^n$$

The general solution to (6.107) can be shown to be

$$G(z, t) = \Psi(z) e^{-(\lambda + \mu)t + (\lambda z + \mu/z)t} \tag{6.108}$$

where $\Psi(z)$ is an arbitrary function.

We now introduce the Bessel function of the first kind of order n, $J_n(x)$, where

$$J_n(x) = x^n \sum_{m=0}^{\infty} \frac{(-1)^m x^{2m}}{2^{2m+n} m! \, \Gamma(n + m + 1)}$$

and the modified Bessel function of the first kind of order n, $I_n(x)$, where

$$I_n(x) = i^{-n}J_n(ix), \quad i = \sqrt{-1}$$

Using the well-known property

$$\left[\exp\frac{1}{2}x\left(y + \frac{1}{y}\right)\right] = \sum_{n=-\infty}^{\infty} y^n I_n(x)$$

we can write

$$e^{(\lambda z + \mu/z)t} = \sum_{n=-\infty}^{\infty} I_n[2t(\lambda\mu)^{1/2}]\left(\frac{\lambda}{\mu}\right)^{n/2} z^n$$

Hence by (6.108)

$$G(z, t) = \Psi(z)e^{-(\lambda+\mu)t} \sum_{n=-\infty}^{\infty} I_n[2t(\lambda\mu)^{1/2}]\left(\frac{\lambda}{\mu}\right)^{n/2} z^n \qquad (6.109)$$

Since $\Psi(z)$ is arbitrary, let us define it as the generating function of some function γ_{-r}. Therefore, (6.109) becomes

$$G(z, t) = \sum_{r=-\infty}^{\infty} \gamma_{-r} z^r e^{-(\lambda+\mu)t} \sum_{n=-\infty}^{\infty} I_n[2t(\lambda\mu)^{1/2}]\rho^{n/2} z^n \qquad (6.110)$$

where $\rho = \lambda/\mu$. Using the relation

$$\sum_{n=-\infty}^{\infty} A_n X^n \sum_{r=-\infty}^{\infty} B_r X^r = \sum_{n=-\infty}^{\infty} \sum_{r=-\infty}^{\infty} A_r B_{n+r} X^n$$

we can rewrite (6.110) as

$$G(z, t) = e^{-(\lambda+\mu)t} \sum_{n=-\infty}^{\infty} \sum_{r=-\infty}^{\infty} \gamma_r \rho^{(n+r)/2} I_{n+r}[2t(\lambda\mu)^{1/2}] z^n$$

Inverting, we obtain

$$P_n(t) = e^{-(\lambda+\mu)t} \sum_{r=-\infty}^{\infty} \gamma_r \rho^{(n+r)/2} I_{n+r}[2t(\lambda\mu)^{1/2}]$$

The initial conditions $P_0(0) = 1$ and $P_n(0) = 0$ for all $n > 0$, along with the properties of the Bessel functions that $I_0(0) = 1$ and $I_n(0) = 0$, $n \neq 0$, combine to give

$$\gamma_0 = 1, \quad \gamma_{-r} = 0, \quad \text{for} \quad r = 1, 2, \ldots$$

Therefore,

$$P_n(t) = e^{-(\lambda+\mu)t}\left[\rho^{n/2}I_n(x) + \sum_{r=1}^{\infty} \gamma_r \rho^{(n+r)/2} I_{n+r}(x)\right] \quad (6.111)$$

where $x = 2t(\lambda\mu)^{1/2}$ and $\rho = \lambda/\mu$.

To determine the remaining value for γ_r, we recall that we originally wished to satisfy the condition $P_0(t) = \rho P_{-1}(t)$. Direct substitution of $n = 0$ into (6.111) followed by the equation of the coefficients of $I_r(x)$ yields

$$\gamma_1 = \rho^{-1}, \quad \gamma_2 = (1-\rho)\rho^{-2}, \quad \gamma_3 = (1-\rho)\rho^{-3}, \quad \ldots$$
$$\gamma_r = (1-\rho)\rho^{-r}$$

Finally, we have the desired result, namely,

$$P_n(t) = e^{-(\lambda+\mu)t}\left[\rho^{n/2}I_n(x) + \rho^{(n-1)/2}I_{n+1}(x) + \rho^n(1-\rho)\sum_{r=2}^{\infty}\rho^{-(n+r)/2}I_{n+r}(x)\right]$$
$$(6.112)$$

For completeness, we include the following result given by Wagner (1969) for the $(M \mid M \mid 1) : (GD \mid N \mid \infty)$ queue:

$$P_n(t) = P_n + \frac{2}{N+1}$$

$$\times \sum_{j=1}^{N} \frac{\exp\left\{-(1+\rho)t + 2t\sqrt{\rho}\cos\left[\frac{j\pi}{(N+1)}\right]\right\}\rho^{(n-i)/2}}{\left(1 - 2\rho\cos\frac{j\pi}{N+1} + \rho\right)}$$

$$\times \left[\sin\frac{ij\pi}{N+1} - \sqrt{\rho}\sin\frac{(i+1)j\pi}{N+1}\right]$$

$$\times \left[\sin\frac{nj\pi}{N+1} - \sqrt{\rho}\sin\frac{(n+1)j\pi}{N+1}\right] \quad (6.113)$$

where N is the maximum number in the system and P_n the steady-state probability of n in the system given by (3.29). The two-state problem treated in the previous section is a special case of (6.113) when $N = 1$.

Equations (6.112) and (6.113) are not inviting. Considering they represent two of the simplest variations of the birth–death process provides motivation for the use of approximate methods for obtaining transient solutions. Transient solutions for variations of the birth–death process can be obtained approximately using Runge–Kutta methods.

Runge–Kutta Methods

Suppose the birth–death equations are given as

$$\frac{d}{dt} P_n(t) = -(\lambda_n(t) + \mu_n(t))P_n(t) + \lambda_{n-1}(t)P_{n-1}(t)$$
$$+ \mu_{n+1}(t)P_{n+1}(t), \quad n \geq 1 \qquad (6.114)$$

$$\frac{d}{dt} P_0(t) = -\lambda_0(t)P_0(t) + \mu_1(t)P_1(t) \qquad (6.115)$$

Notice that here both the mean arrival rate and mean service rate are expressed as functions of n and t. Equations (6.114) and (6.115) can be solved using Runge–Kutta methods.

The Runge–Kutta method was developed to solve first-order differential equations of the form $dy/dx = f(x, y)$ numerically. The method is based on the Taylor series expansion:

$$f(x_0 + h, y_0 + h) = f(x_0, y_0) + df + \frac{1}{2!} d^2f + \frac{1}{3!} d^3f + \cdots$$

where $d^k f$ is the kth total differential of $f(x, y)$. Since the Taylor series may involve the computation of high-order derivatives in order to achieve the desired accuracies, Runge–Kutta methods were developed to avoid this difficulty. In place of the derivatives, additional values of the given function $f(x, y)$ are used in a way that essentially duplicates the accuracy of a Taylor polynomial. The simplicity of the Runge–Kutta methods has made them very popular in carrying out numerical operations.

The most common Runge–Kutta formulas are

$$k_1 = hf(x, y),$$
$$k_2 = hf\left(x + \frac{h}{2}, y + \frac{k_1}{2}\right)$$
$$k_3 = hf\left(x + \frac{h}{2}, y + \frac{k_2}{2}\right),$$
$$k_4 = hf(x + h, y + k_3)$$
$$y(x + h) \sim y(x) + \tfrac{1}{6}(k_1 + 2k_2 + 2k_3 + k_4) \qquad (6.116)$$

Equation (6.116) duplicates the Taylor series through the term in h^4. The Runge–Kutta formulas given above are justified in a number of texts dealing with numerical analysis.

Runge–Kutta methods are used to solve a single equation involving a single independent and a single dependent variable. The birth–death equations representing the queueing system, either single or multiple channel, comprise an infinite system of simultaneous equations involving a single independent variable t, but an infinite number of dependent variables $P_n(t)$, $n = 0, 1, 2, \ldots, \infty$. For solving this system of equations, the Runge–Kutta method can be modified as follows:

The infinite system of equations is truncated to a finite system by choosing an upper limit on n [that is, by assuming $P_n(t) = 0$ if $n > N$]. For values of n from 0 to N, a vector **PA** is created such that

$$\mathbf{PA}_n(t) = P_n(t), \quad n = 0, \ldots, N$$

Similarly, the following vectors represent the multivariate version of the intermediate quantities. Vector **AA** corresponds to quantity k_1:

$$\mathbf{AA}_n = (\Delta t) \cdot \frac{d\mathbf{PA}_n(t)}{dt}$$

$$= \{-[\lambda(t) + \mu(t)] \cdot \mathbf{PA}_n(t) + \lambda(t)\mathbf{PA}_{n-1}(t)$$

$$+ \mu(t) \cdot \mathbf{PA}_{n+1}(t)\} \cdot \Delta t, \quad \text{if} \quad n = 1, \ldots, N$$

or

$$= \{-\lambda(t)\mathbf{PA}_0(t) + \mu(t) \cdot \mathbf{PA}_1(t)\} \cdot \Delta t, \quad \text{if} \quad n = 0$$

Vector **PB** corresponds to quantity $(y + \frac{1}{2}k_1)$:

$$\mathbf{PB}_n(t + \tfrac{1}{2}\Delta t) = \mathbf{PA}_n(t) + \tfrac{1}{2}\mathbf{AA}_n, \quad n = 0, 1, \ldots, N$$

Vector **AB** corresponds to quantity k_2:

$$\mathbf{AB}_n = (\Delta t)\frac{d}{dt}\mathbf{PB}_n(t + \tfrac{1}{2}\Delta t)$$

$$= \{-[\lambda(t + \tfrac{1}{2}\Delta t) + \mu(t + \tfrac{1}{2}\Delta t)] \cdot \mathbf{PB}_n(t + \tfrac{1}{2}\Delta t)$$

$$+ \lambda(t + \tfrac{1}{2}\Delta t)\mathbf{PB}_{n-1}(t + \tfrac{1}{2}\Delta t)$$

$$+ \mu(t + \tfrac{1}{2}\Delta t) \cdot \mathbf{PB}_{n+1}(t + \tfrac{1}{2}\Delta t)\} \Delta t, \quad \text{if} \quad n = 1, \ldots, N$$

or

$$= \{-\lambda(t + \tfrac{1}{2}\Delta t)\mathbf{PB}_0(t + \tfrac{1}{2}\Delta t) + \mu(t + \tfrac{1}{2}\Delta t) \cdot \mathbf{PB}_1(t + \tfrac{1}{2}\Delta t)\} \Delta t,$$

$$\text{if} \quad n = 0$$

Vector **PC** corresponds to quantity $(y + \frac{1}{2}k_2)$:

$$\mathbf{PC}_n(t + \tfrac{1}{2}\Delta t) = \mathbf{PA}_n(t) + \tfrac{1}{2}\mathbf{AB}_n, \quad n = 0, \ldots, N$$

Vector **AC** corresponds to quantity k_3:

$$\mathbf{AC}_n = (\Delta t)\frac{d}{dt}\mathbf{PC}_n(t + \tfrac{1}{2}\Delta t)$$
$$= \{-[\lambda(t + \tfrac{1}{2}\Delta t) + \mu(t + \tfrac{1}{2}\Delta t)]\mathbf{PC}_n(t + \tfrac{1}{2}\Delta t)$$
$$+ \lambda(t + \tfrac{1}{2}\Delta t)\mathbf{PC}_{n-1}(t + \tfrac{1}{2}\Delta t)$$
$$+ \mu(t + \tfrac{1}{2}\Delta t)\mathbf{PC}_{n+1}(t + \tfrac{1}{2}\Delta t)\}\,\Delta t, \quad \text{if}\quad n = 1, \ldots, N$$

or

$$= \{-\lambda(t + \tfrac{1}{2}\Delta t)\mathbf{PC}_0(t + \tfrac{1}{2}\Delta t) + \mu(t + \tfrac{1}{2}\Delta t)\mathbf{PC}_1(t + \tfrac{1}{2}\Delta t)\}\,\Delta t, \quad \text{if}\quad n = 0$$

Vector **PD** corresponds to quantity $y + k_3$:

$$\mathbf{PD}_n(t) = \mathbf{PA}_n(t) + \mathbf{AC}_n$$

Vector **AD** corresponds to quantity k_4:

$$\mathbf{AD}_n = (\Delta t)\frac{d}{dt}\mathbf{PD}_n(t + \Delta t)$$
$$= \{-[\lambda(t) + \mu(t)]\mathbf{PD}_n(t) + \lambda(t)\mathbf{PD}_{n-1}(t)$$
$$+ \mu(t)\mathbf{PD}_{n+1}(t)\}\,\Delta t, \quad \text{if}\quad n = 1, \ldots, N$$

or

$$= \{-\lambda(t)\mathbf{PD}_0(t) + \mu(t)\mathbf{PD}_1(t)\}\,\Delta t, \quad \text{if}\quad n = 0$$

Using these vectors, the probabilities can be computed using the recursive set of equations

$$P_n(t + \Delta t) = P_n(t) + \tfrac{1}{6}(\mathbf{AA}_n + 2\mathbf{AB}_n + 2\mathbf{AC}_n + \mathbf{AD}_n), \quad n = 0, \ldots, N \tag{6.117}$$

A Machine-Release Model

As an illustration of the use of Runge–Kutta methods in obtaining transient solutions to variations of the birth–death equations, a study given by Ghare et al. (1969) will be presented. This study views the job shop as a network of queues with each work center treated as a single- or multichannel service facility.†

In a large number of production systems, units are manufactured or assembled in continuous sequence and a planning horizon may call for quantities of thousands or tens of thousands of such units. Within this framework, detailed and precise standards are possible and desirable. In addition, production costs and schedules are assumed to be essentially constant from one production unit to the next. In the short-term or job-shop system, where comparatively few units of each type

† The following discussion is taken from Ghare et al. (1969, pp. 208–215) by permission of the publisher.

are manufactured, standards will be gross, measurable improvements in the production process may be anticipated, and succeeding units may then have to be projected at less cost and in less time. The rate of improvement may proceed at a decreasing rate and an accurate forecast will have to incorporate some quantitative measure of improvement.

The job shop is likely to incorporate a number of work centers. Each center would consist of a number of similar or identical production stations, for example, machines that operate in parallel and through one of which a production unit must be processed. If improvements are being realized and the unit processing time at each productive station is diminishing, either more units will be able to be processed through a work center over time or the number of machines that are established in parallel within the work center may be able to be reduced over time to maintain a relatively constant throughput. We will be concerned with the second alternative and, toward this goal, will propose a machine-release scheme for the job shop work center. The scheme is based on the analysis of each center as a multichannel queueing system, allowing the machines to be released from a center as soon as the machine utilization at that center falls below some predetermined level.

A job shop can be defined as a network of queues. The production system requires that the flow of work pass through a series of work centers. Each center can be considered a service facility and each production unit can be viewed as a customer demanding this service. When a unit is available for processing through a center (or when a customer arrives), the service facility may or may not be busy processing earlier arrivals. If the center is busy, the unit (customer) has to wait in an in-process inventory queueing system. The essential parameters of this (or any) queueing system are the distributions of service times. From these distributions, the state probabilities $P_n(t)$, which give the probability of n units in the system at time t, can be developed. This, in turn, leads to such relevant estimates as the expected length of the queue, the probability of delay, the expected waiting time, the probability of an available service center, and so forth.

The majority of queueing systems are treated in the steady-state condition. The forms of the arrival and service distributions are assumed to be constant and the system is assumed to have stabilized. In the low-volume job-shop production system, the assumption of an independent service-time distribution ignores manufacturing improvements. If the service rate is actually improving in some measurable quantity from one production unit to the next, this production system must be viewed as a transient or non–steady-state queueing system.

Manufacturing improvements. The manufacturing process (improvement) function is an empirically derived relationship that expresses the reduction in direct labor man-hours necessary to complete an operation as it is repeated over time. The function is based on the assumption that the direct labor man-hours necessary to complete a unit of production will decrease by a constant percentage each time the production quantity is doubled. This phenomenon of learning or manufacturing improvement has been investigated by a number of authors (for example, Fabrycky and Torgersen, 1966). Empirical evidence was first collected by Wright (1936) and expressed in a linear cumulative average formulation; that is, the cumulative average rather than the unit direct labor man-hours reduces by a constant percentage with doubled production quantities. Conway and Schultz have suggested that "since proponents of neither model are able to establish their position by logic, and empirical evidence is far from sufficient to establish the superiority of one alternative, the choice in usage has been largely a matter of computational convenience" (Conway and Schultz, 1959, p. 41). In support of the unit model, Williams reported that "evidence gained from a thorough analysis of historical cost data at United Control Corporation indicates that the linear unit cost curve is more appropriate when applied to short-run, multiproduct industries" (Williams, 1961, p. 108). Since we are concerned with the job-shop or short-run, multiproduct industry, and it is more convenient from a computational standpoint, the unit formulation is accepted.

The learning function can be expressed mathematically as

$$Y_N = KN^\alpha$$

where N is the unit number, Y_N the time required to produce to Nth unit, K the time required to produce the first unit, and $\alpha = \log S/\log 2$, where S is the slope-parameter.

Since the quantity Y_N represents the production time required for the Nth unit of production, the function $\mu_N = 1/Y_N$ yields the production rate in effect for the Nth unit. For a production operation in which the service facility is subject to learning (short-term, low-volume production), the quantity μ_N is established as the mean of the service-rate distribution for the single-channel facility.

The queueing system. The queueing process can be represented by the birth–death equations established in terms of the time rate of change of the probability of a given number n in the system (in service and in the queue) at time t. The birth–death process for the single-channel Poisson input, exponential holding-time situation is described by (6.72) and (6.73). The quantities λ and μ represent the mean arrival rate and mean service rate, respectively. It is from integration of these equations that the state probabilities $P_n(t)$ are obtained.

The queueing system and manufacturing improvements. In the case where no learning is present, the service rate μ remains constant throughout the production since Y_N is constant for all values of N. If, however, there is learning present, the value of μ used in the birth–death equations is a time-dependent function subject to the following constraints:

$$\mu = \begin{cases} \dfrac{1}{K} = \dfrac{1}{Y_1}, & 0 \leq t \leq CT_1 \\ \dfrac{1}{Y_2}, & CT_1 < t \leq CT_2 \\ \dfrac{1}{Y_3}, & CT_2 < t \leq CT_3 \end{cases}$$

or, in general,

$$\mu = \frac{1}{Y_N}, \qquad CT_{N-1} < t \le CT_N$$

The quantity CT_N is the completion time of the Nth item and is calculated by

$$CT_N = \sum_{i=1}^{N} Y_i$$

During the time period between CT_{N-1} and CT_N, the Nth item is being manufactured.

Because of the time-dependent behavior of the mean service rate μ, the birth–death equations yield state probabilities of a transient nature. These quantities are best studied through the use of a numerical solution giving P_n as a function of t.

The method used in this investigation to extract values of $P_n(t)$ from the birth–death equations is a modified version of the Runge–Kutta method of numerical integration. The computation was completed with the use of a computer program in Fortran IV executed on an IBM 7040 machine.

The aim of the computational scheme is to examine the behavior of the queueing system state probabilities under transient conditions over some stipulated time spectrum. The transient conditions are those generated by the time-dependent service rate.

The first step is to define the problem by establishing values for λ, the arrival rate; S, the slope parameter of the appropriate learning rate; and K, the time required to manufacture the first unit of output. Also, the length of time over which the investigation is to be held must be specified.

The initial value of the mean service rate μ is established as the quantity $1/K$. The quantities Y_N and CT_N are easily established, using an iterative subroutine, from the equations that describe the learning function. The values Y_N and CT_N, for all values of N, together determine subsequent values of μ to be used in the integration scheme.

Starting with the initial conditions at $t = 0$, the numerical values of the state probabilities P_n ($n = 0 \to 100$) are established for each successive point in time by step integration of the birth–death equations. The entire procedure was employed several times, each time using different values of the slope parameter S. The results provide a comparison between the effects of the different learning rates on the final numerical values of the state probabilities. These results are illustrated in a series of graphs that are plotted from the data generated by the computer program. The graphs show the values of certain of the state probabilities for the initial 100 time periods. These graphs compare the behavior of P_0 for several learning rates, while others show P_0, P_1, P_2, and P_3 for a single value of the learning rate.

As intuitively expected, the primary effect of learning at the service facility is a general "speeding up" of the manufacturing process and a lessening of queue congestion and waiting time. The actual results validate this intuitive reasoning with a series of charts depicting P_n as a function of time.

As a basis of comparison, Figs. 6.1–6.3 show the normal (no learning, $S = 100\%$) transient behavior of probabilities P_0 through P_4 for three different values of traffic density ρ. In Fig. 6.1, ρ is constant and equal to 0.75, in Figs. 6.2 and 6.3, ρ is likewise constant and equal to 1.00 and 1.25, respectively.

In Figs. 6.4–6.12, the three cases treated in Figs. 6.1–6.3 are each subjected to three different values of the slope parameter S, to show the effects of learning on the transient state probabilities.

Of more practical interest are other quantities derived from the state probabilities. Among these is the probability of delay $P(\text{delay})$, defined as the probability any arriving unit will be required to wait for service. This quantity $P(\text{delay})$ can be used as an indicator of machine utilization. Whenever $P(\text{delay})$ is small, the service facility is underutilized. For the single-channel case, $P(\text{delay})$ is the probability of one or more units in the queueing system. Thus,

$$P(\text{delay}) = 1 - P_0$$

From the same computer program used to calculate the transfer state probabilities. Fig. 6.13 was derived showing $P(\text{delay})$ as a function of learning-curve slope parameter S for the condition where initial traffic density $\rho_0 = 1.0$. From this figure, an evident effect of learning in the service facility is the subsequent reduction in queue congestion.

The multichannel case. The procedure developed for the single-channel queueing case can be adapted to yield transient state probabilities for the multichannel situation. Consider a system of c parallel channels in which a single queue is formed and units enter the facilities as they become vacant. The birth–death equations describing the system are as follows:

$$\frac{dP_0(t)}{dt} = -\lambda P_0(t) + \mu P_1(t), \qquad n = 0$$

$$\frac{dP_n(t)}{dt} = -(\lambda + n\mu)P_n(t) + \lambda P_{n-1}(t) + (n+1)\mu P_{n+1}(t), \qquad 1 \leq n < c$$

$$\frac{dP_n(t)}{dt} = -(\lambda + c\mu)P_n(t) + \lambda P_{n-1}(t) + c\mu P_{n+1}(t), \qquad c \leq n$$

By using these equations in place of those employed for a single channel, the integration scheme will yield transient state probabilities. The application of this multichannel solution can be used to yield valuable results in a "machine-release scheme" for a queueing system of several parallel channels.

A machine-release scheme. If, in a typical multichannel industrial queueing system, the service rate is affected by learning, it is to be expected that the utilization of the service facility will reach a maximum level shortly after startup and then decrease with additional time. One method of evaluating the service facility utilization is by observing the magnitude of $P(\text{delay})$ as it varies with time. When the value of $P(\text{delay})$ is of relatively high magnitude, there is a corresponding high probability that all c channels are busy and that slack time on any of

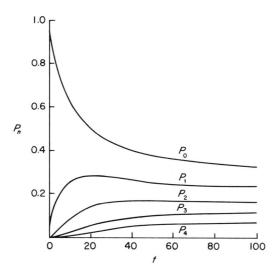

Fig. 6.1 Normal transient behavior, no learning, $\rho < 1.00$.

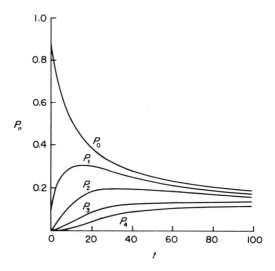

Fig. 6.2 Normal transient behavior, no learning, $\rho = 1.00$.

$(M|M|1):(GD|\infty|\infty)$ Queue

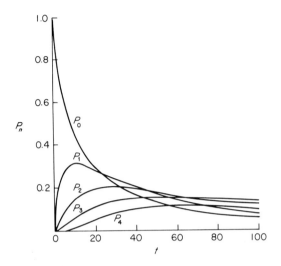

Fig. 6.3 Normal transient behavior, no learning, $\rho > 1.00$.

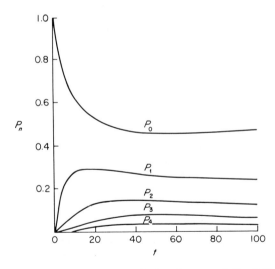

Fig. 6.4 Transient behavior with learning, $\rho_0 < 1$, $S = 90\%$.

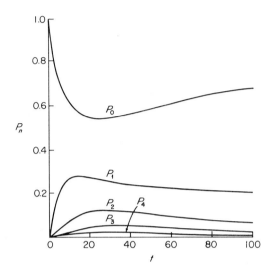

Fig. 6.5 Transient behavior with learning, $\rho_0 < 1$, $S = 80\%$.

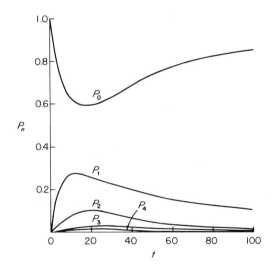

Fig. 6.6 Transient behavior with learning, $\rho_0 < 1$, $S = 70\%$.

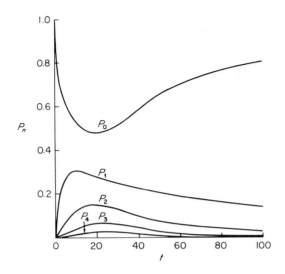

Fig. 6.7 Transient behavior with learning, $\rho_0 = 1$, $S = 90\%$.

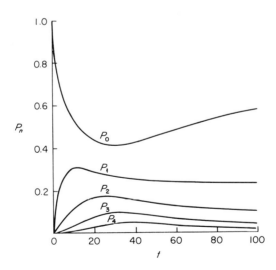

Fig. 6.8 Transient behavior with learning, $\rho_0 = 1$, $S = 80\%$.

288   6. *Transient Analysis and Special Topics*

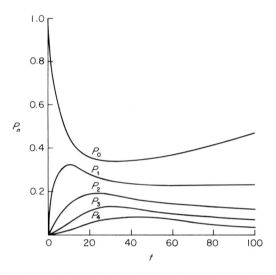

Fig. 6.9 Transient behavior with learning, $\rho_0 = 1$, $S = 70\%$.

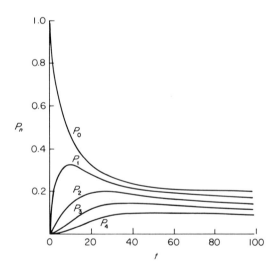

Fig. 6.10 Transient behavior with learning, $\rho_0 > 1$, $S = 90\%$.

$(M|M|1):(GD|\infty|\infty)$ Queue

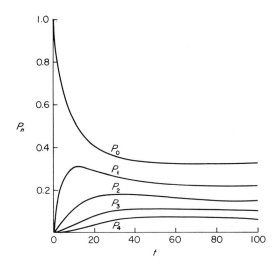

Fig. 6.11 Transient behavior with learning, $\rho_0 > 1$, $S = 80\%$.

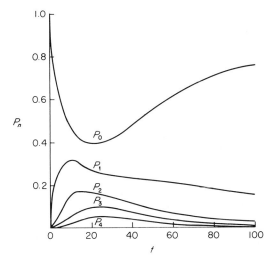

Fig. 6.12 Transient behavior with learning, $\rho_0 > 1$, $S = 70\%$.

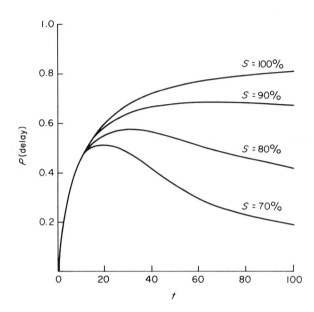

Fig. 6.13 Probability of delay, $\rho_0 = 1$, for various values of S.

the machines is minimal. On the other hand, a very low value of P(delay) indicates that arriving units experience little or no waiting in the queue. However, this lack of congestion, as determined by a small value of P(delay) may also indicate that the service facility is "overdesigned" or, in other words, that the same arrival pattern could be serviced with fewer machines. Where there is a high demand for machine usage for different operations, management may find it advantageous to reassign certain machines to other applications when their utilization falls below some predetermined value.

The purpose of a machine-release scheme is to allow management to make efficient use of each one of the service centers in a queueing system. By employing the quantity P(delay) as the measure of system usage, management can be provided with a forecast of machine-release points. A program would be specifying the initial number of machines, the arrival and initial service rates, the learning rate, and then a minimum desired value of P(delay), referred to as P(critical).

After the system saturates [that is, when P(delay) reaches a maximum and begins to decline], the magnitude of P(delay), calculated from the birth–death equations for each point in time, is compared with the specified value of P(critical). If the present value of P(delay) for the system is less than this critical value, this fact signifies that the service facility utilization has declined below the desirable level, and thus one of the machines in the multichannel system should be released.

Figure 6.14 illustrates a machine-release scheme for a value of P(critical) = 0.3. In this example, the system is initiated with ten machines in parallel. The amount of learning is governed by a value of the slope parameter $S = 80\%$. As shown

Summary

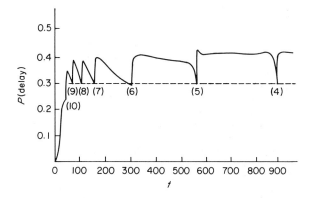

Fig. 6.14 Probability of delay, multichannel machine release system. Initial number of machines = 10. Numbers in parentheses represent machine release points. Arrival rate, $\lambda = 0.5$ units/period, K (each machine) = 20 periods.

in the figure, the value of P(delay) climbs from 0 at $t = 0$ to the value 0.238 at $t = 38.0$ and then begins to level off. At this point, saturation of the system has been achieved and since the value of P(delay) is less than the specified P(critical), the tenth machine is released. As expected, the initial reaction of the system is a substantial increase in P(delay) for the new system of nine machines. The nine-machine system saturates very quickly and P(delay) begins to decline as production continues. When P(delay) reaches a value less than P(critical), the ninth machine is released. As the effects of learning are more pronounced in the initial periods of production, continued output brings increased time intervals between machine-release points. This characteristic is evident from Fig. 6.14 and is explained by the decrease in the service-rate improvement, per time period, as μ tends toward a stable value.

In a practical application of this machine-release scheme, results such as those pictured in Fig. 6.14 would provide management with valuable aids in the analysis of multichannel systems.

SUMMARY

In this chapter several important topics in the study and analysis of queueing systems were presented. The concept of the imbedded Markov process was used to model the $(M|G|1):(GD|\infty|\infty)$ and $(GI|M|1):(GD|\infty|\infty)$ queues. The distribution function of the busy period for the $(M|G|1):(GD|\infty|\infty)$ queue was derived. Priority service disciplines were discussed with both preemptive and nonpreemptive service disciplines being treated. Transient solutions were provided for several simple queueing systems.

REFERENCES

Benes, V. E., On queues with Poisson arrivals, *Ann. Math. Statist.* **28**, 1957.
Conway, R. W., and Schultz, A., The manufacturing progress function, *J. Indust. Eng.* **X** (1), January–February, 1959.
Cooper, R. B., *Queueing Theory*. New York: Macmillan, 1972.
Cox, D. R., and Smith, W. L., *Queues*. London: Methuan, 1961.
Fabrycky, W. J., and Torgersen, P. E., Englewood Cliffs, New Jersey: Prentice-Hall, 1966.
Ghare, P. M., Given, J. M., and Torgersen, P. E. A machine release scheme for the job shop work center, *AIIE Trans.* **1** (3), September, 1969.
Jaiswal, N. K., *Priority Queues*. New York: Academic Press, 1968.
Kendall, D. G., Stochastic processes occurring in the theory of queues and their analysis by means of the imbedded markov chain, *Ann. Math. Statist.* **24**, 1953.
Saaty, T. L., *Elements of Queueing Theory*. New York: McGraw-Hill, 1961.
Takacs, L. *Introduction to the Theory of Queues*. New York: Oxford Univ. Press, 1962.
Takacs, L. *Combinatorial Methods in the Theory of Stochastic Processes*. New York: Wiley, 1967.
Wagner, H. M., *Principles of Operations Research*. Englewood Cliffs, New Jersey: Prentice-Hall, 1969.
Williams, P. F., The application of manufacturing improvement curves in multiproduct industries, *J. Indust. Eng.* **XII** (2), March–April, 1961.
Wright, T. P., Factors affecting the cost of airplanes, *J. Aeronaut. Sci.* **3** (4), February, 1936.

PROBLEMS

1. Consider the $(M \mid G \mid 1) : (GD \mid \infty \mid \infty)$ queue with the following modification: Mr. Slave, the server, serves the queue continuously as long as there is a customer in the system. However, whenever the system becomes empty Mr. Slave takes a coffee break. At the end of the break, Mr. Slave returns to the queue and begins serving those customers who have arrived during his break. If no one is there when Mr. Slave returns, he always begins a new break.

   (a) Show that the generating function of the number of customers left behind by a departing customer is given by

$$G(z) = \frac{(1 - \rho)[F(z) - 1]\mathscr{L}\{f(t_s)\}}{[dF(z)/dz][z - \mathscr{L}\{f(t_s)\}]} \bigg|_{s = \lambda(1 - z)}$$

where $F(z)$ is the generating function of the number of customers waiting for service at the end of a coffee break.

   (b) Compare $G(z)$ with (6.16). Does $G(z)$ reduce to (6.16) if the break ends at the instant an arrival occurs to the empty system? Explain.

2. For the $(M \mid G \mid 1) : (GD \mid \infty \mid \infty)$ queue, let

$$s_j = \int_0^\infty t_s^j f(t_s) \, dt_s, \quad \text{and} \quad b_j = \int_0^\infty x^j b(x) \, dx$$

where $f(t_s)$ and $b(x)$ are the density functions of the service time and busy period, respectively. Show that in steady state,

$$b_1 = \frac{s_1}{1 - \lambda/\mu} = -\frac{d}{ds}\frac{\mathscr{L}\{f(t_s)\}}{1 - \lambda/\mu}\bigg|_{s=0}, \quad b_2 = \frac{s_2}{(1 - \lambda/\mu)^3} = -\frac{d^2}{ds^2}\frac{\mathscr{L}\{f(t_s)\}}{(1 - \lambda/\mu)^3}\bigg|_{s=0}$$

3. Show that the Laplace transform of the density function

$$h(t_q) = \mu\left[1 - \int_0^{t_q} f(t_s)\, dt_s\right]$$

can be written as

$$\mathscr{L}[h(t_q)] = \frac{\mu(1 - \mathscr{L}[f(t_s)])}{s}$$

4. For the $(M|M|1):(GD|\infty|\infty)$ queue, use (6.33) to show that the probability density function of the busy period is given by

$$b(t) = \frac{e^{-(\lambda+\mu)t} I_1[2t(\lambda\mu)^{1/2}]}{t\sqrt{\rho}}$$

where $\rho = \lambda/\mu$ and I_1 is the modified Bessel function of the first kind of order one.

5. Establish the result given in Problem 4 using the relation

$$b(t) = \frac{d}{dt} P_0(t)$$

[Hint:

$$\frac{2}{x} I_1(x) = I_2(x) - I_1(x)$$

where I_1 and I_2 are modified Bessel functions.]

6. Beginning with (6.30), establish (6.31).

7. Customers arrive at a check-out counter at a Poisson rate of 10/hr. Service time is normally distributed with a mean of 2 min and a standard deviation of 0.25 min. Find (a) the generating function for the number of customers who arrive during the service of the nth customer; (b) the steady-state probability that there are n customers at the check-out counter; and (c) the expected duration of time the check-out counter is continuously busy.

8. Arrivals at a cashier's window in a local bank meet the conditions of a Poisson process. The mean interarrival time is 15 min. Service times are gamma distributed with a mean of 6 min and a standard deviation of 2 min. During steady state, determine (a) the expected waiting time per customer; (b) the variance for the number of customers at the window, including the one being served; and (c) the Laplace transform for the amount of time a customer spends in the line, including service time.

9. Consider the emergency-room problem described in Problem 1 of Chapter 4. Find the steady-state probability for the number of customers in the system.

10. Consider Problem 4 of Chapter 4. Find the steady-state probability for the number in the system.

11. Arrivals at a truck weighing station occur at a Poisson rate of 10/hr. The service time is a constant of 5 min/truck. Find (a) the generating function for the number of trucks at the weighing station; and (b) the mean number of trucks at the station.

12. Consider the $(E_k|M|1):(GD|\infty|\infty)$ queue. Use (6.53) to determine the number of customers in the system under steady-state conditions.

13. Using the results of Problem 12, find the steady-state distribution for the number in the system for the $(D|M|1):(GD|\infty|\infty)$ queue.

294   6. Transient Analysis and Special Topics

14. Interarrival time at a hospital emergency room is normally distributed with a mean of 20 min and a standard deviation of 5 min. Time to admit a patient is exponentially distributed with a mean of 5 min. Find (a) the probability that n patients will be at the admittance desk (assume steady-state conditions), and (b) the distribution of the time spent in the admission process for a newly arrived customer.

15. Work Problem 14 for the case where arrivals are uniformly distributed over the interval from 5 through 15 min.

16. A spool of yarn arrives every minute (by conveyor) at the inspection–packaging department. The time required to inspect and package a spool is exponentially distributed with a mean of 2 min. There is one inspector–packer handling spools on a first come–first served basis.

(a) What is the steady-state distribution for the number of spools accumulated on the conveyor at the inspection–packaging department?

(b) What is the expected time a spool must wait on the conveyor (after arriving at the inspection–packaging department) before being removed by the inspector–packer?

17. Consider the $(M \mid M \mid c) : (GD \mid \infty \mid \infty)$ model discussed in Chapter 3 but with the following modification: Upon leaving the server the customer rejoins the queue as a new arrival with probability p and leaves the system with probability $1 - p$. (a) Find the steady-state probability that a new arrival finds all servers busy. (b) Find the mean number of busy servers.

18. Consider a system in which units arrive according to a Poisson distribution with parameter λ. If after a unit arrives the number of units in the waiting line is greater than N, a new channel is added. It may be assumed that there is an infinite number of channels that can be added to the system. Therefore, there can never be more than N units in the waiting line. The service-time distribution for all channels is exponential with parameter μ. When any channel finishes a service it remains available unless there are no units in the waiting line, at which time the channel drops out of the system. At least one channel is always available for service. Let

$P_n(m)$ = probability that n units are in the waiting line and m units are in service
λ = mean arrival rate
μ = mean service rate
$N = 2$
$P(m)$ = probability that m units are in service

Find under steady-state conditions: (a) $P_n(m)$, (b) $P(m)$, (c) the expected number of units in service, and (d) the expected number of available channels.
[*Hint*:

$$P_0(0) = \frac{1 + \rho + \rho^2}{1 + 2\rho + 3\rho^2 e^\rho}$$

where $\rho = \lambda/\mu$.]

19. Let $f(t_s)$ denote the *pdf* of service time. Define

$S_0(t) = P$(service operation takes longer than time t, where time is measured from a point at which last service was completed)

$V_0(t) = P$(service takes longer than time t, where time is measured from a *random* point in time during service)

(a) Show that

$$\mu^{-1} = \int_0^\infty S_0(t)\, dt$$

where μ^{-1} is the mean service time.

(b) Show that

$$V_0(t) = \mu \int_t^\infty S_0(z)\, dz$$

(c) If

$$f(t_s) = \mu e^{-\mu t}$$

show that $V_0(t) = S_0(t)$.

20. Consider the nonpreemptive priority queueing model in which all customers whose service times are less than or equal to "$a\rho$" are classed as 1-customers. All others are classed as 2-customers ($\rho = 1/\mu$ and service is exponential).

   (a) Show that the mean queueing time is given by

$$E(t_q) = \frac{c(1 - \rho + \rho e^{-a})}{2(1 - \rho)(1 - \rho + \rho e^{-a} + \rho a e^{-a})}$$

   (b) Show that $E(t_q)$ can be minimized by selecting "a" such that

$$\frac{1}{\rho} = 1 + \frac{e^{-a}}{a - 1}$$

21. Jobs arrive randomly for processing on a computer. The computer is used to process only one job at a time. It has been found that jobs have one of three priorities: The highest priority jobs have a gamma-distributed service time with a mean of 30 min and a standard deviation of 20 min. The second most important jobs have a service-time distribution that is uniform over the interval from 1 to 5 min. The low-priority jobs (student jobs) have an exponentially distributed service time with a standard deviation of 1 min. All jobs arrive in a Poisson fashion. Priority 3 jobs arrive at an average rate of 26.7/hr. Priority 2 jobs arrive at a mean rate of 3/hr. Top-priority (priority 1) jobs arrive at an average rate of 0.3/hr. The system is operated using a nonpreemptive priority scheme.

   (a) What is the probability that a priority 3 job will be at the head of the waiting line when the computer becomes free for processing a job?
   (b) What is the expected number of 1-customers in the system at a 2-epoch?
   (c) What is the expected number of 3-customers in the system at a 3-epoch?

22. People arrive at the Xerox machine in Norris Hall in such a fashion that the interarrival distribution is given in minutes as $0.1e^{-0.1t}$, $t > 0$. If a person arrives and sees that the machine is in use, he goes elsewhere for Xerox services. Service time (distribution) at the machine is distributed as follows:

$$f(t) = 0.25 e^{-0.25t}, \quad t > 0$$

The room containing the machine is open from 8:00 A.M. to 5:00 P.M., Monday through Friday.

   (a) What is the probability that you will find the machine idle at 8:20 A.M., if the machine is idle at 8:00 A.M.?
   (b) What is the steady-state utilization for the machine?

23. Consider the self-service model $(M \mid M \mid \infty) : (GD \mid \infty \mid \infty)$ with the following change. Arrivals are still Poisson but are influenced by the number of customers in the system. In particular, assume that $\lambda_n = n\lambda$ for all n. Given the initial condition that the system is empty at time $t = 0$, show that the generating function

$$G(z, t) = \sum_{n=0}^\infty P_n(t) z^n$$

of the number in the system at time t is given by

$$G(z, t) = \frac{\mu - \alpha e^{-t(\lambda - \mu)}}{\lambda - \alpha e^{-t(\lambda - \mu)}}$$

where μ is the service rate and $\alpha = (\mu - \lambda z)/(1 - z)$.

24. A collection of m Whambats have been captured in a closed box. The time until death for each Whambat is exponentially distributed with mean time to death μ^{-1}. Let $N(t)$ be the number of Whambats alive at time t. Argue that

$$P\{N(t) = n\} = \binom{m}{n} e^{-n\mu t}(1 - e^{-\mu t})^{m-n}$$

using the premises of the binomial mass function.

25. Good Luck Hospital has facilities sufficient to handle all requests for admittance (i.e., there is, for all practical purposes, an infinite number of beds available). The chief IE has found that the probability of a patient discharge in the interval $(t, t + \Delta t)$ is $\mu \Delta t + 0(\Delta t)$ and that incoming patients occur at the rate of λ with interarrival time exponentially distributed.

(a) Show that the PGF for the number of occupied beds in the hospital is given by

$$G(z, t) = \exp\left\{\frac{-\lambda}{\mu[1 - z][1 - e^{-\mu t}]}\right\}[1 - (1 - z)e^{-\mu t}]^i$$

where i is the number of occupied beds at $t = 0$ (initially).

(b) If $i = 0$, find $P_n(t)$.

26. The transient solution $P_n(t)$ given by (6.112) for the $(M \mid M \mid 1) : (GD \mid \infty \mid \infty)$ queue was based on the initial condition that the system was empty, i.e., $P_0(0) = 1$. Derive an expression for $P_n(t)$ for the case where $P_0(0) = i$. It will be instructive to begin your development from Eq. (6.110).

27. Consider a pure-birth process with the following characteristics:

$$\lambda_n = n\lambda, \qquad \mu_n = 0, \qquad \text{for all} \quad n$$

and initial conditions

$$P_i(0) = 1, \qquad P_n(0) = 0, \qquad \text{for} \quad n \neq i$$

Show that $P_n(t)$ is the negative binomial mass function with parameters $p = e^{-\lambda t}$, $n = i$.

28. Consider a pure-death process where

$$\mu_n = \mu, \qquad \lambda_n = 0, \qquad \text{for all} \quad n = 0, 1, \ldots, m$$

Given the initial condition that there are m customers in the system at time t, show that

$$P_n(t) = \frac{(\mu t)^{m-n}}{(m - n)!} e^{-\mu t}, \quad n = 1, 2, \ldots, m,$$

and

$$P_0(t) = 1 - \sum_{n=1}^{m} P_n(t)$$

Chapter 7

DATA ANALYSIS—ESTIMATION

INTRODUCTION

There are two basic approaches to modeling queueing systems: mathematical modeling and simulation modeling. Thus far we have discussed mathematical modeling. In Chapter 9, we treat simulation modeling. In both of these approaches, certain assumptions must be made about the system to be modeled in order to represent it by mathematical equations or by a simulator. This is particularly true with reference to the random variables involved in the system. For example, we frequently assume that interarrival and service times were exponentially distributed random variables. However, since our attempt is to model physical systems, we would wish to identify these distributions properly rather than simply make assumptions.

In this chapter we will present some of the methods available for the analysis of data prior to model development and for analyzing the sensitivity and output of the model after it has been developed. An area of particular concern in this chapter will be the identification of the probability distribution of a random variable. This is a threefold problem. First, the analyst must collect data that characterize the random variable of concern. Using these data he can then develop and plot a relative-frequency distribution for the random variable. By visually comparing the relative-frequency distribution with known probability mass functions or probability density functions, one is often able to hypothesize that certain families of distribution describe the random variable under study. The selection of these candidate distributions is the first of the three problems mentioned. The second problem is the determination of the numerical values of the parameters of each candidate distribution based on the data collected. Three methods for determining the parameter values will be presented. Once numerical values have been assigned to the parameters of each distribution that the analyst believes may describe the random variable of interest, he is in a position to test the hypothesis that each distribution is, in fact, the true distribution of the random variable under study. There are two widely used tests, called *goodness-of-fit tests*, for determining whether or not a specific distribution describes a given random variable based on data collected for that random variable. These are the chi-square and Kolmogorov–Smirnov tests. Both tests will be presented in this chapter.

To a large extent, distribution identification is of the greatest concern in the analysis of input data, where such random variables as interarrival and service times must be defined. However, when a simulation model is used to represent the system under study, distribution identification may be important in attempting to define random variables such as the number in the system, waiting time, and total time in the system, since analytic expressions describing these random variables cannot be obtained directly from a simulator.

IDENTIFYING THE DISTRIBUTION

As discussed above, there are basically three steps involved in defining the probability distribution of a random variable: (1) obtaining data on the process and making inferences as to how the variable is distributed, (2) estimating the various parameters of the hypothesized distribution, and (3) determining whether or not the collected data can be considered as being obtained from the hypothesized distribution. These three functions can be statistically termed data collection, parameter estimation, and goodness-of-fit testing, respectively. In the next few sections, we will treat each of these topics in some depth.

Data Collection

With few exceptions, one can make a reasonable guess regarding the distribution of a random variable only after data have been collected that can be used as a guide. One commonly used form of summarizing the collected data is the frequency distribution. In the case of discrete variables, we simply record the number of times (frequency) each value was observed. For continuous random variables, we break the range of observed values

Identifying the Distribution

into intervals and record the frequency that occurs within each interval. For example, Table 7.1 lists the frequency distribution for arrivals at a hospital emergency room during an 8-hr period.

TABLE 7.1

Number of arrivals	Frequency	Number of arrivals	Frequency	Number of arrivals	Frequency
6	1	13	9	19	6
7	0	14	11	20	4
8	3	15	11	21	3
9	3	16	10	22	2
10	6	17	8	23	2
11	5	18	6	24	1
12	9				

Once a frequency distribution has been tabularized, as above, a plot of the entries is generally helpful in determining the distribution of the random variable under study. One useful method of displaying these data is to plot the relative-frequency distribution. The relative frequency for each interval—or in case of discrete variables for each observed value—is simply the observed frequency count divided by the total frequency count. The relative-frequency plot of the data listed in Table 7.1 is shown in Fig. 7.1. With this information at hand we can now proceed to select an appropriate probability distribution. Our success at this point is largely a matter of judgment and experience. Comparing Fig. 7.1 with Figs. 7.2–7.4, we might suspect that daily emergency arrivals follow a Poisson, binomial, or even a normal process. The suggestion of a normal fit should not seem too surprising. It is not always necessary to describe a discrete random variable by the probability mass function of the discrete random variable. Often continuous approximations can be made that are suitable. For example, the normal distribution can be used to approximate the binomial distribution when n is large and p is close to 0.5. It also becomes an excellent approximation to the Poisson distribution as the parameter of the Poisson distribution increases.

Once we have identified one or more families of distributions to which the observed data may belong, we must determine estimates for the parameters of the candidate distributions. This becomes the second step in identifying the distribution. Before discussing specific methods by which these estimates can be found, we will find it instructive to discuss some of the basic principles underlying estimation procedures.

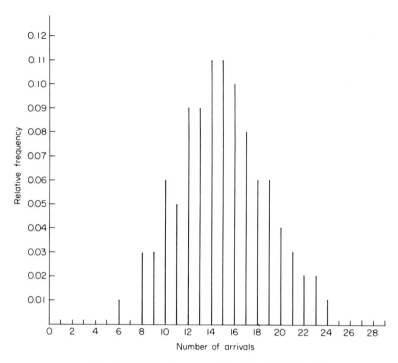

Fig. 7.1 Relative-frequency plot of data in Table 7.1.

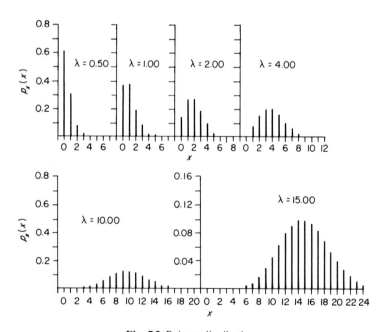

Fig. 7.2 Poisson distribution.

Identifying the Distribution

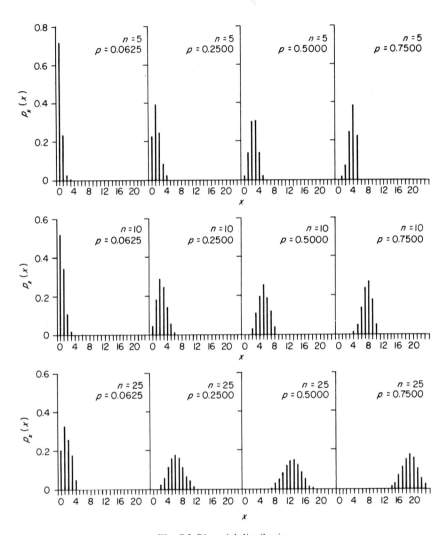

Fig. 7.3 Binomial distribution.

302   7. Data Analysis—Estimation

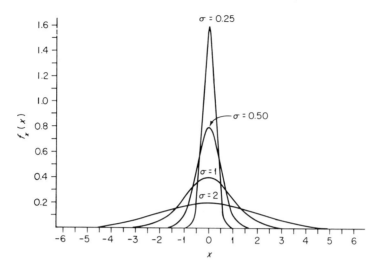

Fig. 7.4 Normal distribution with mean zero.

POINT ESTIMATION

The theory of estimation can be divided into two parts, *point estimation* and *interval estimation*. In *point estimation*, we concern ourselves with the problem of producing a value that, in some sense, represents our best estimate as to the actual value of the parameter of interest. In interval estimation, we are interested in establishing an interval that would contain the true value of the parameter with some given level of probability. Such an interval is called a *confidence interval*. These intervals can also be viewed as possible measures of the precision of a point estimator. This is the view that we will adopt in our discussion of interval estimation in a later section.

The general problem of point estimation can be stated as follows: There exists a random variable X whose distribution function is characterized by some parameter θ that we would like to estimate. A random sample $X_1, X_2, \ldots, X_n$ is to be drawn and a function $\hat{\theta} = \hat{\theta}(X_1, X_2, \ldots, X_n)$ of this sample is formed. The value of $\hat{\theta}$ is then used to estimate the parameter θ. The function $\hat{\theta}$ is referred to as an *estimator* of θ and the value that $\hat{\theta}$ takes on is called the *estimate* of θ. Note that $\hat{\theta}$ is itself a random variable, since it is a function of the random observations $X_1, X_2, \ldots, X_n$.

It should be apparent that there exist many estimators for a parameter θ. For example, if X were normally distributed with known variance and unknown mean m, then several estimators for m come to mind. We might choose

Point Estimation

to estimate m by the sample mean, the sample median, or perhaps by some other function of the sample data such as the smallest of the observed values. Our choice would no doubt be the estimator that provides estimates closest to m. Thus, the problem of point estimation is that of producing an estimate that is close to the true value of the parameter. To resolve this problem, not only must we define precisely what we mean by the ambiguous term "close," but we have to construct estimators that will meet our "closeness" criteria. The first part of the problem amounts to establishing various properties of an estimator that are considered desirable. These properties will be discussed in the next section. The second part of the problem involves developing estimators that would have at least some of these desirable properties. Three such methods—the method of moments, the maximum-likelihood method, and the least-squares method—will be described in later sections.

Properties of Estimators

As noted above, an estimator $\hat{\theta}$ for a parameter θ is itself a random variable and must therefore have a distribution of its own. Obviously we cannot specify this distribution until we have completely described the estimator itself. Even then it may be difficult, if not altogether impossible, to determine the actual distribution of $\hat{\theta}$. We can, however, establish some of the basic characteristics associated with this distribution. In this regard, let us define its mean by $E(\hat{\theta})$, and its variance by

$$\text{Var}(\hat{\theta}) = E[\hat{\theta} - E(\hat{\theta})]^2 = E(\hat{\theta}^2) - [E(\hat{\theta})]^2 \qquad (7.1)$$

The standard deviation of $\hat{\theta}$, defined by $[\text{Var}(\hat{\theta})]^{1/2}$, is called the *standard error* of $\hat{\theta}$. Also of importance are the concepts *sampling error*, *bias*, and *mean squared error*. The sampling error, defined by $\hat{\theta} - \theta$, is simply the difference between the value of the estimator and the true value of the parameter. The bias, defined by $E(\hat{\theta}) - \theta$, is the difference between the expected value of the estimator and the true value of the parameter. Whereas the sampling error may vary from sample to sample, the bias will always remain fixed and may or may not be zero. The third concept, mean-squared error, defined by $E(\hat{\theta} - \theta)^2$, measures the dispersion of an estimator and is therefore similar to the concept of variance—the difference being that while the variance measures the dispersion of $\hat{\theta}$ around its mean $E(\hat{\theta})$, the mean squared error measures the dispersion around the true value of the parameter. If it turns out that $E(\hat{\theta})$ and the true value of the parameter coincide, then the mean squared error and the variance are equivalent. This can be seen by examining Eq. (7.2) below.

The relationship between the mean-squared error (MSE) and the variance is best visualized by considering the following formulation:

$$\begin{aligned}\text{MSE}(\hat{\theta}) &= E(\hat{\theta} - \theta)^2 = E[\hat{\theta} - E(\hat{\theta}) + E(\hat{\theta}) - \theta]^2 \\ &= E\{[\hat{\theta} - E(\hat{\theta})] + [E(\hat{\theta}) - \theta]\}^2 \\ &= E[\hat{\theta} - E(\hat{\theta})]^2 + E[E(\hat{\theta}) - \theta]^2 \\ &\quad + 2E\{[\hat{\theta} - E(\hat{\theta})][E(\hat{\theta}) - \theta]\} \\ &= E[\hat{\theta} - E(\hat{\theta})]^2 + E[E(\hat{\theta}) - \theta]^2\end{aligned}$$

The second term is the expected value of the bias, and since for any estimator $\hat{\theta}$ the bias remains constant, we can write

$$\text{MSE}(\hat{\theta}) = E[\hat{\theta} - E(\hat{\theta})]^2 + [E(\hat{\theta}) - \theta]^2 \tag{7.2}$$

In words, the mean-squared error is equal to the variance of the estimator plus the squared bias.

Having discussed some of the basic characteristics of the distribution of an estimator, let us now examine some estimator properties that are generally considered to be desirable. These properties can be conveniently classed depending on the size of the sample. *Finite-sample* or *small-sample properties* refer to the properties of estimators based on fixed sample sizes. Three such properties are discussed below: (1) unbiasedness, (2) efficiency, and (3) best linear unbiased estimators. On the other hand, *asymptotic* or *large-sample properties* apply when the sample size approaches infinity. The asymptotic properties that we shall subsequently consider are: (1) asymptotic unbiasedness, (2) consistency, and (3) asymptotic efficiency. However, let us first concern ourselves with small-sample properties.

Perhaps the best-known property of an estimator is that of *unbiasedness*. When the bias of an estimator $E(\hat{\theta}) - \theta$ is zero, then the estimator is said to be unbiased. Formally stated, an estimator $\hat{\theta}$ is said to be an *unbiased estimator* for θ if its expected value equals θ, i.e., if $E(\hat{\theta}) = \theta$. Notice that for an unbiased estimator $\text{MSE}(\hat{\theta})$ as given by (7.2) becomes

$$\text{MSE}(\hat{\theta}) = E[\hat{\theta} - E(\hat{\theta})]^2 = \text{Var}(\hat{\theta})$$

Example 7.1 Show that the sample estimators

$$\bar{X} = \sum_{i=1}^{n} \frac{X_i}{n}, \quad S^2 = \frac{\sum_{i=1}^{n}(X_i - \bar{X})^2}{n-1}$$

for the mean m and variance σ^2 of a population are both unbiased estimators.

Consider

$$E(\bar{X}) = E\left[\sum_{i=1}^{n} \frac{X_i}{n}\right] = \frac{1}{n}\sum_{i=1}^{n} E(X_i) = \frac{1}{n}[nE(X)] = E(X) = m$$

i.e., $\bar{X}$ is unbiased.

Now consider

$$E(s^2) = \frac{1}{n-1} \sum_{i=1}^{n} E(X_i - \bar{X})^2$$

$$= \frac{1}{n-1} \sum_{i=1}^{n} E[(X_i - m) - (\bar{X} - m)]^2$$

$$= \frac{1}{n-1} \sum_{i=1}^{n} [E(X_i - m)^2 - E(\bar{X} - m)^2]$$

$$= \frac{1}{n-1} \left\{ \sum_{i=1}^{n} \text{Var}(X_i) - n\,\text{Var}(\bar{X}) \right\}$$

$$= \frac{1}{n-1} [n\,\text{Var}(X) - n\,\text{Var}(\bar{X})]$$

Using the fact that $\text{Var}(\bar{X}) = \text{Var}(X)/n$, we have

$$E(S^2) = \text{Var}(X) = \sigma^2 \qquad (7.3)$$

Thus S^2 is an unbiased estimator for the variance.

It should be recognized that by itself unbiasedness is not a very encompassing property, since it implies nothing about the dispersion of the distribution of the estimator. Thus, an estimator can be unbiased but yet lead to estimates that lie far from the true value of the parameter. On the other hand, a biased estimator even with small variance often can be less useful. Consider the case of an estimator with zero variance. Such an estimator would be any constant, which certainly makes little sense since it completely ignores any sample data. In view of this argument, it would seem desirable to find an estimator that minimizes the mean-squared error. Recall that the mean-squared error is the sum of the variance of $\hat{\theta}$ and its squared bias. Unfortunately, for most distributions there does not exist an estimator that minimizes mean-squared error for all values of θ. One estimator may produce a minimum mean-squared error for some values of θ, while another estimator may produce a minimum mean-squared error for other values of θ.

In view of this difficulty, we are led to the concept of efficiency. An unbiased estimator $\hat{\theta}$ is said to be an *efficient estimator* for θ if $\text{Var}(\hat{\theta}) \leq$

Var(θ^*), where θ^* is any other unbiased estimator for θ. An efficient estimator is frequently called a *minimum-variance unbiased estimator*.

The problem of examining an estimator for efficiency often can be very difficult. To determine whether it is unbiased is rather simple, since all we have to do is compute its expectation. However, to show that an unbiased estimator has minimum variance can be most difficult, since there may be an infinite number of such estimators to consider. Fortunately, the search for a minimum-variance estimator can often be reduced due to the Cramer–Rao inequality, which provides a lower bound on the variance of an unbiased estimator. Thus, if an unbiased estimator can be found that achieves this lower bound, then it will be a minimum-variance unbiased estimator and therefore efficient. The Cramer–Rao inequality can be stated formally as follows:

Theorem 7.1 (Cramer–Rao Inequality) Let X be a random variable with a probability density function $f(x)$ characterized by parameters $\theta_1, \theta_2, \ldots, \theta_k$. Let $\hat{\theta}_i$ be any unbiased estimator of θ_i derived from a sample $X_1, X_2, \ldots, X_n$. Define the logarithmic likelihood function $L = \ln f(x_1, x_2, \ldots, x_n)$, and form the matrix

$$\begin{pmatrix} -E\left[\dfrac{\partial^2 L}{\partial \theta_1^2}\right] & -E\left[\dfrac{\partial^2 L}{\partial \theta_1 \, \partial \theta_2}\right] & \cdots & -E\left[\dfrac{\partial^2 L}{\partial \theta_1 \, \partial \theta_k}\right] \\ -E\left[\dfrac{\partial^2 L}{\partial \theta_2 \, \partial \theta_1}\right] & -E\left[\dfrac{\partial^2 L}{\partial \theta_2^2}\right] & \cdots & -E\left[\dfrac{\partial^2 L}{\partial \theta_2 \, \partial \theta_k}\right] \\ \vdots & \vdots & & \vdots \\ -E\left[\dfrac{\partial^2 L}{\partial \theta_k \, \partial \theta_1}\right] & -E\left[\dfrac{\partial^2 L}{\partial \theta_k \, \partial \theta_2}\right] & \cdots & -E\left[\dfrac{\partial^2 L}{\partial \theta_k^2}\right] \end{pmatrix} \quad (7.4)$$

which is called the information matrix. Now consider the inverse of this matrix and denote the element in the ith row and ith column of the inverse matrix by I_{ii}. Then the Cramer–Rao inequality is

$$\text{Var}(\hat{\theta}_i) \geq I_{ii} \quad (7.5)$$

Unfortunately, an estimator that actually achieves the Cramer–Rao lower bound does not always exist. In fact, it may turn out that we cannot even specify the form of $f(x)$ and hence have no hope of determining the lower bound. For these reasons, we often consider a smaller class of unbiased estimators. If we restrict our attention to the class of unbiased estimators that are formed by a linear combination of the sample observations, i.e. $\hat{\theta} = c_1 X_1 + c_2 X_2 + \cdots + c_n X_n$, where $c_1, c_2, \ldots, c_n$ are constants and $X_1, X_2, \ldots, X_n$ a sample, then a minimum variance estimator can often be

Point Estimation

found. This leads us to the definition of a best linear unbiased estimator. An unbiased estimator $\hat{\theta}$ formed by a linear combination of sample observations is referred to as a *best linear unbiased estimator* (*BLUE*) if $\text{Var}(\hat{\theta}) \leq \text{Var}(\theta^*)$, where θ^* is any other linear unbiased estimator.

It should be noted that the BLUE estimator may not be a minimum variance estimator (efficient). It is only minimum variance relative to other linear unbiased estimators. In such cases we speak of the *relative efficiency* of an estimator. Frequently, however, a BLUE estimator will achieve the Cramer–Rao lower bound and will therefore be efficient.

Example 7.2 Consider the exponential density function

$$f(x) = \begin{cases} \dfrac{1}{\theta} e^{-x/\theta}, & \text{if } x \geq 0 \\ 0, & \text{if } x < 0 \end{cases} \tag{7.6}$$

Show that

$$\hat{\theta} = \frac{\sum_{i=1}^{n} X_i}{n}$$

is a BLUE estimator for the parameter θ and that it is also efficient.

Now

$$E(\hat{\theta}) = E\left(\sum_{i=1}^{n} \frac{X_i}{n}\right) = \frac{1}{n} \sum_{i=1}^{n} E(X_i) = E(X) = \theta \tag{7.7}$$

and therefore $\hat{\theta}$ is an unbiased linear estimator for θ. Now consider the Cramer–Rao lower bound for the special case where $k = 1$. Here the information matrix reduces to the single element whose reciprocal (inverse) is given by

$$I_{11} = -\left\{ E\left[\frac{d^2}{d\theta^2} L\right] \right\}^{-1}$$

where

$$L = \ln \prod_{i=1}^{n} \frac{1}{\theta} \exp -\frac{x_i}{\theta} = \ln \theta^{-n} \exp -\sum_{i=1}^{n} \frac{x_i}{\theta} = -n \ln \theta - \frac{1}{\theta} \sum_{i=1}^{n} x_i$$

Now

$$\frac{d^2}{d\theta^2} L = \frac{n}{\theta^2} - \frac{2}{\theta^3} \sum_{i=1}^{n} x_i, \quad \text{and} \quad E\left[\frac{d^2}{d\theta^2} L\right] = \frac{n - 2n}{\theta^2} = -\frac{n}{\theta^2}$$

Thus

$$I_{11} = \frac{\theta^2}{n}$$

and the Cramer–Rao lower bound is

$$\text{Var}(\hat{\theta}) \geq \frac{\theta^2}{n}$$

Since

$$\hat{\theta} = \sum_{i=1}^{n} \frac{X_i}{n} = \bar{X}, \quad \text{and} \quad \text{Var}(\bar{X}) = \frac{\text{Var}(X)}{n}$$

we have that

$$\text{Var}(\hat{\theta}) = \frac{\theta^2}{n}$$

Thus, $\hat{\theta}$ achieves minimum variance and is therefore efficient and consequently a BLUE estimator as well.

Example 7.3 Show that the elements on the main diagonal of the information matrix can be written as

$$E\left(\frac{\partial L}{\partial \theta_i}\right)^2 \tag{7.8}$$

If we let $\mathbf{X} = (X_1, X_2, \ldots, X_n)$, then from the definition of an expected value we can write

$$E\left(\frac{\partial L}{\partial \theta_i}\right) = \int_{\mathbf{X}} \left[\frac{\partial}{\partial \theta_i} \log f(\mathbf{X})\right] f(\mathbf{X}) \, d\mathbf{X}$$

$$= \int_{\mathbf{X}} \left[\frac{1}{f(\mathbf{X})} \frac{\partial}{\partial \theta_i} f(\mathbf{X})\right] f(\mathbf{X}) \, d\mathbf{X} = \int_{\mathbf{X}} \frac{\partial}{\partial \theta_i} f(\mathbf{X}) \, d\mathbf{X} \tag{7.9}$$

If we can interchange differentiation and integration, then

$$E\left(\frac{\partial L}{\partial \theta_i}\right) = \frac{\partial}{\partial \theta_i} \int_{\mathbf{X}} f(\mathbf{X}) \, d\mathbf{X} = 0$$

since

$$\int_{\mathbf{X}} f(\mathbf{X}) \, d\mathbf{X} = 1$$

Point Estimation

Abbreviating the notation and differentiating (7.9) under the integral, we obtain

$$0 = \int_{\mathbf{X}} \left\{ \frac{1}{f} \frac{\partial f}{\partial \theta} \frac{\partial f}{\partial \theta} + f \frac{\partial}{\partial \theta} \left[\frac{\partial}{\partial \theta} \log f \right] \right\} d\mathbf{X}$$

$$= \int_{\mathbf{X}} \left\{ \left(\frac{1}{f} \frac{\partial f}{\partial \theta} \right)^2 + \frac{\partial^2 L}{\partial \theta^2} \right\} f \, d\mathbf{X} = \int_{\mathbf{X}} \left\{ \left(\frac{\partial L}{\partial \theta} \right)^2 + \frac{\partial^2 L}{\partial \theta^2} \right\} f \, d\mathbf{X}$$

or

$$E\left(\frac{\partial L}{\partial \theta}\right)^2 = -E\left(\frac{\partial^2 L}{\partial \theta^2}\right) \tag{7.10}$$

It is of practical interest to note that for the case where $f(\mathbf{X})$ is characterized by only one unknown parameter, the Cramer–Rao inequality becomes

$$\text{Var}(\hat{\theta}) \geq \left\{ E\left(\frac{\partial L}{\partial \theta}\right)^2 \right\}^{-1} \tag{7.11}$$

This expression can be even further simplified to the computationally convenient form

$$\text{Var}(\hat{\theta}) \geq 1/nE\left[\frac{\partial}{\partial \theta} \log f(\mathbf{X})\right]^2 \tag{7.12}$$

and is left to the reader as an exercise.

We shall now consider the asymptotic properties of estimators. As noted earlier, these properties apply when the sample size is large and approaches infinity. It should be realized that the distribution of a given estimator based on one sample size may be different from the distribution of this same estimator based on a different sample size. Take, for example, the distribution of the sample mean from an exponential distribution. For a sample of size 1, the distribution is exponential. For sample sizes greater than 1, the distribution is gamma, and as n increases the distribution is close to normal, although it is actually still gamma. In fact, the change in the distribution of the sample mean for samples from any distribution is described by the *central limit theorem*. This theorem is one of the most important results in the theory of statistics. It can be stated as follows

Theorem 7.2 (Central Limit Theorem) If X has any distribution with finite mean m and finite variance σ^2, then the distribution of $\bar{X}$ approaches the normal distribution with mean m and variance σ^2/n as the sample size increases.

This theorem says, in essence, that as the sample size increases, the distribution of $\bar{X}$, based on independent samples, tends to be normally distributed regardless of the population from which the sample was taken. In such cases we say that the normal distribution is the limiting or *asymptotic distribution* of the sample mean.

It is important to realize that the term asymptotic distribution does not necessarily apply to the final form that the distribution of the estimator assumes as the sample size approaches infinity. In fact, what often happens as $n \to \infty$ is that the distribution of the estimator collapses on one point, hopefully the true value of the parameter. Such a distribution is referred to as a *degenerate* distribution. To illustrate, consider the statement of the central limit theorem. Clearly, as the sample size n tends to infinity, σ^2/n approaches zero and the distribution will collapse at the population mean m. This is obviously not the asymptotic distribution referred to by the central limit theorem. What is meant by the asymptotic distribution is not the ultimate form of the distribution, which may be degenerate, but the form the distribution assumes just before it collapses, if this occurs.

Asymptotic properties of estimators are generally based on the asymptotic distribution of the estimator. Of particular importance are the mean and variance of this distribution, referred to as the *asymptotic mean* and *asymptotic variance*, respectively. The asymptotic mean can be found simply by determining the limiting value of the finite sample estimator as $n \to \infty$, i.e.,

$$\lim_{n \to \infty} E(\hat{\theta})$$

The asymptotic variance, however, is *not* given by the limiting value of $\text{Var}(\hat{\theta})$ as $n \to \infty$ for rather obvious reasons. Consider those estimators whose variances decrease with an increase in n. Such estimators will become degenerate, implying that the asymptotic variance would be zero. Again, consider the asymptotic distribution of the sample mean $\bar{X}$. By the central limit theorem, this asymptotic distribution has variance σ^2/n, not zero as would be implied if asymptotic variance were defined as the limiting value as $n \to \infty$. Thus, the term asymptotic variance may be misleading if not taken in its proper perspective—as simply an abbreviation for the variance of the asymptotic distribution, and certainly unrelated to the degenerate distribution.

Three desirable asymptotic properties will now be discussed: asymptotic unbiasedness, consistency, and asymptotic efficiency. Let us first consider asymptotic unbiasedness. An estimator $\hat{\theta}$ is said to be an *asymptotically unbiased estimator* of θ if

$$\lim_{n \to \infty} E(\hat{\theta}) = \theta$$

Point Estimation

This definition states that an estimator is asymptotically unbiased if it becomes unbiased as the sample size approaches infinity. Note that if an estimator is unbiased, it is asymptotically unbiased; however, the converse is not necessarily true.

Example 7.4 Show that the sample variance

$$\sigma'^2 = \frac{1}{n} \sum_{i=1}^{n} (X_i - \bar{X})^2 \qquad (7.13)$$

is asymptotically unbiased, i.e., $\sigma'^2 \to \sigma^2$ as $n \to \infty$.

Referring to the estimator S^2 defined in Example 7.1, we have

$$\sigma'^2 = \frac{n-1}{n} S^2$$

Now

$$\lim_{n \to \infty} E(\sigma'^2) = \lim_{n \to \infty} E\left(\frac{n-1}{n} S^2\right) = \lim_{n \to \infty} \frac{n-1}{n} E(S^2) = \lim_{n \to \infty} \frac{n-1}{n} \sigma^2 = \sigma^2$$

i.e., σ'^2 is asymptotically unbiased.

It is interesting to note that σ'^2 is not unbiased, in the finite sense of the definition, but that the expression

$$\frac{n}{n-1} \sigma'^2 \qquad (7.14)$$

is unbiased and therefore also asymptotically unbiased.

The next property to be discussed is consistency. As mentioned earlier, if the asymptotic distribution becomes degenerate at a point, then we hope that this point would be the true value of the parameter θ. In other words, we hope that as $n \to \infty$ the estimator $\hat{\theta}$ would tend toward θ. More precisely, we say that an estimator $\hat{\theta}$ is a *consistent estimator* for θ if given any positive numbers ε and δ, however small, there exists an integer k such that, for $n > k$, $P(|\hat{\theta} - \theta| < \varepsilon) \geq 1 - \delta$. One method of determining whether an estimator is consistent is to examine the bias and the variance as the sample size tends to infinity. If as the sample size increases there is a decrease in both the bias, if any, and the variance and if this continues until both the bias and variance approach zero, then the estimator under consideration is consistent. Since the sum of the squared bias and the variance is equal to the mean-squared error, the convergence of the bias and the variance to zero as $n \to \infty$ is equivalent to the convergence of the mean-squared error to zero.

Formally stated, $\hat{\theta}$ is said to be *squared-error consistent* for θ if $\lim_{n \to \infty} \text{MSE}(\hat{\theta}) = 0$.

It can be shown that if an estimator is squared-error consistent, then it must also be consistent. The reverse, however, is not necessarily true. Also, it is easily seen that if an estimator is squared-error consistent, then it is necessarily asymptotically unbiased. This result follows directly from the definition of mean-squared error.

The last asymptotic property we shall discuss is asymptotic efficiency. Basically this property says that among those estimators that are squared-error consistent, we should concern ourselves only with those whose squared errors approach zero the fastest as $n \to \infty$. This is equivalent to choosing those estimators whose asymptotic distributions have the smallest variance, since the asymptotic distribution represents the last stage before a distribution collapses, and estimators with the smallest variance are closest to assuming this degenerate form. Thus, an estimator $\hat{\theta}$ for θ is said to be *asymptotically efficient* if $\hat{\theta}$ is squared-error consistent and $\text{Var}(\hat{\theta}) \leq \text{Var}(\theta^*)$, where θ^* is any other squared-error consistent estimator.

To determine whether an estimator is squared-error consistent is a relatively simple task. We need only examine the behavior of $\text{MSE}(\hat{\theta})$ as $n \to \infty$. To check, however, for minimum asymptotic variance among all such estimators can be extremely complex. Fortunately, as in the case of efficiency for finite sample sizes, we can determine asymptotic efficiency by comparing the asymptotic variances of consistent estimators to the Cramer–Rao lower bound. If they are equal, then the estimator in question is asymptotically efficient. This procedure is exactly the same as that used for finite sample sizes, the only difference being that now we use the asymptotic variance instead of the variance based on a finite sample size. It should be noted that efficiency implies asymptotic efficiency, although the reverse is not necessarily true.

Example 7.5 Consider the exponential density defined in Example 7.2. A sample of size n is to be drawn and the parameter θ is to be estimated. Two estimators are proposed:

$$\hat{\theta} = \frac{1}{n} \sum_{i=1}^{n} X_i = \bar{X}, \quad \text{and} \quad \tilde{\theta} = \frac{1}{n+1} \sum_{i=1}^{n} X_i \quad (7.15)$$

What are the desirable properties of each of these estimators?

For $\hat{\theta}$. From Example 7.2 $\hat{\theta}$ is unbiased, linear, efficient, and therefore BLUE. Since unbiasedness implies asymptotical unbiasedness and efficiency implies asymptotic efficiency, then $\hat{\theta}$ possesses both of these asymptotic properties and is therefore consistent as well.

For $\tilde{\theta}$. Unbiasedness

$$E(\tilde{\theta}) = \frac{1}{n+1} E\left[\sum_{i=1}^{n} X_i\right] = \frac{n}{n+1} E(X) = \left[\frac{n}{n+1}\right]\theta$$

i.e., $\tilde{\theta}$ is biased.

For $\tilde{\theta}$. Efficiency: Since $\tilde{\theta}$ is biased, it cannot be efficient and certainly not BLUE even though it is a linear combination of X.

For $\tilde{\theta}$. Asymptotic Unbiasedness

$$\lim_{n \to \infty} E(\tilde{\theta}) = \lim_{n \to \infty} \left[\frac{n}{n+1}\right]\theta = \theta$$

i.e., $\tilde{\theta}$ is asymptotically unbiased.

For $\tilde{\theta}$. Consistency

$$\text{MSE}(\tilde{\theta}) = \text{Var}(\tilde{\theta}) + (\text{bias of } \tilde{\theta})^2$$

$$= \text{Var}\left[\frac{1}{n+1}\sum_{i=1}^{n} X_i\right] + \left[\left(\frac{n}{n+1}\right)\theta - \theta\right]^2$$

$$= \left[\frac{1}{n+1}\right]^2 \sum_{i=1}^{n} \text{Var}(X_i) + \left[\frac{1}{n+1}\right]^2 \theta^2$$

$$= \left[\frac{1}{n+1}\right]^2 (n\theta^2 + \theta^2) = \frac{\theta^2}{n+1}$$

and thus

$$\lim_{n \to \infty} \text{MSE}(\tilde{\theta}) = 0$$

i.e., $\tilde{\theta}$ is consistent.

For $\tilde{\theta}$. Asymptotic Efficiency: Since $\tilde{\theta}$ is squared-error consistent, it is a candidate for asymptotic efficiency. Now

$$\text{Var}(\tilde{\theta}) = \left[\frac{1}{n+1}\right]^2 \text{Var}\left(\sum_{i=1}^{n} X_i\right) = \frac{n\theta^2}{(n+1)^2} = \left[\frac{n}{n+1}\right]^2 \frac{\theta^2}{n}$$

and from Example 7.2 the Cramer-Rao lower bound is

$$\text{Var}(\theta^*) \geq \frac{\theta^2}{n}$$

where θ^* denotes any estimator for θ. For large samples, $n/(n+1)$ will be close to 1 so that the asymptotic variance of $\tilde{\theta}$ will be θ^2/n. Since this is the

Cramer–Rao lower bound, $\tilde{\theta}$ is asymptotically efficient. These results are summarized in Table 7.2.

Now that we have discussed several desirable properties of estimators, we come to the problem of devising procedures to obtain estimators that have all or at least some of these properties. Three such methods will be discussed: the *method of moments*, the *maximum-likelihood method*, and the *method of least squares*.

TABLE 7.2 Results of Example 7.5

Properties	Estimator	
	$\hat{\theta}$	$\tilde{\theta}$
Finite sample		
Unbiasedness	yes	no
Efficiency	yes	no
BLUE	yes	no
Asymptotic properties		
Unbiasedness	yes	yes
Consistency	yes	yes
Efficiency	yes	yes

Method of Moments

The method of moments is perhaps the oldest known method for obtaining estimators. It is based on the principle that one should estimate a moment of a population by the corresponding moment of the sample. To illustrate this procedure, consider a population whose density function $f(x)$ is characterized by k parameters $\theta_1, \theta_2, \ldots, \theta_k$, which are to be estimated. Let $X_1, X_2, \ldots, X_n$ be a random sample of size n from the population. The ith sample moment is defined as

$$M_i = \frac{\sum_{j=1}^{n} X_j^k}{n} \tag{7.16}$$

while the ith moment of the parent population, as defined in Chapter 2, is given by $E(X^i)$. Equating the first k population and sample moments, we have

$$M_1 = E(X), \quad M_2 = E(X^2), \quad \ldots, \quad M_k = E(X^k)$$

The solution values $\hat{\theta}_1, \hat{\theta}_2, \ldots, \hat{\theta}_k$ obtained by solving these simultaneous equations are referred to as the method of moment estimators for the k parameters $\theta_1, \theta_2, \ldots, \theta_k$.

Point Estimation

Estimators obtained by the method of moments procedure can be shown to be, under very general conditions, squared-error consistent and their asymptotic distribution is normal. The distribution of an estimator $\hat{\theta}$ is said to be *asymptotically normal* if the random variable $\sqrt{n}(\hat{\theta} - \theta)$ approaches the normal distribution with mean 0 and asymptotic variance $\sigma^2(\theta)$. The symbol $\sigma^2(\theta)$ indicates that the variance depends on θ. Although method of moment estimators are squared-error consistent, and therefore consistent, they are not in general asymptotically efficient.

Example 7.6 Using the method of moments, estimate the parameter θ in the exponential distribution as described in Example 7.2.

The mean of this distribution, its first moment, is given by

$$E(X) = \theta$$

If $X_1, X_2, \ldots, X_n$ represents a sample of size n from this distribution, then the first sample moment, the sample mean, is given by

$$M_1 = \sum_{j=1}^{n} \frac{X_j}{n}$$

Equating, we have as an estimate of θ, the sample mean

$$\hat{\theta} = \sum_{j=1}^{n} \frac{X_j}{n}$$

In this case we note that the method of moments estimator has all the desirable properties as demonstrated in Example 7.5 and listed in Table 7.2. As mentioned above, however, this is not always true.

Example 7.7 Using the method of moments, estimate the parameters k and μ of the Erlang density function

$$f(x) = \frac{(k\mu)^k}{\Gamma(k)} x^{k-1} e^{-k\mu x} \qquad (7.17)$$

The mean and variance of (7.17) are

$$E(X) = \frac{1}{\mu}, \quad \operatorname{Var}(X) = \frac{1}{k\mu^2}$$

Using the relationship

$$E(X^2) = \operatorname{Var}(X) + [E(X)]^2$$

we have, based on a sample of size n, that

$$\sum_{i=1}^{n} \frac{X_i}{n} = \bar{X} = \frac{1}{\mu}$$

and that

$$\sum_{i=1}^{n} \frac{X_i^2}{n} = \text{Var}(X) + \frac{1}{\mu^2}, \quad \text{or} \quad \text{Var}(X) = \sum_{i=1}^{n} \frac{X_i^2}{n} - \left[\sum_{i=1}^{n} \frac{X_i}{n}\right]^2$$

Thus,

$$\hat{\mu} = \frac{1}{\bar{X}}, \quad \text{and} \quad \hat{k} = \frac{1}{\sigma'^2 \mu^2}$$

where $\bar{X}$ and σ'^2 denote the sample mean and variance, respectively. It is worth mentioning that since the variance of X depends on k it is this parameter that determines the dispersion of X. This information is helpful when attempting to fit frequency data to an Erlang distribution.

The reader should notice that although the estimate of k, $\hat{k}$, is the method of moments estimate, it will frequently be infeasible. This will occur whenever $\hat{k}$ is fractional. Recall that the parameter k must be a positive integer. In these instances both the largest integer less than $\hat{k}$, $\hat{k}_l$, and the smallest integer greater than $\hat{k}$, $\hat{k}_g$, should be considered as the possible values of k. Since our main concern is to identify the parent population from which the data have come, we can test separately the hypotheses that the data belong to an Erlang population with parameters $(\hat{\mu}, \hat{k}_l)$ or to an Erlang population with parameters $(\hat{\mu}, \hat{k}_g)$. Procedures for conducting these tests are outlined in detail subsequently.

Maximum-Likelihood Method

The second method of estimation we shall discuss is the maximum-likelihood method, which is based on the idea that different populations generate different samples and that a given sample is more likely to come from some populations than from others. This method is best introduced by a simple example.

Suppose that we were to count the number of arrivals that occur at a hospital emergency room over various nonoverlapping intervals of 10 min. If it can be assumed that the conditions for Poisson arrivals (as listed in Chapter 2) hold, then the probability that exactly X arrivals occur in a 10-min interval is described by the mass function

$$P(x) = \frac{\lambda^x e^{-\lambda}}{x!} \tag{7.18}$$

where λ is the mean number of such arrivals. Suppose that three random observations on the process produce the three sample values 3, 5, and 4. Now our problem is to determine the maximum-likelihood estimate of λ, i.e.,

Point Estimation

the value of λ that most likely created the sample (3, 5, 4). To determine the population that would produce this sample most often, we can consider various values of λ and for these values determine the probability of obtaining the sample. For example, for $\lambda = 2$, the probability of drawing the sample (3, 5, 4) is given by

$$P(3, 5, 4) = P(3)P(5)P(4) = (0.180440)(0.036085)(0.090217) = 0.000587$$

since the sample observations are independent. For other selected values of λ, the results are:

λ	$P(3, 5, 4)$	λ	$P(3, 5, 4)$	λ	$P(3, 5, 4)$
1.0	0.00000	3.0	0.00379	5.0	0.00432
1.5	0.00008	3.5	0.00538	5.5	0.00290
2.0	0.00059	4.0	0.00597	6.0	0.00192
2.5	0.00191	4.5	0.00547		

The function $P(3, 5, 4)$ is referred to as the *likelihood function* for the sample (3, 5, 4). Although we chose values from 1 to 6 in steps of 0.5, we could have chosen any values of λ in the interval 0 to ∞, since the likelihood function is continuous in this interval. Figure 7.5 depicts the likelihood function for the sample (3, 5, 4). Note that this function is at a maximum when $\lambda = 4$; that is, a population with $\lambda = 4$ would generate samples

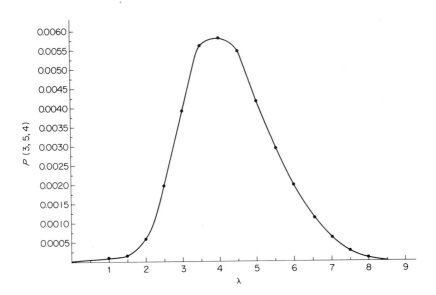

Fig. 7.5 Likelihood function for the sample (3, 5, 4).

(3, 5, 4) more frequently than any other population, and hence $\hat{\lambda} = 4$ is called the maximum-likelihood estimate of λ.

We are now in a position to define the maximum-likelihood estimator formally. Let $l(\theta_1, \theta_2, \ldots, \theta_k) = g(x_1, x_2, \ldots, x_n) = f(x_1)f(x_2)\cdots f(x_n)$ be the likelihood function for a random sample $X_1, X_2, \ldots, X_n$ from a density function $f(x)$ characterized by the parameters $\theta_1, \theta_2, \ldots, \theta_k$. If $\hat{\theta}_1, \hat{\theta}_2, \ldots, \hat{\theta}_k$ are the values of $\theta_1, \theta_2, \ldots, \theta_k$ that maximize $l(\theta_1, \theta_2, \ldots, \theta_k)$, then they are the *maximum-likelihood estimators* for the parameters $\theta_1, \theta_2, \ldots, \theta_k$.

To determine maximum-likelihood estimators we must find those values of the parameters that will maximize the likelihood function. To obtain these estimators we set

$$\frac{\partial l}{\partial \theta_1} = 0, \quad \frac{\partial l}{\partial \theta_2} = 0, \quad \ldots, \quad \frac{\partial l}{\partial \theta_k} = 0 \tag{7.19}$$

and solve this system of k simultaneous first-order equations for the k unknown parameters. Since the first-order condition (7.19) also holds for a minimum of a function, certain second-order conditions must be met before we are assured that our solution is a maximum. These conditions are beyond the scope of our discussion and can be found in Schmidt (1974). It suffices to say that the second-order conditions for a maximum are generally fulfilled for the cases we shall consider.

Frequently, when obtaining maximum-likelihood estimators it is easier to work with the logarithmic likelihood function, log l. This is permissible since the logarithm is a monotonic function and thus achieves a maximum at the same place as the likelihood function. We will denote the logarithm of the likelihood function by L, i.e., $L = \log l$.

Under very general conditions, maximum-likelihood estimators can be shown to be: (1) consistent as well as squared-error consistent, (2) asymptotically efficient, and (3) asymptotically normal. They also possess the important property of *invariance*; that is, if $\hat{\theta}$ is the maximum-likelihood estimator (MLE) of θ, and if $g(\theta)$ is function of θ with a single inverse, then the MLE of $g(\theta)$ is $g(\hat{\theta})$.

Example 7.8 Determine the MLE for the parameter θ of the exponential density function defined in Example 7.2.

Based on a sample of size n, the likelihood function is given by

$$l = \theta^{-n} \exp\left[-\frac{1}{\theta} \sum_{i=1}^{n} x_i\right]$$

Therefore,

$$L = -n \ln \theta - \frac{1}{\theta} \sum_{i=1}^{n} x_i$$

Point Estimation

To find the location of the maximum, we compute

$$\frac{\partial L}{\partial \theta} = \frac{-n}{\theta} + \frac{1}{\theta^2} \sum_{i=1}^{n} x_i$$

equate this result to zero, and solve for θ, obtaining the MLE

$$\hat{\theta} = \frac{1}{n} \sum_{i=1}^{n} x_i \qquad (7.20)$$

Note that $\hat{\theta}$ is the same result obtained using the method of moments.

It is interesting to note that if we make the simple transformation $\lambda = 1/\theta$ in (7.6), the parameter λ of the exponential density function

$$f(x) = \begin{cases} \lambda e^{-\lambda x}, & x > 0 \\ 0, & x \leq 0 \end{cases} \qquad (7.21)$$

has the MLE

$$\hat{\lambda} = \frac{n}{\sum_{i=1}^{n} x_i} = \bar{x}^{-1} \qquad (7.22)$$

This result follows immediately from the invariance property of the MLE.

Example 7.9 Determine the MLE for the parameter λ of the Poisson mass function defined by (7.18).

Let $X_1, X_2, \ldots, X_n$ be a random sample from this population. The likelihood function is given by

$$l = \frac{e^{-n\lambda} \lambda^{\sum_{i=1}^{n} x_i}}{\prod_{i=1}^{n} x_i!}$$

and the logarithmic likelihood function by

$$L = -n\lambda + \sum_{i=1}^{n} x_i \ln \lambda - \sum_{i=1}^{n} \ln x_i$$

Forming $\partial L/\partial \lambda$ and equating to zero, we have

$$\frac{\partial L}{\partial \lambda} = -n + \frac{1}{\lambda} \sum_{i=1}^{n} x_i = 0$$

Thus the MLE is

$$\hat{\lambda} = \sum_{i=1}^{n} \frac{x_i}{n} = \bar{x} \qquad (7.23)$$

Example 7.10 Using the maximum-likelihood method, determine estimates for the parameters k and μ of the Erlang density function defined in (7.17).

We have

$$l = \left[\frac{(k\mu)^k}{\Gamma(k)}\right]^n \prod_{i=1}^n x_i^{k-1} \exp\left[-k\mu \sum_{i=1}^n x_i\right]$$

$$L = nk \ln(k\mu) - n \ln \Gamma(k) + (k-1)\sum_{i=1}^n \ln x_i - k\mu \sum_{i=1}^n x_i$$

$$\frac{\partial L}{\partial k} = n \ln k + n + n \ln \mu - n\Psi(k) + \sum_{i=1}^n \ln x_i - \mu \sum_{i=1}^n x_i = 0 \quad (7.24)$$

$$\frac{\partial L}{\partial \mu} = \frac{nk}{\mu} - k\sum_{i=1}^n x_i = 0 \quad (7.25)$$

where $\Psi(k) = (d/dk) \ln \Gamma(k)$ is called the logarithmic derivative of the Γ-function, or simply the "psi-function." This function is extensively tabulated (see, for example, Jahnke and Emde, 1945). Fortunately, a good approximation to $\Psi(k)$ exists when k is not too small, say $k \geq 2$:

$$\Psi(k) \cong \log\left(k - \frac{1}{2}\right) + \frac{1}{24(k - \frac{1}{2})^2} \quad (7.26)$$

To illustrate, when $k = 2, 3$, the exact values of $\Psi(k)$ are 0.423 and 0.923, respectively, while the values given by the approximation are 0.424 and 0.923, respectively. If we solve (7.25) for μ, obtaining

$$\hat{\mu} = \frac{n}{\sum_{i=1}^n x_i} \quad (7.27)$$

and substitute this expression into (7.24) and simplify, we obtain

$$\hat{k} = \prod_{i=1}^n x_i^{-1/n} \hat{\mu}^{-1} \exp[\Psi(\hat{k})] \quad (7.28)$$

where $\Psi(k)$ is given by (7.26). By inspection of Eq. (7.28), we see that the solution for $\hat{k}$ can only be accomplished through a trial and error procedure.

Again, the reader should recognize that we have a situation similar to that encountered in Example 7.7. The MLE $\hat{k}$ is not necessarily integer valued and therefore can provide an infeasible estimate of k. To alleviate this problem, we can proceed exactly as discussed in Example 7.7, by choosing the nearest integral value on either side of $\hat{k}$ as the possible values of k. However, if a computer is available, the complexity of the problem diminishes since any standard nonlinear root-finding routine such as Newton's method can be employed. We remark that if $\Psi(\hat{k})$ is small then we can use the approximation $e^x \cong 1 + x$ to simplify (7.28). This approximation is highly recommended if a computer is unavailable.

Point Estimation

Example 7.11 Determine the MLEs for the mean m and variance σ^2 of a normal distribution.

Based on a sample of size n, the likelihood function is given by

$$l = \left[\frac{1}{(2\pi)^{1/2}\sigma}\exp-\frac{(x_1-m)^2}{2\sigma^2}\right]\left[\frac{1}{(2\pi)^{1/2}\sigma}\exp-\frac{(x_2-m)^2}{2\sigma^2}\right]$$

$$\cdots \left[\frac{1}{(2\pi)^{1/2}\sigma}\exp-\frac{(x_n-m)^2}{2\sigma^2}\right]$$

$$= \left(\frac{1}{2\pi\sigma^2}\right)^{n/2}\exp\left[-\sum_{i=1}^{n}\frac{(x_i-m)^2}{2\sigma^2}\right]$$

and the logarithmic likelihood function by

$$L = -\frac{n}{2}\ln 2\pi - \frac{n}{2}\ln \sigma^2 - \sum_{i=1}^{n}\frac{(x_i-m)^2}{2\sigma^2}$$

Taking the partial derivatives and setting them equal to zero gives

$$\frac{\partial L}{\partial m} = \sum_{i=1}^{n}\frac{(x_i-m)}{\sigma^2} = 0, \qquad \frac{\partial L}{\partial \sigma^2} = \frac{-n}{2\sigma^2} + \frac{\sum_{i=1}^{n}(x_i-m)^2}{2\sigma^4} = 0$$

Solving these equations simultaneously, we obtain

$$\hat{m} = \sum_{i=1}^{n}\frac{x_i}{n} = \bar{x} \tag{7.29}$$

$$\hat{\sigma}^2 = \sum_{i=1}^{n}\frac{(x_i-\bar{x})^2}{n} = \sigma'^2 \tag{7.30}$$

the maximizing values of m and σ^2, respectively.

Estimation of Time-Dependent Parameters

In our discussion of the sample mean $\bar{X}$ as an estimate of the population mean m, we assumed that the observations $X_1, X_2, \ldots, X_n$, from which $\bar{X}$ was calculated, were drawn from a population with mean m. We were thus able to show that $\bar{X}$ was an unbiased estimate of m. Let us now consider the case where m is not a constant, but rather a function of the point in time at which observations are taken; that is, the mean of the population under study is given by $m(t)$, where t denotes a point in time. Let Y_i be the observed value of a random variable taken at the point in time t_i, $i = 1, 2, \ldots, n$. We can then represent Y_i by

$$Y_i = m(t_i) + \varepsilon \tag{7.31}$$

where ε is a random variable with mean zero. Hence

$$E(Y_i) = m(t_i) \tag{7.32}$$

Unless we are able to obtain several observed values of Y_i at the point in time t_i, our estimate of $m(t_i)$ must be based on one and only one observed value of Y_i. However, we may be able to estimate the behavior of $m(t)$ over time by obtaining observations $Y_1, Y_2, \ldots, Y_n$ at various points in time, $t_1, t_2, \ldots, t_n$.

We will refer to an observation Y_i as a *dependent variable* since its value is dependent on the values of other variables, called *independent variables*, as well as random variation. We shall attempt to obtain an estimate $\hat{Y}_i$ of the mean $m(t_i)$ through a mathematical expression of the type

$$\hat{Y}_i = f(t_i) \tag{7.33}$$

However, we may find that the mean we are attempting to estimate is a function of several independent variables rather than just one. The method of estimation presented below may be applied to functions of a single or several variables.

Least-squares regression analysis. Suppose that the mean of a given random variable is an unknown linear function of the independent variable z_1. Then the mean $m(z_1)$ can be expressed as

$$m(z_1) = \beta_0 + \beta_1 z_1 \tag{7.34}$$

Our problem is to estimate the parameters β_0 and β_1. Let Y_i be an observed value of the random variable with mean $m(z_{1i})$. Then

$$Y_i = m(z_{1i}) + \varepsilon = \beta_0 + \beta_1 z_{1i} + \varepsilon \tag{7.35}$$

where ε is the random component that accounts for the deviation of Y_i from $m(z_{1i})$. Let $Y_1, Y_2, \ldots, Y_n$ be observations corresponding to values of the independent variables $z_{11}, z_{12}, \ldots, z_{1n}$. The values of $z_{11}, z_{12}, \ldots, z_{1n}$ are assumed to be known *without error*. The estimates for β_0 and β_1 are $\hat{\beta}_0$ and $\hat{\beta}_1$ such that the estimated value of Y_i, $\hat{Y}_i$, is given by

$$\hat{Y}_i = \hat{\beta}_0 + \hat{\beta}_1 z_{1i} \tag{7.36}$$

We define $\hat{\beta}_0$ and $\hat{\beta}_1$ such that the sum of the squares of the differences between Y_i and $\hat{\beta}_0 + \hat{\beta}_1 z_{1i}$ is minimized; that is,

$$\sum_{i=1}^{n} (Y_i - \hat{\beta}_0 - \hat{\beta}_1 z_{1i})^2 = \min \tag{7.37}$$

Let

$$E = \sum_{i=1}^{n} (Y_i - \hat{\beta}_0 - \hat{\beta}_1 z_{1i})^2 \tag{7.38}$$

Point Estimation

We minimize E by taking partial derivatives of E with respect to $\hat{\beta}_0$ and $\hat{\beta}_1$, setting each equal to zero, and solving for the resulting values of $\hat{\beta}_0$ and $\hat{\beta}_1$:

$$\frac{\partial E}{\partial \hat{\beta}_0} = -2 \sum_{i=1}^{n} (Y_i - \hat{\beta}_0 - \hat{\beta}_1 z_{1i}) = 0 \qquad (7.39)$$

$$\frac{\partial E}{\partial \hat{\beta}_1} = -2 \sum_{i=1}^{n} (Y_i - \hat{\beta}_0 - \hat{\beta}_1 z_{1i}) z_{1i} = 0 \qquad (7.40)$$

Solving for $\hat{\beta}_0$ and $\hat{\beta}_1$ yields

$$\hat{\beta}_1 = \frac{\sum_{i=1}^{n} Y_i z_{1i} - n\bar{Y}\bar{z}_1}{\sum_{i=1}^{n} z_{1i}^2 - n\bar{z}_1^2} \qquad (7.41)$$

$$\hat{\beta}_0 = \bar{Y} - \hat{\beta}_1 \bar{z}_1 \qquad (7.42)$$

where

$$\bar{Y} = \frac{1}{n} \sum_{i=1}^{n} Y_i \qquad (7.43)$$

$$\bar{z}_1 = \frac{1}{n} \sum_{i=1}^{n} z_{1i} \qquad (7.44)$$

Equation (7.36) is said to be the best simple linear estimating equation for Y_i in z_{1i} in the *least-squares sense*; that is, (7.36) describes the relationship between Y_i and z_{1i} better than any other simple linear relationship, again in the least-squares sense, given the data upon which the estimates $\hat{\beta}_0$ and $\hat{\beta}_1$ are based. However, this does not necessarily mean that the derived simple linear equation *adequately* explains the variation in Y. The actual relationship between Y and z_1 might be quadratic, cubic, logarithmic, etc. Thus, although the derived simple linear equation fits the observed data better than any other simple linear equation, it does not necessarily fit the data better than any other relationship and may actually be quite inadequate.

Example 7.12 Data on the arrival of customers per day have been collected over a 1-yr period. These data are summarized in the following tabulation:

Observation number i	Customer arrivals/day Y_i	Time (years) t_i
1	2	0.01
2	3	0.07
3	3	0.12
4	5	0.27
5	6	0.41
6	8	0.58
7	8	0.63
8	9	0.71
9	11	0.88
10	12	0.95

The collected data indicate that a simple linear relationship may describe the variation of arrival rate per day with time over a 1-yr period. Using (7.41) and (7.42) find the least-squares relationship of the form

$$\hat{Y}_i = \hat{\beta}_0 + \hat{\beta}_1 t_i \tag{7.45}$$

which describes the variation of Y with t.

From (7.41) and (7.42), we must determine the quantities $\sum_{i=1}^{n} t_i Y_i$, $\bar{Y}$, $\bar{t}$, and $\sum_{i=1}^{n} t_i^2$, where $n = 10$. These calculations are summarized as follows:

i	Y_i	t_i	$t_i Y_i$	t_i^2
1	2	0.01	0.02	0.0001
2	3	0.07	0.21	0.0049
3	3	0.12	0.36	0.0144
4	5	0.27	1.35	0.0729
5	6	0.41	2.46	0.1681
6	8	0.58	4.64	0.3364
7	8	0.63	5.04	0.3969
8	9	0.71	6.39	0.5041
9	11	0.88	9.68	0.7744
10	12	0.95	11.40	0.9025
Total	67	4.63	41.55	3.1747

Then

$$\bar{Y} = \frac{1}{10} \sum_{i=1}^{10} Y_i = 6.7000, \qquad \bar{t} = \frac{1}{10} \sum_{i=1}^{10} t_i = 0.4630$$

$$\sum_{i=1}^{10} t_i Y_i = 41.5500, \qquad \sum_{i=1}^{10} t_i^2 = 3.1747$$

and

$$\hat{\beta}_1 = \frac{41.5500 - (10)(6.7000)(0.4630)}{3.1747 - (10)(0.4630)^2} = \frac{10.5290}{1.0310} = 10.2124$$

$$\hat{\beta}_0 = 6.7000 - (10.2124)(0.4630) = 6.7000 - 4.7283 = 1.9717$$

Hence

$$\hat{Y} = 1.9717 + 10.2124t$$

The regression equation and the experimental data are shown graphically in Fig. 7.6.

The method of least squares may be extended not only to polynomials in a single variable but also to more complex models involving several independent variables. For example, suppose that $z_1, z_2, \ldots, z_m$ are independent variables and let $m(z_1, z_2, \ldots, z_m)$ be a parameter that depends on the values

Point Estimation

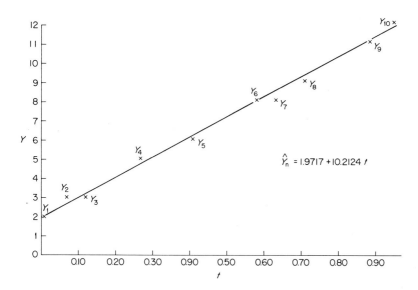

Fig. 7.6 Simple linear regression equation for the data in Example 7.12.

of the independent variables. The relationship describing $m(z_1, z_2, \ldots, z_m)$ can be considered of the form

$$m(z_1, z_2, \ldots, z_m) = \beta_0 + \beta_1 z_1 + \beta_2 z_2 + \cdots + \beta_m z_m \qquad (7.46)$$

Thus an observed value Y_i corresponding to $z_{1i}, z_{2i}, \ldots, z_{mi}$ can be defined as

$$Y_i = m(z_{1i}, z_{2i}, \ldots, z_{mi}) + \varepsilon = \beta_0 + \beta_1 z_{1i} + \beta_2 z_{2i} + \cdots + \beta_m z_{mi} + \varepsilon \qquad (7.47)$$

As before, we will define the estimates of $\beta_0, \beta_1, \ldots, \beta_m, \hat{\beta}_0, \hat{\beta}_1, \ldots, \hat{\beta}_m$, using the least-squares method; that is, we wish to determine the values of $\hat{\beta}_0, \hat{\beta}_1, \ldots, \hat{\beta}_m$ such that

$$E = \sum_{i=1}^{n}(Y_i - \hat{\beta}_0 - \hat{\beta}_1 z_{1i} - \hat{\beta}_2 z_{2i} - \cdots - \hat{\beta}_m z_{mi})^2 \qquad (7.48)$$

is minimized. Taking partial derivatives of E with respect to $\hat{\beta}_0, \hat{\beta}_1, \ldots, \hat{\beta}_m$, we have

$$\frac{\partial E}{\partial \hat{\beta}_0} = -2\sum_{i=1}^{n}(Y_i - \hat{\beta}_0 - \hat{\beta}_1 z_{1i} - \cdots - \hat{\beta}_m z_{mi}) = 0 \qquad (7.49)$$

$$\frac{\partial E}{\partial \hat{\beta}_j} = -2\sum_{i=1}^{n} z_{ji}(Y_i - \hat{\beta}_0 - \hat{\beta}_1 z_{1i} - \hat{\beta}_2 z_{2i} - \cdots - \hat{\beta}_m z_{mi})$$

$$j = 1, 2, \ldots, n \qquad (7.50)$$

Simplifying yields

$$\sum_{i=1}^{n} Y_i = \sum_{i=1}^{n} \sum_{k=0}^{m} \hat{\beta}_k z_{ki} \tag{7.51}$$

$$\sum_{i=1}^{n} Y_i z_{ji} = \sum_{i=1}^{n} \sum_{k=0}^{m} \hat{\beta}_k z_{ki} z_{ji}, \quad j = 1, 2, \ldots, m \tag{7.52}$$

where $z_{0i} = 1$, $i = 1, 2, \ldots, n$. Let

$$Y = \begin{bmatrix} Y_1 \\ Y_2 \\ \vdots \\ Y_n \end{bmatrix} \quad B = \begin{bmatrix} \hat{\beta}_0 \\ \hat{\beta}_1 \\ \vdots \\ \hat{\beta}_m \end{bmatrix} \quad Z = \begin{bmatrix} 1 & z_{11} & z_{21} & \cdots & z_{m1} \\ 1 & z_{12} & z_{22} & \cdots & z_{m2} \\ \vdots & \vdots & \vdots & & \vdots \\ 1 & z_{1m} & z_{2m} & \cdots & z_{mn} \end{bmatrix}$$

Then

$$Z^T Z = \begin{bmatrix} n & \sum_{i=1}^{n} z_{1i} & \sum_{i=1}^{n} z_{2i} & \cdots & \sum_{i=1}^{n} z_{mi} \\ \sum_{i=1}^{n} z_{1i} & \sum_{i=1}^{n} z_{1i}^2 & \sum_{i=1}^{n} z_{1i} z_{2i} & \cdots & \sum_{i=1}^{n} z_{1i} z_{mi} \\ \vdots & \vdots & \vdots & & \vdots \\ \sum_{i=1}^{n} z_{mi} & \sum_{i=1}^{n} z_{1i} z_{mi} & \sum_{i=1}^{n} z_{2i} z_{mi} & \cdots & \sum_{i=1}^{n} z_{mi}^2 \end{bmatrix}$$

and from (7.50) and (7.51)

$$Z^T Y = Z^T B \tag{7.53}$$

Then

$$B = (Z^T Z)^{-1} Z^T Y \tag{7.54}$$

and the approximating regression equation is given by

$$\hat{Y} = \hat{\beta}_0 + \hat{\beta}_1 z_1 + \hat{\beta}_2 z_2 + \cdots + \hat{\beta}_m z_m \tag{7.55}$$

Example 7.13 The mean capacity Y of an air terminal, measured in operations per hour, is assumed to be linearly related to separation on approach z_1 in miles and approach velocity z_2 in hundreds of nautical miles per hour; that is, the hypothesized relationship is

$$Y = \beta_0 + \beta_1 z_1 + \beta_2 z_2 + \varepsilon$$

Point Estimation

The following data have been collected:

Y	z_1	z_2
19	2.0	1.40
27	2.0	1.50
16	2.5	1.40
21	2.5	1.60
17	3.0	1.50
17	3.0	1.60
21	3.0	1.70
11	3.5	1.40
16	3.5	1.60

Using the method of least squares, derive the best-fitting linear relationship of the form

$$\hat{Y} = \hat{\beta}_0 + \hat{\beta}_1 z_1 + \hat{\beta}_2 z_2$$

The matrices Y, B, and Z are given by

$$Y = \begin{bmatrix} 19 \\ 22 \\ 16 \\ 21 \\ 17 \\ 17 \\ 21 \\ 11 \\ 16 \end{bmatrix} \quad B = \begin{bmatrix} \hat{\beta}_0 \\ \hat{\beta}_1 \\ \hat{\beta}_2 \end{bmatrix} \quad Z = \begin{bmatrix} 1 & 2.0 & 1.40 \\ 1 & 2.0 & 1.50 \\ 1 & 2.5 & 1.40 \\ 1 & 2.5 & 1.60 \\ 1 & 3.0 & 1.50 \\ 1 & 3.0 & 1.60 \\ 1 & 3.0 & 1.70 \\ 1 & 3.5 & 1.40 \\ 1 & 3.5 & 1.60 \end{bmatrix}$$

Then

$$Z^T Z = \begin{bmatrix} 9 & 25 & 13.70 \\ 25 & 72 & 38.20 \\ 13.70 & 38.20 & 20.95 \end{bmatrix} \quad Z^T Y = \begin{bmatrix} 160 \\ 434 \\ 245 \end{bmatrix}$$

The determinant of $Z^T Z$, $|Z^T Z|$, is given by

$$|Z^T Z| = 2.01$$

and the adjoint of $Z^T Z$ is

$$\text{adj}(Z^T Z) = \begin{bmatrix} 49.16 & -0.41 & -31.4 \\ -0.41 & 0.86 & -1.3 \\ -31.4 & -1.3 & 23.0 \end{bmatrix}$$

Hence the inverse of $Z^T Z$ is

$$(Z^T Z)^{-1} = \begin{bmatrix} 24.4577 & -0.2040 & -15.6219 \\ -0.2040 & 0.4279 & -0.6468 \\ -15.6219 & -0.6968 & 11.4428 \end{bmatrix}$$

From (7.54),

$$B = (Z^T Z)^{-1} Z^T Y = \begin{bmatrix} 24.4577 & -.2040 & -15.6219 \\ -.2040 & .4279 & -.6468 \\ -15.6219 & -.6468 & 11.4428 \end{bmatrix} \begin{bmatrix} 160 \\ 434 \\ 245 \end{bmatrix}$$

$$= \begin{bmatrix} -2.6695 \\ -5.3974 \\ 23.2708 \end{bmatrix}$$

Therefore, the linear relationship that best fits the data is given by

$$\hat{Y} = -2.6695 - 5.3974 z_1 + 23.2708 z_2$$

In our discussion thus far, we have treated the development of estimating models that were either simple or multiple linear relationships. The method of least-squares regression analysis can be extended to more complex models. For example, suppose that we wish to fit a model of the form

$$\hat{Y} = \hat{\beta}_0 + \hat{\beta}_1 x_1 + \hat{\beta}_2 x_2 + \hat{\beta}_3 x_1 x_2 + \hat{\beta}_4 x_1^2 + \hat{\beta}_5 x_2^2 \qquad (7.56)$$

to a set of experimental data, where x_1 and x_2 are the independent variables. If we define

$$z_1 = x_1, \quad z_2 = x_2, \quad z_3 = x_1 x_2, \quad z_4 = x_1^2, \quad z_5 = x_2^2$$

then the expression in (7.56) reduces to

$$\hat{Y} = \hat{\beta}_0 + \hat{\beta}_1 z_1 + \hat{\beta}_2 z_2 + \hat{\beta}_3 z_3 + \hat{\beta}_4 z_4 + \hat{\beta}_5 z_5 \qquad (7.57)$$

Thus, we can treat, without difficulty, any equation that is linear in the coefficients $\hat{\beta}_i$. Functions that are nonlinear in the coefficients $\hat{\beta}_i$ can also be defined using the method of least squares, although analytic minimization of E may be difficult or impossible in some cases.

Example 7.14 The rate at which a service channel operates, measured in operations per hour, is hypothesized to be a quadratic function of the number of units waiting to be served. The following data were collected, where x is the number of units waiting to be served measured from the average of 3 and Y is the service rate:

Y	x
3	-2
6	-1
8	0
8	1
7	2

Point Estimation

Develop an equation of the form
$$\hat{Y} = \hat{\beta}_0 + \hat{\beta}_1 x + \hat{\beta}_2 x^2$$
which best fits the data using the method of least squares.
  Let
$$z_1 = x, \qquad z_2 = x^2$$
Then
$$\hat{Y} = \hat{\beta}_0 + \hat{\beta}_1 z_1 + \hat{\beta}_2 z_2$$

The matrices Y and Z are given by

$$Y = \begin{bmatrix} 3 \\ 6 \\ 8 \\ 8 \\ 7 \end{bmatrix} \qquad Z = \begin{bmatrix} 1 & -2 & 4 \\ 1 & -1 & 1 \\ 1 & 0 & 0 \\ 1 & 1 & 1 \\ 1 & 2 & 4 \end{bmatrix}$$

and

$$Z^T Z = \begin{bmatrix} 5 & 0 & 10 \\ 0 & 10 & 0 \\ 10 & 0 & 34 \end{bmatrix} \qquad Z^T Y = \begin{bmatrix} 32 \\ 10 \\ 54 \end{bmatrix}$$

The determinant of $Z^T Z$ is
$$|Z^T Z| = 700$$
and the adjoint of $Z^T Z$ is given by
$$\operatorname{adj}(Z^T Z) = \begin{bmatrix} 340 & 0 & -100 \\ 0 & 70 & 0 \\ -100 & 0 & 50 \end{bmatrix}$$

Hence from (7.54)

$$B = \begin{bmatrix} \dfrac{340}{700} & 0 & -\dfrac{100}{700} \\ 0 & \dfrac{70}{700} & 0 \\ -\dfrac{100}{700} & 0 & \dfrac{50}{700} \end{bmatrix} \begin{bmatrix} 32 \\ 10 \\ 54 \end{bmatrix} = \begin{bmatrix} 7.83 \\ 1.00 \\ -0.71 \end{bmatrix}$$

and the estimating equation is given by
$$\hat{Y} = 7.83 + x - 0.71 x^2$$

A note of caution is worth mentioning with regard to the use of a regression equation. The numerical values of the coefficients in a regression equation are based on observed values of the dependent and independent variables. Caution should be exercised in applying the equation beyond the observed range of the independent variables, since there is no evidence available, based on the observed data alone, to indicate that the relationship applies outside this range. Thus, while minor extrapolation beyond the observed range of the independent variables may not introduce serious errors, extensive extrapolation should be avoided.

As for the properties of least-squares estimators, it can be shown that they are unbiased, since they are BLUE, and that they are efficient as well. It can also be shown that the least-squares estimators are the same as the maximum-likelihood estimators and thus possess all the desirable asymptotic properties of these estimators. In particular, they are asymptotically unbiased, consistent, and asymptotically efficient.

Having discussed methods by which point estimates can be obtained, the reader is no doubt concerned with the accuracy of these estimates. Even if an estimator possesses all of the desirable properties discussed, it will rarely produce values that coincide exactly with the true parameter value, since it is based on the incomplete information of a sample. What is needed is some measure of an estimator's precision. This leads quite naturally to a discussion of interval estimation. For the sake of continuity, however, this discussion will be delayed until we have introduced the final step in distribution identification, goodness-of-fit tests.

GOODNESS-OF-FIT TESTS

The third and final step in identifying a distribution is to test the hypothesis that there is no detectable difference between the hypothesized distribution and the sample distribution. Three procedures that can be used for this purpose are described in this section: the Kolmogorov–Smirnov test, the chi-square goodness-of-fit test, and a special test for the Poisson process.

The Kolmogorov–Smirnov Test

The Kolmogorov–Smirnov (K–S) procedure compares the cumulative distribution function for the hypothesized distribution $F(x)$ with the sample cumulative distribution $S_n(x)$. The sample cumulative distribution is defined by $S_n(x) = i/n$, where i is the number of observations less than or equal to x, and n is the sample size. If $S_n(x)$ is "too far" from $F(x)$, then the null hypothesis that there is no detectable difference between $S_n(x)$ and $F(x)$ is

Goodness-of-Fit Tests

rejected. This comparison between $S_n(x)$ and $F(x)$ is based on the absolute value of their difference using the K–S test statistic,

$$D_{\max} = \max_x |F(x) - S_n(x)| \qquad (7.58)$$

The value $D_{\max}$ is called the maximum deviation.

The distribution of $D_{\max}$ is known and can be shown to be independent of $F(x)$. Several values from the distribution have been tabulated as a function of n, the number of observations, and α, the level of significance. Table D in the Appendix lists some of these values for different n and α combinations. The null hypothesis that the data came from the hypothesized distribution is rejected at the α level of significance whenever $D_{\max} \geq D_n^{\alpha}$, where D_n^{α} are the critical values listed in Table D.

As denoted in (7.58) the maximum deviation must be taken over all x. Since $S_n(x)$ plots as a step function, the differences need only be examined at the step points. Figure 7.7 illustrates a typical situation in which $F(x)$ is

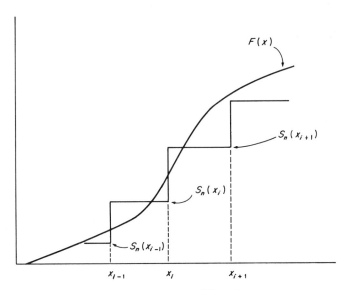

Fig. 7.7 Step points for $F(x)$ continuous.

continuous. It is important to note that two differences must be computed at each step point x_i, $|F(x_i) - S_n(x_i)|$ and $|F(x_i) - S_n(x_{i-1})|$, since one is permitted to choose either the bottom or top of each step.

Although the K–S test is designed specifically to handle continuous functions, it can be successfully applied to discrete functions. The test procedure is exactly the same; however, the level of significance achieved will be less than or equal to the prescribed level. Thus, any bias of the test to discrete data is in the "safe" direction. In other words, if the null hypothesis is rejected by the test, then we have a high degree of confidence in this decision (Siegel, 1956). Another important aspect of the K–S test is that it is more powerful than the chi-square test, which is discussed below. For this reason its use is encouraged.

The K–S test proceeds as follows:

1. Determine $S_n(x)$ from sample data and hypothesize a specified cumulative distribution $F(x)$.
2. Compute $|F(x_i) - S_n(x_i)|$ and $|F(x_i) - S_n(x_{i-1})|$ at each step point x_i, if $F(x)$ is continuous. If $F(x)$ is discrete, only $|F(x_i) - S_n(x_i)|$ need be computed.
3. Determine the maximum value $D_{\max}$ from the calculations performed in Step 2.
4. Select a level of significance α.
5. Locate the critical value D_n^α in Table D of the Appendix.
6. If $D_{\max} \geq D_n^\alpha$, reject the hypothesis that the data could have come from the population described by $F(x)$.

Chi-Square Test

The chi-square test compares the observed frequency in an interval with the expected number of occurrences in that interval if $F(x)$ were the population distribution from which these values were obtained. If the null hypothesis that $F(x) = S_n(x)$ cannot be rejected, then it is assumed that the data may well have come from $F(x)$.

The null hypothesis is tested with the statistic

$$\chi^2 = \sum_{i=1}^{k} \frac{(O_i - E_i)^2}{E_i} \tag{7.59}$$

where O_i is the observed frequency in the ith interval, E_i the expected frequency in the ith interval, and k the number of intervals. If the hypothesis that $F(x) = S_n(x)$ is true, then (7.59) has an approximate chi-square distribution for large n. Hence, we accept the null hypothesis if $\chi^2 \leq \chi^2_{1-\alpha}(d)$, where $\chi^2_{1-\alpha}(d)$ is such that

$$F_{\chi^2}[\chi^2_{1-\alpha}(d)] = 1 - \alpha$$

and $F_{\chi^2}(x)$ is the distribution function of the chi-square random variable with d degrees of freedom. These values can be found in Table C of the Appendix.

If the hypothesized distribution $F(x)$ is completely specified, as discussed above, then the number of degrees of freedom associated with the test is

Goodness-of-Fit Tests

$k - 1$. If, however, the form of $F(x)$ is given but some of its parameters are unspecified, then the number of degrees of freedom is $(k - 1)$ minus the number of parameters that must be estimated from sample data. For example, if the null hypothesis is that the frequency data are from an Erlang distribution (Example 7.10) with parameters μ and k that are to be estimated from sample data, then the number of degrees of freedom associated with the test is $(k - 1) - 2$ or $k - 3$.

In our discussion so far, no suggestions have been given as to how one should choose the value of k in (7.59). A detailed discussion of this problem can be found in the expository article by Cochran (1954). It is generally felt, however, that the number k should be chosen so that the number of occurrences in each interval, both expected and observed, is at least 5. Otherwise a high χ^2 value may result. Also, it is often recommended that the intervals be chosen so the probability of an observation falling into each interval is nearly equal when $F(x)$ is the true population distribution.

One final point should be made in connection with the use of the chi-square test of a continuous distribution with unknown parameters. Specifically, estimates should be of the maximum-likelihood type in order to guarantee an asymptotic chi-square distribution for χ^2. Using sample moments may work well in some cases, but there are others where one can go appreciably astray. The reader is referred to Chernoff and Lehmann (1954) for further discussion.

The chi-square test proceeds as follows:

1. Determine the number of class intervals k such that $O_i > 5$ and $E_i > 5$, for all i.
2. Compute the test statistic χ^2 given by (7.59).
3. Select a level of significance α.
4. Determine the number of degrees of freedom:

$$d = (k - 1) - [\text{number of estimated parameters in the hypothesized distribution } F(x)]$$

5. Locate the critical value $\chi^2_{1-\alpha}(d)$ in Table C of the Appendix.
6. If $\chi^2 > \chi^2_{1-\alpha}(d)$, reject the hypothesis that the data could have come from the population $F(x)$.

Example 7.15 Use the Kolmogorov–Smirnov and chi-square goodness-of-fit tests to identify the parent population for the emergency room arrival data listed in Table 7.1.

Comparing Fig. 7.1 with Figs. 7.2–7.4, we suspect that the arrivals might be generated by a Poisson, binomial, or even a normal process. To test our suspicions we will examine all three candidate distributions.

To determine the sample mean and sample variance of our data we use the relations

$$\bar{x} = \frac{\sum_{i=1}^{k} m_i f_i}{n} \qquad (7.60)$$

and

$$s^2 = \frac{\sum_{i=1}^{k} m_i^2 f_i - n\bar{x}^2}{n-1} \qquad (7.61)$$

where m_i is the midpoint of the ith class interval, f_i the frequency in that interval, n the number of observations, and k the number of intervals. These calculations yield

$$\bar{x} = 14.91, \quad \text{and} \quad s^2 = 14.14$$

The calculations are summarized in Table 7.3.

Kolomogorov–Smirnov Test: From Example 7.9, the maximum-likelihood estimate for the Poisson parameter λ is the sample mean $\bar{X}$. Thus,

TABLE 7.3 Calculations for Example 7.15

Interval midpoint m_i	Frequency f_i	$m_i f_i$	$m_i^2 f_i$
6	1	6	36
7	0	0	0
8	3	24	192
9	3	27	243
10	6	60	600
11	5	55	605
12	9	108	1296
13	9	117	1521
14	11	154	2156
15	11	165	2475
16	10	160	2560
17	8	136	2312
18	6	108	1944
19	6	114	2166
20	4	80	1600
21	3	63	1323
22	2	44	968
23	2	46	1058
24	1	24	576
Total	100	1491	23631

Goodness-of-Fit Tests

we hypothesize that the data may have come from a Poisson mass function with

$$\text{Mean} = 14.91 \cong 15.0, \quad \text{and} \quad \text{Variance} = 14.91 \cong 15.0$$

From Example 7.11, the maximum-likelihood estimators for the mean and variance of a normal distribution are the sample mean $\bar{X}$ and the biased sample variance σ'^2. Thus, we hypothesize that the parent population is normal with

$$\text{Mean} = 14.91 \cong 15.0,$$

and

$$\text{Variance} = \sigma'^2 = \frac{n-1}{n} s^2 = \frac{99}{100}(14.14) = 14.00$$

For the binomial mass function, we have from Table 2.1 that:

$$\text{Mean} = np, \quad \text{Variance} = np(1-p)$$

Using the method of moments,

$$np = 14.91, \quad np(1-p) = 14.14$$

and solving for n and p, we obtain

$$n = 289.97, \quad \text{and} \quad p = 0.0514$$

Since n must be an integer, we hypothesize that the data may well have come from a binomial mass function with parameters $n = 290$, $p = 0.05$.

To summarize, the hypothesized distributions are:

Poisson: $\lambda = 15$

Normal: $m = 15, \quad \sigma^2 = 14.0$ (7.62)

Binomial: $n = 290, \quad p = 0.05$

The calculations relating to the Kolomogorov–Smirnov test are listed in Table 7.4. Letting $\alpha = 0.05$, we find from Table D of the Appendix that $D_{100}^{0.05} = 0.134$. Since $D_{\max} < D_{100}^{0.05}$ for all three distributions, any one of the distributions can be chosen to represent the parent arrival population. The Poisson distribution would most likely be our choice, since we know more about queueing models having Poisson input than any other arrival pattern. It also provides the "best" fit among the candidate distributions.

Chi-Square Test: Again the distributions given by (7.62) are hypothesized. The necessary calculations to test for the Poisson and binomial distributions are summarized in Table 7.5. Notice that the arrival categories 10,

7. Data Analysis—Estimation

TABLE 7.4 Kolmogorov–Smirnov test calculations for Example 7.15

Arrival category	Observed cumulative frequency	Hypothesized cumulative distributions		
		Poisson	Normal	Binomial
6	0.0100	0.0076	0.0080	0.0091
7	0.0100	0.0180	0.0162	0.0214
8	0.0400	0.0374	0.0307	0.0444
9	0.0700	0.0699	0.0548	0.0829
10	0.1300	0.1185	0.0901	0.1383
11	0.1800	0.1848	0.1423	0.2134
12	0.2700	0.2676	0.2119	0.3052
13	0.3600	0.3632	0.2963	0.4085
14	0.4700	0.4657	0.3936	0.5162
15	0.5800	0.5681	0.5000	0.6204
16	0.6800	0.6641	0.6064	0.7147
17	0.7600	0.7489	0.7019	0.7947
18	0.8200	0.8195	0.7881	0.8586
19	0.8800	0.8752	0.8577	0.9067
20	0.9200	0.9170	0.9099	0.9410
21	0.9500	0.9469	0.9452	0.9642
22	0.9700	0.9673	0.9693	0.9792
23	0.9900	0.9805	0.9838	0.9883
24	1.0000	0.9888	0.9920	0.9937
D_{max}		0.0159	0.0800	0.1226

11, and 12 each have fewer than 5 arrivals and have been combined into one class interval. Similarly, arrival categories 19 and 20 and 21, 22, and 23 are also grouped into single-class intervals. The calculations necessary to test for the normal distribution are summarized in Table 7.6.

The test for the Poisson distribution has 11 degrees of freedom, since the parameter λ was estimated from the data. The tests for the normal and binomial distributions each have 10 degrees of freedom, since in both of these cases it was necessary to estimate two parameters from the data. Letting $\alpha = 0.05$, the critical values

$$\chi^2_{0.95}(11) = 19.68, \quad \text{and} \quad \chi^2_{0.95}(10) = 18.31$$

are found in Table C of the Appendix, and the results of the tests are:

Poisson: $\quad \chi^2 < \chi^2_{0.95}(11)$

Normal: $\quad \chi^2 < \chi^2_{0.95}(10)$

Binomial: $\quad \chi^2 < \chi^2_{0.95}(10)$

Goodness-of-Fit Tests

TABLE 7.5 Chi-square test calculations for binomial and Poisson distributions

Number of arrivals	Observed frequency (O_i)	Expected frequencies (E_i) Poisson	Expected frequencies (E_i) Binomial	$\dfrac{(O_i - E_i)^2}{E_i}$ Poisson	$\dfrac{(O_i - E_i)^2}{E_i}$ Binomial
6	1 ⎫	0.48 ⎫	0.58 ⎫		
7	0 ⎬ 7	1.04 ⎬ 6.70	1.23 ⎬ 7.90	0.01	0.10
8	3 ⎪	1.94 ⎪	2.30 ⎪		
9	3 ⎭	3.24 ⎭	3.79 ⎭		
10	6	4.86	5.60	0.27	0.03
11	5	6.63	7.51	0.40	0.84
12	9	8.29	9.18	0.06	0.00
13	9	9.56	10.34	0.06	0.08
14	11	10.24	10.76	0.06	0.10
15	11	10.24	10.42	0.06	0.03
16	10	9.60	9.43	0.02	0.03
17	8	8.47	8.00	0.03	0.00
18	6	7.06	6.39	0.16	0.02
19	6	5.57	4.81	0.03	0.30
20	4 ⎱ 7	4.18 ⎱ 7.17	3.43 ⎱ 5.75	0.00	0.27
21	3 ⎰	2.99 ⎰	2.32 ⎰		
22	2 ⎫	2.04 ⎫	1.49 ⎫		
23	2 ⎬ 5	1.33 ⎬ 4.20	0.92 ⎬ 2.95	0.15	1.42
24	1 ⎭	0.88 ⎭	0.54 ⎭		
				$\chi^2 = 1.31$	$\chi^2 = 3.22$

TABLE 7.6 Chi-square test calculations for normal distribution

Class interval	Observed frequency (O_i)	Normal expected frequency (E_i)	$\dfrac{(O_i - E_i)^2}{E_i}$
< 9.499	7	7.08	0.00
9.5–10.499	6	4.43	0.56
10.5–11.499	5	5.98	0.16
11.5–12.499	9	7.66	0.23
12.5–13.499	9	9.31	0.01
13.5–14.499	11	10.37	0.04
14.5–15.499	11	10.34	0.04
15.5–16.499	10	10.37	0.01
16.5–17.499	8	9.31	0.18
17.5–18.499	6	7.79	0.41
18.5–19.499	6	5.85	0.00
19.5–21.499	7	7.42	0.02
> 21.5	5	4.09	0.20
			$\chi^2 = 1.86$

Thus, any one of the three hypothesized distributions could be chosen as the parent population. Again the Poisson distribution would most likely be chosen.

In practice, only one goodness-of-fit test need be applied to the data. Both tests were used here only for illustrative purposes. Also, it is recommended that the Kolmogorov–Smirnov test generally be used, since statistically it is more powerful than the chi-square test.

Poisson-Process Tests

The majority of queueing models for which we have obtained analytic solutions are based on the assumption of a Poisson process. Therefore we are most anxious to detect the presence of this process when it occurs. Two special tests for the Poisson process are described here. The first can be particularly useful when sample data are sparse, since it does not require any knowledge of the process parameter. This test, however, is not nearly as powerful as the special K–S test for the exponential distribution, which will be discussed later.

Let $t_1, t_2, \ldots, t_n$ denote the times at which n units enter a queueing system during a time interval of length T. If these arrivals are from a Poisson process, then the times are independent and uniformly distributed over the interval 0 to T with mean $T/2$ and variance $T^2/12$. If we form the sum

$$S_n = \sum_{i=1}^{n} \frac{t_i}{n} \qquad (7.63)$$

then by the central limit theorem, for large n, S_n will be normally distributed with mean

$$E(S_n) = \frac{T}{2} \qquad (7.64)$$

and variance

$$\text{Var}(S_n) = \frac{T^2}{12n} \qquad (7.65)$$

Thus, to test the hypothesis that the arriving units are from a Poisson process, we simply compute the normal test statistic

$$Z = \frac{S_n - T/2}{(T^2/12n)^{1/2}} \qquad (7.66)$$

choose a level of significance α, and locate the critical values $Z_{1-\alpha/2}$ and $Z_{\alpha/2}$, in Table A of the Appendix. If $Z < Z_{\alpha/2}$ or $Z > Z_{1-\alpha/2}$, then we reject the null hypothesis that our arrivals are from a Poisson process.

Goodness-of-Fit Tests 339

To summarize, the Poisson test proceeds as follows:
1. Compute the sum S_n given by (7.63).
2. Compute the normal test statistic Z given by (7.66).
3. Select a level of significance α.
4. Locate the critical values $Z_{\alpha/2}$ and $Z_{1-\alpha/2}$ in Table A of the Appendix.
5. If $Z < Z_{\alpha/2}$ or $Z > Z_{1-\alpha/2}$, then we reject the hypothesis that the data could have come from a Poisson process.

Although this test is asymptotically exact, the reader may wonder about its validity for finite sample sizes. Since the test is based on the central limit theorem, as a rule of thumb we can say that the test can be applied safely whenever $n \geq 30$. In fact, tests based on the central limit theorem are more powerful than nonparametric tests such as the Kolmogorov–Smirnov test and the chi-square test.

The above mentioned rule of thumb that n should be greater than 30 is generally suggested when attempting to apply asymptotic properties to finite sample sizes. This rule, although useful as a general guideline, tends to be very conservative. To illustrate that this is indeed the case here, we generated 1000 samples of size 5 and 1000 samples of size 10 from the uniform distribution on the interval 0 to 10. Using these samples we constructed the sampling distributions of $\bar{X}$, S_5, and S_{10}. The results of this experiment are shown in Table 7.7, where they can be compared directly with the values we would expect to receive from the asymptotic normal approximations. Notice that the results given by the normal approximation are very good. In fact, they are in agreement with the sampling results for at least one decimal place and in the majority of cases coincide to two decimal places. This remarkable

TABLE 7.7 Sampling distributions of $\bar{X}$, S_5, and S_{10}

Interval	Sample size, 5		Sample size, 10	
	Sampling distribution	Asymptotic normal approximation	Sampling distribution	Asymptotic normal approximation
0.0– 1.499	0.004	0.003	0.000	0.000
1.5– 2.499	0.025	0.023	0.002	0.003
2.5– 3.499	0.104	0.100	0.044	0.047
3.5– 4.499	0.243	0.225	0.244	0.241
4.5– 5.499	0.307	0.303	0.422	0.418
5.5– 6.499	0.202	0.225	0.238	0.241
6.5– 7.499	0.093	0.097	0.048	0.047
7.5– 8.499	0.022	0.023	0.002	0.003
8.5–10.00	0.000	0.003	0.000	0.000

agreement definitely supports our belief that the Poisson test will be accurate for small sample sizes.

The reader should be careful not to conclude that the rule of thumb should be changed to read $n \geq 5$. The reason why a sample of size 5 worked so well here is due to the symmetry of the parent uniform population. If the parent population were highly skewed, then we should not expect such good results. In general, the rule that $n \geq 30$ is a very good and safe rule.

The second test to be discussed is due to Lilliefors (1969). This test is a special modification of the K–S test and is designed specifically to test for the exponential distribution. The test follows the K–S procedure described earlier, however, the critical values, D_n^α, are now given by Table H of the Appendix. This test compares the cumulative distribution function of the exponential distribution, $1 - e^{-\hat{\lambda}x}$, with the sample cumulative distribution function $S_n(X)$. Here $\hat{\lambda}$ is the MLE of the parameter λ as given by (7.22).

Additional tests for the exponential distribution are available. Although not illustrated here, a very good test is the weighted Cramer–Smirnov–Von Mises test as described by Anderson and Darling (1952). This test is similar in principle to the K–S test. It is based on a weighted deviation of the theoretical distribution from the sampling distribution, as opposed to the K–S test, which is based on maximum deviation. The interested reader is encouraged to consult this work.

Example 7.16 Suppose that over a period of 120 min, the manager of KB's package store recorded the time of arrival of each customer. Starting at $t = 0$, the arrivals occurred at

0.2, 1.1, 4.9, 12.9, 13.0, 21.6, 32.0, 41.1, 42.0, 45.2, 47.2, 52.5, 75.8, 76.4, 82.9, 85.4, 94.1, 96.8, 97.2, 102.6, 104.4, 117.5

Test the hypothesis that the arrivals could have come from a Poisson process.

From (7.64), (7.65), and (7.63), we have

$$E(S_n) = \frac{120}{2} = 60, \quad \text{Var}(S_n) = \frac{(120)^2}{12(22)} = 54.5,$$

and

$$s_n = \frac{\sum t_i}{22} = 56.7$$

Therefore, the computed test statistic is, by (7.66),

$$Z = \frac{56.7 - 60}{(54.5)^{1/2}} = -0.45$$

Goodness-of-Fit Tests

Choosing the level of significance $\alpha = 0.05$, we obtain from Table A of the Appendix the critical value $Z_{0.025} = -1.96$. Since $Z > Z_{0.025}$, we cannot reject the hypothesis that the arrivals are from a Poisson process.

Using the exponential K–S test we find $D_{max} = 0.11$. Since D_{max} is less than 0.21, the critical value $D_{25}^{0.05}$, we can safely conclude that the data are exponential. Note that the value $D_{22}^{0.05}$ is not listed in Table H of the Appendix. However, this value must lie in the region between $D_{22}^{0.05}$ and $D_{25}^{0.05}$. Since values in this region monotonically decrease, we can conclude that if $D_{max} < D_{25}^{0.05}$ then it must be less than $D_{22}^{0.05}$. The reader is encouraged to verify the above values.

Some comments on identifying a Poisson process. As we have seen in earlier chapters queueing models possessing Poisson input and exponential service time have been extensively analyzed. Therefore, we are especially interested in identifying these distributions when they occur. For the arrival population this amounts to verifying that the number of arrivals over a time interval of fixed length has a Poisson distribution or equivalently that the time between successive arrivals has an exponential distribution. This suggests that there are two basic procedures that can be used when collecting data on a given system. The first is related to verifying that the number of arrivals is Poisson distributed, whereas the second relates to verifying that the time between successive arrivals is exponentially distributed.

To verify that the number of arrivals is Poisson, data may be collected in the same fashion as the hospital emergency room data recorded in Table 7.1. With this method, the number of arrivals X is recorded over time intervals of fixed length. In the hospital example, the chosen time interval was 8 hr and 100 observations were made. In other words, the number of arrivals to the emergency room was recorded on an 8-hr basis for a total of 100 days. When data have been collected in this manner an estimate for the mean arrival rate λ is given by

$$\hat{\lambda}^{-1} = \overline{X} = \frac{\sum_{i=1}^{n} X_i}{n}$$

where n is the total number of observations. This result is precisely the maximum-likelihood estimator for the parameter λ of a Poisson mass function, as given by (7.23). To identify that the time between successive arrivals is an exponential random variable, the number of units to be observed, n, is usually fixed and the time t between successive arrivals is recorded until exactly n arrivals have occurred. For this method of data collection, the arrival rate λ can be estimated by

$$\hat{\lambda} = \frac{n}{\sum_{i=1}^{n} t_i}$$

This estimator is the maximum-likelihood estimator for the parameter λ of the exponential density function, defined by (7.21). When data are collected in this manner, it is easiest to record the exact time of each of the arrivals and then later compute their respective interarrival times. If the data are so recorded, then as an alternative to identifying an exponential distribution via a goodness-of-fit test the special Poisson test can be applied to test the hypothesis that the data are from a Poisson process. It is worth mentioning that the data will not be in the proper format for the Poisson process test when collected by the first method, and therefore the nonparametric goodness-of-fit tests should be used for distribution identification.

In the case of service times, one must be very careful to avoid measuring the arrival rate, rather than the service rate. Recall from Chapter 3 that the departure rate from an $(M|M|1): (GD|\infty|\infty)$ queue is λ, rather than μ. Consequently, one cannot simply count the number of departures over a fixed interval of time or measure the cumulative time required to service a given number of customers and use the resulting quantity to estimate μ. Such approaches are appropriate only during a busy period. Therefore, it is recommended that one measure directly the service times for each of n customers and use the sample average as an estimate of $1/\mu$.

It is interesting to note that many practical queueing systems have Poisson input. Even when an attempt is made to schedule the arrivals so as to maintain a uniform loading on the system, the resulting arrival pattern frequently will still be Poisson due to unavoidable deviations from the schedule. A good illustration of this phenomenon occurs in air traffic control when flight controllers attempt to schedule the landing of aircraft. Extensive data gathered at various air terminals tend to indicate that the arrival population is Poisson.

Although Poisson input may be common, generally speaking the same cannot be said of exponential service. Two properties of the exponential distribution often make it an inappropriate choice for the service-time distribution: a strictly monotonically decreasing probability density function, and the lack of memory property, which were discussed in Chapters 2 and 3. A strictly monotonically decreasing density function implies that service times will most likely be short, rather than close to their mean or average value. This fact can be witnessed by observing Fig. 7.8. Certainly if the service operation were of a routine nature, then this distribution would be an inappropriate choice, since service time would be expected to be centered around the mean service time $1/\mu$. In this case, a better choice for the service-time distribution would possibly be the Erlang distribution. On the other hand, if the length of service depends on the arriving unit, then we might well expect service time to be exponentially distributed. For example, consider the service provided by a bank teller. Most customers arriving at

Goodness-of-Fit Tests

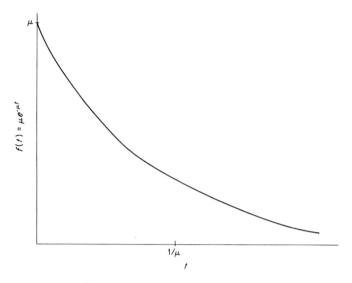

Fig. 7.8 Exponential service-time density.

the teller's window would be expected to have only one transaction and processing would proceed rapidly. However, there would be a few customers requiring several transactions and service time could be prolonged.

The lack of memory property says, in essence, that the probability that service will terminate in a given time period is independent of when it began. Thus, if service has already exceeded its expected time $1/\mu$, then the remaining time until service is completed is still expected to be $1/\mu$. Certainly this property can be unrealistic in a number of service situations. Take, for example, a service that is composed of a sequence of tasks with overall expected service time of $1/\mu$. In this case, if service had progressed through the first few tasks, then we would expect the remaining service time to be dependent on which task in the sequence was currently being performed. A more appropriate model for this situation would be the Erlang distribution. On the other hand, if service requirements differ among the arriving units, then the lack of memory property may be present and an exponential service-time distribution realistic.

As suggested in Chapter 4, if the sample variance for time between consecutive occurrences is significantly less than the square of the sample mean, an Erlang distribution should be considered; if the sample variance is significantly greater than the square of the sample mean, the possibility of a hyperexponential distribution should be explored; and if the sample variance is approximately equal to the square of the sample mean, the presence of an exponential distribution should be tested. Of course, such rules-of-thumb do not guarantee success in identifying the underlying distribution correctly. However, the variety of shapes provided by the Erlang and hyperexponential distributions makes such "quick and dirty" methods particularly inviting.

As a final note on the process of identifying arrival- and service-time distributions, one should gather the data in a form that will allow goodness-of-fit testing for non-Poisson processes. Specifically, in the case of a Poisson process one can gather the data either in the form of number of occurrences per unit of time (Poisson) or in the form of time between consecutive occurrences (exponential), where the occurrences can be either services during a busy period or arrivals. Now, suppose the former approach is employed and the null hypothesis of a Poisson distribution is rejected. What does one do next? Obviously, if the number of occurrences per unit of time is not Poisson distributed, then the distribution of time between consecutive occurrences is not exponential. From Chapters 4, 6, and 9 it is apparent that the distribution of time between consecutive occurrences is the distribution that is needed. Consequently, even though one can employ two methods in collecting data to identify a Poisson process, it is recommended that the data be gathered and analyzed in the form of time between consecutive occurrences.

INTERVAL ESTIMATION

Earlier we discussed methods of obtaining point estimates of unknown parameters along with some of the desirable properties they should possess. Let us now turn our attention to the question of their precision. Suppose we are interested in a population parameter θ for which there is an estimator $\hat{\theta}$. Since $\hat{\theta}$ is based on the incomplete information of a sample, we cannot expect $\hat{\theta}$ to estimate θ precisely. Therefore, when $\hat{\theta}$ is used in place of the true parameter we shall incur some error. What we are concerned with is the size of this error. This is what we mean when we speak of the precision of an estimator. For example, if the population mean were 30.5 and our estimator $\bar{X}$ gave the value 50.1, then our point estimate would not be very precise.

Obviously the precision of an estimator is related to its dispersion. If the estimator is unbiased and the dispersion small, then a large percentage of the estimates will fall very close to the true parameter. If the dispersion is large, then this same percentage will fall in a larger range about the true parameter. Thus the degree of precision of an estimator can be measured by the standard deviation of the estimator, that is, by its standard error.

Interval Estimation

When the distribution of $\hat{\theta}$ is known, the precision of the estimator can be explicitly defined by constructing what is referred to as a confidence interval for the unknown population parameter. These intervals are established by choosing two numbers, say K_1 and K_2, such that

$$P\{K_1 \leq \hat{\theta} \leq K_2\} = 1 - \alpha$$

where $(1 - \alpha)$ is the desired level of confidence, such as 0.95. Then the inequalities are manipulated into the form

$$P\{L \leq \theta \leq U\} = 1 - \alpha \qquad (7.67)$$

where L and U are functions of the sample on which $\hat{\theta}$ is based. Since L and U are functions of the random sample, they too are random variables before experimentation, and thus probability statements such as (7.67) can be made about an interval including the parameter θ. For this reason the interval $[L, U]$ is often referred to as the *interval estimator* for θ. The values L and U are called the $100(1 - \alpha)\%$ *confidence limits*, and the interval between L and U the $100(1 - \alpha)\%$ *confidence interval* (see Table 7.8).

The above interval is called a two-sided confidence interval. Sometimes, however, we are not as interested in the error in one direction as we are in the other. For example, in estimating a standard deviation we may only hope that it is small. In such cases, our only concern is with a probability statement of the form

$$P\{\theta < U\} = 1 - \alpha$$

which is called a $100(1 - \alpha)\%$ *upper one-sided confidence* interval. A statement of the form

$$P\{L < \theta\} = 1 - \alpha$$

is called a $100(1 - \alpha)\%$ *lower one-sided confidence interval*.

When relating a confidence interval, care should be exercised to state precisely what is meant by the interval. It is *not* correct to state that "the probability that θ will lie within the interval $[L, U]$ is $(1 - \alpha)$," since θ is a fixed constant and not a random variable, and is either present in the interval or not. It is the interval endpoints L and U that are the random variables. What is meant by probability statement (7.67) is that if an infinite number of samples were drawn from the population and if for each sample the interval $[L, U]$ were calculated, then $100(1 - \alpha)\%$ of these intervals would contain the true parameter value θ. Thus, a correct statement would be that "we are $100(1 - \alpha)\%$ confident that the interval $[L, U]$ contains the value θ."

TABLE 7.8 Two-sided $100(1-\alpha)\%$ confidence intervals

Parameter estimated	Qualifications	Confidence limits L	Confidence limits U
Mean (m)	Variance (σ^2) known	$\bar{X} - Z_{1-\alpha/2} \dfrac{\sigma}{\sqrt{n}}$	$\bar{X} + Z_{1-\alpha/2} \dfrac{\sigma}{\sqrt{n}}$
Mean (m)	Variance (σ^2) unknown	$\bar{X} - t_{1-\alpha/2}(n-1) \dfrac{S}{\sqrt{n}}$	$\bar{X} + t_{1-\alpha/2}(n-1) \dfrac{S}{\sqrt{n}}$
Variance (σ^2)	—	$\dfrac{(n-1)S^2}{\chi^2_{1-\alpha/2}(n-1)}$	$\dfrac{(n-1)S^2}{\chi^2_{\alpha/2}(n-1)}$
Standard deviation (σ)	—	$\left[\dfrac{(n-1)S^2}{\chi^2_{1-\alpha/2}(n-1)} \right]^{1/2}$	$\left[\dfrac{(n-1)S^2}{\chi^2_{\alpha/2}(n-1)} \right]^{1/2}$
Any population, parameter θ (approximate limits)	Large sample size, $\hat{\theta}$ a MLE, V the Cramer–Rao lower bound with θ replaced by $\hat{\theta}$	$\hat{\theta} - Z_{1-\alpha/2} V^{1/2}$	$\hat{\theta} + Z_{1-\alpha/2} V^{1/2}$

Confidence Interval for the Mean of a Normal Distribution, Known Variance

Let X be a normally distributed random variable with mean m and variance σ^2, and let $X_1, X_2, \ldots, X_n$ denote a random sample from this population. The sample mean

$$\bar{X} = \sum_{i=1}^{n} \frac{X_i}{n}$$

possesses all the desirable properties of a point estimator for m and is known to be normally distributed with mean m and variance σ^2/n; that is, the probability density function of $\bar{X}$ is given by

$$f(\bar{x} \mid m) = \frac{\sqrt{n}}{\sigma(2\pi)^{1/2}} \exp - \frac{n(\bar{x} - m)^2}{2\sigma^2}, \quad -\infty < \bar{x} < \infty \quad (7.68)$$

If $\bar{X}$ is transformed into the standard normal random variable Z by

$$Z = \frac{\sqrt{n}(\bar{X} - m)}{\sigma} \quad (7.69)$$

then the probability density function of Z is given by

$$g(z) = \frac{1}{(2\pi)^{1/2}} \exp - \frac{1}{2} z^2 \quad (7.70)$$

As a result, the following probability statement can be made:

$$P\left\{Z_{\alpha/2} \leq \frac{\sqrt{n}(\bar{X} - m)}{\sigma} \leq Z_{1-\alpha/2}\right\} = 1 - \alpha \quad (7.71)$$

The points $Z_{\alpha/2}$ and $Z_{1-\alpha/2}$ are values of the standard normal distribution defined by

$$F_Y(z_{\alpha/2}) = \frac{\alpha}{2}, \quad \text{and} \quad F_Y(z_{1-\alpha/2}) = 1 - \frac{\alpha}{2} \quad (7.72)$$

where $F_Y(z)$ is the normal distribution function of Z. Since the standard normal density is symmetric about 0, then

$$Z_{\alpha/2} = -Z_{1-\alpha/2} \quad (7.73)$$

and the probability statement (7.71) becomes

$$P\left\{-Z_{1-\alpha/2} \leq \frac{\sqrt{n}(\bar{X} - m)}{\sigma} \leq Z_{1-\alpha/2}\right\} = 1 - \alpha$$

or
$$P\left\{-Z_{1-\alpha/2}\frac{\sigma}{\sqrt{n}} \leq \bar{X} - m \leq Z_{1-\alpha/2}\frac{\sigma}{\sqrt{n}}\right\} = 1 - \alpha$$
and finally
$$P\left\{\bar{X} - Z_{1-\alpha/2}\frac{\sigma}{\sqrt{n}} \leq m \leq \bar{X} + Z_{1-\alpha/2}\frac{\sigma}{\sqrt{n}}\right\} = 1 - \alpha \quad (7.74)$$
Thus,
$$L = \bar{X} - Z_{1-\alpha/2}\frac{\sigma}{\sqrt{n}}, \quad U = \bar{X} + Z_{1-\alpha/2}\frac{\sigma}{\sqrt{n}}$$
and the $100(1 - \alpha)\%$ confidence for m is
$$\bar{X} \pm Z_{1-\alpha/2}\frac{\sigma}{\sqrt{n}}$$

The value of the point $Z_{1-\alpha/2}$ can be found for a given value of α in Table A of the Appendix. A lower one-sided confidence interval can be obtained by setting $U = \infty$ and replacing $Z_{1-\alpha/2}$ by $Z_{1-\alpha}$ in (7.74). This yields the probability statement

$$P\left\{\bar{X} - Z_{1-\alpha}\frac{\sigma}{\sqrt{n}} \leq m\right\} = 1 - \alpha$$

which gives $\bar{X} - Z_{1-\alpha}\sigma/\sqrt{n}$ as the $100(1 - \alpha)\%$ lower one-sided confidence interval. The $100(1 - \alpha)\%$ upper one-sided confidence interval $\bar{X} + Z_{1-\alpha}\sigma/\sqrt{n}$ can be similarly obtained by setting $L = -\infty$ in (7.74).

The construction of the confidence interval for m is based on the point estimator $\bar{X}$ that possesses all of the desirable point estimation properties we have defined. No doubt these properties also have great influence on the behavior of the confidence intervals themselves. To illustrate that this is indeed the case, consider the properties of unbiasedness and squared-error consistency. If an unbiased estimator is squared-error consistent, its variance approaches 0 as n approaches infinity. Thus the confidence interval will reduce as the sample size increases, and finally will collapse at the true value of the parameter. This behavior can be witnessed by examining the confidence interval for m, given by (7.74) as $n \to \infty$. Now consider the property of efficiency. Confidence intervals based on unbiased estimators that are efficient will have smaller intervals than can otherwise be obtained. This follows from the fact that the variance of an inefficient estimator is larger than that of the efficient estimator, and therefore the interval endpoints are spread farther apart. Thus, the desirable properties of point estimators also have a significant effect on their precision.

Interval Estimation

Example 7.17 An automatic bottling process is used in an attempt to place 12 oz. of liquid into each bottle leaving the assembly line. However, after operating for a period of time this average amount will change significantly and the process will need adjustment. It is known that the liquid level is normally distributed with a variance of 0.04 oz. regardless of the process average setting. A sample of 9 bottles produces $\bar{X} = 11.6$ as a point estimate of the process average. Determine 95 and 99% confidence intervals for the current process average m.

For a 95% confidence interval, $\alpha = 0.05$ and from Table A of the Appendix $Z_{0.975} = 1.96$. The standard error of $\bar{X}$ is

$$s_{\bar{X}} = \frac{\sigma}{\sqrt{n}} = \frac{0.2}{3}$$

Thus

$$L = 11.6 - \frac{0.2}{3}(1.96) = 11.46, \quad \text{and} \quad U = 11.6 + \frac{0.2}{3}(1.96) = 11.74$$

For a 99% confidence interval, $\alpha = 0.01$ and from Table A of the Appendix $Z_{0.995} = 2.57$. Thus

$$L = 11.6 - \frac{0.2}{3}(2.57) = 11.42, \quad \text{and} \quad U = 11.6 + \frac{0.2}{3}(2.57) = 11.78$$

To summarize, we may state that we are 95% confident that the process average is at present in the interval [11.46, 11.74] and 99% confident that it is in the interval [11.42, 11.78].

Confidence Interval for the Mean of a Normal Distribution, Unknown Variance

When the variance of the normal distribution is unknown, confidence intervals for the mean must be based on the t distribution. Simply replacing the population standard deviation σ in

$$Z = \frac{\sqrt{n}(\bar{X} - m)}{\sigma}$$

by the sample standard deviation

$$S = \left[\frac{\sum_{i=1}^{n}(X_i - \bar{X})^2}{n - 1}\right]^{1/2}$$

and using the confidence interval given by (7.74) can lead to serious error, particularly when n is small. This results from the fact that S may not be an accurate estimate of σ. Fortunately, however, the random variable

$$T = \frac{\sqrt{n}(\bar{X} - m)}{S} \qquad (7.75)$$

can be shown to have a t distribution with $n - 1$ degrees of freedom. Stated formally, a random variable

$$T = \frac{U\sqrt{d}}{V} \qquad (7.76)$$

is said to have a t distribution with d degrees of freedom, if U is normally distributed with mean 0 and variance 1 and V^2 is distributed as a chi-square random variable with d degrees of freedom and U and V are independent.

Since Z is normally distributed with mean 0 and variance 1, it possesses the properties of the random variable U. We can also make the following statement:

If X is normally distributed with mean m and variance σ^2 and $X_1, X_2, \ldots, X_n$ is a random sample of size n, then the random variable

$$\chi^2 = \frac{(n-1)S^2}{\sigma^2} \qquad (7.77)$$

has a chi-square distribution with $n - 1$ degrees of freedom, where S^2 is the unbiased estimator for the variance σ^2. Thus the random variable χ^2 possesses the properties of V^2 described above. Hence, the random variable

$$T = \frac{[\sqrt{n}(\bar{X} - m)/\sigma](n-1)^{1/2}}{(n-1)^{1/2}/\sigma} = \frac{\sqrt{n}(\bar{X} - m)}{S}$$

has a t distribution with $n - 1$ degrees of freedom. The probability density function of T is given by

$$f(t) = \frac{\Gamma(n/2)}{[2(n-1)]^{1/2}\Gamma[(n-1)/2]} \left(1 + \frac{t^2}{n-1}\right)^{-n/2}, \quad -\infty < t < \infty$$

(7.78)

and values of the t distribution $t_p(n - 1)$, defined by

$$F[t_p(n - 1)] = \int_{-\infty}^{t_p(n-1)} f(t)\, dt = p \qquad (7.79)$$

can be found in Table B of the Appendix.

Interval Estimation

Based on the above discussion, the following probability statement can be made:

$$P\left\{t_{\alpha/2}(n-1) \leq \frac{\sqrt{n}(\bar{X}-m)}{S} \leq t_{1-\alpha/2}(n-1)\right\} = 1 - \alpha \qquad (7.80)$$

Since the t distribution is symmetric about 0,

$$t_{\alpha/2}(n-1) = -t_{1-\alpha/2}(n-1) \qquad (7.81)$$

and (7.80) can be written as

$$P\left\{-t_{1-\alpha/2}(n-1) \leq \frac{\sqrt{n}(\bar{X}-m)}{S} \leq t_{1-\alpha/2}(n-1)\right\} = 1 - \alpha$$

which after simplifying becomes

$$P\left\{\bar{X} - t_{1-\alpha/2}(n-1)\frac{S}{\sqrt{n}} \leq m \leq \bar{X} + t_{1-\alpha/2}(n-1)\frac{S}{\sqrt{n}}\right\} = 1 - \alpha$$

$$(7.82)$$

Thus

$$L = \bar{X} - t_{1-\alpha/2}(n-1)\frac{S}{\sqrt{n}}, \quad U = \bar{X} + t_{1-\alpha/2}(n-1)\frac{S}{\sqrt{n}} \qquad (7.83)$$

and the $100(1-\alpha)\%$ confidence interval for m is given by

$$\bar{X} \pm t_{1-\alpha/2}(n-1)\frac{S}{\sqrt{n}} \qquad (7.84)$$

The $100(1-\alpha)\%$ upper and lower one-sided confidence intervals for m are given by

$$\bar{X} + t_{1-\alpha}(n-1)\frac{S}{\sqrt{n}}, \quad \text{and} \quad \bar{X} - t_{1-\alpha}(n-1)\frac{S}{\sqrt{n}} \qquad (7.85)$$

respectively

Example 7.18 The time to perform a polishing task on a part is known to be normally distributed. A random sample of 6 performance times was recorded as 2.6, 2.9, 2.5, 2.3, 2.6, and 2.7. Estimate the average polishing time for the part and give a confidence statement.

The point estimate for the average polishing time is given by $\bar{X} = 2.6$. Choosing $\alpha = 0.05$, we have from Table B of the Appendix that $t_{0.975}(5) = 2.571$, and from (7.83) that

$$L = 2.6 - 2.571 s_{\bar{X}} = 2.6 - 2.571(0.082) = 2.389$$

and

$$U = 2.6 + 2.571 s_{\bar{X}} = 2.6 + 2.571(0.082) = 2.811$$

Therefore, we are 95% confident that the average polishing time is contained in the time interval [2.389, 2.811].

Suppose now that we were interested in knowing something about how small we can expect the polishing time to be; that is, we desire a one-sided confidence interval on the average polishing time. Using (7.85) we have

$$P\{\bar{X} - t_{0.95}(5)s_{\bar{X}} \le m\} = 0.95, \quad \text{or} \quad P\{2.6 - 2.015(0.082) \le m\} = 0.95$$

Thus, we are 95% confident that the average polishing time is above 2.435 time units.

Confidence Interval for the Standard Deviation of a Normal Distribution

In the preceding section a confidence interval for the mean of the normal distribution was obtained using the sampling standard deviation S as an estimate of the population standard deviation σ. Since we rarely know σ with certainty, we would also be interested in a confidence statement for this parameter. Such a statement can be obtained from the sampling distribution of the random variable χ^2, defined by (7.77). Since this variable has a chi-square distribution, the following probability statement can be made:

$$P\left\{\chi^2_{\alpha/2}(n-1) \le \frac{(n-1)S^2}{\sigma^2} \le \chi^2_{1-\alpha/2}(n-1)\right\} = 1 - \alpha \quad (7.86)$$

The points $\chi^2_{\alpha/2}(n-1)$ and $\chi^2_{1-\alpha/2}(n-1)$ of the chi-square distribution are defined by

$$F_{\chi^2}[\chi^2_p(n-1)] = p \quad (7.87)$$

where $F_{\chi^2}(x)$ is the distribution function of the chi-square random variable with $n - 1$ degrees of freedom. Values for these points can be found in Table C of the Appendix. Now probability statement (7.86) is equivalent to

$$P\left\{\frac{1}{\chi^2_{1-\alpha/2}(n-1)} \le \frac{\sigma^2}{(n-1)S^2} \le \frac{1}{\chi^2_{\alpha/2}(n-1)}\right\} = 1 - \alpha \quad (7.88)$$

or

$$P\left\{S\left[\frac{n-1}{\chi^2_{1-\alpha/2}}\right]^{1/2} \le \sigma \le S\left[\frac{n-1}{\chi^2_{\alpha/2}(n-1)}\right]^{1/2}\right\} = 1 - \alpha \quad (7.89)$$

Interval Estimation

Thus

$$L = S\left[\frac{n-1}{\chi^2_{1-\alpha/2}(n-1)}\right]^{1/2}, \quad \text{and} \quad U = S\left[\frac{n-1}{\chi^2_{\alpha/2}(n-1)}\right]^{1/2} \quad (7.90)$$

are the lower and upper limits of the $100(1-\alpha)\%$ confidence interval for the population standard deviation σ. The $100(1-\alpha)\%$ lower and upper one-sided confidence intervals L and U can be obtained by replacing $\alpha/2$ by α in (7.90).

Example 7.19 Using the data listed in Example 7.18, obtain a 95% confidence interval for the population standard deviation σ.

From the calculations of Example 7.18, $s = 0.2$, and from Table C of the Appendix $\chi^2_{0.025}(5) = 0.831$ and $\chi^2_{0.975}(5) = 12.8$. Therefore,

$$L = 0.2\left(\frac{5}{12.8}\right)^{1/2} = 0.12, \quad \text{and} \quad U = 0.2\left(\frac{5}{0.831}\right)^{1/2} = 0.49$$

and the 95% confidence interval for the variation in polishing time is given by $[0.12, 0.49]$.

Approximate Confidence Intervals

As stated previously, maximum-likelihood estimators are, under very general conditions, asymptotically normal; that is, if $\hat{\theta}$ is the maximum-likelihood estimator for θ, based on a large sample size n, then $\hat{\theta}$ will tend to be normally distributed with mean θ and asymptotic variance $\sigma^2(\theta)$. Since maximum-likelihood estimators are also asymptotically efficient, the asymptotic variance of $\hat{\theta}$, $\sigma^2(\theta)$, is the Cramer–Rao lower bound, (7.5). Therefore, by transforming $\hat{\theta}$ into a standard normal random variable by

$$Z = \frac{\hat{\theta} - \theta}{\sigma(\theta)}$$

a $100(1-\alpha)\%$ confidence interval for θ is given by

$$\hat{\theta} \pm Z_{1-\alpha/2}\sigma(\theta) \quad (7.91)$$

This result follows immediately from the steps leading to the establishment of (7.74). The asymptotic variance $\sigma^2(\theta)$, which is generally a function of θ and sometimes of other parameters as well, can be evaluated by replacing the parameters by their respective estimates. Denoting this result by V, we obtain the approximate confidence interval

$$\hat{\theta} \pm Z_{1-\alpha/2}V^{1/2} \quad (7.92)$$

Thus the interval $\hat{\theta} \pm Z_{1-\alpha/2}V^{1/2}$ contains the true parameter θ with probability approximately equal to $(1-\alpha)$.

Example 7.20 In Example 7.9, the MLE for the parameter λ of the Poisson distribution was shown to be the sample mean $\bar{X}$. Find an approximate 95% confidence interval for λ.

From (7.12), the Cramer–Rao inequality can be written as

$$\text{Var}(\hat{\lambda}) \geq \frac{1}{nE[(\partial/\partial \lambda) \ln f(x)]^2} \tag{7.93}$$

Here

$$f(x) = \frac{e^{-\lambda}\lambda^x}{x!}, \quad \text{and} \quad \ln f(x) = -\lambda + x \ln \lambda - \ln(x!)$$

Taking the derivative with respect to λ gives

$$\frac{d}{d\lambda} \ln f(x) = -1 + \frac{x}{\lambda} = \frac{x - \lambda}{\lambda}$$

Now

$$E\left[\frac{d}{d\lambda} \ln f(x)\right]^2 = E\left[\frac{X - \lambda}{\lambda}\right]^2 = E\left[\frac{X^2 - 2X\lambda + \lambda^2}{\lambda^2}\right]$$

$$= \frac{E(X^2) - 2E(X) + \lambda^2}{\lambda^2} = \frac{E(X^2) - \lambda^2}{\lambda^2}$$

$$\tag{7.94}$$

since $E(X) = \lambda$ (Table 2.1). Substituting the relation

$$\text{Var}(X) = E(X^2) - E(X)^2$$

into (7.94), we obtain

$$E\left[\frac{d}{d\lambda} \ln f(x)\right]^2 = \frac{\text{Var}(X)}{\lambda^2} = \frac{1}{\lambda}$$

Thus, the Cramer–Rao lower bound becomes

$$\text{Var}(\hat{\lambda}) \geq \frac{\lambda}{n} \tag{7.95}$$

Replacing λ by its MLE estimate $\hat{\lambda}$, we obtain from (7.92) the approximate 95% confidence interval for λ,

$$\bar{X} \pm Z_{0.975}\left(\frac{\bar{X}}{n}\right)^{1/2}, \quad \text{or} \quad \bar{X} \pm 1.96\left(\frac{\bar{X}}{n}\right)^{1/2} \tag{7.96}$$

It should be noted that

$$\text{Var}(\hat{\lambda}) = \text{Var}(\bar{X}) = \frac{\text{Var}(X)}{n} = \frac{\lambda}{n} \tag{7.97}$$

and hence the MLE estimator $\hat{\lambda}$ achieves the Cramer–Rao lower bound, i.e., it is efficient.

SUMMARY

The accurate modelling of any queueing system is significantly dependent on one's ability to analyze data. This chapter, therefore, has been entirely devoted to this important topic. Methods of distribution identification and parameter estimation have been considered in detail. The methods presented were chosen to provide the analyst with a wide assortment of proven techniques. Armed with these techniques, the analyst is now ready to tackle the problems of data collection and identification. The success of model performance will significantly depend on his ability to apply these methods correctly.

REFERENCES

Anderson, T. W., and Darling, D. A. Asymptotic theory of certain "goodness-of-fit" criteria based on stochastic processes, *Ann. Math. Statist.* **23**, 193–212, 1952.
Chernoff, H., and Lehmann, E. L., The use of maximum likelihood estimates in χ^2 tests of goodness of fit, *Ann. Math. Statist.* **25**, 573, 1954.
Cochran, W. G., Some methods for strengthening the common tests, *Biometrics* **10**, 417–51, 1954.
Jahnke, E., and Emde, F., *Table of Functions*, 4th edition. New York: Dover, 1945.
Lilliefors, H. W., On the Kolmogorov–Smirnov test for the exponential distribution with mean unknown, *J. Amer. Statist. Assoc.* **64**, 387–389, 1969.
Schmidt, J. W., *Mathematical Foundations for Management Science and Systems Analysis*. New York: Academic Press, 1974.
Siegel, S., *Nonparametric Statistics*. New York: McGraw-Hill, 1956.

PROBLEMS

1. Consider the Poisson distribution

$$P(x) = \frac{e^{-\lambda}\lambda^x}{x!}, \quad x = 0, 1, \ldots, \quad \lambda > 0$$

with unknown parameter λ. Show that $\sum_{i=1}^{n} x_i/n$ is a BLUE estimator for the parameter λ and that it is also efficient.

2. Consider the binomial distribution with unknown parameter p given by

$$P(x) = \binom{n}{x} p^x (1-p)^{n-x}, \quad x = 0, 1, 2, \ldots, n, \quad 0 \le p \le 1$$

Find the maximum-likelihood estimator for p based on n independent observations.

3. Consider the following density function:

$$f(x) = \lambda^2 x e^{-\lambda x}, \quad \lambda > 0, \quad x \ge 0$$

The parameter λ is to be estimated by

$$\hat{\lambda} = \frac{2n-1}{\sum_{i=1}^{n} x_i}$$

which is based on a sample of size n. (a) Is $\hat{\lambda}$ an unbiased and consistent estimate of λ? (b) Is $\hat{\lambda}$ an efficient estimator for λ? (c) Is $\hat{\lambda}$ asymptotically efficient?

4. Consider the following density function:

$$f(x) = (1 + \lambda)x^\lambda, \quad 0 < x < 1$$

Show that the maximum-likelihood estimator for λ based on a sample of size n is given by

$$\hat{\lambda} = -\left(1 + \frac{n}{\sum_{i=1}^{n} \ln x_i}\right)$$

5. Consider the uniform density function

$$f(x) = \frac{1}{\beta}, \quad 0 \le x \le \beta$$

Find the maximum-likelihood estimator for β, given a sample of size n.

6. The time T to failure of an electronic component has been shown to have a Weibull distribution with parameters α and β, i.e.,

$$f(t) = \alpha \beta t^{\beta - 1} \exp{-\alpha t^\beta}, \quad t \ge 0, \quad \alpha, \beta > 0$$

Assume that β is a known constant and find the maximum-likelihood estimator for α.

7. We have noted that there has been a significant growth of aircraft operations (takeoffs and landings) at the High-Flight air terminal over the last several years. This growth seems to be exponential. It is impossible to say exactly what the operation rates have been due to insufficient data but we have the following estimates:

Year	$\hat{\lambda}$
1970	0.5
1971	1.2
1972	2.4
1973	3.5
1974	5.2
1975	7.3

What is the least-squares estimate for the arrival rate in 1976 and 1977? [*Hint:* Transform to linear regression.]

8. The number of cars that turn away from a car wash operation is hypothesized to be a quadratic function of the number of units waiting to be served. The following data were

collected, where x is the number of cars waiting to be serviced, measured from an average of 5, and y is the number of cars observed to turn away:

y	x
1	-4
6	-3
9	-2
12	-1
13	0
13	1
13	2
12	3
9	4

Develop an equation of the form

$$\hat{y} = \hat{\beta}_0 + \hat{\beta}_1 x + \hat{\beta}_2 x^2$$

that best fits the data using the method of least squares.

9. The average capacity y of an air terminal, measured in operations per hour, is assumed to be linearly related to aircraft separation on approach x_1 in miles and approach velocity x_2 in hundreds of nautical miles per hour. The following data have been collected:

y	x_1	x_2
19	2.0	1.4
26	3.3	1.6
18	2.5	1.4
27	3.0	1.7
16	3.0	1.4
15	2.5	1.3
21	3.0	1.7

Fit these data using the method of least squares.

10. Over a period of 60 min data were collected on automobile traffic at a particular intersection by a traffic engineer. Starting at time $t = 0$, the arrivals occurred at

0.1, 1.3, 5.6, 8.2, 13.4, 21.6, 34.0, 40.1, 43.3, 51.5, 52.4, 58.4

Test the hypothesis that the arrivals could have come from a Poisson process.

11. Data on the number of parts arriving at a work station were collected over a period of 100 days and are given below:

Number of parts	Frequency	Number of parts	Frequency
50	1	57	16
51	2	58	12
52	4	59	8
53	7	60	3
54	11	61	2
55	15	62	1
56	17	63	1

Using the chi-square test, fit these data to a Poisson distribution.

12. Below you are given some interarrival data. Use either method of moments or maximum-likelihood estimators to fit the data to Erlang, exponential, and hyperexponential distributions:

0.754	0.533	0.192	0.544	0.127
0.365	0.222	0.097	0.293	0.095
0.125	0.130	0.695	0.420	0.285
0.101	0.297	0.074	0.226	0.371
0.425	0.496	0.517	0.489	0.585

13. Referring to Problem 12, what was your prior guess about which of the above distributions would best fit the data? Use both a chi-square test and a K–S test to test the goodness of fit for each of the three families. Is it your posterior opinion that your prior guess was best for these data?

14. After answering Problems 12 and 13, you are probably able to deal with the following service-time data more efficiently. Fit the data to a distribution and test the goodness of fit:

0.112	0.102	0.063	0.004	0.019
0.041	0.643	0.005	0.029	0.148
0.008	0.004	0.238	0.047	0.025
0.052	0.030	0.058	0.023	0.052
0.029	0.018	0.168	0.027	0.027

15. Between the Kolmogorov–Smirnov test and the chi-square test, which is uniformly most powerful for continuous distributions?

16. Given the following data

Random variable	Frequency of occurrence
$-\infty < x \leq 20$	8
$20 < x \leq 30$	93
$30 < x \leq 50$	405
$50 < x \leq 70$	390
$70 < x \leq 80$	95
$80 < x \leq \infty$	9
	1000

(a) From what family of distributions did the data come?
(b) What are the values of the parameters of the best distribution in the family?
(c) Perform a goodness-of-fit test on the answers to (a) and (b).

17. Write a computer program for computing the K–S goodness-of-fit test for (a) normal distributions, (b) gamma distributions, and (c) hyperexponential distributions.

18. Write a computer program for computing the chi-square goodness-of-fit statistic.

19. A computer simulation model has been developed that is to be used to estimate the mean annual cost of a production system. After simulating 10 years of operation of the system, the sample mean and standard deviation of cost per year are estimated to be

$$\bar{x} = \$430{,}000, \quad s = \$16{,}000$$

If the cost of each year of simulation C_R is $10.00, how many additional years of simulation are necessary to minimize the expected cost of error in estimation $|\bar{x} - \mu|$, plus the cost of simulation $C_R n$; that is, the expression to be minimized is given by

$$C_E = C_R n + E[|\bar{x} - \mu|]$$

where μ is the mean annual cost of operation. Note that

$$\int_0^\infty z \frac{1}{(2\pi)^{1/2}} \exp -\tfrac{1}{2}z^2 \, dz = 0.282$$

20. Referring to Problem 19, if mean annual cost of operation is to be estimated using a 95% confidence interval and

$$\frac{U - L}{\bar{x}} = 0.01$$

how many additional years of simulation should be carried out, where U is the upper confidence limit and L the lower confidence limit?

21. The optimum number of replications is to be determined for the estimation of the annual cost of a given operation. The cost of each replication (year of simulation) is C_R. The sample mean $\bar{x}$ is to be used to estimate the expected annual cost of operation μ. The cost of the error in estimation is given by $(\bar{x} - \mu)^2$. The expression to be minimized is the sum of the cost of simulation and the expected cost of error in estimation, $E[(\bar{x} - \mu)^2]$. Derive an expression for the optimum number of replications.

22. Assume you have six independently and identically distributed exponential variables with unknown parameter λ. Find the 95% confidence interval for λ of the form $0 \le \lambda \le \bar{x}_i c$, where c is to be determined and $\bar{x}_i$ is the sample mean.

23. Is it possible to find a confidence interval for the single-parameter exponential distribution of the form

$$\bar{x} \pm c_{\alpha/2} s^2$$

24. Derive Eq. (7.12) from Eq. (7.11).

Chapter 8

DATA ANALYSIS— HYPOTHESIS TESTING

INTRODUCTION

Quite often we are interested not only in the calculation of the estimate of a parameter but also in the conclusions that can be drawn from the estimator regarding changes that may have taken place. To illustrate, suppose that the mean arrival rate λ of customers was estimated at time t_0 to be $\hat{\lambda}_0$. At time t_1 we calculate another estimate of λ, $\hat{\lambda}_1$, where $\hat{\lambda}_1 \neq \hat{\lambda}_0$. Do we conclude that the mean arrival rate of customers has changed between t_0 and t_1? Even if the mean arrival rate had not changed between t_0 and t_1, we would not be surprised if $\hat{\lambda}_0$ and $\hat{\lambda}_1$ were not equal since both are random variables; that is, a slight difference between $\hat{\lambda}_0$ and $\hat{\lambda}_1$ may not indicate a change in the mean arrival rate. However, if the difference between $\hat{\lambda}_0$ and $\hat{\lambda}_1$ is sufficiently large, we might reasonably conclude that the mean arrival rate has in fact changed.

In this chapter we will present several statistical tests that can be used to test hypotheses about the means and variances of one or two populations. In the illustration given above, we would have been interested in testing the hypothesis that the mean arrival rates at t_0 and t_1 were equal. In this case, we would be comparing the means of two populations: customer arrivals at t_0 and customer arrivals at t_1. However, we may not be interested in population means only. Suppose that the variance of service time σ^2 must be less than σ_0^2 if the service system is to operate effectively. Here we would be testing the hypothesis that $\sigma^2 \leq \sigma_0^2$ and our statistical test is about the parameter of a single population.

NULL AND ALTERNATIVE HYPOTHESES

As implied in the above discussion, before conducting tests of hypotheses, we must first establish the hypothesis to be tested. The hypothesis tested is generally referred to as the null hypothesis and denoted by H_0. If the result of the statistical test is to reject the null hypothesis, then the alternative to the null hypothesis is assumed to be true. The alternative to the null hypothesis is simply called the alternative hypothesis and denoted H_1. For example, if we wish to compare the means, m_1 and m_2, of two populations to determine whether or not $m_1 = m_2$, the null and alternatives are expressed by

$$H_0: \quad m_1 = m_2, \quad H_1: \quad m_1 \neq m_2$$

If we are to determine whether or not the variance σ^2 of a single population is equal to some value σ_0^2, then

$$H_0: \quad \sigma^2 = \sigma_0^2, \quad H_1: \quad \sigma^2 \neq \sigma_0^2$$

Finally, if our test is to determine whether or not the population mean m is greater than or equal to some value m_0, we have

$$H_0: \quad m \geq m_0, \quad H_1: \quad m < m_0$$

As we shall see, rejection or acceptance of the null hypothesis is based in part on the values obtained from statistical estimates. Since these estimates are random variables, one may incorrectly reject or accept a null hypothesis. Thus one cannot prove that the null hypothesis is true or false as a result of a statistical test. However, in many cases, if the test is properly designed one can limit the probabilities of falsely rejecting or falsely accepting the null hypothesis to reasonable values.

TYPE I AND TYPE II ERRORS

Prior to conducting a statistical test the analyst knows that one of three results may occur after the test is completed. He may draw the correct conclusion about the null hypothesis, he may reject the null hypothesis when it is true, or he may accept the null hypothesis when it is false. Rejection of the null hypothesis when it is true is called a type I error and acceptance of the null hypothesis when it is false is referred to as a type II error. Of course, after the experiment has been completed, the required analysis conducted, and the null hypothesis either accepted or rejected, only one of these errors can occur.

Before the experiment is carried out, one usually attempts to design it so that the probability of occurrence of each of these errors is restricted to prescribed values. Suppose that the null hypothesis H_0 is actually true. The only error possible in this case is the type I error. The probability that this error will occur is defined as α. Hence

$$\alpha = P(\text{type I error}) = P(H_0 \text{ rejected} \mid H_0 \text{ true}) \quad (8.1)$$

Now let us assume that the null hypothesis is false. In this case, only the type II error may occur. The probability that a type II error occurs depends on the specific case of the alternative hypothesis that is true. To illustrate, suppose that we wish to test the hypothesis

$$H_0: \quad m = m_0$$

where m is the population mean and m_0 a constant. If H_0 is false, then m is equal to some value other than m_0, say m_1. Then the probability β of a type II error, given $m = m_1$, is

$$\beta = P(\text{type II error} \mid m = m_1) = P(H_0 \text{ accepted} \mid m = m_1)$$

As the absolute difference between m_1 and m_0 increases, β decreases; that is, as the true situation deviates further from the situation described by the null hypothesis, the less likely it is that the statistical test will indicate that the null hypothesis is true. Thus, if H'_1 is a specific case of the alternative hypothesis, then

$$\beta = P(\text{type II error} \mid H'_1 \text{ true}) = P(H_0 \text{ accepted} \mid H'_1 \text{ true}) \quad (8.2)$$

Hence, β can be used as a measure of the ability of the statistical test to detect a specific deviation from the null hypothesis, H'_1. Another similar measure of performance of the test is the power defined as

$$\text{power} = P(H_0 \text{ rejected} \mid H'_1 \text{ true}) = 1 - \beta \quad (8.3)$$

SAMPLE SIZE

To conduct a statistical test, one must collect data to estimate the parameter or parameters included in the null hypothesis. As the sample size increases, the value of β, for a specific H'_1, will decrease for a given value of α. Normally the analyst specifies the value of α to be used in the test. He may then specify an arbitrary sample size for the test. However, a more realistic approach is to define the case of the alternative hypothesis H'_1 that is critical in the sense that if H'_1 is true, H_0 should be rejected with probability $1 - \beta$. That is, we define a critical condition H'_1, which will be rejected with a specified probability $1 - \beta$. By defining α, H'_1, $1 - \beta$, in some situations, and, estimating the variance of the population or populations involved we are able to define the sample size necessary to meet these conditions. Thus,

Tests for a Single Parameter

our approach to hypothesis testing will be to specify the null and alternative hypotheses, identify a critical case of the alternative hypothesis, specify acceptable values of α and β, estimate the variances required, and then define the sample size necessary to conduct the statistical test. This is the design phase of the experiment. The remaining steps include collecting data, calculating the test statistic, testing the null hypothesis, and deciding whether to accept or reject the null hypothesis.

TESTS FOR A SINGLE PARAMETER

Test about a Single Mean, Known Variance

In this section we are concerned with testing one of the following null hypotheses

$$H_0: \quad m = m_0, \qquad H_0: \quad m \leq m_0, \qquad H_0: \quad m \geq m_0$$

against the corresponding alternative

$$H_1: \quad m \neq m_0, \qquad H_1: \quad m > m_0, \qquad H_1: \quad m < m_0$$

where m is the population mean. To conduct the necessary test, a sample of n observations must be drawn at random from the population having mean m. Let $X_1, X_2, \ldots, X_n$ be this set of observations. Since $X_1, X_2, \ldots, X_n$ are selected at random, they may be assumed to be independent and identically distributed with mean m and some variance σ^2. We will assume that the value of σ^2 is known for the test to be developed here, although such an assumption is usually unrealistic. However, the concepts presented here will serve as an introduction to subsequent developments. We will relax this assumption in subsequent sections. From the set of observations $X_1, X_2, \ldots, X_n$ we calculate the sample mean $\bar{X}$.

Let us assume that we wish to test the hypothesis that $m = m_0$. Then

$$H_0: \quad m = m_0, \qquad H_1: \quad m \neq m_0$$

If the value of $\bar{X}$, $\bar{x}$, is close to m_0 we might be justified in accepting the null hypothesis. On the other hand, if $\bar{x}$ is far from m_0 we might be led to the conclusion that $m \neq m_0$, or that the null hypothesis should be rejected. The critical question is how far should $\bar{x}$ be from m_0 before we reject the null hypothesis? To answer this question we must examine the distribution of $\bar{X}$.

From the central limit theorem, $\bar{X}$ can be considered to have an approximate normal distribution with mean m and variance σ^2 if the sample size n is large. Hence, if the mean is m, the probability density function of $\bar{X}$ is given by (7.68).

Now let us consider the limits for $\bar{x}$ and the errors we wish to allow in the test. We know that if $\bar{x}$ is either too large or too small we should reject the null hypothesis, although we have not defined what too large and too small are. Let L and U ($L < U$) be limits such that the null hypothesis is accepted if $L \leq \bar{x} \leq U$ and rejected otherwise. Now let α be the probability of rejecting the null hypothesis if $m = m_0$. Then

$$\alpha = 1 - \int_L^U f(\bar{x} \mid m_0) \, d\bar{x}, \quad \text{or} \quad \int_L^U f(\bar{x} \mid m_0) \, d\bar{x} = 1 - \alpha \quad (8.4)$$

We will define L and U such that

$$\int_{-\infty}^L f(\bar{x} \mid m_0) \, d\bar{x} = \int_U^\infty f(\bar{x} \mid m_0) \, d\bar{x} \quad (8.5)$$

which leads to the condition that

$$m_0 - L = U - m_0 \quad (8.6)$$

Transforming $\bar{X}$ to a standard normal random variable Z by

$$Z = \frac{\sqrt{n}(\bar{X} - m_0)}{\sigma}$$

we have

$$\int_{\sqrt{n}(L-m_0)/\sigma}^{\sqrt{n}(U-m_0)/\sigma} g(z) \, dz = 1 - \alpha \quad (8.7)$$

where $g(z)$ is the probability density function of the standard normal random variable as given by (7.70). Let z_p, $0 < p < 1$, be a value of the standard normal random variable such that the distribution function of Z, $F_Y(z)$, is

$$F_Y(z_p) = p \quad (8.8)$$

From (8.7),

$$1 - \alpha = \int_{-\infty}^{\sqrt{n}(U-m_0)/\sigma} g(z) \, dz - \int_{-\infty}^{\sqrt{n}(L-m_0)/\sigma} g(z) \, dz$$

$$= F_Y\left(\frac{U - m_0}{\sigma/\sqrt{n}}\right) - F_Y\left(\frac{L - m_0}{\sigma/\sqrt{n}}\right) \quad (8.9)$$

Since $m_0 - L = U - m_0$,

$$F_Y\left(\frac{U - m_0}{\sigma/\sqrt{n}}\right) = F_Y(z_{1-\alpha/2}) \quad (8.10)$$

and

$$F_Y\left(\frac{L - m_0}{\sigma/\sqrt{n}}\right) = F_Y(z_{\alpha/2}) \quad (8.11)$$

Tests for a Single Parameter

or

$$\frac{U - m_0}{\sigma/\sqrt{n}} = Z_{1-\alpha/2} \qquad (8.12)$$

and

$$\frac{L - m_0}{\sigma/\sqrt{n}} = Z_{\alpha/2} \qquad (8.13)$$

Therefore, the limits L and U are given by

$$L = m_0 + \frac{\sigma}{\sqrt{n}} Z_{\alpha/2} \qquad (8.14)$$

$$U = m_0 + \frac{\sigma}{\sqrt{n}} Z_{1-\alpha/2} \qquad (8.15)$$

Since the standard normal probability density function is symmetric about zero by (7.73), we have

$$L = m_0 - \frac{\sigma}{\sqrt{n}} Z_{1-\alpha/2} \qquad (8.16)$$

$$U = m_0 + \frac{\sigma}{\sqrt{n}} Z_{1-\alpha/2} \qquad (8.17)$$

Before we develop a method for calculating the sample size for the test, let us consider an illustrative example.

Example 8.1 The average time required by an employee to perform a certain task is supposed to be 2.00 hr. A new employee has been trained for this task and a test is to be run to determine whether or not he meets the established standard. A history of data on the time X required to complete the task indicates that the variance of X, σ^2, is 0.25 or $\sigma = 0.5$ hr. The new employee performs the task 100 times, the time required to complete the task being recorded in each case. The average observed time to complete the task is 2.05 hr. Would you say that this employee conforms to the established standard if $\alpha = 0.10$?

In the problem statement we are asked to draw a conclusion as to whether or not the new employee meets the established standard; that is, we wish to know whether or not the true mean time m to perform the task is actually 2.00 hr. Hence the null hypothesis is

$$H_0: \quad m = 2.00$$

and the alternative is

$$H_1: \quad m \neq 2.00$$

From the data given in the problem

$$\sigma = 0.5, \quad \alpha = 0.10, \quad n = 100, \quad \bar{x} = 2.05$$

The limits for the test are given in (8.16) and (8.17). Thus

$$U = m_0 + \frac{\sigma}{\sqrt{n}} Z_{1-\alpha/2} = 2.00 + \frac{0.50}{10} Z_{0.95}$$

and

$$L = m_0 - \frac{\sigma}{\sqrt{n}} Z_{1-\alpha/2} = 2.00 - \frac{0.5}{10} Z_{0.95}$$

From Table A of the Appendix

$$Z_{0.95} = 1.65$$

Therefore, the limits for the test are

$$U = 2.00 + 0.08 = 2.08, \quad L = 2.00 - 0.08 = 1.92$$

Since $1.92 \leq \bar{x} \leq 2.08$, we accept the hypothesis that the new employee conforms to the established standard.

When the null hypothesis is of the form

$$H_0: \quad m \leq m_0$$

the hypothesis can be rejected only when $\bar{x}$ is too large. Hence we need only one limit for the test, an upper limit U. For any α, U is given by

$$U = m_0 + \frac{\sigma}{\sqrt{n}} Z_{1-\alpha} \tag{8.18}$$

If the null hypothesis is

$$H_0: \quad m \geq m_0$$

H_0 can be rejected only in the case where $\bar{x}$ is small, and we need only the lower limit L for this case, where

$$L = m_0 - \frac{\sigma}{\sqrt{n}} Z_{1-\alpha} \tag{8.19}$$

When both an upper and lower limit for $\bar{x}$ are needed to carry out a statistical test, the test is called a two-tailed test. Otherwise the test is called a single- or one-tailed test.

Example 8.2 Over the past several years the mean number of customers served per week has been 50 and the variance 9. However, the manager of

Tests for a Single Parameter

the service center suspects that the service rate has dropped based on data from the past 25 weeks. The average service rate over this period was observed to be 45 services/week. Letting $\alpha = 0.15$, determine whether the manager's claim is valid.

From the data given in the problem

$$\alpha = 0.15, \quad \sigma = 3, \quad m_0 = 50, \quad n = 25, \quad \bar{x} = 45$$

If the manager's claim is valid, then the hypothesis that $m \geq 50$ should be rejected. Hence, the null and alternative hypotheses are

$$H_0: \quad m \geq 50, \quad H_1: \quad m < 50$$

Since the null hypothesis can be rejected only when $\bar{x}$ is small, the statistical test is a one-tailed test requiring a lower limit L, where

$$L = m_0 - \frac{\sigma}{\sqrt{n}} Z_{1-\alpha} = 50.00 - \frac{3}{5}(1.04) = 50.00 - 0.62 = 49.38$$

Since $\bar{x} < 49.38$, H_0 is rejected and we conclude that the manager's claim is valid.

In our development of the statistical tests presented above, we arbitrarily specified the sample size and did not include in our analysis consideration of the type II error. The reader will recall that we accepted the null hypothesis in Example 8.1. Hence, we may have made a type II error. Suppose that if the mean time required to complete the task by the new employee differed from 2.00 hr by as much as 0.02 hr we would like the statistical test to reject the null hypothesis with probability 0.90. Does the statistical test designed in Example 8.1 satisfy this requirement? To answer this question we must determine the power of the test when $m = 1.98$ and $m = 2.02$. Now

$$\text{power} = 1 - \beta = 1 - P(\text{type II error} \mid m = m_1)$$

where $m_1 = 1.98$ and 2.02. Let $m_1 = 1.98$. Then

$$\text{power} = 1 - P(\text{accept } H_0 \mid m = 1.98) = 1 - \int_L^U f(\bar{x} \mid m_1) \, d\bar{x}$$

$$= 1 - F\left(\frac{U - m_1}{\sigma/\sqrt{n}}\right) + F\left(\frac{L - m_1}{\sigma/\sqrt{n}}\right) = 1 - F(2.00) + F(-1.20)$$

$$= 1 - 0.977 + 0.115 = 0.138$$

The power is identical when $m_1 = 2.02$. Thus the probability of detecting the critical values of m_1 is not what we would like it to be. The reason that this requirement is not reflected in the test designed is that β and m_1 were not included in the analysis that led to the development of the test.

Suppose that we are to test for the null hypothesis

$$H_0: \quad m \leq m_0$$

where

$$H_1: \quad m > m_0$$

Again we assume that the variance σ^2 is known and that α is defined. In addition, we define a value of m, m_1 ($m_1 > m_0$) such that if $m = m_1$ the null hypothesis is to be rejected with probability $1 - \beta$ or accepted with probability β. Now

$$P(H_0 \text{ rejected} \mid m = m_0) = \alpha \tag{8.20}$$

and

$$P(H_0 \text{ accepted} \mid m = m_1) = \beta \tag{8.21}$$

Since the test required is one-tailed with an upper limit U,

$$1 - \alpha = \int_{-\infty}^{U} f(\bar{x} \mid m_0) \, d\bar{x} \tag{8.22}$$

and

$$\beta = \int_{-\infty}^{U} f(\bar{x} \mid m_1) \, dx \tag{8.23}$$

Letting Z_1 and Z_2 be standard normal random variables defined by

$$Z_1 = \frac{\sqrt{n}(\bar{X} - m_0)}{\sigma}, \quad Z_2 = \frac{\sqrt{n}(\bar{X} - m_1)}{\sigma}$$

we have

$$1 - \alpha = \int_{-\infty}^{\sqrt{n}(U - m_0)/\sigma} g(z_1) \, dz_1 \tag{8.24}$$

$$\beta = \int_{-\infty}^{\sqrt{n}(U - m_1)/\sigma} g(z_2) \, dz_2 \tag{8.25}$$

Thus

$$Z_{1-\alpha} = \frac{U - m_0}{\sigma/\sqrt{n}} \tag{8.26}$$

and

$$Z_\beta = \frac{U - m_1}{\sigma/\sqrt{n}} \tag{8.27}$$

Tests for a Single Parameter

Solving (8.26) for U yields

$$U = m_0 + \frac{\sigma}{\sqrt{n}} Z_{1-\alpha} \qquad (8.28)$$

which is identical to (8.18). Substituting (8.28) into (8.27) and solving for n yields

$$n = \left(\frac{Z_\beta - Z_{1-\alpha}}{m_0 - m_1}\right)^2 \sigma^2 \qquad (8.29)$$

For a lower-tailed test, n is also given by (8.29). For a two-tailed test

$$n = \left(\frac{Z_\beta - Z_{1-\alpha/2}}{\delta}\right)^2 \sigma^2 \qquad (8.30)$$

where δ is the critical deviation of m_1 from m_0 in either direction; that is,

$$m_1 = m_0 \pm \delta \qquad (8.31)$$

Example 8.3 Determine the number of observations required in Example 8.1 if $\delta = 0.02$ and $\beta = 0.20$.

From Example 8.1 and the above data,

$$\alpha = 0.10, \qquad \beta = 0.20, \qquad \sigma = 0.5, \qquad \delta = 0.02$$

Therefore,

$$Z_{1-\alpha/2} = Z_{0.95} = 1.65, \qquad Z_\beta = Z_{0.20} = -0.84$$

Since a two-tailed test is required, n is given by (8.30):

$$n = \left(\frac{-0.84 - 1.65}{0.02}\right)^2 0.25 = 3875$$

Tests about a Single Mean, Unknown Variance

As in the previous section we wish to find a limit or limits for $\bar{x}$ such that conditions for acceptance or rejection of the null hypothesis are established. Let us first consider the two-tailed test, where the null and alternative hypotheses are

$$H_0: \quad m = m_0, \qquad H_1: \quad m \neq m_0$$

and the lower and upper limits for the statistical test are L and U, respectively. In order to formulate this test, we estimate the population variance σ^2 by the sample variance S^2. Then from (7.75), the random variable

$$T = \frac{\sqrt{n}(\bar{X} - m)}{S}$$

has a t distribution with probability density function given by (7.78) and

$$P(L \leq \bar{x} \leq U) = P\left(\frac{L - m_0}{s/\sqrt{n}} \leq t \leq \frac{U - m_0}{s/\sqrt{n}}\right)$$

$$= P\left(t \leq \frac{U - m_0}{s/\sqrt{n}}\right) - P\left(t \leq \frac{L - m_0}{s/\sqrt{n}}\right)$$

$$= F_t\left(\frac{U - m_0}{s/\sqrt{n}}\right) - F_t\left(\frac{L - m_0}{s/\sqrt{n}}\right) \tag{8.32}$$

where $F_t(x)$ is the distribution function of the t random variable. If the probability of occurrence of the type I error is to be α, then

$$F_t\left(\frac{U - m_0}{s/\sqrt{n}}\right) - F_t\left(\frac{L - m_0}{s/\sqrt{n}}\right) = 1 - \alpha \tag{8.33}$$

If L and U are chosen equidistant from m_0, then

$$F_t\left(\frac{U - m_0}{s/\sqrt{n}}\right) = 1 - \frac{\alpha}{2} \tag{8.34}$$

$$F_t\left(\frac{L - m_0}{s/\sqrt{n}}\right) = \frac{\alpha}{2} \tag{8.35}$$

Thus

$$\frac{U - m_0}{s/\sqrt{n}} = t_{1-\alpha/2}(n - 1) \tag{8.36}$$

and

$$\frac{L - m_0}{s/\sqrt{n}} = -t_{1-\alpha/2}(n - 1) \tag{8.37}$$

since the density function of the t random variable is symmetric about zero. The limits L and U are then given by

$$U = m_0 + \frac{s}{\sqrt{n}} t_{1-\alpha/2}(n - 1) \tag{8.38}$$

$$L = m_0 - \frac{s}{\sqrt{n}} t_{1-\alpha/2}(n - 1) \tag{8.39}$$

If $L \leq \bar{x} \leq U$, H_0 is accepted, and is rejected otherwise.
    For the null hypothesis

$$H_0: \quad m \leq m_0$$

Tests for a Single Parameter

an upper-tailed test is required and the limit for the test is

$$U = m_0 + \frac{s}{\sqrt{n}} t_{1-\alpha}(n-1) \tag{8.40}$$

If the null hypothesis is

$$H_0: \quad m \geq m_0$$

a lower-tailed test is required and the lower limit is given by

$$L = m_0 - \frac{s}{\sqrt{n}} t_{1-\alpha}(n-1) \tag{8.41}$$

Example 8.4 An expected-cost model has been developed for a maintenance facility. The model is based on an equipment failure rate λ of 2.00/day and the optimum solution to the model has been found to be sensitive to changes in λ; that is, if λ decreases or increases it may be necessary to change the optimum solution. Over the past 25 days, data have been collected on the number of equipment failures per day. The results of the data collection are as follows:

$$\bar{x} = 2.21, \quad s = 0.53$$

Test the hypothesis that the mean failure rate λ has not changed using $\alpha = 0.10$.

The null hypothesis for the test is

$$H_0: \quad \lambda = 2.00$$

and the alternative hypothesis is given by

$$H_1: \quad \lambda \neq 2.00$$

Since a two-tailed test is necessary, the limits for the test are given in (8.38) and (8.39). From Table B of the Appendix

$$t_{1-\alpha/2}(n-1) = t_{0.90}(24) = 1.318$$

Hence

$$L = 2.00 - \frac{0.53}{5.00}(1.318) = 1.860$$

and

$$U = 2.00 + \frac{0.53}{5.00}(1.318) = 2.140$$

Since 2.21 > U, the null hypothesis is rejected and consideration should be given to altering the solution deemed optimal.

As in the previous section, the number of observations required for the *t*-test can be determined once we have defined a critical change in the mean and the associated value of β. Suppose we wish to test the null hypothesis

$$H_0: \quad m \leq m_0$$

against the alternative

$$H_1: \quad m > m_0$$

Let β be the probability of accepting H_0 if $m = m_1 > m_0$. Then

$$\beta = P(H_0 \text{ accepted} \mid m = m_1)$$

$$= 1 - P\left[\bar{x} \geq m_0 + \frac{s}{\sqrt{n}} t_{1-\alpha}(n-1) \mid m = m_1\right]$$

$$= 1 - P\left[\bar{x} - \frac{s}{\sqrt{n}} t_{1-\alpha}(n-1) \geq m_0 \mid m = m_1\right] \qquad (8.42)$$

Now $\bar{x} - (s/\sqrt{n})t_{1-\alpha}(n-1)$ has an approximate normal distribution with mean $m_1 - (\sigma/\sqrt{n})t_{1-\alpha}(n-1)$ and variance

$$(\sigma^2/n)[1 + (t_{1-\alpha}^2(n-1)/2(n-1))]$$

where $\sigma^2 = E(s^2)$. Therefore

$$\frac{[\bar{x} - (s/\sqrt{n})t_{1-\alpha}(n-1)] - [m_1 - (\sigma/\sqrt{n})t_{1-\alpha}(n-1)]}{(\sigma/\sqrt{n})\{1 + [t_{1-\alpha}^2(n-1)/2(n-1)]\}^{1/2}}$$

has an approximate standard normal distribution. β can then be approximated by (Hald, 1952)

$$\beta = \int_{-\infty}^{V} \frac{1}{(2\pi)^{1/2}} \exp\left[-\frac{1}{2} z^2\right] dz \qquad (8.43)$$

where

$$V = \frac{(m_0 - m_1) + t_{1-\alpha}(n-1)(\sigma/\sqrt{n})}{(\sigma/\sqrt{n})\{1 + [t_{1-\alpha}^2(n-1)/2(n-1)]\}^{1/2}}$$

Let

$$\delta = \frac{(m_1 - m_0)}{\sigma/\sqrt{n}} \qquad (8.44)$$

Tests for a Single Parameter

Then

$$\frac{(m_0 - m_1) + t_{1-\alpha}(n-1)(\sigma/\sqrt{n})}{(\sigma/\sqrt{n})\{1 + [t_{1-\alpha}^2(n-1)/2(n-1)]\}^{1/2}}$$

$$= \frac{t_{1-\alpha}^2(n-1) - \delta}{\{1 + [t_{1-\alpha}^2(n-1)/2(n-1)]\}^{1/2}} = U(t, \alpha, n) \qquad (8.45)$$

and

$$\beta = \int_{-\infty}^{U(t,\alpha,n)} \frac{1}{(2\pi)^{1/2}} \exp\left[-\frac{1}{2}z^2\right] dz = F[U(t, \alpha, n)] \qquad (8.46)$$

or

$$Z_\beta = U(t, \alpha, n) \qquad (8.47)$$

The required value of n is the one that satisfies (8.47), where $U(t, \alpha, n)$ is a function of σ^2. Since σ^2 is not known, it must be replaced by an estimate obtained from data already available or from a pilot study.

Example 8.5 Optimal operation of a particular service system is dependent on knowledge of the rate at which arrivals occur to the system. The system operation is presently based on the assumption that the arrival rate is 100 units/day. If the arrival rate increases to 105 units/day, the operation of the system should be altered if optimal operation is to be maintained. To monitor the arrival rate, a statistical test is to be designed such that the null hypothesis that the arrival rate has not increased will be rejected with probability 0.05 if the arrival rate has, in fact, not increased. However, if the arrival rate increases to 105 units/day the null hypothesis should be rejected with probability 0.90. Historical data indicate that the standard deviation of arrivals per day is approximately 8. How many days data should be taken to run the required test?

As given in the statement of the problem, the null and alternative hypotheses are given by

$$H_0: \quad m \le 100, \qquad H_1: \quad m > 100$$

From (8.44),

$$\delta = \frac{105 - 100}{8}\sqrt{n} = 0.625\sqrt{n}$$

and

$$U(t, \alpha, n) = \frac{t_{0.95}^2(n-1) - 0.625\sqrt{n}}{\{1 + [t_{0.95}^2(n-1)/2(n-1)]\}^{1/2}}$$

374   8. Data Analysis—Hypothesis Testing

Finally
$$Z_{0.10} = -1.28$$
since
$$\beta = 1 - 0.90$$

We find the desired value of n by trial and error starting with $n = 100$ (see Table 8.1). Hence, the required sample size is 24.

TABLE 8.1 Summary of the trial and error procedure to find n for Example 8.5

n	$t_{0.95}(n-1)$	$0.625\sqrt{n}$	$\left[1 + \dfrac{t_{0.95}^2(n-1)}{2(n-1)}\right]^{1/2}$	$U(t, \alpha, n)$	$Z_{0.10}$
100	1.66	6.25	1.007	-4.56	-1.28
49	1.68	4.38	1.015	-2.66	-1.28
25	1.71	3.13	1.030	-1.38	-1.28
16	1.75	2.50	1.050	-0.71	-1.28
23	1.71	2.99	1.033	-1.24	-1.28
24	1.71	3.06	1.055	-1.28	-1.28

When the null and alternative hypotheses are given by
$$H_0: \quad m \geq m_0, \qquad H_1: \quad m < m_0$$
β can be approximated by
$$\beta = \int_{L(t,\alpha,n)}^{\infty} \frac{1}{(2\pi)^{1/2}} \exp\left[-\frac{1}{2}z^2\right] dz = 1 - F[L(t, \alpha, n)] \quad (8.48)$$
where
$$L(t, \alpha, n) = \frac{-t_{1-\alpha}(n-1) - \delta}{\{1 + [t_{1-\alpha}^2(n-1)/2(n-1)]\}^{1/2}} \quad (8.49)$$
and
$$\delta = \frac{m_1 - m_0}{\sigma/\sqrt{n}} \quad (8.50)$$
for $m_1 < m_0$. Thus, the desired value of n is that satisfying
$$Z_{1-\beta} = L(t, \alpha, n) \quad (8.51)$$

For the two-tailed test, where
$$H_0: \quad m = m_0, \qquad H_1: \quad m \neq m_0$$
β is given by
$$\beta = \int_{-\infty}^{U(t,\alpha/2,n)} \frac{1}{(2\pi)^{1/2}} \exp\left[-\frac{1}{2}z^2\right] dz = F[U(t, \tfrac{1}{2}\alpha, n)] \quad (8.52)$$

Tests for a Single Parameter

where $F[L(t, \frac{1}{2}\alpha, n)]$ is small compared to β, and

$$Z_\beta = U(t, \tfrac{1}{2}\alpha, n) \tag{8.53}$$

for $m_1 > m_0$; that is, m_1 is the critical value of m such that $m_1 > m_0$. A similar expression for β can be obtained using $L(t, \frac{1}{2}\alpha, n)$ but will yield a result identical to that given in (8.52) for the case where $m_1 < m_0$.

Example 8.6 Suppose that in the problem given in Example 8.5 it is critical to detect both positive and negative shifts in the arrival rate; that is, it is critical to detect an arrival rate as low as 95 or as great as 105. How many observations are necessary for the test?

Again we will solve for n by trial and error, where n must be such that either

$$Z_\beta = U(t, \tfrac{1}{2}\alpha, n), \quad \text{or} \quad Z_\beta = L(t, \tfrac{1}{2}\alpha, n)$$

A summary of the trial and error solution is given in Table 8.2, where the starting value of n is 24. Thus, the necessary sample size is 29.

TABLE 8.2 Summary of the trial and error solution for n for Example 8.6

n	$t_{0.975}(n-1)$	$0.625\sqrt{n}$	$\left[1 + \dfrac{t_{0.975}^2(n-1)}{2(n-1)}\right]^{1/2}$	$U(t, \tfrac{1}{2}\alpha, n)$	$Z_{0.10}$
24	2.07	3.06	1.046	-0.99	-1.28
36	2.03	3.75	1.030	-1.67	-1.28
27	2.06	3.25	1.040	-1.14	-1.28
30	2.05	3.44	1.035	-1.34	-1.28
29	2.05	3.36	1.037	-1.26	-1.28

Tests about a Single Variance

The reader will recall that in both of the statistical tests about a single mean, knowledge of the variance of the random variable in question was important in the design of the tests. At times we are as interested in changes in the variance of a random variable as we are in changes in the mean. For example, suppose that service time for a given service channel is gamma distributed. Mean service time might remain constant while the variance changed. If the variance increased appreciably, the number of people spending long periods of time in the service channel would also increase and these people might decide to take their business elsewhere in the future.

Let X be a random variable that is normally distributed with mean m and variance σ^2. Let S^2 be the sample variance obtained from n observed values of X. Then by (7.77) the random variable $(n-1)S^2/\sigma^2$ has a chi-square distribution with $n - 1$ degrees of freedom. Knowledge of the distribution of $(n - 1)S^2/\sigma^2$ allows us to develop tests for hypotheses about σ^2. Assume that we wish to test the null hypothesis

$$H_0: \quad \sigma^2 = \sigma_0^2$$

against the alternative

$$H_1: \quad \sigma^2 \neq \sigma_0^2$$

where σ_0^2 is a constant. We establish limits L and U such that if $L \leq s^2 \leq U$, the null hypothesis is accepted; otherwise it is rejected. The limits can be determined once we have specified a value for α. However, as in the tests for the mean, the sample size required for the test cannot be defined until critical values of σ^2 have been established and the associated value of β identified. Now

$$\alpha = 1 - P(L \leq s^2 \leq U \mid \sigma^2 = \sigma_0^2) \tag{8.54}$$

or

$$1 - \alpha = P(L \leq s^2 \leq U \mid \sigma^2 = \sigma_0^2)$$

$$= P\left(\frac{n-1}{\sigma_0^2} L \leq \frac{(n-1)s^2}{\sigma_0^2} \leq \frac{(n-1)}{\sigma_0^2} U \mid \sigma^2 = \sigma_0^2\right) \tag{8.55}$$

Since S^2 is an unbiased estimate of σ^2 and therefore σ_0^2, $(n - 1)S^2/\sigma_0^2$ has a chi-square distribution with $n - 1$ degrees of freedom. Then

$$1 - \alpha = P\left(\chi^2 \leq \frac{(n-1)}{\sigma_0^2} U \mid \sigma^2 = \sigma_0^2\right) - P\left(\chi^2 \leq \frac{(n-1)}{\sigma_0^2} L \mid \sigma^2 = \sigma_0^2\right)$$

$$\tag{8.56}$$

The chi-square probability density function is not symmetrical as was the case with the normal and t density functions. Thus, if the limits L and U are defined in a manner similar to that given by (8.5), they will not be equidistant from σ_0^2. If

$$P\left(\chi^2 > \frac{(n-1)}{\sigma_0^2} U \mid \sigma^2 = \sigma_0^2\right) = P\left(\chi^2 \leq \frac{(n-1)}{\sigma_0^2} L \mid \sigma^2 = \sigma_0^2\right) \tag{8.57}$$

then

$$P\left(\chi^2 > \frac{(n-1)}{\sigma_0^2} U \mid \sigma^2 = \sigma_0^2\right) = \frac{\alpha}{2} \tag{8.58}$$

Tests for a Single Parameter

and

$$P\left(\chi^2 \le \frac{(n-1)}{\sigma^2} L \mid \sigma^2 = \sigma_0^2\right) = \frac{\alpha}{2} \tag{8.59}$$

From (7.87),

$$\frac{(n-1)}{\sigma_0^2} U = \chi_{1-\alpha/2}^2(n-1) \tag{8.60}$$

and

$$\frac{(n-1)}{\sigma_0^2} L = \chi_{\alpha/2}^2(n-1) \tag{8.61}$$

or

$$U = \frac{\sigma_0^2}{n-1} \chi_{1-\alpha/2}^2(n-1) \tag{8.62}$$

$$L = \frac{\sigma_0^2}{n-1} \chi_{\alpha/2}^2(n-1) \tag{8.63}$$

To test the null hypothesis

$$H_0: \quad \sigma^2 \le \sigma_0^2$$

against the alternative

$$H_1: \quad \sigma^2 > \sigma_0^2$$

we require a one-tailed test where

$$U = \frac{\sigma_0^2}{n-1} \chi_{1-\alpha}^2(n-1) \tag{8.64}$$

For the case where

$$H_0: \quad \sigma^2 \ge \sigma_0^2, \qquad H_1: \quad \sigma^2 < \sigma_0^2$$

we have a lower-tailed test where

$$L = \frac{\sigma_0^2}{n-1} \chi_{\alpha}^2(n-1) \tag{8.65}$$

Example 8.7 The number of jet aircraft arriving at a terminal during the peak afternoon hour is normally distributed with mean 30. If the standard deviation of the number of arrivals per hour exceeds 3, a standby controller should be added to the active controllers to help handle aircraft during periods of high-intensity traffic. Over the past thirty days the number of arrivals during the peak hour has been recorded. Based upon these data the sample variance was calculated and found to be 11.6. Is a standby controller necessary? Let $\alpha = 0.10$.

The null and alternative hypotheses for this problem are

$$H_0: \quad \sigma^2 \leq 9, \qquad H_1: \quad \sigma^2 > 9$$

Since the test is upper-tailed, we need only calculate U from (8.64):

$$U = \frac{\sigma_0^2}{n-1}\chi^2_{1-\alpha}(n-1) = \frac{9}{29}\chi^2_{0.90}(29)$$

From Table C of the Appendix, by interpolation,

$$\chi^2_{0.90}(29) = 39.09$$

Hence

$$U = \tfrac{9}{29}(39.09) = 12.15$$

since $s^2 < 12.15$, the null hypothesis is accepted and a standby controller is not necessary.

We will now turn our attention to the determination of the number of observations necessary for the test if a given case of the alternative hypothesis is to be detected with a specified power. Consider the null and alternative hypotheses given by

$$H_0: \quad \sigma^2 \leq \sigma_0^2, \qquad H_1: \quad \sigma^2 > \sigma_0^2$$

Suppose that the null hypothesis is to be rejected with probability $1 - \beta$ if $\sigma^2 = \sigma_1^2$ ($\sigma_1^2 > \sigma_0^2$). Now

$$P(H_0 \text{ rejected} \,|\, \sigma^2 = \sigma_1^2) = P\left[s^2 > \frac{\sigma_0^2}{n-1}\chi^2_{1-\alpha}(n-1) \,\Big|\, \sigma^2 = \sigma_1^2\right] = 1 - \beta \tag{8.66}$$

Since S^2 is an unbiased estimate of $\sigma^2 = \sigma_1^2$, $(n-1)S^2/\sigma^2$ is not chi-square distributed. However, $(n-1)S^2/\sigma_1^2$ is chi-square distributed, since S^2 is an unbiased estimate of σ_1^2. Thus

$$\begin{aligned}1 - \beta &= P\left[s^2 > \frac{\sigma_0^2}{n-1}\chi^2_{1-\alpha}(n-1) \,\Big|\, \sigma^2 = \sigma_1^2\right] \\ &= P\left[\frac{(n-1)s^2}{\sigma_1^2} > \frac{\sigma_0^2}{\sigma_1^2}\chi^2_{1-\alpha}(n-1) \,\Big|\, \sigma^2 = \sigma_1^2\right] \end{aligned} \tag{8.67}$$

Tests for a Single Parameter

Since $(n-1)S^2/\sigma_1^2$ is chi-square distributed with $n-1$ degrees of freedom, $(\sigma_0^2/\sigma_1^2)\chi_{1-\alpha}^2(n-1)$ is the value of the chi-square random variable such that

$$F_{\chi^2}\left[\frac{\sigma_0^2}{\sigma_1^2}\chi_{1-\alpha}^2(n-1)\right]=\beta \qquad (8.68)$$

Hence

$$\frac{\sigma_0^2}{\sigma_1^2}\chi_{1-\alpha}^2(n-1)=\chi_\beta^2(n-1) \qquad (8.69)$$

Thus, by trial and error we can find the value of n satisfying Eq. (8.69).

Example 8.8 Find the number of observations necessary if the null hypothesis in Example 8.7 is to be rejected with probability 0.75 if the variance is as large as 12.

From Example 8.7 and the above statement,

$$\sigma_0^2=9, \qquad \sigma_1^2=12$$

Thus we must find n such that

$$\tfrac{9}{12}\chi_{0.90}^2(n-1)=\chi_{0.25}^2(n-1)$$

The summary of the trial and error solution for n is given in Table 8.3. As shown in this table, $n \simeq 91$.

TABLE 8.3 Summary of the trial and error solution for n for the problem in Example 8.8

n	$\chi_{0.25}^2(n-1)$	$\chi_{0.90}^2(n-1)$	$\tfrac{9}{12}\chi_{0.90}^2(n-1)$
10	5.90	14.68	11.01
51	42.94	63.17	47.37
101	90.13	118.50	88.89
91	80.62	107.57	80.67

The number of observations n necessary to test the null hypothesis

$$H_0: \quad \sigma^2 \geq \sigma_0^2$$

against the alternative hypothesis

$$H_1: \quad \sigma^2 < \sigma_0^2$$

is that value of n such that

$$\frac{\sigma_0^2}{\sigma_1^2}\chi_\alpha^2(n-1)=\chi_{1-\beta}^2(n-1) \qquad (8.70)$$

where $\sigma_1^2 < \sigma_0^2$. When the null and alternative hypotheses are given by

$$H_0: \quad \sigma^2 = \sigma_0^2, \qquad H_1: \quad \sigma^2 \neq \sigma_0^2$$

the sample size is given approximately as that value of n such that

$$\frac{\sigma_0^2}{\sigma_1^2} \chi_{1-\alpha/2}^2(n-1) = \chi_\beta^2(n-1) \tag{8.71}$$

when $\sigma_1^2 > \sigma_0^2$ and when

$$P\left[\frac{(n-1)s^2}{\sigma_1^2} \leq \frac{\sigma_0^2}{\sigma_1^2} \chi_{\alpha/2}^2(n-1)\right]$$

is small compared to β, or such that

$$\frac{\sigma_0^2}{\sigma_1^2} \chi_{\alpha/2}^2(n-1) = \chi_{1-\beta}^2(n-1) \tag{8.72}$$

when $\sigma_1^2 < \sigma_0^2$ and when

$$P\left[\frac{(n-1)s^2}{\sigma_1^2} \geq \frac{\sigma_0^2}{\sigma_1^2} \chi_{1-\alpha/2}^2(n-1)\right]$$

is small compared to β. The reader will note that (8.71) and (8.72) may lead to different sample sizes for the same test. Thus, the analyst would usually calculate the values of n satisfying each of the equations and choose the largest as the sample size.

TESTS FOR THE COMPARISON OF TWO PARAMETERS

In the tests presented thus far we were concerned with testing the hypotheses that a given parameter was equal to, less than or equal to, and greater than or equal to a specific numerical value. However, these tests do not apply when we are faced with a situation where we have to compare two unknown parameters. For example, we might be interested in whether the mean arrival rate of customers is the same during two different periods. Now, it is unlikely that we would know the true mean arrival rate in either period. Thus, we would have to compare sample means from the two periods and, based on their respective values, draw a conclusion as to whether or not the difference in the observed means indicates that the true means are different.

In this section we will present tests for the comparison of two means and two variances. The test for the comparison of two means is given for the case where the values of the variances of the respective populations are unknown, since this is usually the case. For a treatment of the corresponding test when the population variances are known the reader should see the work of Ostle (1963). The test for the comparison of two means treated here depends on whether or not the respective population variances are equal or not. As we shall see, a t-test, although only approximate in one case, is used whether or not the variances are equal. However, the computational requirements are different in the two cases.

In order to determine which of the above mentioned t-tests to use in comparing two population means, we must be able to determine whether or not the variances of the two populations are equal. Since we will not assume that the population variances are known, we will first need a test that allows us to compare population variances. Of course, a test for the comparison of two variances may be of interest whether or not we are interested in comparing the corresponding means.

Test for the Comparison of Two Variances

Let σ_1^2 and σ_2^2 be the variances of two normally distributed random variables and let S_1^2 and S_2^2 be the corresponding sample variances, where

$$S_i^2 = \frac{1}{n_i - 1} \sum_{i=1}^{n_i} (X_i - \bar{X}_i)^2, \quad i = 1, 2 \quad (8.73)$$

Then $(n_1 - 1)S_1^2/\sigma_1^2$ and $(n_2 - 1)S_2^2/\sigma_2^2$ are chi-square random variables with $n_1 - 1$ and $n_2 - 1$ degrees of freedom, respectively. Now, if Y_1 and Y_2 are chi-square random variables with r_1 and r_2 degrees of freedom, respectively, then $(Y_1/r_1)/(Y_2/r_2)$ is F distributed with degrees of freedom r_1 and r_2. Hence, $(S_1^2/\sigma_1^2)/(S_2^2/\sigma_2^2)$ is F distributed with $n_1 - 1$ and $n_2 - 1$ degrees of freedom.

Now suppose that we are to test the null hypothesis

$$H_0: \quad \sigma_1^2 = \sigma_2^2$$

against the alternative

$$H_1: \quad \sigma_1^2 \neq \sigma_2^2$$

Under the null hypothesis, $\sigma_1^2 = \sigma_2^2$ and S_1^2/S_2^2 is F distributed with $n_1 - 1$ and $n_2 - 1$ degrees of freedom. If the probability of a type I error is to be α, then

$$P\left(L \leq \frac{S_1^2}{S_2^2} \leq U \,\middle|\, \sigma_1^2 = \sigma_2^2\right) = 1 - \alpha \quad (8.74)$$

Assuming that L and U are to be defined such that

$$P\left(\frac{s_1^2}{s_2^2} < L \mid \sigma_1^2 = \sigma_2^2\right) = P\left(\frac{s_1^2}{s_2^2} > U \mid \sigma_1^2 = \sigma_2^2\right) \tag{8.75}$$

we have

$$P\left(\frac{s_1^2}{s_2^2} < L \mid \sigma_1^2 = \sigma_2^2\right) = \frac{\alpha}{2} \tag{8.76}$$

and

$$P\left(\frac{s_1^2}{s_2^2} < U \mid \sigma_1^2 = \sigma_2^2\right) = \frac{1-\alpha}{2} \tag{8.77}$$

Let F_p be the value of the F random variable such that

$$F_F(F_p) = p \tag{8.78}$$

Then L is the value of the F random variable such that

$$F_F(L) = \frac{\alpha}{2} \tag{8.79}$$

and

$$F_F(U) = \frac{1-\alpha}{2} \tag{8.80}$$

Therefore, L and U are given by

$$L = F_{\alpha/2}(n_1 - 1, n_2 - 1) \tag{8.81}$$
$$U = F_{1-\alpha/2}(n_1 - 1, n_2 - 1) \tag{8.82}$$

Values of $F_p(r_1, r_2)$ can be found in most texts on statistical methods (see, for example, Bennett and Franklin, 1954; Bowker and Lieberman, 1959; Dixon and Massey, 1957; Ostle, 1963).

For the lower-tailed test defined by the null and alternative hypotheses

$$H_0: \quad \sigma_1^2 \geq \sigma_2^2, \qquad H_1: \quad \sigma_1^2 < \sigma_2^2$$

the lower limit for the test is

$$L = F_\alpha(n_1 - 1, n_2 - 1) \tag{8.83}$$

For the upper-tailed test defined by

$$H_0: \quad \sigma_1^2 \leq \sigma_2^2, \qquad H_1: \quad \sigma_1^2 > \sigma_2^2$$

the upper limit for the test is given by

$$U = F_{1-\alpha}(n_1 - 1, n_2 - 1) \tag{8.84}$$

Tests for the Comparison of Two Parameters

Example 8.9 A manufacturer is dissatisfied with a particular piece of extrusion equipment. Specifically, the length of the extruded product is too variable. Based on the measurement of 100 units of product, the sample variance was found to be 0.82. A vendor claims that his extrudor is less variable. A sample of 200 units is selected from the vendor's equipment and the variance is found to be 0.69. Would you accept the vendor's claim if $\alpha = 0.01$?

Let σ_1^2 be the variance of the present equipment, and σ_2^2 the variance of the vendor's equipment. The null and alternative hypotheses are

$$H_0: \quad \sigma_1^2 \leq \sigma_2^2, \qquad H_1: \quad \sigma_1^2 > \sigma_2^2$$

The null hypothesis is constructed on the assumption that the vendor's claim is not justified until it is established by rejection of the null hypothesis. For the test,

$$U = F_{0.99}(99, 199) = 1.48$$

Now

$$\frac{s_1^2}{s_2^2} = \frac{0.82}{0.69} = 1.19$$

Since $s_1^2/s_2^2 < 1.19$, the null hypothesis is not rejected and the vendor's claim is not verified.

The values of n_1 and n_2 necessary to test the null hypothesis

$$H_0: \quad \sigma_1^2 \leq \sigma_2^2$$

against the alternative

$$H_1: \quad \sigma_1^2 > \sigma_2^2$$

when H_0 is to be rejected with probability $1 - \beta$ when $\sigma_1^2/\sigma_2^2 = \lambda > 1$ are those n_1 and n_2 satisfying the equation

$$F_\beta(n_1 - 1, n_2 - 1) = \frac{1}{\lambda^2} F_{1-\alpha}(n_1 - 1, n_2 - 1) \qquad (8.85)$$

To test the null hypothesis

$$H_0: \quad \sigma_1^2 \geq \sigma_2^2$$

against the alternative

$$H_1: \quad \sigma_1^2 < \sigma_2^2$$

when H_0 is to be rejected with probability $1 - \beta$ when $\sigma_1^2/\sigma_2^2 = \lambda < 1$, the number of observations n_1 and n_2 necessary are those values of n_1 and n_2 satisfying

$$F_{1-\beta}(n_1 - 1, n_2 - 1) = \frac{1}{\lambda^2} F_\alpha(n_1 - 1, n_2 - 1) \tag{8.86}$$

For the two-tailed procedure where the null and alternative hypotheses are given by

$$H_0: \quad \sigma_1^2 = \sigma_2^2, \qquad H_1: \quad \sigma_1^2 \neq \sigma_2^2$$

the required values of n_1 and n_2 can be defined in approximate form. Specifically, the values of n_1 and n_2 required to reject H_0 with probability $1 - \beta$ when $\sigma_1^2/\sigma_2^2 = \lambda > 1$ satisfy the relationship

$$F_\beta(n_1 - 1, n_2 - 1) = \frac{1}{\lambda^2} F_{1-\alpha/2}(n_1 - 1, n_2 - 1) \tag{8.87}$$

when

$$P\left[\frac{s_1^2/\sigma_1^2}{s_2^2/\sigma_2^2} \leq \frac{1}{\lambda^2} F_{\alpha/2}(n_1 - 1, n_2 - 1)\right]$$

is small compared to β and by

$$F_{1-\beta}(n_1 - 1, n_2 - 1) = \frac{1}{\lambda^2} F_{\alpha/2}(n_1 - 1, n_2 - 1) \tag{8.88}$$

when $\sigma_1^2/\sigma_2^2 = \lambda < 1$ and when

$$P\left[\frac{s_1^2/\sigma_1^2}{s_2^2/\sigma_2^2} \geq \frac{1}{\lambda^2} F_{1-\alpha/2}(n_1 - 1, n_2 - 1)\right]$$

is small compared to β.

In determining the sample sizes n_1 and n_2 for the two-tailed test, one would usually specify two ratios σ_1^2/σ_2^2, one less than and one greater than unity, such that the null hypothesis will be rejected with probability $1 - \beta$ if either ratio is in fact true. These two ratios are likely to result in two different sets of values of n_1 and n_2. One method that can be used to select between these two sets is to select that set of values of n_1 and n_2 such that the probability of a type II error on both tails is no greater than the specified value of β.

Since most tables of the cumulative F distribution are not extensive and as a result do not give probabilities for all combinations of degrees of freedom, an approximation for the F distribution function is useful to calculate these probabilities when extensive tables are not available. Such an approximation is given as follows (Hald, 1952). Let Z be a standard normal random

Tests for the Comparison of Two Parameters

variable and X an F random variable with r_1 and r_2 degrees of freedom. Then

$$P(x \le F) = P\left\{z \le \frac{[(2r_2 - 1)(r_1/r_2)F]^{1/2} - (2r_1 - 1)^{1/2}}{[1 + (r_1/r_2)F]^{1/2}}\right\} \quad (8.89)$$

Let

$$\lambda_1 = \frac{\sigma_1^2}{\sigma_2^2} < 1, \quad \text{and} \quad \lambda_2 = \frac{\sigma_1^2}{\sigma_2^2} > 1$$

and let β_1 be the probability of a type II error when $\sigma_1^2/\sigma_2^2 = \lambda_1$ and β_2 the probability of a type II error when $\sigma_1^2/\sigma_2^2 = \lambda_2$. Then

$$\beta_1 = P\left[\lambda_1 F_{\alpha/2}(n_1 - 1, n_2 - 1) \le \frac{s_1^2/\sigma_1^2}{s_2^2/\sigma_2^2} \le \lambda_1 F_{1-\alpha/2}(n_1 - 1, n_2 - 1)\right] \quad (8.90)$$

and

$$\beta_2 = P\left[\lambda_2 F_{\alpha/2}(n_1 - 1, n_2 - 1) \le \frac{s_1^2/\sigma_1^2}{s_2^2/\sigma_2^2} \le \lambda_2 F_{1-\alpha/2} F(n_1 - 1, n_2 - 1)\right] \quad (8.91)$$

Letting

$$A[F_p(r_1, r_2)] = \frac{[(2r_2 - 1)(r_1/r_2)F_p(r_1, r_2)]^{1/2} - (2r_1 - 1)^{1/2}}{[1 + (r_1/r_2)F_p(r_1, r_2)]^{1/2}} \quad (8.92)$$

we have

$$\beta_1 = P\{z \le A[\lambda_1 F_{1-\alpha/2}(n_1 - 1, n_2 - 1)]\}$$
$$\quad - P\{z \le A[\lambda_1 F_{\alpha/2}(n_1 - 1, n_2 - 1)]\} \quad (8.93)$$
$$\beta_2 = P\{z \le A[\lambda_2 F_{1-\alpha/2}(n_1 - 1, n_2 - 1)]\}$$
$$\quad - P\{z \le A[\lambda_1 F_{\alpha/2}(n_1 - 1, n_2 - 1)]\} \quad (8.94)$$

Example 8.10 The variances σ_1^2 and σ_2^2 of service time for two processes are to be compared using the F test with $\alpha = 0.05$. If $\sigma_1^2/\sigma_2^2 = 0.25$ or $\sigma_1^2/\sigma_2^2 = 4.00$, the null hypothesis that $\sigma_1^2 = \sigma_2^2$ should be rejected with probability at least 0.90. How many observations should be taken from each process if the number of observations from each is to be the same?

Let

$$\lambda_1^2 = \frac{\sigma_1^2}{\sigma_2^2} = 0.25, \quad \text{and} \quad \lambda_2^2 = \frac{\sigma_1^2}{\sigma_2^2} = 4.00$$

From the statement of the problem

$$\alpha = 0.05, \quad \beta = 0.10, \quad \text{and} \quad n_1 = n_2$$

Let $n = n_1 = n_2$. Define β_1 and β_2 as

$$\beta_1 = P\left[\frac{1}{\lambda_1^2}F_{\alpha/2}(n-1, n-1) \leq \frac{s_1^2/\sigma_1^2}{s_2^2/\sigma_2^2} \leq \frac{1}{\lambda_1^2}F_{1-\alpha/2}(n-1, n-1)\right]$$

$$= P\left[0.25F_{0.025}(n-1, n-1) \leq \frac{s_1^2/\sigma_1^2}{s_2^2/\sigma_2^2} \leq 0.25F_{0.975}(n-1, n-1)\right]$$

and

$$\beta_2 = P\left[4.0F_{0.025}(n-1, n-1) \leq \frac{s_1^2/\sigma_1^2}{s_2^2/\sigma_2^2} \leq 4.0F_{0.975}(n-1, n-1)\right]$$

For λ_1, n is given by

$$F_{0.90}(n-1, n-1) = 4F_{0.025}(n-1, n-1)$$

and for λ_2

$$F_{0.10}(n-1, n-1) = 0.25F_{0.975}(n-1, n-1)$$

The trial and error procedure to find n is summarized in Table 8.4, where (8.92) and (8.93) were used to calculate β_1 and β_2. Therefore by selecting samples of size 25 from each process the desired power is achieved.

TABLE 8.4 Trial and error solution for n in Example 8.10

n	$F_{0.90}$ $(n-1, n-1)$	$4F_{0.025}$ $(n-1, n-1)$	$F_{0.10}$ $(n-1, n-1)$	$0.25F_{0.975}$ $(n-1, n-1)$	β_1	β_2
10	2.44	1.00	0.41	1.01	0.49	0.50
15	2.02	1.36	0.50	0.75	0.30	0.29
21	1.79	1.64	0.56	0.62	0.15	0.14
25	1.70	1.80	0.59	0.57	0.08	0.09

Test for the Comparison of Two Means, Unknown but Equal Variances

Let X_1 and X_2 be normal random variables with means m_1 and m_2 and variances σ_1^2 and σ_2^2, respectively. If $\bar{X}_1$ and $\bar{X}_2$ are sample means estimating m_1 and m_2, respectively, and if $\sigma_1^2 = \sigma_2^2$, then

$$\frac{\bar{X}_1 - \bar{X}_2}{S_p\left(\dfrac{1}{n_1} + \dfrac{1}{n_2}\right)^{1/2}}$$

Tests for the Comparison of Two Parameters

is t distributed with $n_1 + n_2 - 2$ degrees of freedom if $m_1 = m_2$, where S_p^2 is the pooled estimate of variance given by

$$S_p^2 = \frac{(n_1 - 1)S_1^2 + (n_2 - 1)S_2^2}{n_1 + n_2 - 2} \tag{8.95}$$

and

$$S_i^2 = \frac{1}{n_i - 1} \sum_{j=1}^{n_i} (X_{ij} - \bar{X}_i)^2 \tag{8.96}$$

Since $\sigma_1^2 = \sigma_2^2 = \sigma^2$, S_1^2 and S_2^2 are estimates of the same variance σ^2. Therefore, the pooled estimate, S_p^2, is a more precise estimate of σ^2 than either S_1^2 or S_2^2. To verify the assumptions that $\sigma_1^2 = \sigma_2^2$, the test in the previous section should be applied prior to using the test given in this section.

To test the null hypothesis

$$H_0: \quad m_1 = m_2$$

against the alternative

$$H_1: \quad m_1 \neq m_2$$

the limits L and U must be defined such that

$$P(L \leq \bar{x}_1 - \bar{x}_2 \leq U \mid m_1 = m_2) = 1 - \alpha \tag{8.97}$$

Then

$$P\left[\frac{L}{S_p\left(\frac{1}{n_1} + \frac{1}{n_2}\right)^{1/2}} \leq \frac{\bar{x}_1 - \bar{x}_2}{S_p\left(\frac{1}{n_1} + \frac{1}{n_2}\right)^{1/2}} \leq \frac{U}{S_p\left(\frac{1}{n_1} + \frac{1}{n_2}\right)^{1/2}} \,\bigg|\, m_1 = m_2\right]$$
$$= 1 - \alpha \tag{8.98}$$

If

$$P\left[\frac{\bar{x}_1 - \bar{x}_2}{S_p\left(\frac{1}{n_1} + \frac{1}{n_2}\right)^{1/2}} \leq \frac{U}{S_p\left(\frac{1}{n_1} + \frac{1}{n_2}\right)^{1/2}} \,\bigg|\, m_1 = m_2\right]$$

$$= P\left[\frac{\bar{x}_1 - \bar{x}_2}{S_p\left(\frac{1}{n_1} + \frac{1}{n_2}\right)^{1/2}} \geq \frac{L}{S_p\left(\frac{1}{n_1} + \frac{1}{n_2}\right)^{1/2}} \,\bigg|\, m_1 = m_2\right]$$

we have

$$\frac{U}{S_p\left(\frac{1}{n_1} + \frac{1}{n_2}\right)^{1/2}} = t_{1-\alpha/2}(n_1 + n_2 - 2),$$

and

$$\frac{L}{s_p\left(\frac{1}{n_1} + \frac{1}{n_2}\right)^{1/2}} = t_{\alpha/2}(n_1 + n_2 - 2)$$

and

$$L = s_p\left(\frac{1}{n_1} + \frac{1}{n_2}\right)^{1/2} t_{\alpha/2}(n_1 + n_2 - 2) \tag{8.99}$$

$$U = s_p\left(\frac{1}{n_1} + \frac{1}{n_2}\right)^{1/2} t_{1-\alpha/2}(n_1 + n_2 - 2) \tag{8.100}$$

To test the null hypothesis

$$H_0: \quad m_1 \leq m_2$$

against

$$H_1: \quad m_1 > m_2$$

the limit for the test is given by

$$U = s_p\left(\frac{1}{n_1} + \frac{1}{n_2}\right)^{1/2} t_{1-\alpha}(n_1 + n_2 - 1) \tag{8.101}$$

For the null and alternative hypotheses

$$H_0: \quad m_1 \geq m_2, \qquad H_1: \quad m_1 < m_2$$

the limit for the test is

$$L = s_p\left(\frac{1}{n_1} + \frac{1}{n_2}\right)^{1/2} t_{\alpha}(n_1 + n_2 - 1) \tag{8.102}$$

Example 8.11 During the first five weeks (25 days) of operation of a particular production process, the average number of units manufactured per day $\bar{x}_1$ was found to be 23.6 and the standard deviation s_1 of the number of units produced per day was 4.7. During the next two weeks (10 days) the average number of units produced per day $\bar{x}_2$ was 21.2 and the standard deviation s_2 was 5.1. Letting $\alpha = 0.10$, test the hypothesis that the mean daily production rate has not decreased.

The null and alternative hypotheses are

$$H_0: \quad m_1 \leq m_2, \qquad H_1: \quad m_1 > m_2$$

To determine whether or not the t-test presented above can be used, we must first determine whether or not $\sigma_1^2 = \sigma_2^2$. Letting $\alpha = 0.10$, we have

$$H_0: \quad \sigma_1^2 = \sigma_2^2, \qquad H_1: \quad \sigma_1^2 \neq \sigma_2^2$$

Tests for the Comparison of Two Parameters

From (8.81) and (8.82)

$$L = F_{0.05}(24, 9) = 0.34, \quad U = F_{0.95}(24, 9) = 2.90$$

Now

$$\frac{s_1^2}{s_2^2} = \left(\frac{4.7}{5.1}\right)^2 = 0.85$$

Since $0.34 < s_1^2/s_2^2 < 2.90$, we accept the hypothesis that $\sigma_1^2 = \sigma_2^2$, and the t-test presented above can be used.

From (8.101)

$$U = s_p \left(\frac{1}{n_1} + \frac{1}{n_2}\right)^{1/2} t_{1-\alpha}(n_1 + n_2 - 2)$$

Now

$$s_p^2 = \frac{24(4.7) + 9(5.1)}{33} = 4.81$$

from (8.95). Then

$s_p = 2.2$, and $U = 2.2(\frac{1}{25} + \frac{1}{10})^{1/2} t_{0.90}(33) = 2.2(0.374)(1.31) = 1.07$

Now

$$\bar{x}_1 - \bar{x}_2 = 23.6 - 21.2 = 2.4$$

Therefore, the null hypothesis is rejected, since $\bar{x}_1 - \bar{x}_2 > U$, and we conclude that the mean production rate has decreased.

If we wish to find the sample sizes n_1 and n_2 such that the null hypothesis

$$H_0: \quad m_1 \leq m_2$$

is accepted with probability β if $m_1 - m_2 = M > 0$, then

$$\beta = P(\bar{X}_1 - \bar{X}_2 \leq U | m_1 - m_2 = M)$$

$$= P\left[\frac{\bar{X}_1 - \bar{X}_2}{s_p\left(\frac{1}{n_1} + \frac{1}{n_2}\right)^{1/2}} \leq \frac{U}{s_p\left(\frac{1}{n_1} + \frac{1}{n_2}\right)^{1/2}} \bigg| m_1 - m_2 = M\right] \quad (8.103)$$

but

$$U = s_p \left(\frac{1}{n_1} + \frac{1}{n_2}\right)^{1/2} t_{1-\alpha}(n_1 + n_2 - 2)$$

Hence

$$\beta = P\left[\frac{\bar{x}_1 - \bar{x}_2}{s_p\left(\frac{1}{n_1} + \frac{1}{n_2}\right)^{1/2}} \le t_{1-\alpha}(n_1 + n_2 - 2) \,\middle|\, m_1 - m_2 = M\right]$$

$$= P\left[(\bar{x}_1 - \bar{x}_2) - s_p\left(\frac{1}{n_1} + \frac{1}{n_2}\right)^{1/2} t_{1-\alpha}(n_1 + n_2 - 2) \le 0 \,\middle|\, m_1 - m_2 = M\right]$$
(8.104)

Now

$$(\bar{X}_1 - \bar{X}_2) - S_p\left(\frac{1}{n_1} + \frac{1}{n_2}\right)^{1/2} t_{1-\alpha}(n_1 + n_2 - 2)$$

has an approximate normal distribution with mean

$$M - \sigma\left(\frac{1}{n_1} + \frac{1}{n_2}\right)^{1/2} t_{1-\alpha}(n_1 + n_2 - 2)$$

and variance

$$\frac{(n_1 + n_2)\sigma^2}{n_1 n_2}\left[1 + \frac{t_{1-\alpha}^2(n_1 + n_2 - 2)}{2(n_1 + n_2 - 2)}\right]$$

Then

$$\frac{(\bar{X}_1 - \bar{X}_2) - s_p\left(\frac{1}{n_1} + \frac{1}{n_2}\right)^{1/2} t_{1-\alpha}(n_1 + n_2 - 2) - M + \sigma\left(\frac{1}{n_1} + \frac{1}{n_2}\right)^{1/2} t_{1-\alpha}(n_1 + n_2 - 2)}{[(n_1 + n_2)/n_1 n_2]^{1/2} \sigma\{1 + [t_{1-\alpha}^2(n_1 + n_2 - 2)/2(n_1 + n_2 - 2)]\}^{1/2}}$$

has an approximate standard normal distribution. Let us define $U(t, \alpha, n_1, n_2)$ as

$$U(t, \alpha, n_1, n_2) = \frac{t_{1-\alpha}(n_1 + n_2 - 2) - \delta}{\{1 + [t_{1-\alpha}^2(n_1 + n_2 - 2)/2(n_1 + n_2 - 2)]\}^{1/2}}$$
(8.105)

where

$$\delta = \frac{M}{\sigma[(n_1 + n_2)/n_1 n_2]^{1/2}}$$
(8.106)

Hence

$$\beta = \int_{-\infty}^{U(t, \alpha, n_1, n_2)} \frac{1}{(2\pi)^{1/2}} \exp\left[-\frac{1}{2}z^2\right] dz = P[z \le U(t, \alpha, n_1, n_2)]$$
(8.107)

and the required values of n_1 and n_2 are those satisfying

$$Z_\beta = U(t, \alpha, n_1, n_2)$$
(8.108)

If the null hypothesis

$$H_0: \quad m_1 \geq m_2$$

is to be accepted with probability β when $m_1 - m_2 = M < 0$, then the required values of n_1 and n_2 are those satisfying the relationship

$$Z_{1-\beta} = L(t, \alpha, n_1, n_2) \tag{8.109}$$

where

$$L(t, \alpha, n_1, n_2) = \frac{-t_{1-\alpha}(n_1 + n_2 - 2) - \delta}{\{1 + [t_{1-\alpha}^2(n_1 + n_2 - 2)/2(n_1 + n_2 - 2)]\}^{1/2}} \tag{8.110}$$

and

$$\delta = \frac{M}{\sigma[(n_1 + n_2)/n_1 n_2]^{1/2}} \tag{8.111}$$

If the null hypothesis

$$H_0: \quad m_1 = m_2$$

is to be accepted with probability β when $|m_1 - m_2| = M$, then n_1 and n_2 satisfy the equation

$$Z_\beta = U(t, \tfrac{1}{2}\alpha, n_1, n_2) \tag{8.112}$$

if $F[L(t, \tfrac{1}{2}\alpha, n_1, n_2)]$ is small compared to β.

Test for the Comparison of Two Means, Unknown and Unequal Variances

When we are not able to verify the assumption that $\sigma_1^2 = \sigma_2^2$, the t-test presented above should not be used. However, if S_1^2 and S_2^2 are estimates of σ_1^2 and σ_2^2 and if $m_1 = m_2$, then

$$\frac{\bar{X}_1 - \bar{X}_2}{(S_1^2/n_1) + (S_2^2/n_2)}$$

has an approximate t distribution with r degrees of freedom given by

$$r = \frac{(S_1^2/n_1 + S_2^2/n_2)^2}{[(S_1^2/n_1)^2/(n_1 + 1)] + [(S_2^2/n_2)^2/(n_2 + 1)]} - 2 \tag{8.113}$$

Hence, to test the null hypothesis

$$H_0: \quad m_1 = m_2$$

the limits for the test are

$$L = \left(\frac{s_1^2}{n_1} + \frac{s_2^2}{n_2}\right)^{1/2} t_{\alpha/2}(r) \tag{8.114}$$

$$U = \left(\frac{s_1^2}{n_1} + \frac{s_2^2}{n_2}\right)^{1/2} t_{1-\alpha/2}(r) \tag{8.115}$$

For the lower-tailed test of the null hypothesis

$$H_0: \quad m_1 \geq m_2$$

the lower limit L is defined by

$$L = \left(\frac{s_1^2}{n_1} + \frac{s_2^2}{n_2}\right)^{1/2} t_{\alpha}(r) \tag{8.116}$$

The upper limit for testing the null hypothesis

$$H_0: \quad m_1 \leq m_2$$

is given by

$$U = \left(\frac{s_1^2}{n_1} + \frac{s_2^2}{n_2}\right)^{1/2} t_{1-\alpha}(r) \tag{8.117}$$

The probability distribution of

$$\frac{\bar{X}_1 - \bar{X}_2}{(S_1^2/n_1) + (S_2^2/n_2)}$$

has not been defined for $m_1 \neq m_2$. Therefore, relationships for defining n_1 and n_2 for β corresponding to a given difference $m_1 - m_2$ will not be given.

Example 8.12 In Example 8.11 suppose that

$$s_1 = 4.7, \quad s_2 = 2.4$$

Using $\alpha = 0.10$, test the hypothesis that the mean daily production rate has not changed.
From Example 8.11,

$$\bar{x}_1 = 23.6, \quad \bar{x}_2 = 21.2$$

We first test the null hypothesis that $\sigma_1^2 = \sigma_2^2$ using $\alpha = 0.10$, where the lower and upper limits for the test are

$$L = 0.34, \quad U = 2.90$$

Now

$$\frac{s_1^2}{s_2^2} = \left(\frac{4.7}{2.4}\right)^2 = 3.84$$

Since $s_1^2/s_2^2 > 2.90$, we conclude that $\sigma_1^2 \neq \sigma_2^2$. Thus to test the null hypothesis

$$H_0: \quad m_1 \leq m_2$$

we will use the approximate t-test just presented, where the limit for the test is

$$U = \left(\frac{s_1^2}{n_1} + \frac{s_2^2}{n_2}\right)^{1/2} t_{0.90}(r)$$

where r is given in (8.113). Thus

$$r = \frac{[(12.09/25) + (5.76/10)]^2}{\dfrac{(12.09/25)^2}{26} + \dfrac{(5.76/10)^2}{11}} - 2$$

and

$$U = (0.48 + 0.58)^{1/2} t_{0.90}(26) = 1.40$$

Since

$$\bar{x}_1 - \bar{x}_2 = 2.4$$

the null hypothesis is rejected and we conclude that the mean production rate has decreased.

EFFECTS OF NONNORMALITY

In all of the statistical tests presented in this section we have made the assumption that the underlying distribution of the observations on which all estimates were based was normal. In general, moderate departures from normality will not cause significant errors for tests concerning the means (Bennett and Franklin, 1954; Duncan, 1965; Hald, 1952). However, tests about variances seem to be sensitive to departures from normality (Bennett and Franklin, 1954). Should serious departures from normality occur, as indicated by the results of a goodness-of-fit test, the analyst has two avenues of approach available. First, he can develop the necessary statistical test, basing the development on the distributions he believes to be appropriate. Since the development of a statistical test based on a distribution other than the normal is sometimes complex, the analyst may wish to apply an appropriate nonparametric statistical test. Generally, nonparametric statistical tests make no assumptions about the underlying distributions of the random variables whose values form the data for the test. However, nonparametric tests are less powerful than the corresponding parametric tests when the assumptions of the parametric tests are satisfied.

OTHER STATISTICAL TESTS

In this section we have presented certain basic statistical tests that may be of use in analyzing experimental data drawn from queueing systems. However, there are many other tests that are also of general interest, although their treatment is beyond the scope of this text.

We have discussed tests of hypothesis about a single population mean and the means of two populations. There are instances where we might wish to compare several means simultaneously. For example, suppose that we were concerned with whether or not the mean daily arrival rate of customers was the same on each day of the week. Assuming a five-day week, we would require a test for the null hypothesis

$$H_0: \quad m_1 = m_2 = m_3 = m_4 = m_5$$

This problem can be resolved through a testing procedure known as the analysis of variance (Bennett and Franklin, 1954; Brownlee, 1960; Edwards, 1950; Graybill, 1961). While the analysis of variance is a parametric testing procedure, requiring assumptions of normality and homogeneity of variance, nonparametric analogs are also available when these assumptions are seriously violated.

In a previous section of this chapter we discussed estimation of parameters that are dependent on one or more independent variables. Our concern was to define a relationship of the form

$$\hat{Y} = \hat{\beta}_0 + \hat{\beta}_1 X_1 + \hat{\beta}_2 X_2 + \cdots + \hat{\beta}_m X_m \qquad (8.118)$$

which was used as an estimating equation for

$$E(Y) = \beta_0 + \beta_1 X_1 + \beta_2 X_2 + \cdots + \beta_m X_m \qquad (8.119)$$

The adequacy of an approximating equation such as that given in (8.118) can be assessed through the use of the index of multiple determination and a test for lack of fit (Graybill, 1961; Ostle, 1963). The index of multiple determination, usually denoted by R^2, provides an estimate of the proportion of the total variation in Y explained by the estimating regression equation. The test for lack of fit determines whether or not the fitted regression equation adequately explains the variation in Y; that is, if the lack-of-fit test indicates that the equation does not fit the data, then one might attempt to develop another regression equation that better represents the variation in Y. On the other hand, if the test for lack of fit indicates that the regression equation adequately explains the variation in Y, then one can conclude that any unexplained variation is due to random error. Ideally, one would attempt to find a regression equation that adequately explains the variation in the dependent variable and that has an index of determination near unity. However, in many practical situations we must be satisfied with a regression equation that satisfies only one of these criteria.

Just as we may be interested in comparing more than two means, we may also be concerned with comparing more than two variances. Bartlett's test (Bennett and Franklin, 1954; Brownlee, 1960; Ostle, 1963) can be used for this purpose. However, Bartlett's test assumes that the underlying populations are all normal and, as in the case of the previously mentioned tests for variances, is sensitive to departures from normality (Bennett and Franklin, 1954) and when such departures are evident the analyst may wish to apply a nonparametric test (Siegel and Tukey, 1960).

CHARTING TECHNIQUES

Control Charts and Their Use

In the preceding section we presented several statistical tests, their design, and their application to the analysis of queueing systems through examples. In some situations the analyst may be concerned with the validity of a given hypothesis at a single point in time. On the other hand, one might be interested in testing a hypothesis repeatedly over time, to determine those points in time at which the hypothesis is invalid. To illustrate, suppose that the optimum operation of a single-channel queueing system depends on control of the average service rate. Specifically, let the optimum service rate be μ_0. However, if the service rate falls to $\mu_1 < \mu_0$ or increases to $\mu_2 > \mu_0$, the operation of the system becomes costly enough that the service rate should be adjusted to the desired value μ_0. Using the methods already presented, we can design a statistical test that will determine when such departures from the desired value μ_0 occur. Thus, we can use either the normal or the t-test, depending on knowledge of the variance, to monitor variation in the service rate. This is accomplished by repeatedly testing the null hypothesis

$$H_0: \quad \mu = \mu_0$$

at successive points in time. Of course, tests about the variance can be used in a similar manner to monitor changes in the variance or the relationship between two variances.

A convenient method for keeping track of the results of the monitoring effort is the control chart. Control charts were introduced by Shewhart (1926) to control the quality of manufactured product. To illustrate the construction and use of control charts, assume that the mean service time for a particular service channel is supposed to be $m_0 = 1.0$ hr. If the mean service time falls to $m_1 = 0.9$ hr or increases to $m_2 = 1.1$ hr, the cost of operation of the channel becomes excessive and operation of the service system should be interrupted to find the cause of the deviation from m_0 and correct it. The standard deviation of service time is known to be 0.2 hr. The probability of rejecting the null hypothesis that $m = m_0$ when $m = m_1$ or $m = m_2$ is to be $1 - \beta = 0.95$ and $\alpha = 0.01$.

The required sample size is approximately 72 and the limits for the test are

$$L = 0.939, \quad U = 1.061$$

Thus, we record the service time for 72 consecutive units, calculate the sample mean $\bar{x}$ and compare it with the limits L and U. If $\bar{x} < L$ or $\bar{x} > U$, adjustment of the process is called for; otherwise, operation of the service system continues uninterrupted. The process is then repeated for the next 72 units, again for the third set of 72 units, etc.

The control chart for this example is composed of a graph consisting of three horizontal lines representing m_0, L, and U as shown in Fig. 8.1. The sample mean $\bar{x}$ for each sample of 72 units is plotted on the control chart as shown in Fig. 8.1. Any time a value of $\bar{x}$ falls below L or above U on the chart, the conclusion that $m \neq m_0$ is drawn and the process is adjusted. As shown in Fig. 8.1, we would conclude that $m \neq m_0$ for samples 5 and 8. For all other samples, the process would appear to be in control, that is, $m = m_0$.

The reader will note that the limits L and U were constant for each sample in the illustrative problem presented above. This was due to the fact that the variance was assumed to be known and constant. However, this is not always the case as illustrated in the following example.

Example 8.13 Two service channels are supposed to have the same mean service times m_1 and m_2. It has been established that the variances of service time for the two channels are equal. To monitor differences in the two means, 18 consecutive units are to be selected from each channel and the service time for each recorded. The data in Table 8.5 were recorded for the first 8 tests. All data are in hours and $\alpha = 0.10$. Sketch the control chart and determine at which points the means are significantly different.

Charting Techniques

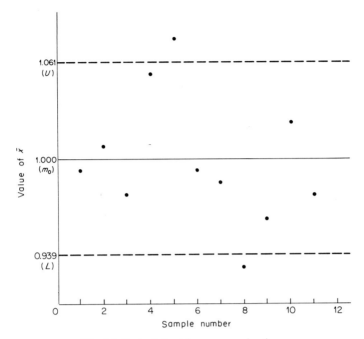

Fig. 8.1 Control chart for mean service time.

TABLE 8.5 Data for Example 8.13

Sample number	Sample mean		Sample standard deviation	
	Channel 1	Channel 2	Channel 1	Channel 2
1	0.511	0.485	0.050	0.080
2	0.519	0.533	0.070	0.060
3	0.531	0.538	0.050	0.050
4	0.521	0.508	0.040	0.030
5	0.533	0.570	0.030	0.070
6	0.531	0.537	0.080	0.040
7	0.581	0.552	0.060	0.050
8	0.514	0.544	0.070	0.080

The lower and upper limits for testing the null hypothesis

$$H_0: \quad m_1 = m_2$$

are given by (8.101) and (8.102) since the variances of service time of the two channels are assumed equal. The pooled estimate of the standard deviation s_p, the lower and upper limits, and $\bar{x}_1 - \bar{x}_2$ for each test are given in Table 8.6. The control chart and the location of $\bar{x}_1 - \bar{x}_2$ for each sample number

8. Data Analysis—Hypothesis Testing

TABLE 8.6

Sample number	s_p	L	U	$\bar{x}_1 - \bar{x}_2$
1	0.067	−0.037	0.037	0.026
2	0.066	−0.037	0.037	−0.014
3	0.050	−0.028	0.028	−0.007
4	0.036	−0.020	0.020	0.013
5	0.054	−0.030	0.030	−0.037
6	0.063	−0.035	0.035	−0.006
7	0.056	−0.031	0.031	0.029
8	0.075	−0.042	0.042	−0.030

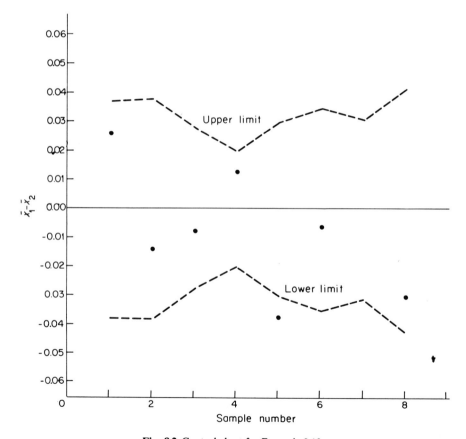

Fig. 8.2 Control chart for Example 8.13.

are shown in Fig. 8.2. Since the limits for the test are dependent on the pooled estimate of variance s_p^2, which is a random variable, the limits are not constant from one sample to the next, as shown in Fig. 8.2. For seven of the eight samples, we are led to the conclusion that $m_1 = m_2$. However, at sample 5 there is reason to believe that the mean service times for the two channels are unequal. At that point, maintenance on the two channels should have been carried out to identify the indicated cause of the discrepancy and correct the problem. However, it is possible that such a maintenance check would indicate that there is no explainable cause of the observed variation. In such a case we would conclude that the observed variation is due to random variation alone, although situations of this type should occur in only 10% of the samples.

Example 8.14 The variance of the total time spent by a unit in the system is supposed to be 10 min or less. Total time in the system can be assumed to be a random variable that has an approximate normal distribution. If the variance increases to 20 min, the hypothesis that $\sigma^2 \leq 10$ should be rejected with probability 0.90 and the service rates of the channels adjusted to reduce σ^2. The value of α is to be 0.05. Develop a control chart for σ^2. If the following sample variances were observed in the first 15 tests, each including n observed values of total time in the system, for which tests should the hypothesis that $\sigma^2 \leq 10$ be rejected?

$$s_1^2 = 10.87, \quad s_6^2 = 14.01, \quad s_{11}^2 = 16.33$$
$$s_2^2 = 9.71, \quad s_7^2 = 8.22, \quad s_{12}^2 = 11.19$$
$$s_3^2 = 7.16, \quad s_8^2 = 10.54, \quad s_{13}^2 = 8.45$$
$$s_4^2 = 12.08, \quad s_9^2 = 12.62, \quad s_{14}^2 = 9.71$$
$$s_5^2 = 15.77, \quad s_{10}^2 = 13.11, \quad s_{15}^2 = 13.27$$

Since we are interested only in increases in the variance of total time in the system, we require an upper tailed test for σ^2. The upper limit for the test is given by (8.64) and is

$$U = \frac{10}{n-1} \chi_{0.95}^2(n-1)$$

To find the sample size such that $\beta = 0.10$ when $\sigma^2 = 20$, we will use (8.71); that is, n must satisfy the relationship

$$0.5\chi_{0.95}^2(n-1) = \chi_{0.10}^2(n-1)$$

A summary of the trial and error solution for n is given in Table 8.7,

TABLE 8.7 Summary of the trial and error solution for n for Example 8.14

n	$\chi^2_{0.10}(n-1)$	$\chi^2_{0.95}(n-1)$	$0.5\chi^2_{0.95}(n-1)$
10	4.17	16.92	8.46
20	11.65	30.14	15.07
30	19.77	42.56	21.28
40	28.20	54.57	27.29
35	23.95	48.60	24.30
37	25.64	51.00	25.50
36	24.80	49.80	24.90

which shows that the required sample size is 36. Therefore,

$$U = \frac{10}{35}(49.80) = 14.23$$

Since U is a constant for all tests, the control chart for the variance σ^2 is as shown in Fig. 8.3.

Each sample includes the observed total time spent in the system by each of 36 units. Thus, each value of s_i^2, $i = 1, 2, \ldots, 15$, given in the statement of the problem is based on 36 observations. Accordingly, there are two cases

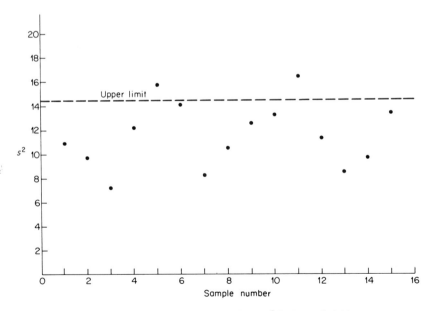

Fig. 8.3 Control chart for the variance σ^2 in Example 8.14.

Charting Techniques

where the process requires adjustment, namely, at the fifth and eleventh samples.

Extensions

Consider the control of the mean of a process where the variance is unknown. If m_0 is the value of the mean when the process is under proper control, then the lower and upper limits for testing the null hypothesis

$$H_0: \quad m = m_0$$

are given by

$$L = m_0 - \frac{s}{\sqrt{n}} t_{1-\alpha/2}(n-1), \quad U = m_0 + \frac{s}{\sqrt{n}} t_{1-\alpha/2}(n-1)$$

Suppose the control chart for the first 25 samples is as shown in Fig. 8.4. As indicated in the figure, the results shown do not indicate that the hypothesis $m = m_0$ should be rejected for any of the 25 samples. However, the reader will note that the sample mean lies above m_0 for 22 of the 25 samples. One might wonder whether or not such an occurrence is likely if, in fact, $m = m_0$.

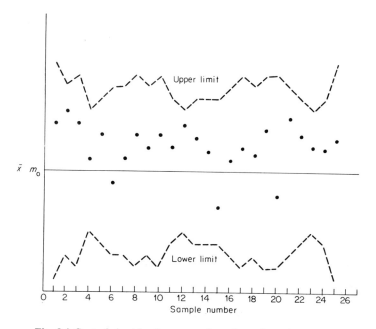

Fig. 8.4 Control chart for the mean, where the variance is unknown.

Now
$$P(\bar{x} \geq m_0 \mid m = m_0) = 0.5$$
Let k be the number of sample means falling above m_0. If $m = m_0$, then k has a binomial distribution with $p = 0.5$ and $n = 25$. Then

$P(22$ or more sample means above $m_0 \mid m = m_0)$

$= P(3$ or less sample means below $m_0 \mid m = m_0)$

$$= \sum_{k=0}^{3} \frac{25!}{k!(25-k)!} (0.5)^{25} = 0.0001$$

Thus the event observed in Fig. 8.4 is highly unlikely if $m = m_0$.

Let us consider another example. Suppose that the 25 sample means had the values shown in Fig. 8.5. Again, we have no reason to believe that

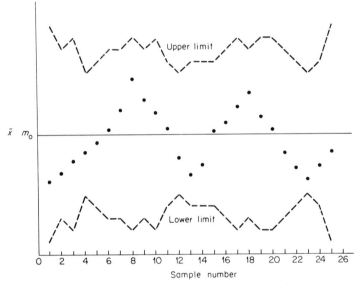

Fig. 8.5 Control chart for the mean, where the variance is unknown.

$m \neq m_0$ based on the individual hypothesis tests at each sample point. However, based on the cyclic nature of the data, we might well question the validity of the hypothesis that $m = m_0$ throughout the sampling period.

Binomial Test

As mentioned above, if m_0 is the true mean throughout N samples, then

$$P(\bar{x} \geq m_0 \mid m = m_0) = P(\bar{x} \leq m_0 \mid m = m_0) = 0.5$$

Charting Techniques

Thus, the number of sample means falling above m_0, k, is binomially distributed with parameters N and 0.5, or

$$p(k) = \frac{N!}{k!(N-k)!}(0.5)^N \tag{8.120}$$

We wish to test the null hypothesis

$$H_0: \quad m = m_0$$

Since we tend to reject this hypothesis if k is either too small or too large, a two-tailed test is required. Hence, we must find limits L and U such that

$$P(k < L \mid N, m = m_0) = \frac{\alpha}{2} \tag{8.121}$$

$$P(k > U \mid N, m = m_0) = \frac{\alpha}{2} \tag{8.122}$$

Since k is integer valued, we may not be able to find values of L and U such that (8.121) and (8.122) are satisfied exactly. Thus, we seek integer-valued limits L and U that satisfy (8.121) and (8.122) as nearly as possible. Equations (8.121) and (8.122) are then expressed as

$$\sum_{i=0}^{L-1} \frac{N!}{L!(N-L)!}(0.05)^n \simeq \frac{\alpha}{2} \tag{8.123}$$

$$\sum_{i=U+1}^{N} \frac{N!}{L!(N-L)!}(0.05)^N \simeq \frac{\alpha}{2} \tag{8.124}$$

If $L \leq k \leq U$, the null hypothesis is accepted and is rejected otherwise.

Example 8.15 After the selection of 10 samples and the calculation of $\bar{x}$ for each, the hypothesis that $m = m_0$ is to be tested using the binomial test if each of the 10 values of $\bar{x}$ fall within the control limits. The test is to be run using $\alpha = 0.10$. Develop the limits L and U for the test. If 6 of the 10 values of $\bar{x}$ fall above m_0, should the hypothesis that the mean is m_0 be accepted for all 10 samples?

Now

$$p(i) = \frac{10!}{i!(10-i)!}(0.5)^{10} \simeq 0.001 \frac{10!}{i!(10-i)!}$$

The limits L and U are determined by trial and error as shown in Table 8.8. The resulting values of L and U are

$$L = 3, \quad U = 7$$

TABLE 8.8 Summary of the determination of L and U by trial and error for Example 8.15

L	U	$\sum_{i=0}^{L-1} \binom{10}{i}(0.5)^N$	$\sum_{i=U+1}^{N} \binom{10}{i}(0.5)^N$
1	9	0.001	0.001
2	8	0.011	0.011
3	7	0.056	0.056
4	6	0.170	0.170

Since $k = 6$ and $L < 6 < U$, the hypothesis that $m = m_0$ throughout the 10 samples is accepted.

When $p = 0.5$, the binomial probability mass function is symmetric about its mean, Np. In such cases

$$U = N - L$$

Where a control chart is used for the mean, the assumption that $p = 0.5$ is usually valid due to the central limit theorem. However, such an assumption does not hold when the control chart is used to detect departures of the variance σ^2 from the hypothesized value σ_0^2.

Example 8.16 A control chart is to be used to monitor the validity of the hypothesis

$$H_0: \quad \sigma^2 = \sigma_0^2$$

where $\sigma_0^2 = 5$. The value of s^2 is plotted on the control chart based on samples of size 13. After 12 values of s^2 are plotted, the number of values s^2 falling above σ_0^2, k, is recorded. If k is either too large or too small the hypothesis that $\sigma^2 = \sigma_0^2$ throughout the sampling period is rejected. The limits for this test are to be L and U such that $\alpha = 0.05$. Define the limits L and U.

If we assume that $(n - 1)S^2/\sigma_0^2$ is chi-square distributed with $n - 1$ degrees of freedom, then

$$P(s^2 > \sigma_0^2 \mid \sigma^2 = \sigma_0^2) = P\left[\frac{(n-1)s^2}{\sigma_0^2} > (n-1) \mid \sigma^2 = \sigma_0^2\right]$$

$$= P\left[\frac{12s^2}{\sigma_0^2} > 12 \mid \sigma^2 = \sigma_0^2\right] \simeq 0.45$$

Hence L and U are defined such that

$$\sum_{i=0}^{L-1} \frac{12!}{i!(12-i)!} (0.45)^i (0.55)^{12-i} = 0.025$$

$$\sum_{i=U+1}^{12} \frac{12!}{i!(12-i)!} (0.45)^i (0.55)^{12-i} = 0.025$$

The trial and error solution for L and U is given in Table 8.9. As shown in this table, the values of L and U are given by

$$L = 2, \quad U = 10$$

TABLE 8.9 Trial and error solution for L and U for Example 8.16

L	U	$\sum_{i=0}^{L-1} \frac{12!}{i!(12-i)!}(0.45)^i(0.55)^{12-i}$	$\sum_{i=U+1}^{12} \frac{12!}{i!(12-i)!}(0.45)^i(0.55)^{12-i}$
1	11	0.005	0.010
2	10	0.029	0.023
3	9	0.095	0.049

In Examples 8.15 and 8.16 the values of L and U were found using the cumulative distribution function of the binomial random variable. For large values of n and values of p near 0.5, the distribution function of the binomial random variable can be accurately approximated by the distribution function of the normal random variable (Brownlee, 1960). This approximation can be quite useful since the computational effort required to calculate values of the binomial distribution function can be extensive.

Runs Test

The binomial test presented in the preceding section is designed to test the hypothesis that a set of observations is randomly distributed about a given value. However, this test would not be appropriate in attempting to detect behavior such as that illustrated in Fig. 8.5. The pattern of variation illustrated in Fig. 8.5 would suggest that successive values of $\bar{x}$ are not independent. We can test for this type of behavior by using the runs test.

Consider the sequence of fifteen numbers given by

28 34 41 16 10 12 17 29 30 37 44 32 27 26 18

A run up is defined as a sequence of numbers in which each number is followed by a larger number. A run down is a sequence of numbers in which each number is followed by a smaller number. To identify runs, we record a plus for each number preceded by a smaller number and a minus for each number preceded by a larger number, starting with the second number in the sequence. Thus, for the above sequence we have

$$+ + - - + + + + + + - - - -$$

A run up is then an unbroken succession of pluses and a run down is an unbroken succession of minuses. Therefore, for the sequence given above we have two runs up and two down. The total number of runs k in a sequence of N numbers is the sum of the runs up and the runs down.

In a sequence of N values, the minimum number of runs is one and the maximum number of runs is $N - 1$. Either too many or too few runs may indicate that the variable of concern is not randomly distributed. For example, consider the control chart shown in Fig. 8.6. In this case there is only

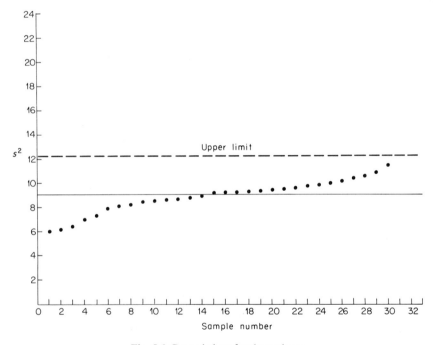

Fig. 8.6 Control chart for the variance.

one run, an upward run, indicating that the variance is increasing with each sample. On the other hand, if we consider the results shown in Fig. 8.7 for the same problem, we note that the sample variance increases and decreases successively, again indicating a predictable or nonrandom pattern of variation.

Charting Techniques

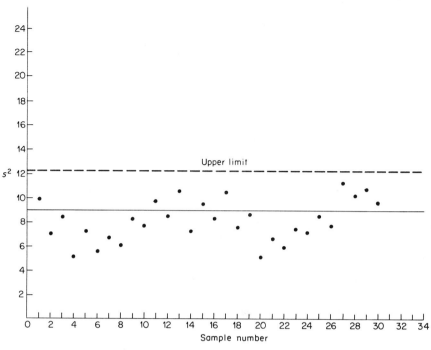

Fig. 8.7 Control chart for the variance.

For $N > 20$, the number of runs k in a sequence has an approximate normal distribution with mean m_k and variance σ_k^2, given by

$$m_k = \frac{2N - 1}{3} \tag{8.125}$$

$$\sigma_k^2 = \frac{16N - 29}{90} \tag{8.126}$$

Thus to test the hypothesis that the values of the variable of interest are randomly distributed using the runs test requires lower and upper limits as follows:

$$L = \frac{2N - 1}{3} - \frac{16N - 29}{90} Z_{1-\alpha/2} \tag{8.127}$$

$$U = \frac{2N - 1}{3} + \frac{16N - 29}{90} Z_{1-\alpha/2} \tag{8.128}$$

Example 8.17 Two service channels are available to arriving customers. It is hypothesized that arrivals choose the channel they will enter in a random

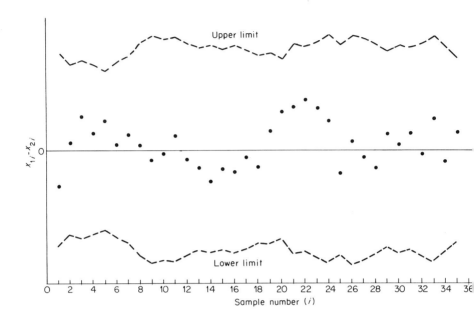

Fig. 8.8 Control chart for the difference between two means for Example 8.17.

manner. For a period of 35 days, a study was conducted recording the average number of arrivals per hour for each channel. Let $\bar{x}_{1i}$ and $\bar{x}_{2i}$ be the sample mean for the number of arrivals on the ith day for the first and second channels, respectively. Control limits for $\bar{x}_{1i} - \bar{x}_{2i}$ were calculated for $i = 1, 2, \ldots, 35$, with the results shown in Fig. 8.8. From the information given in Fig. 8.8, would you conclude that there is a nonrandom pattern of variation to the difference between the number of arrivals per hour to the two channels? Let $\alpha = 0.10$.

From the statement of the problem, $N = 35$ and

$$m_k = \frac{2(35) - 1}{3} = 23, \quad \text{and} \quad \sigma_k^2 = \frac{16(35) - 29}{90} = 5.9$$

Hence, the lower and upper limits for the test are

$$L = 23 - \sqrt{5.9}\, Z_{0.95} = 18.99, \quad U = 23 + \sqrt{5.9}\, Z_{0.95} = 27.01$$

Recording a plus for any i for which $\bar{x}_{1i} - \bar{x}_{2i} > \bar{x}_{1,i-1} - \bar{x}_{2,i-1}$ and a minus for any i for which $\bar{x}_{1i} - \bar{x}_{2i} < \bar{x}_{1,i-1} - \bar{x}_{2,i-1}$, $i = 2, 3, \ldots, 35$, we have

$$+ + - + - + - - + + - - - - + - + - + + + + $$
$$- - - + - - + - + - + - +$$

There are 12 runs up and 11 down, and the total number of runs k is 23. Since $18.99 < k < 27.01$, we cannot conclude that the pattern of variation of $\bar{x}_{1i} - \bar{x}_{2i}$ is nonrandom.

SUMMARY

The purpose of this chapter has been to present some of the statistical testing procedures that can be quite useful in analyzing data describing a queueing system. We have presented those testing procedures that should be most generally useful to the analyst. The reader will note that we have placed emphasis on the design of the statistical tests presented in this chapter. The importance of a proper experimental design lies in the fact that it allows the analyst to specify the reliability of the tests a priori. For a comprehensive introductory treatment of statistical methods the reader should see Ostle (1963) or Bowker and Lieberman (1959). An introductory discussion of nonparametric methods is presented in Siegel (1956). An excellent discussion of the power of statistical tests and their associated design is presented in Hald (1952).

REFERENCES

Abramowitz, M., and Stegun, I. A. (eds.), *Handbook of Mathematical Functions*, Appl. Math. Ser. **54**, Washington: National Bureau of Standards, 1964.
Bennett, C. A., and Franklin, N. L., *Statistical Analysis in Chemistry and the Chemical Industry*. New York: Wiley, 1954.
Bowker, A. H., and Lieberman, G. J., *Engineering Statistics*. Englewood Cliffs, New Jersey: Prentice-Hall, 1959.
Brownlee, K. A., *Statistical Theory and Methodology in Science and Engineering*. New York: Wiley, 1960.
Dixon, W. J., and Massey, F. J., *Introduction to Statistical Analysis*. New York: McGraw-Hill, 1957.
Duncan, A. J., *Quality Control and Industrial Statistics*. Homewood, Illinois: Irwin, 1965.
Edwards, A. L., *Experimental Design in Psychological Research*. New York: Rinehart, 1950.
Freeman, H., *Introduction to Statistical Inference*. Reading, Massachusetts: Addison-Wesley, 1963.
Graybill, F. A., *An Introduction to Linear Statistical Models*, Volume I. New York: McGraw-Hill, 1961.
Hald, A., *Statistical Theory with Engineering Applications*. New York: Wiley, 1952.
Jahnke, E., and Emde, F., *Tables of Functions*. New York: Dover, 1945.
Kamenta, J., *Elements of Econometrics*, New York: MacMillan, 1971.
Ostle, B., *Statistics in Research*. Ames, Iowa: Iowa State Univ. Press, 1963.
Shewhart, W. A., *Finding Causes of Quality Variations*. Manufacturing Industries, 1926.
Siegel, S., *Nonparametric Statistics for the Behavioral Sciences*. New York: McGraw-Hill, 1956.

Siegel, S., and Tukey, J., A nonparametric sum of ranks procedure for relative spread in unpaired samples, *J. Amer. Statist. Assoc.* **55**, 1960.

Wadsworth, G. P., and Bryan, J. G., *Introduction to Probability and Random Variables.* New York: McGraw-Hill, 1960.

PROBLEMS

1. Derive Eq. (8.19).
2. Derive Eq. (8.32).
3. Derive Eq. (8.41).
4. Derive Eq. (8.72).
5. Derive Eqs. (8.87) and (8.88).
6. A mathematical model has been developed that estimates the mean daily cost of operation of a particular multiple-channel queueing system. The optimum conditions for operation of the system have been identified and implemented. However, the optimum operating conditions have been shown to be sensitive to the daily arrival rate of customers. The arrival rate is currently 1000 units/day and the standard deviation of arrivals per day is 100. If the arrival rate changes by as much as 25 units/day, the current "optimum solution" is no longer optimal. If such a change is to be detected with probability 0.90 and $\alpha = 0.10$, how many days' data will be necessary before one could test the null hypothesis that the arrival rate has not changed?

7. The weekly cost of operation of a service facility has been predicted to be $6835. The following weekly costs have been observed:

Week	Cost ($)
1	6017
2	6043
3	7200
4	6411
5	6559
6	7275
7	7528
8	6345
9	6257
10	6420

If $\alpha = 0.10$, should the prediction be accepted as satisfactory?

8. Based on the information in Problem 7, how many observations would be necessary to identify an error in the prediction of $50 with probability 0.95?

9. In order for a single-channel queueing system to operate with stability, it is necessary that $\lambda < \mu$, where λ is the arrival rate and μ the service rate. Ten periods of time of length one hour each were chosen at random. During each one-hour period the service channel operated continuously. The number of arrivals and services during these ten periods are given below. It is hypothesized that the service rate is greater than the arrival rate. Would you accept this hypothesis if $\alpha = 0.10$?

Hour	Arrivals	Services
1	22	24
2	22	27
3	20	19
4	25	22
5	23	24
6	19	22
7	27	27
8	27	21
9	15	26
10	19	25

10. The mean arrival rates λ_1 and λ_2 for two service channels are to be compared. The standard deviations of both arrival rates are known to be 25/hr. The null hypothesis

$$H_0: \quad \lambda_1 = \lambda_2$$

is to be tested using 20 observations on each arrival rate and $\alpha = 0.05$. If the null hypothesis should be rejected with probability 0.80 if $|\lambda_1 - \lambda_2| \geq 5$, is the design specified appropriate?

11. Parts are scheduled to arrive at an assembly station at a mean rate λ of 60/hr. A preliminary analysis indicates that the standard deviation is approximately 10/hr. A t-test is to be used to test the null hypothesis

$$H_0: \quad \lambda = 60$$

If $|\lambda - 60| \geq 10$, the null hypothesis should be rejected with probability 0.90. If $\alpha = 0.10$, how many hours of observation are necessary to conduct the test?

12. The following data have been collected on the number of units reneging per hour from two different queueing channels. Letting $\alpha = 0.05$, determine whether or not the variances of the number of units reneging from the two channels are significantly different.

Channel 1	Channel 2
5	8
3	1
8	4
4	0
6	3
2	7
7	
6	
1	
4	

13. It has been noted that the efficiency of a particular operator decreases as the variability of service time increases. A standard deviation of 3 min is considered acceptable. Given the following service times, test the hypothesis

$$H_0: \quad \sigma^2 \leq 9$$

using α = 0.15:

2.5, 7.2, 3.4, 3.9, 1.4, 8.5, 7.7, 6.2, 9.3, 10.2

14. In Problem 12, determine if there is a significant difference between the mean number of customers reneging per hour for channels 1 and 2 if α = 0.15.

15. In Problem 11, a test is to be conducted to determine whether or not the variance of arrivals per hour is greater than 60. Specifically, if the variance is as great as 80, the hypothesis that the variance is 60 or less is to be rejected with probability 0.80. If α = 0.10, how many hours of data will be necessary to conduct the test?

16. In Problem 13, what is the probability that the null hypothesis will be rejected if $\sigma^2 = 16$?

17. In Problem 9, suppose that the null hypothesis should be rejected with probability 0.95 if σ^2 is as large as 16. How many observations are necessary to conduct the test under the specified criteria?

18. The time required to serve customers has been shown to be normally distributed. The standard deviation of service time is 0.4 hr for the average worker. The following data have been collected on service time, in hours, for a new employee.

Service time (hours)	
0.98	1.23
1.32	0.84
1.40	1.04
1.52	0.76
1.53	0.68
0.76	0.52
1.08	0.64
1.20	1.28
1.48	0.44
1.45	1.12
0.56	1.72

If α = 0.15, would you accept the hypothesis that the new employee is an average worker with respect to variability?

19. Given the data in Problem 18, would you conclude that the observations are randomly distributed about the sample mean using the runs test and letting α = 0.10?

20. The time to perform a given task is exponentially distributed. The mean time to complete the service is supposed to be 8 min. The hypothesis that the mean is actually 8 min is to be tested using three observations:

(a) Find the exact distribution of the sample mean.

(b) If α = 0.05, develop exact limits L and U for the test of the exact distribution in (a).

(c) Compare the limits in (b) to those that would be used if the sample mean were assumed to be normally distributed.

21. The distribution of fabrication time for a particular product is normal with mean m and variance σ^2. Derive the distribution of the sample standard deviation.

22. Develop limits for the two-tailed test for the standard deviation using the result given in Problem 21.

Problems

23. Four statistical tests are to be run on a given process each at $\alpha = 0.10$. What is the probability that a type I error will be made somewhere in the analysis if the null hypothesis is, in fact, true in all four cases?

24. The daily costs of lost customers, provision of service, and waiting time have been predicted to be $433, $5836 and $566, respectively. The following ten days of data have been collected:

	Cost ($)		
Day	Lost customers	Service	Waiting time
1	427	5100	490
2	433	5400	210
3	410	6300	490
4	451	5400	560
5	439	5700	420
6	435	6000	840
7	408	6000	1120
8	435	5700	210
9	437	5400	420
10	440	5700	280

The probability of making a type I error somewhere is to be 0.10. Test the hypothesis that each of the predictions is satisfactory.

25. For the situation discussed in Problem 11, plot the power of the test $(1 - \beta)$ as a function of λ.

26. Using the runs test, test the hypothesis that the sample means given in Fig. 8.4 are randomly distributed, where $\alpha = 0.01$.

27. Using the binomial test, test the hypothesis that the mean of the process is m_0 for the 25 samples shown in Fig. 8.4. Let $\alpha = 0.15$.

28. Using the binomial test, determine whether or not the data shown in Fig. 8.8 indicate that $E(\bar{X}_1 - \bar{X}_2) = 0$ over the sampling period. Use $\alpha = 0.05$.

29. The number of units processed per day is assumed to be randomly distributed about its mean. The following data were collected on this system and represent the number of units processed for a twenty-day period.

Day:	1	2	3	4	5	6	7	8	9	10
Units processed:	59	12	19	5	59	58	83	18	36	61

Day:	11	12	13	14	15	16	17	18	19	20
Units processed:	47	24	41	42	98	23	67	84	43	29

Using $\alpha = 0.05$, would you conclude that the number of units processed per day is randomly distributed about the mean?

Chapter 9

SIMULATION OF QUEUEING SYSTEMS

INTRODUCTION

Simulation, in the general sense, can be thought of as performing experiments on a model of a system. Generally then, we can consider the manipulation carried out on the mathematical models developed in preceding chapters to be simulations of the systems represented by those models. The difference between simulation as we shall use the term in this chapter and simulation as presented in preceding chapters lies in the type of model used to represent the system under study. We will refer to these models as simulation models to distinguish them from mathematical models.

Before we discuss simulation models and their development, we should give some justification for the use of simulation as a tool for analyzing queueing systems. First, if we can model a queueing system mathematically, why should we consider simulation at all? In general, one should not simulate a system that he can model mathematically. However, when the system to be analyzed is complex, involving many interactive elements, mathematical analysis may be beyond the capabilities of the analyst involved, or any analyst for that matter. System complexity does not usually present an insurmountable problem when the analysis is to be carried out through simulation.

The preceding discussion brings to light the principle advantages of simulation. The first is versatility; that is, simulation models can be

Simulation Modeling

developed for a wide variety of systems, including systems that cannot be properly modeled mathematically. The second advantage rests on the fact that successful simulation modeling can usually be achieved with less background in mathematical analysis and probability theory than is generally required for a similar analysis through mathematical modeling.

Although ease of application is a major advantage of simulation, it can also be a disadvantage. Many analysts apply simulation rather indiscriminately, using it where mathematical techniques can be easily applied. The problem with such applications of simulation is that systems analysis through simulation often requires a significant amount of time on a digital computer and is therefore expensive.

SIMULATION MODELING

Perhaps the best way to explain modeling through simulation is to compare the operation of the simulation model with that of the system simulated. Consider a simple single-channel service system where units arrive and are serviced one at a time, on a first come–first served basis, $(G\,|\,G\,|\,1):(FCFS\,|\,\infty\,|\,\infty)$. We start observing the system at a random point in time at which there are n units in the system. The status of the system will remain unchanged until a service or arrival occurs. Let us define the following notation:

t = present time

t_δ = duration of time for which we intend to observe the system

τ_s = service time

t_s = time of next service as measured from $t = 0$
 $= t + \tau_s$

τ_a = time until next arrival

t_a = time of next arrival as measured from $t = 0$
 $= t + \tau_a$

t_e = time of next event
 $= \min(t_s, t_a, t_\delta)$

If we assume that the population of arrivals is infinite, then each arrival will be followed by another. Thus each time an arrival occurs, we can consider that nature generates a time interval until the occurrence of the next arrival. Similarly, each time a unit enters the service channel, a time interval for that service is generated by nature. As we observe the system, the next event that occurs in the future will be either a service completion or an arrival, except in the case where both the next service and the next arrival occur after the end of the period of observation. The operation of the system and the associated changes in system status can be summarized as shown in the flowchart in Fig. 9.1.

Figure 9.1 indicates that observation of the system occurs only at the time of an arrival, the time of a service completion, and at the termination of the period of observation. However, for the purpose of computing pertinent statistics regarding the system, observation is necessary at these points only, since these points in time include all points at which system status changes occur. For example, let t_i be the time of occurrence of the ith event, N_{si} the number in the system at the ith event, and N_{qi} the number in the queue at the ith event, $i = 0, 1, 2, \ldots, m$, where t_0 is the start of the period of observation ($t_0 = 0$) and $t_m = t_s$. Therefore, the period of observation includes a combined total of $m - 1$ arrivals and services. Then

$$\text{average number in system} = \frac{1}{t_m} \sum_{i=0}^{m-1} N_{si}(t_{i+1} - t_i) \tag{9.1}$$

$$\text{average number in queue} = \frac{1}{t_m} \sum_{i=0}^{m-1} N_{qi}(t_{i+1} - t_i) \tag{9.2}$$

Let k be the number of units spending time in the system between t_0 and t_m. Then

$$\text{average time in system} = \frac{1}{k} \sum_{i=0}^{m-1} N_{si}(t_{i+1} - t_i) \tag{9.3}$$

$$\text{average time in queue} = \frac{1}{k} \sum_{i=0}^{m-1} N_{qi}(t_{i+1} - t_i) \tag{9.4}$$

To calculate the probability that there are i units in the system, we record the period of time T_{si} for which there were i units in the system. Then

$$P(i \text{ units in system}) = \frac{T_{si}}{t_m} \tag{9.5}$$

If T_{qi} is the total time for which there are i units in the queue, the probability that there are i units in the queue is given by

$$P(i \text{ units in queue}) = \frac{T_{qi}}{t_m} \tag{9.6}$$

The statistics given in (9.1)–(9.6) as well as other pertinent information can be obtained directly from data recorded during the period of observa-

Simulation Modeling

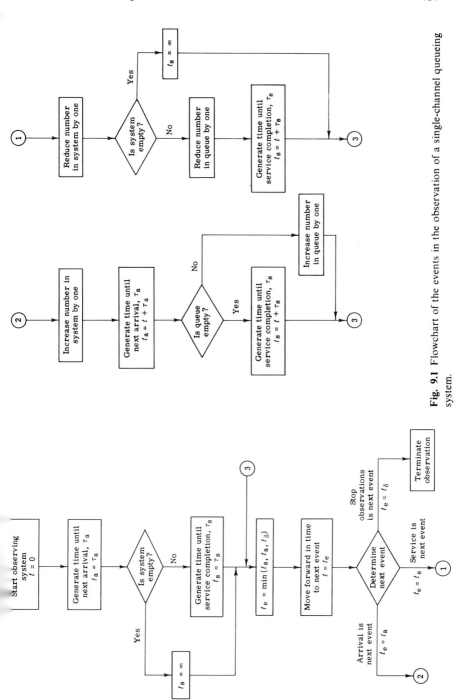

Fig. 9.1 Flowchart of the events in the observation of a single-channel queueing system.

tion. To obtain these data, it is necessary to observe the system only at those times at which changes in system status occur and at the beginning and end of the observation period.

To simulate the system described above (or any system for that matter), those events generated by nature must be generated synthetically; that is, we must capture the essential characteristics of natural phenomena artificially. Therefore, to simulate the system described in Fig. 9.1 we must generate service times and interarrival times synthetically but in a manner such that the generated times are representative of those that would be found in the physical system. For example, if service time is exponentially distributed with parameter μ and interarrival time is exponentially distributed with parameter λ, then we must devise a means of generating these random variables artificially. If this can be accomplished, then the simulated system will be representative of the physical system.

MONTE CARLO METHOD

Let X be a discrete random variable with probability mass function

$$p(x) = \begin{cases} \dfrac{x}{10}, & x = 1, 2, 3, 4 \\ 0, & \text{otherwise} \end{cases} \quad (9.7)$$

The cumulative distribution function of X, $F(x)$, is given by

$$F(x) = \begin{cases} 0.0, & x < 1 \\ 0.1, & 1 \leq x < 2 \\ 0.3, & 2 \leq x < 3 \\ 0.6, & 3 \leq x < 4 \\ 1.0, & 4 \leq x < \infty \end{cases} \quad (9.8)$$

$F(x)$ is shown graphically in Fig. 9.2. The values that $F(x)$ can assume range between 0 and 1 inclusive. Now suppose that we were to select a value of $F(x)$ between 0.00 and 1.00 in a manner such that each possible value of $F(x)$ had an equal and independent chance of occurrence. Corresponding to each value of $F(x)$ is a unique value of X as indicated in (9.8); that is, if $0.00 \leq F(x) \leq 0.10$, then the associated value of X is 1. Similarly, if $0.11 \leq F(x) \leq 0.30$, the associated value of X is 2. According to our procedure for selecting values of $F(x)$, the probability of selecting a value of $F(x)$ between 0.00 and 0.10 inclusive is 0.1. Hence, the probability of obtaining a value of $X = 1$ is 0.10. By a similar argument, the probability of obtaining a value of $F(x)$

Monte Carlo Method

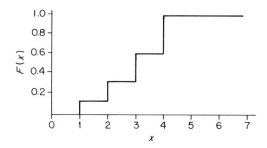

Fig. 9.2 Cumulative distribution of X.

between 0.11 and 0.30, and therefore a value of $X = 2$, is 0.20. Continuing in this fashion, the probability of obtaining a value of $X = 3$ is 0.30 and the probability of obtaining a value of $X = 4$ is 0.40.

If we were to generate a large number of values of X in the manner just described, we would find that of the total number generated approximately 10% would be 1's, 20% would be 2's, 30% would be 3's, and 40% would be 4's. Thus we have developed a method for generating random variables from the distribution specified by the probability mass function in (9.7). Randomly selected values of $F(x)$ can be obtained from a table of random numbers. To illustrate the process of generating random variables, consider the following example.

Example 9.1 Generate 100 values of X, where X has the probability mass function given in (9.7). Compare the observed occurrence of 1's, 2's, 3's, and 4's with what would be expected from the probability mass function of X.

The results of the generation process are summarized in Table 9.1. From this table it can be seen that 12 ones, 23 twos, 26 threes, and 39 fours were generated. The expected number of ones, twos, threes, and fours is 10, 20, 30, and 40, respectively, based on the probability mass function of X. In general, as the number of values of the random variable generated is increased, the observed probability distribution will more closely approximate the theoretical distribution; that is, if we were to increase the number of trials in this experiment from 100 to 1000, we would expect the proportion of 1's, 2's, 3's, and 4's more closely to approximate the expected proportions given by the probability mass function of X.

The process just described for generating random variables is called the Monte Carlo method. Essentially, the Monte Carlo method of generating random variables consists of establishing a relationship between the random variable to be generated and one or more random variables that are uniformly and independently distributed on the unit interval. The reader will notice that this is precisely what we did in Example 9.1; that is, we first

TABLE 9.1 Generation of 100 values of X, where X has the probability mass function given in Eq. (9.7)

Trial number	Random number	F(x)	X	Trial number	Random number	F(x)	X	Trial number	Random number	F(x)	X	Trial number	Random number	F(x)	X
1	86	0.86	4	26	17	0.17	2	51	89	0.89	4	76	38	0.38	3
2	21	0.21	2	27	23	0.23	2	52	64	0.64	4	77	71	0.71	4
3	49	0.49	3	28	72	0.72	4	53	72	0.72	4	78	24	0.24	2
4	22	0.22	2	29	30	0.30	2	54	07	0.07	1	79	79	0.79	4
5	55	0.55	3	30	20	0.20	2	55	13	0.13	2	80	20	0.20	2
6	04	0.04	1	31	79	0.79	4	56	59	0.59	2	81	61	0.61	4
7	61	0.61	4	32	58	0.58	3	57	18	0.18	2	82	54	0.54	3
8	55	0.55	3	33	13	0.13	2	58	31	0.31	3	83	08	0.08	1
9	92	0.92	4	34	07	0.07	1	59	29	0.29	2	84	31	0.31	3
10	88	0.88	4	35	06	0.06	1	60	63	0.63	4	85	72	0.72	4
11	53	0.53	3	36	68	0.68	4	61	69	0.69	4	86	46	0.46	3
12	19	0.19	2	37	03	0.03	1	62	56	0.56	3	87	24	0.24	2
13	42	0.42	3	38	04	0.04	1	63	48	0.48	3	88	32	0.32	3
14	20	0.20	2	39	68	0.68	4	64	06	0.06	1	89	21	0.21	2
15	51	0.51	3	40	20	0.20	2	65	73	0.73	4	90	80	0.80	4
16	76	0.76	4	41	83	0.83	4	66	91	0.91	4	91	21	0.21	2
17	08	0.08	1	42	92	0.92	4	67	70	0.70	4	92	63	0.63	4
18	98	0.98	4	43	76	0.76	4	68	10	0.10	1	93	91	0.91	4
19	46	0.46	3	44	84	0.84	4	69	38	0.38	3	94	85	0.85	4
20	42	0.42	3	45	20	0.20	2	70	64	0.64	4	95	82	0.82	4
21	40	0.40	3	46	99	0.99	4	71	67	0.67	4	96	52	0.52	3
22	49	0.49	3	47	52	0.52	3	72	69	0.69	4	97	04	0.04	1
23	59	0.59	3	48	92	0.92	4	73	29	0.29	2	98	87	0.87	4
24	32	0.32	3	49	14	0.14	2	74	39	0.39	3	99	19	0.19	2
25	58	0.58	3	50	08	0.08	1	75	87	0.87	4	100	73	0.73	4

Generation of Uniformly Distributed Random Numbers

selected a random number, placed a decimal point before the first digit of the number, thus restricting it to values between zero and unity, and established a relationship between the random number and a value of the random variable to be generated. In Example 9.1 this relationship was established in a tabular fashion through (9.8). The relationship through which values of the random variable to be generated are associated with values of uniformly distributed random numbers is called a process generator.

GENERATION OF UNIFORMLY DISTRIBUTED RANDOM NUMBERS

As indicated in the preceding discussion, one must first select a uniformly distributed random number before a value of the random variable of interest can be generated. If the simulation is to be carried out by hand, as in the case of Example 9.1, random numbers can be drawn from tables of random numbers (see, for example, Rand Corporation, 1955). However, since most simulations of the type discussed here are executed on a digital computer, it is usually necessary to have at hand a routine through which these numbers can be obtained on such a machine. FORTRAN function subprograms for the generation of uniformly distributed random numbers between 0.00 and 1.00 are given in Figs. 9.3 and 9.4. For the random number generator given

```
      FUNCTION RANDU (IX, IY)
      IY = IX*03125
      IF (IY) 5, 6, 6
    5 IY = IY + 2**35
    6 YFL = IY
      RANDU = YFL*2.0**(-35)
      IX = IY
      RETURN
      END
```

```
      FUNCTION RANDU (IX, IY)
      IY = IX*65539
      IF (IY) 5, 6, 6
    5 IY = IY + 2147483647 + 1
    6 YFL = IY
      RANDU = YFL*.4656613E-9
      IX = IY
      RETURN
      END
```

Fig. 9.3 FORTRAN IV random-number generator for a 35-bit word binary computer.

Fig. 9.4 System/360 random-number generator.

in Fig. 9.3, an initial value for IX must be defined and should be an odd integer containing five digits. Succeeding values of IX are defined as shown in Fig. 9.3. For the random-number generator given in Fig. 9.4, the initial value of IX must also be an odd integer but may contain up to nine digits. In both cases, the random number returned by the subprogram lies between zero and one.

Whatever method is used to generate random numbers, there are certain properties that those numbers should possess (Schmidt and Taylor, 1970). First, a sequence of generated numbers should be uniformly distributed on the interval zero to one. This property may be checked through a goodness-of-fit test. Second, the numbers occurring in the sequence should be randomly arranged. Runs tests and tests for autocorrelation are frequently used to determine whether or not a sequence of numbers is randomly arranged. Finally, it is sometimes necessary to determine whether or not the digits occurring within the individual random numbers are randomly arranged. Statistical tests such as the gap, poker, and Yule tests are often used to test this property.

PROCESS GENERATION OF CONTINUOUS RANDOM VARIABLES WITH KNOWN DENSITY FUNCTIONS

Let $F(x)$ be the cumulative distribution function of the continuous random variable X and r a uniformly distributed random number on the interval $(0, 1)$. As mentioned earlier, we choose values of $F(x)$ at random. This random selection can be obtained by setting $F(x) = r$. Therefore,

$$r = F(x) \quad (9.9)$$

Solving for x in terms of r, we have

$$x = \phi(r) \quad (9.10)$$

and $\phi(r)$ is the desired process generator for X.

Example 9.2 Let X be exponentially distributed with parameter μ. Determine a process generator for X.

The distribution function of X is given by

$$F(x) = \int_0^x \mu e^{-\mu y}\, dy = 1 - e^{-\mu x} \quad (9.11)$$

Setting $F(x)$ equal to the random number r, we have

$$r = 1 - e^{-\mu x}, \quad \text{and} \quad x = -\frac{1}{\mu}\ln(1 - r) \quad (9.12)$$

is the desired process generator for an exponential random variable with parameter μ.

Example 9.3 Let X be a Weibull random variable with probability density function given by

$$f(x) = \frac{a}{b-c}\left(\frac{x-c}{b-c}\right)^{a-1}\exp\left[-\left(\frac{x-c}{b-c}\right)^a\right], \quad c < x < \infty \quad (9.13)$$

Derive a process generator for X.

Process Generation of Continuous Random Variables

The distribution function for X is given by

$$F(x) = \int_c^x \frac{a}{b-c}\left(\frac{y-c}{b-c}\right)^{a-1} \exp\left[-\left(\frac{y-c}{b-c}\right)^a\right] dy$$

$$= 1 - \exp\left[-\left(\frac{x-c}{b-c}\right)^a\right], \quad c < x < \infty \quad (9.14)$$

Then

$$r = 1 - \exp\left[-\left(\frac{x-c}{b-c}\right)^a\right], \quad \text{and} \quad x = c + (b-c)[-\ln(1-r)]^{1/a}$$

$$(9.15)$$

The reader should note that if r is uniformly distributed on the interval $(0, 1)$, then so is $1 - r$. Therefore, it is customary to replace $1 - r$ by r in (9.12) and (9.15).

Although the approach taken in Examples 9.2 and 9.3 can frequently be employed to develop process generators for continuous random variables, this method introduces computational difficulties for some continuous random variables. In Examples 9.2 and 9.3 we were able to eliminate the integral operator in the expressions for $F(x)$; that is, we were able to evaluate $\int_0^x f(y)\, dy$ analytically. However, this is not always possible, as in the case of the normal random variable. When this problem arises, the analyst may wish to resort to approximation techniques or he may be able to generate the random variable of interest by noting its relationship to other random variables for which he can obtain exact process generators (Schmidt and Taylor, 1970).

Further problems may arise even when an analytic expression for $F(x)$ can be obtained; that is, we may be able to determine an analytic expression for $F(x)$, but may not be able to determine the inverse relationship between x and r. This problem arises in the case of the Erlang random variable. To illustrate, let X be an Erlang random variable with probability density function given by

$$f(x) = \lambda^2 x e^{-\lambda x}, \quad 0 < x < \infty \quad (9.16)$$

Then

$$F(x) = 1 - e^{-\lambda x} - \lambda x e^{-\lambda x}, \quad 0 < x < \infty \quad (9.17)$$

Although we have obtained a rather simple expression for $F(x)$, we are unable to define an exact relationship between x and a uniformly distributed random number r. Here again we might resort to approximation techniques, although a preferred approach would be to generate values of X by noting its relationship to the exponential random variable.

Erlang Process Generator

Let X be an Erlang random variable with probability density function given by

$$f(x) = \frac{(n\lambda)^n}{(n-1)!} x^{n-1} e^{-n\lambda x}, \quad 0 < x < \infty \tag{9.18}$$

We have already noted the problems that arise in attempting to generate this random variable by the method described in Examples 9.2 and 9.3. However, in Chapter 2 we noted that the Erlang random variable could be expressed as the sum of independent, identically distributed exponential random variables. Let $Y_1, Y_2, \ldots, Y_n$ be independent exponential random variables each with parameter λ. Then

$$X = \sum_{i=1}^{n} Y_i \tag{9.19}$$

Now Y_i can be generated by

$$Y_i = -\frac{1}{n\lambda} \ln(1 - r_i)$$

from Example 9.2, where r_i is a uniformly distributed random number on the interval (0, 1). Then

$$x = \sum_{i=1}^{n} Y_i = -\frac{1}{n\lambda} \ln\left[\prod_{i=1}^{n} (1 - r_i)\right] \tag{9.20}$$

Since r_i and $1 - r_i$ are identically distributed, the process generator for the Erlang random variable becomes

$$x = -\frac{1}{n\lambda} \ln\left(\prod_{i=1}^{n} r_i\right) \tag{9.21}$$

Thus, to generate an Erlang random variable with parameters λ and n, we generate n random numbers r_i, $i = 1, 2, \ldots, n$, take the natural logarithm of their product, and multiply the result by $-1/n\lambda$.

Normal Process Generator

Let r_1 and r_2 be two random numbers on the interval (0, 1). Then

$$x = [-2 \ln(r_1)]^{1/2} \cos(2\pi r_2) \tag{9.22}$$

is a standard normal random variable (Abramowitz and Stegun, 1964). To generate a normal random variable with mean m and variance σ^2 we modify (9.22) as follows:

$$x = m + \sigma[-2 \ln(r_1)]^{1/2} \cos(2\pi r_2) \tag{9.23}$$

Chi-Square Process Generator

A chi-square random variable can be generated by noting its relationship to the standard normal random variable. Let $Z_1, Z_2, \ldots, Z_n$ be n independent standard normal random variables. Then

$$X = \sum_{i=1}^{n} Z_i^2 \tag{9.24}$$

is a chi-square random variable with n degrees of freedom.

Uniform Process Generator

Let X be a uniformly distributed random variable with probability density function

$$f(x) = \frac{1}{b-a}, \quad a < x < b \tag{9.25}$$

Then

$$F(x) = \int_a^x \frac{1}{b-a} dy = \frac{x-a}{b-a}, \quad a < x < b \tag{9.26}$$

and

$$x = a + (b-a)r \tag{9.27}$$

is the required process generator.

Beta Process Generator

Let X be a beta random variable with probability density function

$$f(x) = \frac{\Gamma(a+b)}{\Gamma(a)\Gamma(b)} x^{a-1}(1-x)^{b-1}, \quad 0 < x < 1 \tag{9.28}$$

Then

$$x = \frac{r_1^{1/a}}{r_1^{1/a} + r_2^{1/b}} \tag{9.29}$$

is beta distributed with parameters a and b if $r_1^{1/a} + r_2^{1/b} \leq 1$ (Berman, 1971). If $r_1^{1/a} + r_2^{1/b} > 1$, another set of random numbers r_1 and r_2 on the interval (0, 1) is generated and the magnitude of $r_1^{1/a} + r_2^{1/b}$ is again checked. This process is repeated until the condition $r_1^{1/a} + r_2^{1/b} \leq 1$ is satisfied.

PROCESS GENERATORS FOR DISCRETE RANDOM VARIABLES WITH KNOWN PROBABILITY MASS FUNCTIONS

The general approach to the generation of discrete random variables is similar to that taken for continuous random variables; that is, we can sometimes define a relatively simple expression for $F(x)$, set this expression equal to a random number r on the interval (0, 1), and then express x as a function of r. However, as in the case of continuous random variables we often have trouble either determining a usable expression for $F(x)$ or, if a simple expression for $F(x)$ can be determined, inverting this expression to obtain a useful equation for x in terms of r. These problems can be overcome by noting a relationship between the random variable to be generated and other random variables that can be simply generated or generated by approximation procedures (Schmidt and Taylor, 1970). The use of the distribution function in generating discrete random variables is illustrated in the following example.

Example 9.4 If X is a discrete random variable with probability mass function

$$p(x) = \frac{1}{k+1}, \quad x = 0, 1, \ldots, k \qquad (9.30)$$

find a process generator for X.

The distribution function of X is given by

$$F(x) = \sum_{y=0}^{x} \frac{1}{k+1} = \frac{x+1}{k+1}, \quad x = 0, 1, \ldots, n \qquad (9.31)$$

Now X assumes the value x whenever

$$\frac{x}{k+1} < F(x) \leq \frac{x+1}{k+1} \qquad (9.32)$$

Therefore, replacing $F(x)$ by r, we have

$$\frac{x}{k+1} < r \leq \frac{x+1}{k+1}, \quad \text{or} \quad x < (k+1)r \leq x+1 \qquad (9.33)$$

Hence, for any value of r, x is integer valued such that

$$(k+1)r - 1 \leq x < (k+1)r \qquad (9.34)$$

Bernoulli Process Generator

If X is a Bernoulli random variable, then

$$p(x) = p^x(1-p)^{1-x}, \quad x = 0, 1 \qquad (9.35)$$

The distribution function of X is given by

$$F(x) = \begin{cases} 0, & x < 0 \\ 1 - p, & 0 \le x < 1 \\ 1, & x \ge 1 \end{cases} \qquad (9.36)$$

Setting $F(x)$ equal to the random number r, we have

$$x = \begin{cases} 0, & r > p \\ 1, & r \le p \end{cases} \qquad (9.37)$$

Binomial Process Generator

From Chapter 2 we know that a binomial random variable can be expressed as the sum of independent, identically distributed Bernoulli random variables. Thus, if $Y_1, Y_2, \ldots, Y_n$ are independent Bernoulli random variables each with parameter p, then

$$X = \sum_{i=1}^{n} Y_i \qquad (9.38)$$

is binomially distributed with parameters p and n.

Geometric Process Generator

The geometric random variable can be generated using the method described in Example 9.4. The distribution function for the geometric random variable X is given by

$$F(x) = 1 - (1 - p)^x, \quad x = 1, 2, \ldots, \qquad \text{or} \qquad r = 1 - (1 - p)^x \qquad (9.39)$$

Then X has the value x whenever

$$1 - (1 - p)^{x-1} < r \le 1 - (1 - p)^x,$$

or

$$(1 - p)^x \le 1 - r < (1 - p)^{x-1}$$

Taking the natural logarithm, we have

$$x \log(1 - p) \le \log(1 - r) < (x - 1) \log(1 - p)$$

which leads to

$$0 \ge \frac{\log(1 - r)}{\log(1 - p)} - x > -1$$

Therefore, for any uniformly distributed random number r on the interval $(0, 1)$, the value of X is an integer x such that

$$\frac{\log(1-r)}{\log(1-p)} \leq x < \frac{\log(1-r)}{\log(1-p)} + 1 \qquad (9.40)$$

Poisson Process Generator

To generate a random variable that is Poisson distributed, we will utilize the relationship between the Poisson and exponential random variables established in Chapter 2; that is, if the number of events occurring in a fixed time interval T is Poisson distributed, then the time between successive events is exponentially distributed. If Y_1 is the time until the first event, Y_2 the time between the first and second events, ..., Y_x the time between the $(x-1)$th and xth events, and if $Y_1, Y_2, \ldots, Y_x$ are independent and identically distributed exponential random variables with parameter λ, then the number of events occurring in T is Poisson distributed with parameter λT and has the value $x - 1$ if and only if

$$\sum_{i=1}^{x-1} Y_i < T \leq \sum_{i=1}^{x} Y_i \qquad (9.41)$$

Hence, to generate a Poisson random variable X with parameter λT, we generate an exponential random variable Y_1 with parameter λ, and compare Y_1 with T. If $Y_1 > T$, then the value of X is zero. If $Y_1 < T$, we generate another exponential random variable Y_2, and compare $Y_1 + Y_2$ with T. If $Y_1 + Y_2 > T$, the value of X is one. If $Y_1 + Y_2 < T$, a third exponential random variable Y_3 is generated and $Y_1 + Y_2 + Y_3$ is compared with T. This process continues until the sum of the exponential random variables generated exceeds T. At this point the value of the Poisson random variable is the number of exponential random variables included in the sum minus one.

Tables 9.2 and 9.3 present process generators for several well-known continuous and discrete random variables. For a more complete discussion of the development of process generators for random variables having known distributions, the reader should see Schmidt and Taylor (1970).

EMPIRICAL PROCESS GENERATORS

The foregoing discussion of process generators assumed that the random variable to be generated could be identified as one having a well-known probability density or mass function. Unfortunately, it is not always possible so to categorize random variables in practice. In such cases, it is frequently possible to develop a process generator directly from observed data.

Empirical Process Generators

We will first discuss the development of empirical process generators for discrete random variables. Let i be any value that the random variable may assume and let f_i be the observed frequency of occurrence of the value i. Furthermore, let

$$F_i = \sum_{j=m}^{i} f_j \qquad (9.42)$$

where it is assumed that $i = m$ is the minimum value that the random variable may assume. If M is the maximum observed value of the random variable, then F_i/F_M, $i = m, m+1, \ldots, M$, defines the observed cumulative distribution function of the random variable. If r is a random number, then the associated value of the random variable is the value of i such that $F_{i-1} < r < F_i$, $i = m, m+1, \ldots, M$.

Example 9.5 The data in Table 9.4 have been collected on the number of emergency patients treated per day at a hospital. Develop an empirical process generator for the number of emergency patients treated per day.

For the data given, $m = 7$, and $M = 21$, the values of f_i, F_i, and F_i/F_M are given in the Table 9.5. The process generator is thus given by

$$i = \begin{cases} 7, & 0 \leq r \leq 0.033 \\ 8, & 0.033 < r \leq 0.067 \\ 9, & 0.067 < r \leq 0.134 \\ 10, & 0.134 < r \leq 0.166 \\ 11, & 0.166 < r \leq 0.200 \\ 12, & 0.200 < r \leq 0.333 \\ 13, & 0.333 < r \leq 0.433 \\ 14, & 0.433 < r \leq 0.500 \\ 15, & 0.500 < r \leq 0.600 \\ 16, & 0.600 < r \leq 0.700 \\ 17, & 0.700 < r \leq 0.767 \\ 18, & 0.767 < r \leq 0.867 \\ 19, & 0.867 < r \leq 0.933 \\ 20, & 0.933 < r \leq 0.967 \\ 21, & 0.967 < r \leq 1.000 \end{cases}$$

TABLE 9.2 Process generators for continuous random variables

Distribution	Density function	Process generator	Comments
Uniform	$f(x) = \dfrac{1}{b-a},\ a < x < b$	$x = a + (b-a)r$	$r \sim U(0, 1)$
Exponential	$f(x) = \lambda e^{-\lambda x},\ 0 < x < \infty$	$x = -\dfrac{1}{\lambda}\ln(r)$	$r \sim U(0, 1)$
Erlang	$f(x) = \dfrac{n\lambda}{(n-1)!} x^{n-1} e^{-n\lambda x},\ 0 < x < \infty$	$x = -\dfrac{1}{n\lambda}\ln\left[\prod_{i=1}^{n} r_i\right]$	$r_i \sim U(0, 1),\ i = 1, 2, \ldots, n$
Beta	$f(x) = \dfrac{\Gamma(a+b)}{\Gamma(a)\Gamma(b)} x^{a-1}(1-x)^{b-1},\ 0 < x < 1$	$x = \dfrac{r_1^{1/a}}{r_1^{1/a} + r_2^{1/b}}$	$r_i \sim U(0, 1),\ i = 1, 2,\ r_1^{1/a} + r_2^{1/b} \leq 1$
Normal	$f(x) = \dfrac{1}{\sigma(2\pi)^{1/2}}\exp\left[\dfrac{-(x-m)^2}{2\sigma^2}\right],$ $-\infty < x < \infty$	$x = m + \sigma[-2\ln(r_1)]^{1/2}\cos(2\pi r_2)$	$r_i \sim U(0, 1),\ i = 1, 2$
Chi-square	$f(x) = \dfrac{x^{(n/2)-1}}{2^{n/2}\Gamma(n/2)} e^{-x/2},\ 0 < x < \infty$	$x = \sum_{i=1}^{n} Z_i^2$	$Z_i \sim N(0, 1),\ i = 1, 2, \ldots, n$
Weibull	$f(x) = \dfrac{a}{b-c}\left(\dfrac{x-c}{b-c}\right)^{a-1}\exp\left[-\left(\dfrac{x-c}{b-c}\right)^a\right],$ $-c < x < \infty$	$x = c + (b-c)[-\ln(r)]^{1/a}$	$r \sim U(0, 1)$
Hyperexponential	$f(x) = p\lambda_1 e^{-\lambda_1 x} + (1-p)\lambda_2 e^{-\lambda_2 x},$ $0 < x < \infty$	$x = \begin{cases} -\dfrac{1}{\lambda_1}\ln(r_2), & r_1 \leq p \\ -\dfrac{1}{\lambda_2}\ln(r_2), & r_1 > p \end{cases}$	$r_i \sim U(0, 1),\ i = 1, 2$

TABLE 9.3 Process generators for discrete random variables

Distribution	Probability mass function	Process generator	Comments
Bernoulli	$p(x) = p^x(1-p)^{1-x}, x = 0, 1$	$x = \begin{cases} 0, & r > p \\ 1, & r \leq p \end{cases}$	$r \sim U(0, 1)$
Binomial	$p(x) = \binom{n}{x} p^x (1-p)^{n-x}, x = 0, 1, \ldots, n$	$x = \sum_{i=1}^{n} y_i$	y_i is Bernoulli distributed with parameter p, $i = 1, 2, \ldots, n$
Geometric	$p(x) = p(1-p)^{x-1}, x = 1, 2, \ldots$	$\dfrac{\ln(1-r)}{\ln(1-p)} \leq x < \dfrac{\ln(1-r)}{\ln(1-p)} + 1$	$r \sim U(0, 1)$
Rectangular	$p(x) = \dfrac{1}{b-a+1}, x = a, a+1, \ldots, b$	$(b-a+1)r + a - 1 \leq x < (b-a+1)r + a$	$r \sim U(0, 1)$
Poisson	$p(x) = \dfrac{(\lambda T)^x}{x!} e^{-\lambda T}, x = 0, 1, \ldots$	$x = \begin{cases} 0, & y_1 > T \\ x, & \sum_{i=1}^{x} y_i < T \leq \sum_{i=1}^{x+1} y_i \end{cases}$	y_i is exponentially distributed with parameter λ, $i = 1, 2, \ldots, x+1$
Negative binomial	$p(x) = \binom{x-1}{r-1} p^r (1-p)^{x-r}, x = r, r+1, \ldots$	$x = \sum_{i=1}^{r} y_i$	y_i is geometrically distributed with parameter p, $i = 1, 2, \ldots, r$

TABLE 9.4 Data for Example 9.5

Day of the month	Patients treated	Day of the month	Patients treated	Day of the month	Patients treated
1	18	11	15	21	13
2	12	12	15	22	8
3	7	13	12	23	10
4	9	14	19	24	9
5	13	15	18	25	20
6	21	16	11	26	17
7	14	17	12	27	13
8	17	18	19	28	18
9	16	19	16	29	12
10	14	20	15	30	16

TABLE 9.5 Observed cumulative distribution function for the problem in Example 9.5

i	f_i	F_i	F_i/F_M
7	1	1	0.033
8	1	2	0.067
9	2	4	0.134
10	1	5	0.166
11	1	6	0.200
12	4	10	0.333
13	3	13	0.433
14	2	15	0.500
15	3	18	0.600
16	3	21	0.700
17	2	23	0.767
18	3	26	0.867
19	2	28	0.933
20	1	29	0.967
21	1	30	1.000

Empirical process generators for continuous random variables can be developed in a manner quite similar to that presented for discrete random variables. Again we first form a frequency distribution from observed data. In this case, let x_m and x_M be the minimum and maximum values of the random variable under study. The interval (x_m, x_M) is divided into n intervals, each of width Δx, where

$$\Delta x = \frac{x_M - x_m}{n} \quad (9.43)$$

Empirical Process Generators

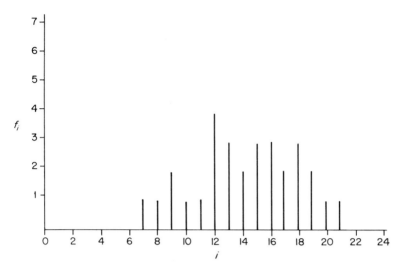

Fig. 9.5 Frequency distribution of the random variable X.

The frequency count in the ith interval, f_i, is then the number of observed values of the random variable lying between $x_m + (i - 1)\,\Delta x$ and $x_m + i\,\Delta x$. The resulting frequency distribution can then be represented graphically as shown in Fig. 9.5, where m_i is the midpoint of the ith interval. The corresponding observed cumulative distribution function of the random variable can then be represented as shown in Fig. 9.6, where

$$F_i = \sum_{j=1}^{i} f_j \tag{9.44}$$

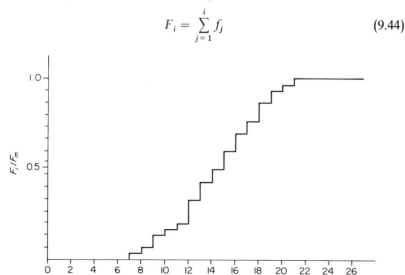

Fig. 9.6 Observed cumulative distribution function for the random variable X.

To generate values of the random variable X, we might generate a random number r and let $x = m_i$, where m_i is the midpoint of the interval for which $F_{i-1}/F_n < r \leq F_i/F_n$. However, this procedure allows us to generate only a limited set of discrete values of X. To circumvent this problem, we connect consecutive points $(F_{i-1}/F_n, m_{i-1})$ and $(F_i/F_n, m_i)$ by a straight line, yielding a smoothed observed cumulative distribution function as

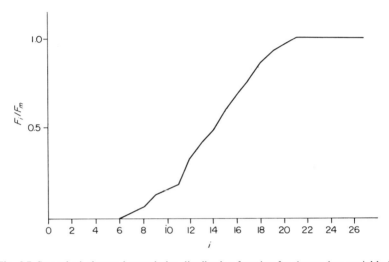

Fig. 9.7 Smoothed observed cumulative distribution function for the random variable X.

shown in Fig. 9.7. The value of the random variable X corresponding to a random number r is then given by

$$x = m_{i-1} + (m_i - m_{i-1}) \frac{F_n r - F_{i-1}}{F_i - F_{i-1}} \quad (9.45)$$

where $F_{i-1}/F_n < r \leq F_i/F_n$.

Example 9.6 Given the information presented in Table 9.6 on the random variable X, find the value of X corresponding to the random numbers 0.28 and 0.71.

From (9.45)

$$x = m_{i-1} + (m_i - m_{i-1}) \frac{F_n r - F_{i-1}}{F_i - F_{i-1}}$$

Since $0.081 < r < 0.324$ for $r = 0.28$, we have

$$x = 1.00 + (2.00 - 1.00) \frac{(74.00)(0.28) - 6.00}{24.00 - 6.00} = 1.82$$

Simulation of a Single-Channel Queueing System

TABLE 9.6 Observed cumulative distribution function for the problem in Example 9.6

i	Interval	M_i	f_i	F_i	F_i/F_n
1	0.51–1.50	1.00	6	6	0.081
2	1.51–2.50	2.00	18	24	0.324
3	2.51–3.50	3.00	7	31	0.419
4	3.51–4.50	4.00	1	32	0.432
5	4.51–5.50	5.00	4	36	0.486
6	5.51–6.50	6.00	8	44	0.596
7	6.51–7.50	7.00	12	56	0.757
8	7.51–8.50	8.00	10	66	0.892
9	8.51–9.50	9.00	6	72	0.973
10	9.51–10.50	10.00	2	74	1.000

For $r = 0.71$, $0.596 < r < 0.757$, and

$$x = 6.00 + (7.00 - 6.00)\frac{(74.00)(0.71) - 44.00}{56.00 - 44.00} = 6.71$$

SIMULATION OF A SINGLE-CHANNEL QUEUEING SYSTEM $(G|G|1):(FCFS|\infty|\infty)$

As we have already mentioned, to simulate the operation of any system we must generate artificially those events occurring in nature. In the case of the single-channel queueing system discussed earlier in this chapter, this would mean that we must generate interarrival and service times. To capture these phenomena realistically, it is necessary to identify the probability density functions of these random variables before attempting to simulate the operation of the system.

To illustrate, let us assume that interarrival and service times are exponentially distributed. Let the arrival rate be 10/yr and the service rate 20/yr. Assume that the system is empty at time $t = 0$ and that the period of simulation is 1 yr. To start the simulation we must generate the time until the first arrival using (9.12). We will not generate the time of completion of service for this unit until it arrives and enters the service channel. Thus, at time $t = 0$, the time of completion of service has not been generated and the simulator looks for the next event to occur. The events that can occur in the course of the simulation are an arrival, a service, or the end of the period of simulation. The simulator determines the next event occurring by comparing the time of the next arrival, the time of the next service completion, and the time of the termination of the simulation. Using the notation in Fig. 9.1, the next event is defined by $\min(t_s, t_a, t_\delta)$. Thus, t_s, t_a, and t_δ must be defined at all times in the simulation. At time $t = 0$, t_a is defined, since we have generated its value. Similarly, t_δ was defined prior to the start of the simulation. However, at time $t = 0$, t_s is not defined. Thus at time $t = 0$ we set $t_s = \infty$ so that $\min(t_s, t_a, t_\delta)$ must be either t_a or t_δ.

Assume that the first event is an arrival at t_a; that is, $t_a < t_\delta$. The simulator moves forward in time to t_a, sets present time to t_a, $t = t_a$, records the arrival, places the arrival in the service channel since the system is empty, and generates the service time τ_s for this unit. Then

$$t_s = t + \tau_s \tag{9.46}$$

Finally, it generates the interarrival time τ_a for the next arrival, and

$$t_a = t + \tau_a \tag{9.47}$$

At this point the simulator determines $\min(t_s, t_a, t_\delta)$ again. This time let

$$t_s = \min(t_s, t_a, t_\delta) \tag{9.48}$$

Therefore, the simulator sets $t = t_s$, removes the unit in the service channel from the system, and sets $t_s = \infty$, since there are no units in the system after the completion of the present service. Had there been units in the waiting line at the time of completion of the above service, the simulator would have brought the unit at the head of the waiting line into the service channel, generated a service time τ_s for that unit, and reduced the waiting line length by one. In this case

$$t_s = t + \tau_s \tag{9.49}$$

To summarize, starting at $t = 0$, the simulator moves from one event to the next, recording the changes in the system at each event. After each arrival the time for the next arrival is generated, as well as the time for service of the present arrival if it enters the service channel. After each service the time of the next service is generated; in some cases this is equal to infinity. This process continues until the next event to occur is the end of the simulation; that is,

$$t_\delta = \min(t_s, t_a, t_\delta) \tag{9.50}$$

At this point the simulator calculates those statistics of interest to the analyst.

An illustrative summary of the simulation of one year of operation of the $(M \mid M \mid 1) : (FCFS \mid \infty \mid \infty)$ queueing system is shown in Table 9.7.

Example 9.7 Based on the results given in Table 9.7, estimate the probability mass function of the number of units in the system, the mean of the

Simulation of a Single-Channel Queueing System

TABLE 9.7 Summary of the simulation of an $(M|M|1)$ queueing system for one year[a]

Event number i	t	Service Random number	τ_s	t_s	Arrival Random number	τ_a	t_a	$\min(t_s, t_a, t_\delta)$	Number in system, N_{si}	Number in queue, N_{qi}	$N_{si}(t_{i+1} - t_i)$	$N_{qi}(t_{i+1} - t_i)$
0	0.00	—	∞	∞	0.34	0.12	0.12	$t_a = 0.12$	0	0	0.00	0.00
1	0.12	0.77	0.01	0.13	0.43	0.08	0.20	$t_s = 0.13$	1	0	0.01	0.00
2	0.13	—	∞	∞			0.20	$t_a = 0.20$	0	0	0.00	0.00
3	0.20	0.40	0.05	0.25	0.22	0.15	0.35	$t_s = 0.25$	1	0	0.05	0.00
4	0.25	—	∞	∞			0.35	$t_a = 0.35$	0	0	0.00	0.00
5	0.35	0.04	0.16	0.51	0.12	0.21	0.56	$t_s = 0.51$	1	0	0.16	0.00
6	0.51	—	∞	∞			0.56	$t_a = 0.56$	0	0	0.00	0.00
7	0.56	0.02	0.19	0.75	0.88	0.01	0.57	$t_a = 0.57$	1	0	0.01	0.00
8	0.57			0.75	0.68	0.04	0.61	$t_a = 0.61$	2	1	0.08	0.04
9	0.61			0.75	0.20	0.16	0.77	$t_s = 0.75$	3	2	0.42	0.28
10	0.75	0.41	0.04	0.79			0.77	$t_a = 0.77$	2	1	0.04	0.02
11	0.77			0.79	0.88	0.01	0.78	$t_a = 0.78$	3	2	0.03	0.02
12	0.78			0.79	0.51	0.07	0.85	$t_s = 0.79$	4	3	0.04	0.03
13	0.79	0.70	0.02	0.81			0.85	$t_s = 0.81$	3	2	0.06	0.04
14	0.81	0.29	0.06	0.87			0.85	$t_a = 0.85$	2	1	0.08	0.04
15	0.85			0.87	0.26	0.13	0.98	$t_s = 0.87$	3	2	0.06	0.04
16	0.87	0.53	0.03	0.90			0.98	$t_s = 0.90$	2	1	0.06	0.03
17	0.90	0.48	0.04	0.94			0.98	$t_s = 0.94$	1	0	0.04	0.00
18	0.94	—	∞	∞			0.98	$t_a = 0.98$	0	0	0.00	0.00
19	0.98			∞	0.55	0.06	1.04	$t_\delta = 1.00$	1	0	0.02	0.00
20	1.00			∞			1.04		1	0	—	—

[a] For $\lambda = 10$, $\mu = 20$, $t_1 = 1.00$.

number in the system, the mean of the number in the queue, the mean time a unit spends in the system, and the mean time a unit spends in the queue.

To calculate the empirical probability mass function of the number of units in the system, we will use (9.5). Let $\hat{P}_n$ be the probability that there are n units in the system as estimated through simulation, and t_i the time of occurrence of the ith event. Then

$$\hat{P}_0 = \frac{T_{s0}}{t_m}$$

$$= \frac{(t_1 - t_0) + (t_3 - t_2) + (t_5 - t_4) + (t_7 - t_6) + (t_{19} - t_{18})}{t_m}$$

$$= \frac{0.12 + 0.07 + 0.10 + 0.05 + 0.04}{1.00} = 0.38$$

$$\hat{P}_1 = \frac{T_{s1}}{t_m}$$

$$= \frac{\begin{array}{c}(t_2 - t_1) + (t_4 - t_3) + (t_6 - t_5) + (t_8 - t_7) \\ + (t_{18} - t_{17}) + (t_{20} - t_{19})\end{array}}{t_m}$$

$$= \frac{0.01 + 0.05 + 0.16 + 0.01 + 0.04 + 0.02}{1.00} = 0.29$$

$$\hat{P}_2 = \frac{T_{s2}}{t_m}$$

$$= \frac{(t_9 - t_8) + (t_{11} - t_{10}) + (t_{15} - t_{14}) + (t_{17} - t_{16})}{t_m}$$

$$= \frac{0.04 + 0.02 + 0.04 + 0.03}{1.00} = 0.13$$

$$\hat{P}_3 = \frac{T_{s3}}{t_m}$$

$$= \frac{(t_{10} - t_9) + (t_{12} - t_{11}) + (t_{14} - t_{13}) + (t_{16} - t_{15})}{t_m}$$

$$= \frac{0.14 + 0.01 + 0.02 + 0.02}{1.00} = 0.19$$

$$\hat{P}_4 = \frac{T_{s4}}{t_m} = \frac{(t_{13} - t_{12})}{t_m}$$

$$= \frac{0.01}{1.00} = 0.01$$

Simulation of a Single-Channel Queueing System

where $t_m = t_\delta$ and $m = 20$. For $n > 4$, $\hat{P}_n = 0$. The average number in the system is given by (9.1). Therefore,

$$\text{average number in system} = \frac{1}{t_m} \sum_{i=0}^{m-1} N_{si}(t_{i+1} - t_i) = \frac{1}{1.00}(1.16) = 1.16$$

The average number in the queue is given by

$$\text{average number in queue} = \frac{1}{t_m} \sum_{i=0}^{m-1} N_{qi}(t_{i+1} - t_i) = \frac{1}{1.00}(0.54) = 0.54$$

The average time in the system and in the queue are given by (9.3) and (9.4), respectively, and lead to

$$\text{average time in system} = \frac{1}{k} \sum_{i=0}^{m-1} N_{si}(t_{i+1} - t_i) = \frac{1}{10}(1.15) = 0.115$$

$$\text{average time in queue} = \frac{1}{k} \sum_{i=0}^{m-1} N_{qi}(t_{i+1} - t_i) = \frac{1}{10}(0.54) = 0.054$$

where k is the number of units spending time in the system and is 10 in this case.

TABLE 9.8 Comparison of the simulated and theoretical probability mass functions for the number in the system for the $(M|M|1):(FCFS|\infty|\infty)$ queue[a]

Number in the system n	Theoretical P_n	Simulated $\hat{P}_n$
0	0.500	0.380
1	0.250	0.290
2	0.125	0.130
3	0.063	0.190
4	0.032	0.010

[a] For $\lambda = 10.00$, $\mu = 20.00$, $t_1 = 1.00$.

For the $(M|M|1):(FCFS|\infty|\infty)$ queueing system, we know that the probability mass function of the number of units in the system is given by

$$P_n = (1 - \rho)\rho^n \tag{9.51}$$

where $\rho = \lambda/\mu$. A comparison of the empirical (simulated in Example 9.7) and the exact probability mass functions for the number in the system is given in Table 9.8. Some discrepancy between the simulated and theoretical probability mass functions, $\hat{P}_n$ and P_n, is apparent. However, the reader should remember that the queueing system was simulated for only one year. Since this period included only ten arrivals and nine services, the length of the simulation would be considered rather short for reliable results. To extend the period of the simulation, we would be well advised to program the simulation for execution on a digital computer.

Example 9.8 Simulate the operation of an $(M|M|1): (FCFS|\infty|)$ queueing system for a period of 20 yr. Let the mean arrival rate be 100 units/yr and let the mean service rate be 200 units/yr. Estimate:

(a) the probability mass function of the number of units in the system,
(b) the probability mass function of the number of units in the queue,
(c) the mean time a unit spends in the system,
(d) the mean time a unit spends in the queue,
(e) the mean number of units in the system,
(f) the mean number of units in the queue,
(g) the utilization of the server.

Compare the estimates indicated above with the corresponding expected values.

To simulate this system we will develop a FORTRAN IV computer program. The program will consist of three parts: The first part is the main program, which establishes the conditions prevailing at the start of the simulation, determines the next event to occur in the simulation, modifies the status of the queueing system at the occurrence of each event, and calculates the required estimates. The second part of the program is a subroutine subprogram that accumulates the information necessary to calculate the required estimates at the end of the simulation. The last part of the program is a function subprogram, which generates uniformly distributed random numbers. The program is written for execution on an IBM System/360 computer.

For the purposes of this example the following FORTRAN variables are defined:

$$\text{TS(I)} = T_{si}, \quad \text{I} = i - 1, \quad \text{I} = 1, 2, \ldots, 100$$

$$\text{TQ(I)} = T_{qi}, \quad \text{I} = i - 1, \quad \text{I} = 1, 2, \ldots, 100$$

SYSNT = total number of units spending time in the system

QLEN = current length of the queue

$$\text{TSYS} = \sum_{i=0}^{m-1} N_{si}(t_{i+1} - t_i)$$

$$\text{TQUEUE} = \sum_{i=0}^{m-1} N_{qi}(t_{i+1} - t_i)$$

SYSN = current number in the system

TLAST = time of occurrence of the last or most recent event

TNEXT = time of occurrence of the next event

ALAM = λ

AMU = μ

IX = random-number seed

$$\text{TIM(I)} = \begin{cases} \text{time of the next service,} & I = 1 \\ \text{time of the next arrival,} & I = 2 \\ \text{duration of the simulation,} & I = 3 \end{cases}$$

ASYS = average number in the system

AQUEUE = average number in the queue

UTIL = utilization of the server

$$\text{NEXT} = \begin{cases} 1, & \text{next event is a service} \\ 2, & \text{next event is an arrival} \\ 3, & \text{next event is termination of the simulation} \end{cases}$$

The flowchart for the main program is given in Fig. 9.8 and is similar to that given in Fig. 9.7. The initial conditions for the system assume that at time zero the system is empty. The reader will notice that the time of the next service is set equal to 10^{30} instead of infinity when the system is empty, since infinity is an undefinable quantity. Actually, the time of the next service could have been set equal to any number greater than the simulation period, since this would guarantee that the next event detected in the simulation could not be a service.

In the main program, subroutine UPDATE is called after the next event and its associated times of occurrence are determined, but before the execution of that event. In this subroutine, the quantities T_{si}, T_{qi}, $\sum_{i=0}^{m-1} N_{si}(t_{i+1} - t_i)$, and $\sum_{i=0}^{m-1} N_{qi}(t_{i+1} - t_i)$ are accumulated. The flowchart for subroutine UPDATE is given in Fig. 9.9. The coded FORTRAN IV program is given in Fig. 9.10.

The results of the 20-yr simulation are summarized in Tables 9.9 and 9.10.

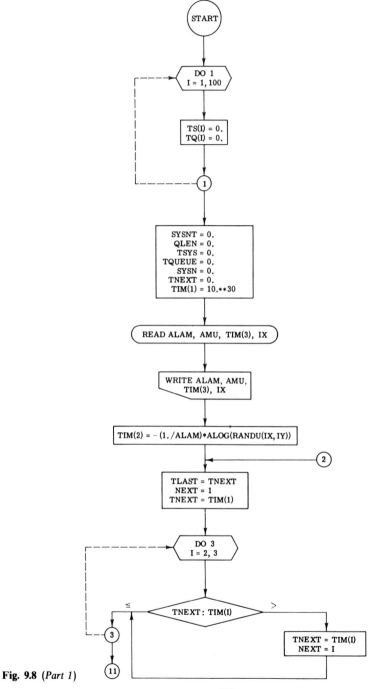

Fig. 9.8 (*Part 1*)

Simulation of a Single-Channel Queueing System

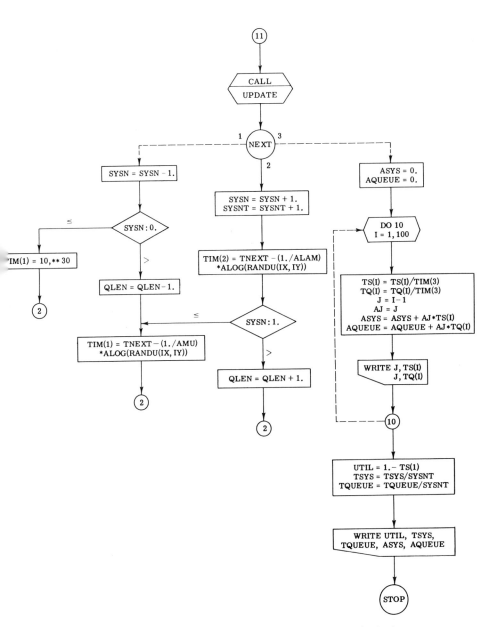

Fig. 9.8 Main program for the simulation of an $(M \mid M \mid 1) : (FCFS \mid \infty \mid \infty)$ queue.

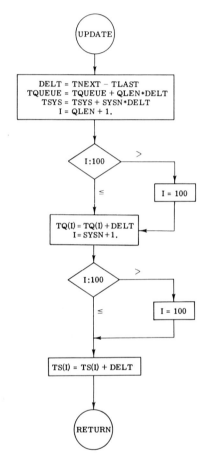

Fig. 9.9 Subroutine UPDATE for the $(M \mid M \mid 1) : (FCFS \mid \infty \mid \infty)$ queueing simulator.

Simulation of a Single-Channel Queueing System

TABLE 9.9 Simulated and theoretical probability mass functions (pmf) for number in the system and number in the queue for Example 9.8

	Number in the system			Number in the queue	
Number	Theoretical pmf	Simulated pmf	Number	Theoretical pmf	Simulated pmf
0	0.5000	0.4901	0	0.7500	0.7458
1	0.2500	0.2557	1	0.1250	0.1380
2	0.1250	0.1380	2	0.0625	0.0683
3	0.0625	0.0683	3	0.0313	0.0328
4	0.0313	0.0328	4	0.0156	0.0101
5	0.0156	0.0101	5	0.0078	0.0032
6	0.0078	0.0032	6	0.0039	0.0014
7	0.0039	0.0014	7	0.0019	0.0002
8	0.0019	0.0002	8	0.0010	0.0002
9	0.0010	0.0002	9	0.0005	0.0000
10	0.0005	0.0000	10	0.0003	0.0000
11	0.0003	0.0000	11	0.0002	0.0000
12	0.0002	0.0000			

TABLE 9.10 Comparison of simulated statistics with expected values for Example 9.8

Quantity	Expected value	Simulated value
Mean time in system	0.0100 yr	0.0096 yr
Mean time in queue	0.0050 yr	0.0044 yr
Mean number in system	1.0000	0.9506
Mean number in queue	0.5000	0.4407
Server utilization	0.5000	0.5098

In the examples discussed thus far, we have centered interest on the steady-state analysis of single-channel queueing systems; that is, we were interested in the behavior of the system only after the steady state was achieved. Although steady-state behavior is usually of primary interest, situations arise where the analysis of transient-state behavior is also important. In the following example, we will attempt to illustrate how the behavior of a system in the transient state can be analyzed through simulation.

Example 9.9 Arrivals to a single-channel queueing system are Poisson distributed with a mean arrival rate of 49 units/day. Service time is exponentially distributed with a mean service rate of 51 units/day. The capacity of the system is such that no waiting line is allowed; that is, if an arrival occurs

```
      DIMENSION TS(100),TQ(100),TIM(3)
      COMMON TS,TQ,TLAST,TNEXT,QLEN,TQUEUE,SYSN,TSYS
C INITIALIZE VARIABLES AND ACCUMULATORS AND READ IN INPUT DATA ****************
      DO 1 I=1,100
      TS(I)=0
    1 TQ(I)=0
      SYSNT=0
      QLEN=0
      TSYS=0
      TQUEUE=0
      SYSN=0
      ASYS=0
      AQUEUE=0
      TNEXT=0
      TIM(1)=10.**30
      READ(5,100) ALAM,AMU,TIM(3),IX
      WRITE(6,100) ALAM,AMU,TIM(3),IX
      TIM(2)=-(1./ALAM)*ALOG(RANDU(IX,IY))
C ********************************************************************
C DETERMINE THE NEXT EVENT TO OCCUR AND THE TIME OF ITS OCCURRENCE ###########
    2 TLAST=TNEXT
      NEXT=1
      TNEXT=TIM(1)
      DO 3 I=2,3
      IF(TNEXT.LE.TIM(I)) GO TO 3
      TNEXT=TIM(I)
      NEXT=I
    3 CONTINUE
C ####################################################################
C UPDATE ALL ACCUMULATORS AND BRANCH TO THE ROUTINE FOR THE NEXT EVENT <<<<<<<<
      CALL UPDATE
      GO TO (4,7,8),NEXT
C <<<<<<<<<<<<<<<<<<<<<<<<<<<<<<<<<<<<<<<<<<<<<<<<<<<<<<<<<<<<<<<<<<<<
C SERVICE ROUTINE *****************************************************
    4 SYSN=SYSN-1.
      IF(SYSN.GT.0.) GO TO 5
      TIM(1)=10.**30
      GO TO 2
    5 QLEN=QLEN-1.
    6 TIM(1)=TNEXT-(1./AMU)*ALOG(RANDU(IX,IY))
      GO TO 2
C ********************************************************************
C ARRIVAL ROUTINE #####################################################
    7 SYSN=SYSN+1.
      SYSNT=SYSNT+1.
      TIM(2)=TNEXT-(1./ALAM)*ALOG(RANDU(IX,IY))
      IF(SYSN.LE.1.) GO TO 6
      QLEN=QLEN+1.
      GO TO 2
C ####################################################################
C TERMINATION ROUTINE <<<<<<<<<<<<<<<<<<<<<<<<<<<<<<<<<<<<<<<<<<<<<<<<<
    8 WRITE(6,200)
C CALCULATE THE PROBABILITY MASS FUNCTION OF THE NUMBER IN THE SYSTEM, THE
C PROBABILITY MASS FUNCTION OF THE NUMBER IN THE QUEUE, THE AVERAGE NUMBER IN
C THE SYSTEM,  THE   AVERAGE NUMBER IN THE QUEUE, AND WRITE OUT EACH OF THESE
C STATISTICS **********************************************************
      DO 10 I=1,100
      TS(I)=TS(I)/TIM(3)
      TQ(I)=TQ(I)/TIM(3)
      J=I-1
      AJ=J
      ASYS=ASYS+AJ*TS(I)
      AQUEUE=AQUEUE+AJ*TQ(I)
      IF(J.EQ.99) GO TO 9
      WRITE(6,300) J,TS(I),J,TQ(I)
      GO TO 10
    9 WRITE(6,400) J,TS(I),J,TQ(I)
   10 CONTINUE
C ********************************************************************
C CALCULATE AND WRITE OUT THE CHANNEL UTILIZATION, THE AVERAGE TIME IN THE
C SYSTEM, AND THE AVERAGE TIME IN THE QUEUE ###########################
      UTIL=1.-TS(1)
      TSYS=TSYS/SYSNT
      TQUEUE=TQUEUE/SYSNT
      WRITE(6,500) UTIL,TSYS,TQUEUE,ASYS,AQUEUE
C ####################################################################
C <<<<<<<<<<<<<<<<<<<<<<<<<<<<<<<<<<<<<<<<<<<<<<<<<<<<<<<<<<<<<<<<<<<<
  100 FORMAT(3F10.2,I5)
  200 FORMAT(1X,'NO. IN SYSTEM',3X,'PROBABILITY',5X,'NO. IN QUEUE',3X,'P
     1ROBABILITY')
  300 FORMAT(1X,I3,13X,E11.5,5X,I3,12X,E11.5)
  400 FORMAT(1X,I3,1X,'OR MORE',5X,E11.5,5X,I3,1X,'OR MORE',4X,E11.5///)
  500 FORMAT(1X,'UTILIZATION=',F10.4,2X,'AVG. TIME IN SYSTEM=',F10.4,2X,
     1'AVG. TIME IN QUEUE=',F10.4/1X,'AVG. NO. IN SYSTEM=',F10.4,2X,'AVG
     2. NO. IN QUEUE=',F10.4)
      STOP
      END
```

Fig. 9.10 FORTRAN IV program for an $(M \mid M \mid 1):(FCFS \mid \infty \mid \infty)$ queueing simulator.

Simulation of a Single-Channel Queueing System

```
      SUBROUTINE UPDATE
      DIMENSION TS(100),TQ(100)
      COMMON TS,TQ,TLAST,TNEXT,QLEN,TQUEUE,SYSN,TSYS
C ACCUMULATE UNIT-TIME IN THE QUEUE AND UNIT-TIME IN THE SYSTEM ****************
      DELT=TNEXT-TLAST
      TQUEUE=TQUEUE+QLEN*DELT                                                  *
      TSYS=TSYS+SYSN*DELT                                                      *
C *******************************************************************************
C ACCUMULATE THE TOTAL TIME FOR WHICH THERE WERE I-1 UNITS IN THE QUEUE AND FOR
C WHICH THERE WERE I-1 UNITS IN THE SYSTEM #####################################
      I=QLEN+1.
      IF(I.GT.100) I=100                                                       #
      TQ(I)=TQ(I)+DELT                                                         #
      I=SYSN+1.                                                                #
      IF(I.GT.100) I=100                                                       #
      TS(I)=TS(I)+DELT                                                         #
C #################################################################################
      RETURN
      END

      FUNCTION RANDU(IX,IY)
      IY=IX*65539
      IF(IY)5,6,6
    5 IY=IY+2147483647+1
    6 YFL=IY
      RANDU=YFL*.4656613E-9
      IX=IY
      RETURN
      END
```

Fig. 9.10 (continued)

when there is a unit in the service channel, that arrival is rejected by the system. Develop a simulator for this system that will record the probability that the system is empty in periods of length 0.001 days starting at time zero. Assume that the system is empty at the start of the simulation.

Let $\Delta t = 0.001$. We wish to calculate the probability that the system is empty between t_i and t_{i+1}, where

$$t_{i+1} - t_i = \Delta t, \quad \text{and} \quad t_0 = 0 \tag{9.52}$$

Let δ_i be the total time in the interval t_i to t_{i+1} during which the system is empty. Then

$$P(\text{system empty between } t_i \text{ and } t_{i+1}^*) = \frac{\delta_i}{\Delta t} = P_0(t - \Delta t, t) \tag{9.53}$$

To determine an estimate of $\delta_i/\Delta t$ through simulation, we simulate n days of operation, starting each day with zero units in the system. For each interval t_i to t_{i+1} within a given day, we record the observed value of δ_i. Let δ_{ij} be the total time in the interval t_i to t_{i+1} during which the system is empty for the jth day simulated. Then

$$P_0(t - \Delta t, t) = \frac{1}{n\Delta} \sum_{j=1}^{n} \delta_{ij} \tag{9.54}$$

The simulator for the system described is given in Fig. 9.11, where

ALAM = mean daily arrival rate

AMU = mean daily service rate

DELT = Δt

IX = random-number seed

NSIM = n

TIM(1) = time of the next service

TIM(2) = time of the next arrival

TIM(3) = time at which the next period of length Δt begins

TIM(4) = time of the end of the simulation for one day

TNEXT = time of occurrence of the next event

TLAST = time of occurrence of the last event

SYSN = current number in the system (0 or 1)

$$\text{NEXT} = \begin{cases} 1, & \text{next event is a service completion} \\ 2, & \text{next event is an arrival} \\ 3, & \text{next event is the end of a period of length } \Delta t \\ 4, & \text{next event is the end of one day's simulation} \end{cases}$$

ISTAT = Indicator defining the current period of length Δt

$\text{CUMPO(I)} = \sum_{j=1}^{n} \delta_{1j}$, prior to the end of the n replications of the simulation

$= \dfrac{1}{n\Delta} \sum_{j=1}^{n} \delta_{1j}$, after completion of n replications

The results of the simulation are shown graphically in Fig. 9.12, where the number of days replicated was 10,000. If $P_0(t)$ is the transient-state probability of zero in the system at time t, then it can be shown that

$$P_0(t) = \frac{\mu}{\lambda + \mu} + \frac{\lambda}{\lambda + \mu} e^{-(\lambda + \mu)t} \tag{9.55}$$

The probability of zero in the system between $t - \Delta t$ and t is then given by

$$P_0(t - \Delta t, t) = \frac{1}{\Delta t} \int_{t - \Delta t}^{t} P_0(x) \, dx \tag{9.56}$$

The results given in Fig. 9.12 can be shown to agree quite well with those obtained from Eq. (9.56).

Simulation of a Single-Channel Queueing System

```
      DIMENSION TIM(4),CUMPO(1000),T(1000)
C     READ INPUT DATA AND SET INITIAL CONDITIONS FOR NSIM REPLICATIONS OF
C     THE SIMULATION ***************************************************
      READ(5,100) ALAM,AMU,TIM(4),DELT,IX,NSIM                         *
      WRITE(6,100) ALAM,AMU,TIM(4),DELT,IX,NSIM                        *
      INT=TIM(4)/DELT                                                  *
      INT=INT+1                                                        *
      DO 1 I=1,INT                                                     *
    1 CUMPO(I)=0.                                                      *
C     ******************************************************************
C     REPLICATE THE SIMULATION OF ONE DAY NSIM TIMES ..................
      DO 9 ISIM=1,NSIM                                                 .
C     SET INITIAL CONDITIONS FOR THE ITH REPLICATE OF THE SIMULATION *********.
      ISTAT=1                                                          * .
      SYSN=0                                                           * .
      TNEXT=0                                                          * .
      TIM(1)=10.**30                                                   * .
      TIM(3)=DELT                                                      * .
C     ****************************************************************** .
C     GENERATE THE TIME OF THE NEXT ARRIVAL --------------------------- .
    2 TIM(2)=TNEXT-(1./ALAM)*ALOG(RANDU(IX,IY))                        - .
C     ---------------------------------------------------------------- .
C     DEFINE THE TIME OF THE LAST EVENT AND FIND THE TIME AND NATURE OF THE .
C     NEXT EVENT ***************************************************** .
    3 TLAST=TNEXT                                                      * .
      NEXT=1                                                           * .
      TNEXT=TIM(1)                                                     * .
      DO 4 I=2,4                                                       * .
      IF(TNEXT.LE.TIM(I)) GO TO 4                                      * .
      TNEXT=TIM(I)                                                     * .
      NEXT=I                                                           * .
    4 CONTINUE                                                         * .
C     ****************************************************************** .
C     BRANCH TO THE ROUTINE WHICH ALTERS THE SYSTEM FOR THE NEXT EVENT ------- .
      GO TO (5,6,7,8),NEXT                                             - .
C     ---------------------------------------------------------------- .
C     SERVICE ROUTINE ************************************************** .
    5 SYSN=SYSN-1.                                                     * .
      TIM(1)=10.**30                                                   * .
      GO TO 3                                                          * .
C     ****************************************************************** .
C     ARRIVAL ROUTINE ------------------------------------------------- .
    6 IF(SYSN.EQ.0.) CUMPO(ISTAT)=CUMPO(ISTAT)+(TNEXT-TLAST)           - .
      IF(SYSN.GT.0.) GO TO 2                                           - .
      SYSN=SYSN+1.                                                     - .
      TIM(2)=TNEXT-(1./ALAM)*ALOG(RANDU(IX,IY))                        - .
      TIM(1)=TNEXT-(1./AMU)*ALOG(RANDU(IX,IY))                         - .
      GO TO 3                                                          - .
C     ---------------------------------------------------------------- .
C     ROUTINE FOR CHANGING THE PERIOD OF LENGTH DELT ********************* .
    7 TIM(3)=TNEXT+DELT                                                * .
      IF(SYSN.EQ.0.) CUMPO(ISTAT)=CUMPO(ISTAT)+(TNEXT-TLAST)           * .
      ISTAT=ISTAT+1                                                    * .
      GO TO 3                                                          * .
C     ****************************************************************** .
C     ROUTINE FOR TERMINATING ONE REPLICATION OF THE SIMULATION ------------- .
    8 IF(SYSN.EQ.0.) CUMPO(ISTAT-1)=CUMPO(ISTAT-1)+(TNEXT-TLAST)       - .
    9 CONTINUE                                                         - .
C     ---------------------------------------------------------------- .
C     ..................................................................
C     CALCULATE AND WRITE OUT THE PROBABILITY OF ZERO UNITS IN THE SYSTEM FOR
C     PERIOD OF LENGTH DELT ********************************************
      ASIM=NSIM                                                        *
      ADELT=DELT*ASIM                                                  *
      DO 10 I=1,INT                                                    *
      AI=I                                                             *
      T(I)=AI*DELT                                                     *
   10 CUMPO(I)=CUMPO(I)/ADELT                                          *
      WRITE(6,200) (T(I),CUMPO(I),I=1,INT)                             *
      ******************************************************************
      STOP
  100 FORMAT(1X,4F10.4,2I5)
  200 FORMAT(1X,'PROBABILITY OF ZERO UNITS IN THE SYSTEM IN INTERVAL TO'
     1,F10.6,1X,'=',3X,F10.6)
      END

      FUNCTION RANDU(IX,IY)
      IY=IX*65539
      IF(IY)5,6,6
    5 IY=IY+2147483647+1
    6 YFL=IY
      RANDU=YFL*.4656613E-9
      IX=IY
      RETURN
      END
```

Fig. 9.11 Simulator for the transient-state analysis in Example 9.9.

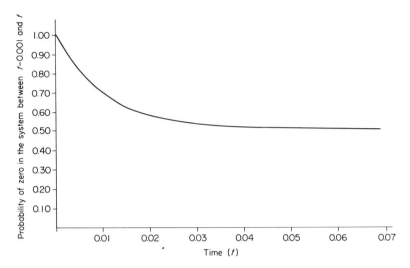

Fig. 9.12 Transient-state probability of zero in the system between $t - 0.001$ and t for the system described in Example 9.9.

MULTIPLE CHANNELS IN PARALLEL $(G\,|\,G\,|\,C):(FCFS\,|\,\infty\,|\,\infty)$

The simulator for a queueing system with several channels in parallel does not differ greatly from that given for a single channel. One difference arises in attempting to define the next event to occur in the simulation. In the single-channel simulator developed in Example 9.8, the next event was given by $\min(t_s, t_a, t_\delta)$. However, if there are c channels in parallel, then a service in any one of the c channels could be the next event. Let t_{si} be the time of the next service in the ith channel. Then the next event to occur in the simulation is given by $\min(t_{s1}, t_{s2}, \ldots, t_{sc}, t_a, t_\delta)$. Thus, to determine the next event, we must compare $c + 2$ event times.

In the single-channel simulator, when the next event was the completion of a service, the number in the system was reduced by one and the length of the queue was compared with zero to determine whether or not another unit was to be placed in the service channel. The same operations are carried out in the multiple-channel simulator. If the number of units in the queue is greater than zero, the queue length is reduced by one, the time of service is generated, and the time of service completion is calculated; that is, if the present service occurs in channel i and the queue length is greater than zero, then the service time τ_{si} for the unit entering channel i is generated and the time of the next service completion in this channel is given by

$$t_{si} = t + \tau_{si} \tag{9.57}$$

Multiple Channels in Parallel

If the queue length is zero at the time of the current service completion in channel i, then we assign an arbitrarily large value to t_{si}, 10^{30} here, since that channel is now empty and a service in channel i cannot be the next event. The reader will recall that the same manipulations were carried out for the single-channel system. However, in the case of multiple channels in parallel we must keep track of service times for each individual channel.

When an arrival occurs in a multiple-channel queueing system, the number in the system is increased by one, the cumulative number of entries to the system is increased by one, and the time of the next arrival is generated. All of these operations are identical to those carried out in the single-channel simulator. The next operation in the single-channel simulator was to compare the number in the system, including the current arrival, with one. If the number in the system was one, the arrival entered the service channel and the time of service completion was generated. If the number in the system was greater than one, the queue length was increased by one. Similar operations are carried out in the multiple-channel simulator. However, in this case the number in the system, including the current arrival, is compared with the number of channels c, since the arrival will enter a service channel if any one of the channels is empty. Therefore, if the number in system is greater than c, the queue length is increased by one. On the other hand, if the number in the system is c or less the arrival enters one of the service channels, and the queue length remains zero. We will assume that the channel the arrival enters is selected at random from those that are currently empty. Suppose that there are j channels empty and that these are channels $i_1, i_2, \ldots, i_j$, where $1 \leq i_1 < i_2 < \cdots < i_j \leq c$. The probability of selecting any channel is given by $1/j$. To select a channel at random from those that are empty, we generate a random number r and compare it with $1/j$. If $r \leq 1/j$, the arrival enters channel i_1, a service time τ_{si} is generated for this channel, and

$$t_{si_1} = t + \tau_{si_1} \tag{9.58}$$

If $1/j < r \leq 2/j$, the arrival enters channel i_2 and

$$t_{si_2} = t + \tau_{si_2} \tag{9.59}$$

Therefore, if $(k-1)/j < r \leq k/j$, $1 \leq k \leq j$, the arrival enters channel i_k and t_{si_k} is generated.

To determine which channels are empty, we compare t_{si}, $i = 1, 2, \ldots, c$, with 10^{30}. If $t_{si} < 10^{30}$, then a service completion time for the ith channel has been recorded and that channel is busy. If $t_{si} = 10^{30}$, then the ith channel is empty. Therefore, looking at each channel in order starting with $i = 1$, i_1 is the first value of i for which $t_{si} = 10^{30}$. Similarly, i_2 is the second value of *i* for which $t_{si} = 10^{30}$, i_3 is the third value of *i* for which $t_{si} = 10^{30}$, and so forth.

When the next event to occur is the termination of the simulation run, the simulator calculates the statistics of interest to the analyst. For the simulator presented here, the statistics calculated will be the same as those given in the single-channel simulator developed in Example 9.8. With one exception, these calculations are identical to those for the single-channel simulator. In the multiple-channel simulator, we will calculate the utilization for each individual channel. If U_i is the utilization for the *i*th channel, then

$$U_i = \frac{\text{time of occupancy of the } i\text{th channel}}{\text{duration of the simulation}} \qquad (9.60)$$

The reader will recall that the single-channel simulator was developed for the case of exponential interarrival and service times. This restriction will not be placed on the multiple-channel queueing simulator. Specifically, the simulator allows the user to generate interarrival and service times from the Erlang, normal, beta, or uniform distribution. The parameters of the distributions used are defined by the user. For example, suppose a queueing system with three channels in parallel is to be simulated for a period of 10 yr, where interarrival times are normally distributed with mean 0.10 yr and standard deviation 0.01 yr. Service time in the first channel is uniformly distributed on the interval (0.01, 0.08), service time in the second channel is beta distributed with parameters 100 and 1, and service time in the third channel is beta distributed with $\lambda = 200$ and $n = 4$. By appropriately defining the input variables, the simulator developed here can be used to analyze this system.

Definition of Variables

The variables TS(I), TQ(I), SYSNT, QLEN, TSYS, TQUEUE, SYSN, TLAST, TNEXT, IX, ASYS, and AQUEUE are defined as in Example 9.8. In addition, the following variables will be used:

IC = number of channels in parallel

UTIL(I) = utilization of the Ith channel, I = 1, 2, ..., IC

TIM(I) = $\begin{cases} \text{time of the next service in channel I, I} = 1, 2, \ldots, \text{IC} \\ \text{time of the next arrival, I} = \text{IC} + 1 \\ \text{duration of the simulation, I} = \text{IC} + 2 \end{cases}$

A, B = parameters of the interarrival-time distribution
C(I), D(I) = parameters of the service-time distribution for the Ith channel

The variable DIST(I) defines the distribution from which a given random variable is to be generated. For I = 1, 2, ..., IC,

$$\text{DIST(I)} = \begin{cases} 1, & \text{service time in the Ith channel is Erlang} \\ & \text{distributed with } \lambda = C(I), n = D(I) \\ 2, & \text{service time in the Ith channel is normally} \\ & \text{distributed with } m = C(I), \sigma = D(I) \\ 3, & \text{service time in the Ith channel is beta} \\ & \text{distributed with } a = C(I), b = D(I) \\ 4, & \text{service time in the Ith channel is uniformly} \\ & \text{distributed with } a = C(I), b = D(I) \\ 5, & \text{service time in the Ith channel is constant with} \\ & \text{value } C(I) \end{cases}$$

For I = IC + 1, DIST(I) specifies the distribution of interarrival time, and

$$\text{DIST(I)} = \begin{cases} 1, & \text{interarrival time is Erlang distributed with} \\ & \lambda = A, n = B \\ 2, & \text{interarrival time is normally distributed with} \\ & m = A, \sigma = B \\ 3, & \text{interarrival time is beta distributed with } a = A, \\ & b = B \\ 4, & \text{interarrival time is uniformly distributed with} \\ & a = A, b = B \\ 5, & \text{interarrival time is constant with value A} \end{cases}$$

The parameters given in the definitions for DIST(I) are as given in Table 9.2.

The flowcharts for the main program and subroutine UPDATE are given in Figs. 9.13 and 9.14. The FORTRAN IV program for this simulator is given in Fig. 9.15. All interarrival and service times are generated in the function subprogram RNVAR. Thus, whenever a service time or interarrival time is generated, FUNCTION RNVAR must be referenced. The flowchart for FUNCTION RNVAR is given in Fig. 9.16. The arguments IX and IC are as already defined. The argument I is used to determine whether the time to be generated is for the next arrival or for a service in one of the channels; that is, if I = IC + 1, the time to be generated is for the next arrival, and the distribution is of type DIST(IC + 1) with parameters A and B. If I = 1, 2, ..., IC, the time to be generated is for a service in channel I. In this case service time is a random variable that belongs to the family defined by DIST(I) and has parameters C(I) and D(I).

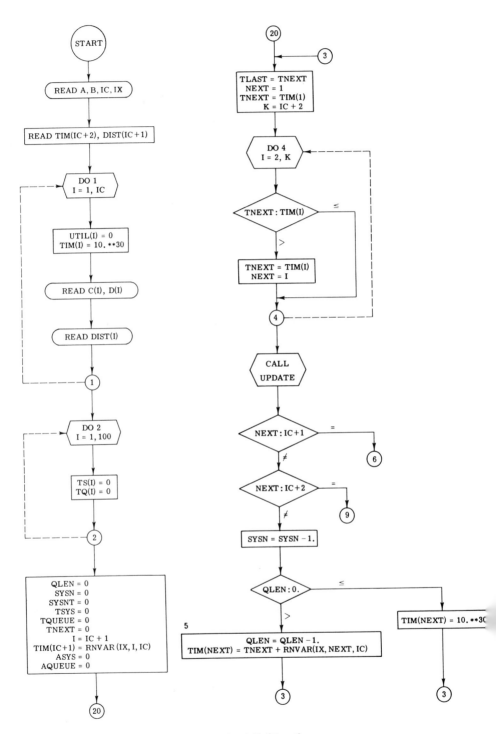

Fig. 9.13 (*Part 1*)

Multiple Channels in Parallel

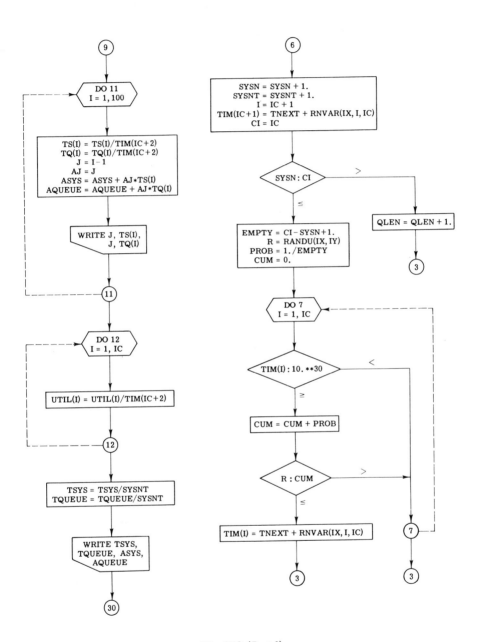

Fig. 9.13 (*Part 2*)

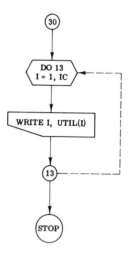

Fig. 9.13 Flowchart of the main program for a multiple-parallel-channel queueing simulator.

Multiple Channels in Parallel

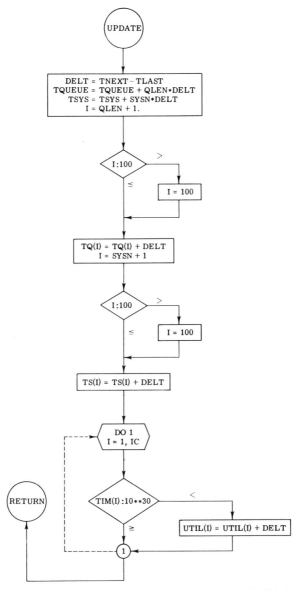

Fig. 9.14 Flowchart for subroutine UPDATE for the multiple-parallel-channel queueing simulator.

```
      DIMENSION TS(100),TQ(100),TIM(50),UTIL(50),C(50),D(50),DIST(51)
      COMMON/BLOKA/TS,TQ,UTIL,TIM,TLAST,TNEXT,QLEN,TQUEUE,SYSN,TSYS,IC
      COMMON/BLOKB/A,B,C,D,DIST
C READ IN INPUT DATA, SET INITIAL CONDITIONS, AND INITIALIZE ACCUMULATORS AND
C AND COUNTERS *************************************************************
      READ(5,100)A,B,IC,IX
      WRITE(6,100)A,B,IC,IX
      WRITE(6,100) A,B,IC,IX
      READ(5,100) TIM(IC+2),DIST(IC+1)
      WRITE(6,100) TIM(IC+2),DIST(IC+1)
      DO 1 I=1,IC
      UTIL(I)=0
      TIM(I)=10.**30
      READ(5,100) C(I),D(I)
      WRITE(6,100) C(I),D(I)
      READ(5,100) DIST(I)
    1 WRITE(6,100) DIST(I)
      DO 2 I=1,100
      TS(I)=0
    2 TQ(I)=0
      QLEN=0
      SYSN=0
      SYSNT=0
      TSYS=0
      TQUEUE=0
      TNEXT=0
      ASYS=0
      AQUEUE=0
      I=IC+1
      TIM(IC+1)=RNVAR(IX,I,IC)
C ****************************************************************************
C IDENTIFY THE NEXT EVENT AND THE TIME OF ITS OCCURRENCE ####################
    3 TLAST=TNEXT
      NEXT=1
      TNEXT=TIM(1)
      K=IC+2
      DO 4 I=2,K
      IF(TNEXT.LE.TIM(I)) GO TO 4
      TNEXT=TIM(I)
      NEXT=I
    4 CONTINUE
C ############################################################################
C UPDATE STATISTICS AND BRANCH TO THE NEXT EVENT <<<<<<<<<<<<<<<<<<<<<<<<<<<<<
      CALL UPDATE
      IF(NEXT.EQ.IC+1) GO TO 6
      IF(NEXT.EQ.IC+2) GO TO 9
C <<<<<<<<<<<<<<<<<<<<<<<<<<<<<<<<<<<<<<<<<<<<<<<<<<<<<<<<<<<<<<<<<<<<<<<<<<<
C SERVICE ROUTINE ***********************************************************
      SYSN=SYSN-1.
      IF(QLEN.GT.0.) GO TO 5
      TIM(NEXT)=10.**30
      GO TO 3
    5 QLEN=QLEN-1.
      TIM(NEXT)=TNEXT+RNVAR(IX,NEXT,IC)
      GO TO 3
C ****************************************************************************
C ARRIVAL ROUTINE ############################################################
    6 SYSN=SYSN+1.
      SYSNT=SYSNT+1.
      TIM(IC+1)=TNEXT+RNVAR(IX,NEXT,IC)
      CI=IC
      IF(SYSN.GT.CI) GO TO 8
C SELECT THE CHANNEL ENTERED BY THE ARRIVAL AT RANDOM ***********************
      EMPTY=CI-SYSN+1.
      R=RANDU(IX,IY)
      PROB=1./EMPTY
      CUM=0
      DO 7 I=1,IC
      IF(TIM(I).LT.10.**30) GO TO 7
      CUM=CUM+PROB
      IF(R.GT.CUM) GO TO 7
      TIM(I)=TNEXT+RNVAR(IX,I,IC)
      GO TO 3
    7 CONTINUE
C ****************************************************************************
      GO TO 3
    8 QLEN=QLEN+1.
      GO TO 3
C ############################################################################
C ROUTINE FOR TERMINATION OF THE SIMULATION <<<<<<<<<<<<<<<<<<<<<<<<<<<<<<<<<<
    9 WRITE(6,200)
      DO 11 I=1,100
      TS(I)=TS(I)/TIM(IC+2)
      TQ(I)=TQ(I)/TIM(IC+2)
      J=I-1
      AJ=J
      ASYS=ASYS+AJ*TS(I)
      AQUEUE=AQUEUE+AJ*TQ(I)
      IF(J.EQ.99) GO TO 10
      WRITE(6,300) J,TS(I),J,TQ(I)
      GO TO 11
   10 WRITE(6,400) J,TS(I),J,TQ(I)
```

Fig. 9.15 FORTRAN IV program for the multiple–parallel-channel simulator.

```
   11 CONTINUE
      DO 12 I=1,IC
   12 UTIL(I)=UTIL(I)/TIM(IC+2)
      TSYS=TSYS/SYSNT
      TQUEUE=TQUEUE/SYSNT
      WRITE(6,500) TSYS,TQUEUE,ASYS,AQUEUE
      DO 13 I=1,IC
   13 WRITE(6,600) I,UTIL(I)
C <<<<<<<<<<<<<<<<<<<<<<<<<<<<<<<<<<<<<<<<<<<<<<<<<<<<<<<<<<<<<<<<<
  100 FORMAT(2F10.4,2I5)
  200 FORMAT(1X,'NO. IN SYSTEM',3X,'PROBABILITY',5X,'NO. IN QUEUE',3X,'P
     1ROBABILITY')
  300 FORMAT(1X,I3,13X,E11.5,5X,I3,12X,E11.5)
  400 FORMAT(1X,I3,1X,'OR MORE',5X,E11.5,5X,I3,1X,'OR MORE',4X,E11.5///)
  500 FORMAT(1X,'AVERAGE TIME IN SYSTEM=',F10.4/1X,'AVERAGE TIME IN QUEU
     1E=',F10.4/1X,'AVERAGE NUMBER IN SYSTEM=',F10.4/1X,'AVERAGE NUMBER
     2IN QUEUE=',F10.4)
  600 FORMAT(1X,'UTIL(',I2,')=',F10.4)
      STOP
      END

      SUBROUTINE UPDATE
      DIMENSION TS(100),TQ(100),TIM(50),UTIL(50)
      COMMON/BLOKA/TS,TQ,UTIL,TIM,TLAST,TNEXT,QLEN,TQUEUE,SYSN,TSYS,IC
C ACCUMULATE UNIT-TIME IN THE QUEUE AND UNIT-TIME IN THE SYSTEM ***************
      DELT=TNEXT-TLAST                                              *
      TQUEUE=TQUEUE+QLEN*DELT                                       *
      TSYS=TSYS+SYSN*DELT                                           *
C *****************************************************************
C ACCUMULATE THE TIME FOR WHICH THERE WERE I-1 UNITS IN THE QUEUE AND THE TIME
C FOR WHICH THERE WERE I-1 UNITS IN THE SYSTEM ###############################
      I=QLEN+1.                                                     #
      IF(I.GT.100) I=100                                            #
      TQ(I)=TQ(I)+DELT                                              #
      I=SYSN+1.                                                     #
      IF(I.GT.100) I=100                                            #
      TS(I)=TS(I)+DELT                                              #
C ############################################################################
C ACCUMULATE THE UTILIZATION TIME FOR EACH CHANNEL <<<<<<<<<<<<<<<<<<<<<<<<<<<<
      DO 1 I=1,IC                                                   <
      IF(TIM(I).GE.10.**30) GO TO 1                                 <
      UTIL(I)=UTIL(I)+DELT                                          <
    1 CONTINUE                                                      <
C <<<<<<<<<<<<<<<<<<<<<<<<<<<<<<<<<<<<<<<<<<<<<<<<<<<<<<<<<<<<<<<<<<<<<<<<<<<<
      RETURN
      END

      FUNCTION RNVAR(IX,I,IC)
      DIMENSION C(50),D(50),DIST(51)
      COMMON/BLOKB/A,B,C,D,DIST
C IDENTIFY DISTRIBUTION AND PARAMETERS ****************************
      M=DIST(I)                                                     *
      IF(I.EQ.IC+1) GO TO 1                                         *
      A1=C(I)                                                       *
      A2=D(I)                                                       *
      GO TO 2                                                       *
    1 A1=A                                                          *
      A2=B                                                          *
C *****************************************************************
C BRANCH TO APPROPRIATE PROCESS GENERATOR #########################
    2 GO TO (3,5,6,7,8),M                                           #
C #################################################################
C PROCESS GENERATOR FOR ERLANG RANDOM VARIABLES <<<<<<<<<<<<<<<<<<<
    3 IA=A2                                                         <
      RNVAR=0                                                       <
      DO 4 J=1,IA                                                   <
    4 RNVAR=RNVAR-(1./A1)*ALOG(RANDU(IX,IY))                        <
      RETURN                                                        <
C <<<<<<<<<<<<<<<<<<<<<<<<<<<<<<<<<<<<<<<<<<<<<<<<<<<<<<<<<<<<<<<<<
C PROCESS GENERATOR FOR NORMAL RANDOM VARIABLES ********************
    5 R1=RANDU(IX,IY)                                               *
      R2=RANDU(IX,IY)                                               *
      RNVAR=A1+A2*SQRT(-2.*ALOG(R1))*COS(6.28*R2)                   *
      RETURN                                                        *
C *****************************************************************
C PROCESS GENERATOR FOR BETA RANDOM VARIABLES ######################
    6 R1=RANDU(IX,IY)                                               #
      R2=RANDU(IX,IY)                                               #
      A3=1./A1                                                      #
      A4=1./A2                                                      #
      RNVAR=(R1**A3)/(R1**A3+R2**A4)                                #
      IF(R1**A3+R2**A4.GT.1.) GO TO 6                               #
      RETURN                                                        #
C #################################################################
C PROCESS GENERATOR FOR UNIFORM RANDOM VARIABLES <<<<<<<<<<<<<<<<<<
    7 RNVAR=A1+(A2-A1)*RANDU(IX,IY)                                 <
      RETURN                                                        <
C <<<<<<<<<<<<<<<<<<<<<<<<<<<<<<<<<<<<<<<<<<<<<<<<<<<<<<<<<<<<<<<<<
C PROCESS GENERATOR FOR A CONSTANT *********************************
    8 RNVAR=A1                                                      *
      RETURN                                                        *
C *****************************************************************
      END

      FUNCTION RANDU(IX,IY)
      IY=IX*65539
      IF(IY)5,6,6
    5 IY=IY+2147483647+1
    6 YFL=IY
      RANDU=YFL*.4656613E-9
      IX=IY
      RETURN
      END
```

Fig. 9.15 (*continued*)

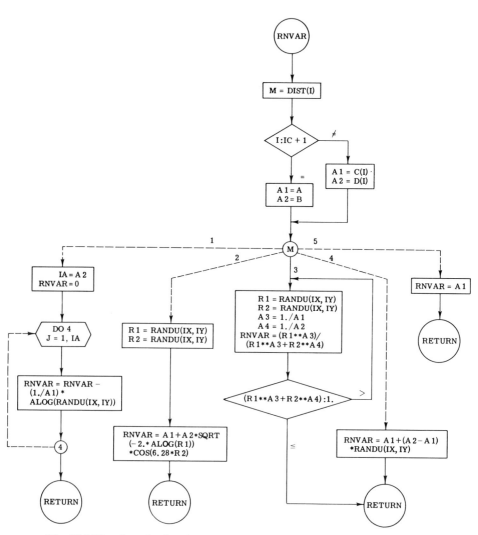

Fig. 9.16 Flowchart for function RNVAR for the multiple-parallel-channel queueing simulator.

Multiple Channels in Parallel

Example 9.10 Simulate the operation of a queueing system with five channels in parallel for a period of 20 yr. Interarrival time is exponentially distributed with a mean arrival rate of 100/yr. Service time in each channel is exponentially distributed with a mean service rate of 40/yr. Calculate the following for the steady state:

(a) the probability mass function of the number in the system,
(b) the probability mass function of the number in the queue,
(c) the average number in the system,
(d) the average number in the queue,
(e) the average time in the system,
(f) the average time in the queue.

Compare the above results with those that would be expected from a mathematical analysis of the system.

The input data for the simulator given in Fig. 9.16 are

$$A = 100.$$
$$B = 1.$$
$$IC = 5$$
$$IX = \text{five-digit odd integer}$$
$$TIM(IC + 2) = 20.$$
$$DIST(IC + 1) = 1.$$
$$C(I) = 40.0, \quad I = 1, 2, 3, 4, 5$$
$$D(I) = 0.0, \quad I = 1, 2, 3, 4, 5$$
$$DIST(I) = 1.0, \quad I = 1, 2, 3, 4, 5$$

The results of 20 yr of simulation of the multiple–parallel-channel queueing system are given in Tables 9.11 and 9.12.

In our treatment of single– and multiple–parallel-channel queueing simulators, we have discussed the estimation of system parameters such as the mean number in the system and the queue, and the mean time in the system and the queue. In each case we used a sample statistic, the sample mean, as the estimator of the corresponding population parameter. However, each of these estimators is a point estimate for the corresponding population parameter, and we have provided no information regarding the precision with which these statistics estimate the corresponding population parameters. We may measure this precision by means of a confidence interval.

TABLE 9.11 Simulated and theoretical probability mass functions (pmf) for the number in the system and the number in the queue for Example 9.10

	Number in the system			Number in the queue	
Number	Simulated pmf	Theoretical pmf	Number	Simulated pmf	Theoretical pmf
0	0.0964	0.0801	0	0.9515	0.9348
1	0.2166	0.2003	1	0.0230	0.0163
2	0.2620	0.2503	2	0.0117	0.0082
3	0.2010	0.2086	3	0.0098	0.0041
4	0.1231	0.1304	4	0.0024	0.0020
5	0.0524	0.0652	5	0.0012	0.0010
6	0.0230	0.0163	6	0.0004	0.0005
7	0.0117	0.0082	7	0.0000	0.0003
8	0.0098	0.0041	8	0.0000	0.0001
9	0.0024	0.0020	9	0.0000	0.0001
10	0.0012	0.0010	10	0.0000	0.0000
11	0.0004	0.0005	11	0.0000	0.0000

TABLE 9.12 Comparison of simulated statistics with expected values for Example 9.10

Quantity	Simulated value	Expected value
Mean time in system	0.0252 yr	0.0263 yr
Mean time in queue	0.0010 yr	0.0013 yr
Mean number in system	2.4338	2.6304
Mean number in queue	0.0934	0.1304

For example, suppose that the random variable X is the time a unit spends in the system. Let $\bar{X}$ be the sample mean of n observations of X. Then

$$E(\bar{X}) = m \qquad (9.61)$$

and a $(1 - \alpha)\, 100\%$ confidence interval for m is given by

$$(L, U) = \left[\bar{x} - \frac{s}{\sqrt{n}} t_{1-\alpha/2}(n-1),\ \bar{x} + \frac{s}{\sqrt{n}} t_{1-\alpha/2}(n-1) \right] \qquad (9.62)$$

where L and U are the upper and lower confidence limits for m, s is the sample standard deviation of X given by

$$s = \left[\frac{\sum_{i=1}^{n} x_i^2 - n\bar{x}^2}{n-1} \right]^{1/2} \qquad (9.63)$$

Multiple Channels in Parallel

and $t_{1-\alpha/2}(n-1)$ is the tabular value of the t random variable with $n-1$ degrees of freedom such that

$$P[t(n-1) \le t_{1-\alpha/2}(n-1)] = 1 - \frac{\alpha}{2} \qquad (9.64)$$

We have already discussed the calculation of $\bar{X}$ for single- and multiple-parallel-channel queueing simulators. We will illustrate the calculation of s by showing how the standard deviation of time in the system can be calculated in a queueing simulator. Let x_i be the time the ith unit spends in the system. If t_{ei} is the time the ith unit enters the system and t_{li} is the time unit leaves the system, then

$$x_i = t_{li} - t_{ei} \qquad (9.65)$$

If there are k entries to the system during the simulation, then†

$$s^2 = \frac{\sum_{i=1}^{k}(t_{li} - t_{ei})^2 - k\bar{x}^2}{k-1} \qquad (9.66)$$

From (9.66), it can be seen that the time of entry to the system and the time at which each unit leaves the system must be recorded by the simulator to calculate the variance or standard deviation of time in the system. Similar calculations are necessary to estimate the variance or standard deviation of time in the queue.

The modifications necessary to compute the standard deviation of time in the system and the standard deviation of time in the queue for the multiple-parallel-channel queueing simulator are given in Fig. 9.17, which shows the main programs for the multiple-parallel-channel queueing simulator only, since the subprograms UPDATE, RNVAR, and RANDU require no modification. The reader will note that the main program shown in Fig. 9.17 is identical to that given in Fig. 9.15 except for the addition of those statements necessary to compute the standard deviation of time in the system and the standard deviation of time in the queue. These statements are

---

† The estimator given in (9.66) is an unbiased estimate of σ^2 if k includes only those units that both enter and leave the system during the simulation period. However, if k includes all entries to the system and if j units are left in the system when the simulation ends, then x_i can be defined by

$$x_i = t_\delta - t_{ei}, \qquad i = j, j+1, \ldots, k$$

In this case s^2 will be a biased estimate of σ^2. However, the effect of this bias is likely to be insignificant unless the duration of the simulation period is short. The latter approach is used in this chapter.

```
      DIMENSION TS(100),TQ(100),TIM(50),UTIL(50),C(50),D(50),DIST(51)
C.............................................................
      DIMENSION TINS(100),TINQ(100)
C.............................................................
      COMMON/BLOKA/TS,TQ,UTIL,TIM,TLAST,TNEXT,QLEN,TQUEUE,SYSN,TSYS,IC
      COMMON/BLOKB/A,B,C,D,DIST
C READ IN INPUT DATA, SET INITIAL CONDITIONS, AND INITIALIZE ACCUMULATORS AND
C AND COUNTERS ****************************************************
      READ(5,100)A,B,IC,IX
      WRITE(6,100) A,B,IC,IX
      READ(5,100) TIM(IC+2),DIST(IC+1)
      WRITE(6,100) TIM(IC+2),DIST(IC+1)
      DO 1 I=1,IC
      UTIL(I)=0
      TIM(I)=10.**30
      READ(5,100) C(I),D(I)
      WRITE(6,100) C(I),D(I)
      READ(5,100) DIST(I)
    1 WRITE(6,100) DIST(I)
      DO 2 I=1,100
      TS(I)=0
    2 TQ(I)=0
      QLEN=0
      SYSN=0
      SYSNT=0
      TSYS=0
      TQUEUE=0
      TNEXT=0
      ASYS=0
      AQUEUE=0
C.............................................................
      SSTSYS=0
      SSTQU=0
C.............................................................
      I=IC+1
      TIM(IC+1)=RNVAR(IX,I,IC)
C ****************************************************************
C IDENTIFY THE NEXT EVENT AND THE TIME OF ITS OCCURRENCE **********
    3 TLAST=TNEXT
      NEXT=1
      TNEXT=TIM(1)
      K=IC+2
      DO 4 I=2,K
      IF(TNEXT.LE.TIM(I)) GO TO 4
      TNEXT=TIM(I)
      NEXT=I
    4 CONTINUE
C ****************************************************************
C UPDATE STATISTICS AND BRANCH TO THE NEXT EVENT <<<<<<<<<<<<<<<<<<
      CALL UPDATE
      IF(NEXT.EQ.IC+1) GO TO 6
      IF(NEXT.EQ.IC+2) GO TO 9
C <<<<<<<<<<<<<<<<<<<<<<<<<<<<<<<<<<<<<<<<<<<<<<<<<<<<<<<<<<<<<<<<
C SERVICE ROUTINE *************************************************
      SYSN=SYSN-1.
C.............................................................
      SSTSYS=SSTSYS+(TINS(NEXT)-TNEXT)**2
C.............................................................
      IF(QLEN.GT.0.) GO TO 5
      TIM(NEXT)=10.**30
      GO TO 3
    5 QLEN=QLEN-1.
      TIM(NEXT)=TNEXT+RNVAR(IX,NEXT,IC)
C.............................................................
      SSTQU=SSTQU+(TINQ(1)-TNEXT)**2
      TINS(NEXT)=TINS(IC+1)
      IF(QLEN.LE.0.) GO TO 3
      K=QLEN
      DO 7777 I=1,K
      TINS(IC+I)=TINS(IC+I+1)
 7777 TINQ(I)=TINQ(I+1)
C.............................................................
      GO TO 3
C ****************************************************************
C ARRIVAL ROUTINE *************************************************
    6 SYSN=SYSN+1.
      SYSNT=SYSNT+1.
      TIM(IC+1)=TNEXT+RNVAR(IX,NEXT,IC)
      CI=IC
      IF(SYSN.GT.CI) GO TO 8
C SELECT THE CHANNEL ENTERED BY THE ARRIVAL AT RANDOM *************
      EMPTY=CI-SYSN+1.
      R=RANDU(IX,IY)
      PROB=1./EMPTY
      CUM=0
      DO 7 I=1,IC
      IF(TIM(I).LT.10.**30) GO TO 7
      CUM=CUM+PROB
      IF(R.GT.CUM) GO TO 7
      TIM(I)=TNEXT+RNVAR(IX,I,IC)
C.............................................................
      TINS(I)=TNEXT
```

```
C.........................................................................  * #
            GO TO 3                                                          * #
         7 CONTINUE                                                          * #
C ##########################################################################   #
            GO TO 3                                                            #
         8 QLEN=QLEN+1.                                                        #
C...........................................................................  #
            K=QLEN                                                             #
            TINS(IC+K)=TNEXT                                               .   #
            TINQ(K)=TNEXT                                                  .   #
C...........................................................................  #
            GO TO 3                                                            #
C ##############################################################################
C ROUTINE FOR TERMINATION OF THE SIMULATION <<<<<<<<<<<<<<<<<<<<<<<<<<<<<<<<<<<
         9 WRITE(6,200)                                                      <
            DO 11 I=1,100                                                    <
            TS(I)=TS(I)/TIM(IC+2)                                            <
            TQ(I)=TQ(I)/TIM(IC+2)                                            <
            J=I-1                                                            <
            AJ=J                                                             <
            ASYS=ASYS+AJ*TS(I)                                               <
            AQUEUE=AQUEUE+AJ*TQ(I)                                           <
            IF(J.EQ.99) GO TO 10                                             <
            WRITE(6,300) J,TS(I),J,TQ(I)                                     <
            GO TO 11                                                         <
        10 WRITE(6,400) J,TS(I),J,TQ(I)                                      <
        11 CONTINUE                                                          <
            DO 12 I=1,IC                                                     <
        12 UTIL(I)=UTIL(I)/TIM(IC+2)                                         <
            TSYS=TSYS/SYSNT                                                  <
            TQUEUE=TQUEUE/SYSNT                                              <
            WRITE(6,500) TSYS,TQUEUE,ASYS,AQUEUE                             <
            DO 13 I=1,IC                                                     <
        13 WRITE(6,600) I,UTIL(I)                                            <
C.......................................................................    <
            IF(SYSN.EQ.0.) GO TO 9999                                    .   <
            K=SYSN                                                       .   <
            DO 8888 I=1,K                                                .   <
            AQ=I                                                         .   <
            IF(QLEN.LT.AQ) GO TO 8888                                    .   <
            SSTQU=SSTQU+(TNEXT-TINQ(I))**2                               .   <
      8888 SSTSYS=SSTSYS+(TNEXT-TINS(I))**2                              .   <
      9999 SSTSYS=SQRT((SSTSYS-SYSNT*(TSYS**2))/(SYSNT-1.))               .   <
            SSTQU=SQRT((SSTQU-SYSNT*(TQUEUE**2))/(SYSNT-1.))             .   <
            WRITE(6,700) SSTSYS,SSTQU,SYSNT                              .   <
C..........................................................................  <
C <<<<<<<<<<<<<<<<<<<<<<<<<<<<<<<<<<<<<<<<<<<<<<<<<<<<<<<<<<<<<<<<<<<<<<<<<<<<<
       100 FORMAT(2F10.4,2I5)
       200 FORMAT(1X,'NO. IN SYSTEM',3X,'PROBABILITY',5X,'NO. IN QUEUE',3X,'P
           1ROBABILITY')
       300 FORMAT(1X,I3,13X,E11.5,5X,I3,12X,E11.5)
       400 FORMAT(1X,I3,1X,'OR MORE',5X,E11.5,5X,I3,1X,'OR MORE',4X,E11.5///)
       500 FORMAT(1X,'AVERAGE TIME IN SYSTEM=',F10.4/1X,'AVERAGE TIME IN QUEU
           1E=',F10.4/1X,'AVERAGE NUMBER IN SYSTEM=',F10.4/1X,'AVERAGE NUMBER
           2IN QUEUE=',F10.4)
       600 FORMAT(1X,'UTIL(',I2,')=',F10.4)
       700 FORMAT(1X,'STANDARD DEVIATION OF TIME IN THE SYSTEM=',F10.4/1X,'ST
           1ANDARD DEVIATION OF TIME IN THE QUEUE=',F10.4/1X,'NUMBER OF ENTRIE
           2S TO THE SYSTEM=',F10.4)
            STOP
            END
```

Fig. 9.17 Main program for the multiple-parallel-channel queueing simulator, where the standard deviation of time in the system and the standard deviation of time in the queue are estimated.

enclosed by rows of periods. The following variables have been added in Fig. 9.17 to the main program given in Fig. 9.15 to allow for the calculation of the standard deviation of time in the system and the standard deviation of time in the queue:

$$\text{TINS(I)} = \text{time unit I entered the system}$$
$$\text{TINQ(I)} = \text{time unit I entered the queue}$$
$$\text{SSTSYS} = \text{sum of squares of time in the system}$$
$$\text{SSTQU} = \text{sum of squares of time in the queue}$$

Example 9.11 Simulate the queueing system described in Example 9.10 for a period of 40 yr and calculate 99% confidence intervals for the mean time in the system and the mean time in the queue.

The input data for this problem are the same as that given in Example 9.10 except that TIM(IC + 2) = 40. in this case. During the simulation period, there were 3947 entries to the system with the following results:

$$\text{average time in the system} = 0.0260 \text{ yr}$$
$$\text{standard deviation of time in the system} = 0.0258 \text{ yr}$$
$$\text{average time in the queue} = 0.0012 \text{ yr}$$
$$\text{standard deviation of time in the queue} = 0.0045 \text{ yr}$$

Let L_s, L_q and U_s, U_q be the upper and lower 95% confidence limits for mean time in the system and mean time in the queue, respectively. Then

$$L_s = 0.0260 - t_{0.975}(3946)\frac{0.0258}{(3947)^{1/2}} = 0.0260 - 2.58\frac{0.0258}{62.8} = 0.249 \text{ yr}$$

$$U_s = 0.0260 + t_{0.975}(3946)\frac{0.0258}{(3947)^{1/2}} = 0.0271 \text{ yr}$$

$$L_q = 0.0012 - t_{0.975}(3946)\frac{0.0045}{(3947)^{1/2}} = 0.0012 - 2.58\frac{0.0045}{62.8} = 0.0010 \text{ yr}$$

$$U_q = 0.0012 + t_{0.975}(3946)\frac{0.0045}{(3947)^{1/2}} = 0.0014 \text{ yr}$$

From the mathematical theory of queues, the expected time in the system is 0.0263 yr, while the expected time in the queue is 0.0013 yr. Although it is not surprising that the calculated confidence intervals (L_s, U_s) and (L_q, U_q)

Multiple Channels in Parallel

contain the true mean times spent in the system and in the queue, respectively, it should be remembered that there was a 5% chance that each calculated interval would not contain the true means prior to running the simulation.

Example 9.12 The problems discussed in Examples 9.10 and 9.11 dealt with a multiple–parallel service system, where the service time in each of the five channels was exponentially distributed with mean rate 40 units/yr. Therefore, when all service channels are operating, the mean total service rate is 200 units/yr. Now consider a system identical to that discussed in Examples 9.10 and 9.11 except that the mean service rates in the service channels are 5, 5, 5, 5, and 180 units/yr. Thus, when all channels are operating, the mean total service rate is 200 as it was in Examples 9.10 and 9.11. By simulating this system for a period of 5 yr, determine whether the mean time in the system has significantly changed from the value of 0.0263 yr, which was the mean time in the system for the case discussed in the preceding two examples; that is, test the hypothesis

$$H_0: \quad E(x) = 0.0263, \quad H_1: \quad E(x) \neq 0.0263$$

using $\alpha = 0.10$, where $\bar{x}$ is the total time a unit spends in the system.

Let $\bar{x}$ be the estimated mean time a unit spends in the system based on a 5-yr simulation and let s_x be the estimated standard deviation of time a unit spends in the system. Then

$$\frac{\bar{x} - E(x)}{s/\sqrt{n}} \sim t(n-1) \tag{9.67}$$

where n is the number of entries to the system. If

$$\left| \frac{\bar{x} - E(x)}{s/\sqrt{n}} \right| > t_{0.95}(n-1)$$

the null hypothesis is rejected.

The input data for the problem are as follows:

$$A = 100.$$
$$B = 0.$$
$$IC = 5$$
$$IX = \text{five-digit odd integer}$$
$$TIM(IC + 2) = 5.$$
$$DIST(IC + 1) = 1.$$

$$C(I) = \begin{cases} 180., & I = 1 \\ 5., & I = 2 \\ 5., & I = 3 \\ 5., & I = 4 \\ 5., & I = 5 \end{cases}$$

$$D(I) = 0., \qquad I = 1, 2, 3, 4, 5$$

$$DIST(I) = 1., \qquad I = 1, 2, 3, 4, 5$$

After 5 yr of simulation, the estimated mean and standard deviation of time in the system were

$$\bar{x} = 0.0466, \qquad s = 0.1219$$

There were 494 entries to the system during the simulation period. Therefore,

$$\left| \frac{\bar{x} - E(x)}{s/\sqrt{n}} \right| = \left| \frac{0.0466 - 0.0263}{0.1219/\sqrt{494}} \right| = 3.757$$

Since

$$t_{0.95}(493) = 1.648$$

we have

$$\left| \frac{\bar{x} - E(x)}{s/\sqrt{n}} \right| > t_{0.95}(n - 1)$$

and the null hypothesis is rejected; that is, the results of a 5-yr simulation lead us to the conclusion that the alteration of the system has led to a significant change in the mean time a unit spends in the system.

Example 9.13 A system of n service channels operates in parallel. However, the parallel arrangement of channels is not fed by a single waiting line, but rather by individual waiting lines as shown in Fig. 9.18. Units in each waiting line are served on a first come–first served basis. When a unit arrives it surveys the n queueing subsystems to determine the subsystem i_m that possesses the least number of units. However, the arriving unit does not necessarily enter subsystem i_m, but may leave (balk). Specifically, the probability that the unit enters i_m is given by

$$P(\text{unit enters system } i_m) = \frac{1}{1 + An_{i_m}} \qquad (9.68)$$

Multiple Channels in Parallel

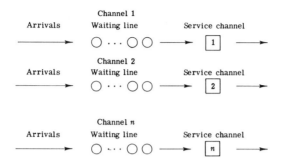

Fig. 9.18 Multiple-service-channel system for Example 9.13.

where A is a positive constant and n_{i_m} the number of units in subsystem i_m. If the unit does not leave the system, it enters the waiting line for subsystem i_m or the service channel if it is not busy.

Units that enter a given subsystem will not remain in the waiting line indefinitely. After a certain period of time the unit becomes impatient and will look for a shorter line or may decide to leave (renege). The density function of the time, w, until a unit becomes impatient is given by

$$h(w) = \frac{1}{a_2(2\pi)^{1/2}} \exp\left[-\frac{(w-a_1)^2}{2a_2^2}\right], \qquad -\infty < w < \infty \qquad (9.69)$$

that is, the time a unit will wait before becoming impatient is normally distributed with mean a_1 and standard deviation a_2. Once a unit becomes impatient it looks for the subsystem i_m that possesses the least number of units. The probability that it enters or jockeys to that subsystem is given by (9.68). If the unit does not enter the subsystem i_m, it leaves or reneges.

Presently two channels are open throughout the day. However, a maximum of five channels could be opened. The entire system operates 8 hr/day and channels can be opened or closed on the hour. When a channel closes it will serve the remaining customers in its waiting line but will not allow entry of new arrivals. Interarrival and service times are exponentially distributed, but the mean arrival rate changes throughout the day and the service rate is a function of the channel operating as given in Tables 9.13 and 9.14.

Management is concerned about the number of customers lost due to balking and reneging. In particular, it would like to keep the number of lost customers to approximately 5/day on the average.

Develop a simulator that will estimate the average number of customers lost per day under the present operating policy. Then vary the number of channels provided per hour of the day in an attempt to determine an allocation of channels that will yield an average number of lost customers per day of five or less.

TABLE 9.13

Time of day	Arrival rate in customers per day
9:00–10:00	50
10:00–11:00	30
11:00–12:00	220
12:00–1:00	70
1:00–2:00	200
2:00–3:00	10
3:00–4:00	50
4:00–5:00	10

TABLE 9.14

Channel number	Service rate (customers/day)
1	42
2	40
3	38
4	35
5	30

To determine the effect of using a given number of channels in a given 1-hr period, the simulator must record the behavior of the system during each 1-hr period. Thus, the simulator must be constructed to record the number of balks, reneges, and units jockeying during each 1-hr period. In addition, the simulator should record the number of arrivals and services per hour to indicate the amount of carryover customers waiting from one hour to the next. This information is important, since the number of units to be served in a given hour is a function of the number of units arriving in that hour and the number of units arriving but not serviced in previous hours. To estimate the mean number of units balking, reneging, and jockeying, the simulation of each day's operation is repeated or replicated NSIM times.

The FORTRAN IV program for the simulator required for this problem is presented in Figs. 9.19–9.24. Function RANDU is the same as presented in the preceding simulators. All variables are defined in the main program. The main program, Fig. 9.11, reads the input information, initializes all variables, counters, and indices for each replicate of the simulation, determines the time and nature of the next event, calls the routine that updates statistical accumulators, and calls the routines that alter the status of the system whenever an event occurs. The main program also alters the number of channels to be available during each hour as defined in the input data.

Subroutine ARRIV is called when the next event is an arrival and determines the subsystem possessing the least number of units, generates the time of the next arrival, and determines whether the arriving unit enters the system or balks using (9.68). If the unit balks, the number of units balking during the present 1-hr period is increased by one. If the unit enters the subsystem possessing the least number of units, then the number in that subsystem is increased by one, the queue length is increased by one if the service channel is busy, or the unit is placed in the service channel if the server is not busy. If the unit enters the waiting line, the time at which that unit will become impatient is generated.

Subroutine SERV is called whenever the next event is a service completion. The number in the subsystem where the service completion occurs is decreased by one. If there are no units in the waiting line, the time of the next service completion in that channel is set equal to the time limit of one replication of the simulation plus one. If the waiting line is not zero, the unit at the head of the waiting line is brought into the service channel, the time of service completion for that unit is generated, and the queue length is reduced by one. Finally, all units, if any, remaining in the waiting line are moved up one position.

Subroutine CHANGE is called whenever the next event in the simulation is impatience on the part of a unit in one of the waiting lines. A unit becomes impatient only when it is in a waiting line; that is, a unit will not renege or jockey once it is in a service channel. Subroutine CHANGE first determines whether the impatient unit can improve its position by moving to another waiting line. If there is no subsystem to which the unit can move and improve its position, then the unit either stays where it is or reneges. If the unit can improve its position by jockeying to another channel, then it either does so or reneges. In either case, Eq. (9.68) is used to determine whether the unit reneges or continues to wait for service. If the unit reneges, the number of units reneging during the current period is increased by one. If the unit does not renege, then the time at which the unit will become impatient again is generated. If the unit jockeys to another waiting line, the number of units jockeying during the current period is increased by one, the number in the waiting line and in the subsystem to which it moves is increased by one if that subsystem is not empty. If the unit jockeys and the subsystem that it enters is empty, the jockeying unit enters the service channel, the number in that subsystem is increased by one, and the time of the next service completion for that service channel is generated. If the unit either reneges or jockeys, the units remaining in the waiting line that the reneging or jockeying unit leaves are moved up one position. If the unit neither reneges or jockeys, then the status of all subsystems remains unchanged.

```
C     NMAX=MAXIMUM NUMBER OF SERVERS AVAILABLE
C     IX=RANDOM NUMBER SEED
C     NPER=NUMBER OF PERIODS DURING WHICH DIFFERENT ARRIVAL RATES OCCUR
C     NSIM=NUMBER OF REPLICATIONS OF THE SIMULATION
C     MPER=NUMBER OF TIMES THE NUMBER OF CHANNELS AVAILABLE FOR SERVICE CHANGE
C         DURING ONE REPLICATION OF THE SIMULATION
C     IPRINT=0, STATUS OF THE SYSTEM IS NOT PRINTED OUT BEFORE EACH EVENT
C            1, STATUS OF THE SYSTEM IS PRINTED OUT BEFORE EACH EVENT
C     AMU(J)=AVERAGE SERVICE RATE FOR THE JTH CHANNEL
C     ALAM(I)=AVERAGE ARRIVAL RATE DURING THE ITH PERIOD
C     A=CONSTANT WHICH DETERMINES WHETHER AN ARRIVING UNIT BALKS OR WHETHER A
C         UNIT ALREADY IN THE SYSTEM RENIGS
C     A1,A2=MEAN AND STANDARD DEVIATION OF THE NORMAL RANDOM VARIABLE USED TO
C         DETERMINE THE LENGTH OF TIME A UNIT REMAINS IN THE WAITING LINE
C         BEFORE HE DECIDES TO JOCKEY OR RENIG
C     TIM(NMAX+3,1)=LENGTH OF EACH REPLICATION OF THE SIMULATION
C     TSER(I)=TIME AT WHICH THE NUMBER OF SERVICE CHANNELS CHANGES DURING A
C         REPLICATION OF THE SIMULATION
C     NSER(I)=NUMBER OF SERVICE CHANNELS AVAILABLE BEFORE THE ITH CHANGE IN THE
C         NUMBER AVAILABLE
C     ARRV(I)=NUMBER OF ARRIVALS DURING THE ITH PERIOD
C     ASERV(I)=NUMBER OF SERVICES DURING THE ITH PERIOD
C     BALK(I)=NUMBER OF BALKS DURING THE ITH PERIOD
C     REN(I)=NUMBER OF RENIGS DURING THE ITH PERIOD
C     AJOC(I)=NUMBER OF UNITS JOCKEYING DURING THE ITH PERIOD
C     NSYS=NUMBER OF CHANNELS CURRENTLY AVAILABLE
C     IPER=CURRENT PERIOD
C     JPER=NUMBER OF THE NEXT CHANGE IN THE NUMBER OF CHANNELS AVAILABLE
C     SYSN(I)=NUMBER OF UNITS CURRENTLY IN THE ITH SERVICE SYSTEM
C     QLEN(I)=NUMBER OF UNITS IN THE ITH WAITING LINE
C     TIM(1,1)=TIME OF THE NEXT ARRIVAL
C     TIM(2,J)=TIME OF THE NEXT SERVICE IN THE JTH CHANNEL
C     TIM(I,J)=TIME AT WHICH THE UNIT IN POSITION J OF WAITING LINE I-2 WILL
C         JOCKEY OR RENIG
C     TNEXT=TIME OF THE NEXT EVENT
C     NEXT=1, THE NEXT EVENT IS AN ARRIVAL
C         =2, THE NEXT EVENT IS A SERVICE
C         =3, THE NEXT EVENT IS A JOCKEY OR A RENIG
C         =4, THE NEXT EVENT IS A CHANGE IN THE NUMBER OF CHANNELS AVAILABLE
C         =5, THE NEXT EVENT IS THE END OF A REPLICATION OF THE SIMULATION
C     JNEXT=CHANNEL IN WHICH THE NEXT SERVICE TAKES PLACE OR THE WAITING LINE IN
C         WHICH THE NEXT JOCKEY TAKES PLACE
C     JPOS=POSITION OF THE NEXT UNIT TO JOCKEY
      DIMENSION TIM(10,20),AMU(10),ALAM(12),SYSN(10),QLEN(10)
      DIMENSION REN(12),BALK(12),AJOC(12),TSER(12),NSER(12),ARRV(12)
      DIMENSION ASERV(12)
      COMMON/BLOCKA/IX
      COMMON/BLOCKB/TIM,TNEXT,NSYS,NMAX
      COMMON/BLOCKC/SYSN,QLEN
      COMMON/BLOCKD/ARRV,ASERV,NPER,NSIM,NEXT,IK
      COMMON/BLOCKE/A
      COMMON/BLOCKF/A1,A2
      COMMON/BLOCKG/REN,AJOC
      COMMON/BLOCKH/AMU
      COMMON/BLOCKI/ALAM
      COMMON/BLOCKJ/BALK
      COMMON/BLOCKK/JPOS
      COMMON/BLOCKL/IPER
C     READ INPUT DATA, INITIALIZE INDICATORS AND ACCUMULATORS ******************
      READ(5,130) NMAX,IX,NPER,NSIM,MPER,IPRINT                          *
      WRITE(6,100)NMAX,IX,NPER,NSIM,MPER,IPRINT                          *
      READ(5,200) (AMU(J),J=1,NMAX),A,A1,A2,TIM(NMAX+3,1)                *
      WRITE(6,200)(AMU(J),J=1,NMAX),A,A1,A2,TIM(NMAX+3,1)                *
      READ(5,200) (ALAM(I),I=1,NPER)                                     *
      WRITE(6,200) (ALAM(I),I=1,NPER)                                    *
      READ(5,300) (TSER(I),NSER(I),I=1,MPER)                             *
      WRITE(6,300)(TSER(I),NSER(I),I=1,MPER)                             *
      NPAR=NMAX+2                                                        *
      DO 1 I=1,NPER                                                      *
      ARRV(I)=0                                                          *
      ASERV(I)=0                                                         *
      BALK(I)=0                                                          *
      REN(I)=0                                                           *
    1 AJOC(I)=0                                                          *
C     ****************************************************************************
C     REPLICATE THE SIMULATION NSIM TIMES ........................................
      DO 14 II=1,NSIM                                                    .
C     INITIALIZE SYSTEM VARIABLES, INDICATORS AND THE TIME OF NEXT SERVICE IN .
C     EACH CHANNEL, AND GENERATE THE FIRST ARRIVAL TIME FOR THE IITH     .
C     REPLICATION OF THE SIMULATION -------------------------------------- .
      NSYS=NSER(1)                                                       - .
      TPER=TSER(1)                                                       - .
      JPER=1                                                             - .
      IPER=1                                                             - .
      DO 2 J=1,NMAX                                                      - .
      SYSN(J)=0                                                          - .
      QLEN(J)=0                                                          - .
      TIM(2,J)=TIM(NMAX+3,1)+1.                                          - .
    2 CONTINUE                                                           - .
      TNEXT=0                                                            - .
      TIM(1,1)=-(1./ALAM(1))*ALOG(RANDU(IX,IY))                          - .
C     ---------------------------------------------------------------------- .
```

```
C       DETERMINE THE TIME AND NATURE OF THE NEXT EVENT TO OCCUR IN THE
C       SIMULATION *********************************************************
      3 NEXT=1
C       WRITE OUT THE STATUS OF THE SYSTEM AT THE PRESENT TIME IF IPRINT IS
C       GREATER THAN 0 -----------------------------------------------------
        IF(IPRINT.GT.0) WRITE(6,400) JPER,NSYS,TPER
        IF(IPRINT.GT.0) WRITE(6,600) (BALK(K1),REN(K1),AJOC(K1),K1=1,NPER)
        IF(IPRINT.GT.0) WRITE(6,1000) TIM(1,1)
        TNEXT=TIM(1,1)
        DO 4 J=1,NMAX
        IF(IPRINT.GT.0) WRITE(6,500) TIM(2,J),SYSN(J),QLEN(J),J
        IF(SYSN(J).EQ.0.) GO TO 4
        IF(TIM(2,J).GT.TNEXT) GO TO 4
        TNEXT=TIM(2,J)
        JNEXT=J
        NEXT=2
      4 CONTINUE
        DO 6 I=3,NPAR
        IF(SYSN(I-2).LE.1.) GO TO 6
        NPOS=SYSN(I-2)
        DO 5 J=2,NPOS
        IF(IPRINT.GT.0) WRITE(6,800) TIM(I,J),I,J
C       -------------------------------------------------------------------
        IF(TIM(I,J).GT.TNEXT) GO TO 5
        TNEXT=TIM(I,J)
        NEXT=3
        JNEXT=I-2
        JPOS=J
      5 CONTINUE
      6 CONTINUE
        IF(TPER.GE.TNEXT) GO TO 7
        TNEXT=TPER
        NEXT=4
      7 IF(TIM(NMAX+3,1).GE.TNEXT) GO TO 8
        TNEXT=TIM(NMAX+3,1)
        NEXT=5
C       ********************************************************************
C       CALL UPDATING ROUTINE AND WRITE OUT STATUS OF THE SYSTEM IF IPRINT IS
C       GREATER THAN 0 -----------------------------------------------------
      8 CALL UPDATE(II)
        IF(IK.EQ.1) GO TO 3
        IF(IPRINT.GT.0) WRITE(6,900) IPER
        IF(IPRINT.GT.0) WRITE(6,700) TNEXT,NEXT
C       -------------------------------------------------------------------
C       BRANCH TO NEXT EVENT ROUTINE ***************************************
        GO TO (9,10,11,12,14),NEXT
      9 CALL ARRIV
        ARRV(IPER)=ARRV(IPER)+1.
        GO TO 3
     10 CALL SERV(JNEXT)
        ASERV(IPER)=ASERV(IPER)+1.
        GO TO 3
     11 CALL CHANGE(JNEXT)
        GO TO 3
     12 JPER=JPER+1
        IF(JPER.GT.MPER) GO TO 13
        TPER=TSER(JPER)
        NSYS=NSER(JPER)
        GO TO 3
     13 TPER=TIM(NMAX+3,1)+1.
        GO TO 3
     14 CONTINUE
C       ********************************************************************
        STOP
    100 FORMAT(1X,6I5)
    200 FORMAT(1X,7F10.4)
    300 FORMAT(1X,F10.4,I5)
    400 FORMAT(1X,'JPER=',I5,1X,'NSYS=',I5,1X,'TPER=',F10.4)
    500 FORMAT(1X,'TIM(2,J)=',F14.7,2X,'SYSN(J)=',F14.7,2X,'QLEN(J)=',E14.
       17,2X,'J=',I5)
    600 FORMAT(1X,'BALK=',F10.4,2X,'REN=',F10.4,2X,'AJOC=',F10.4)
    700 FORMAT(1X,'TNEXT=',E14.7,2X,'NEXT=',I5)
    800 FORMAT(1X,'TIM(I,J)=',F14.7,2X,'I=',I5,2X,'J=',I5)
    900 FORMAT(1X,'IPER=',I5)
   1000 FORMAT(1X,'TIM(1,1)=',F14.7)
        END
```

Fig. 9.19 Main program for the simulator in Example 9.13.

```
      SUBROUTINE ARRIV
      DIMENSION SYSN(10),TIM(10,20),ALAM(12),BALK(12),QLEN(10)
      DIMENSION AMU(10)
      COMMON/BLOKA/IX
      COMMON/BLOKB/TIM,TNEXT,NSYS,NMAX
      COMMON/BLOKC/SYSN,QLEN
      COMMON/BLOKE/A
      COMMON/BLOKF/A1,A2
      COMMON/BLOKH/AMU
      COMMON/BLOKI/ALAM
      COMMON/BLOKJ/BALK
      COMMON/BLOKL/IPER
C     FIND THE SUBSYSTEM POSSESSING THE MINIMUM NUMBER OF UNITS ***************
      AMIN=SYSN(1)                                                             *
      JMIN=1                                                                   *
      IF(NSYS.EQ.1) GO TO 2                                                    *
      DO 1 I=2,NSYS                                                            *
      IF(AMIN.LT.SYSN(I)) GO TO 1                                              *
      AMIN=SYSN(I)                                                             *
      JMIN=I                                                                   *
    1 CONTINUE                                                                 *
C     *************************************************************************
C     GENERATE THE TIME OF ARRIVAL OF THE NEXT UNIT ...........................
    2 TIM(1,1)=TNEXT-(1./ALAM(IPER))*ALOG(RANDU(IX,IY))                        .
C     .........................................................................
C     DETERMINE WHETHER OR NOT THE ARRIVING UNIT REMAINS IN THE SYSTEM OR
C     BALKS -------------------------------------------------------------------
      IF(RANDU(IX,IY).LT.1./(1.+A*AMIN)) GO TO 3                               -
C     -------------------------------------------------------------------------
C     INCREASE THE NUMBER OF BALKS IN PERIOD IPER BY ONE *********************
      BALK(IPER)=BALK(IPER)+1.                                                 *
C     *************************************************************************
      RETURN
C     ADJUST THE NUMBER IN THE SUBSYSTEM AND, IF NECESSARY, THE NUMBER IN THE
C     QUEUE FOR SUBSYSTEM JMIN ................................................
    3 SYSN(JMIN)=SYSN(JMIN)+1.                                                 .
      J=SYSN(JMIN)                                                             .
      IF(SYSN(JMIN).LE.1.) GO TO 4                                             .
      QLEN(JMIN)=QLEN(JMIN)+1.                                                 .
C     .........................................................................
C     GENERATE THE TIME AT WHICH THE ARRIVING UNIT BECOMES IMPATIENT ----------
      TIM(JMIN+2,J)=TNEXT+TCHANG(A1,A2)                                        -
C     -------------------------------------------------------------------------
      RETURN
C     GENERATE THE TIME OF SERVICE COMPLETION FOR THE ARRIVING UNIT ***********
    4 TIM(2,JMIN)=TNEXT-(1./AMU(JMIN))*ALOG(RANDU(IX,IY))                      *
C     *************************************************************************
      RETURN
      END
```

Fig. 9.20 Subroutine ARRIV for the simulator in Example 9.13.

```
      SUBROUTINE SERV(J)
      DIMENSION SYSN(10),TIM(10,20),QLEN(10),AMU(10)
      COMMON/BLOKA/IX
      COMMON/BLOKB/TIM,TNEXT,NSYS,NMAX
      COMMON/BLOKC/SYSN,QLEN
      COMMON/BLOKH/AMU
C     REDUCE THE NUMBER IN SUBSYSTEM J BY ONE AND, IF THERE IS NO WAITING LINE,
C     SET THE TIME OF THE NEXT SERVICE IN SUBSYSTEM J TO A NUMBER GREATER THAN
C     THE PERIOD OF ONE REPLICATION OF THE SIMULATION *************************
      SYSN(J)=SYSN(J)-1.                                                       *
      IF(SYSN(J).GT.0.) GO TO 1                                                *
      TIM(2,J)=TIM(NMAX+3,1)+1.                                                *
C     *************************************************************************
      RETURN
C     REDUCE THE NUMBER IN QUEUE J BY ONE, GENERATE THE TIME OF THE NEXT
C     SERVICE COMPLETION IN CHANNEL J, AND, IF THE RESULTING WAITING LINE IS OF
C     LENGTH GREATER THAN 0, MOVE EACH UNIT IN WAITING LINE J UP ONE POSITION..
    1 QLEN(J)=QLEN(J)-1.                                                       .
      TIM(2,J)=TNEXT-(1./AMU(J))*ALOG(RANDU(IX,IY))                            .
      IF(QLEN(J).EQ.0.) RETURN                                                 .
      K=SYSN(J)                                                                .
      DO 2 K1=2,K                                                              .
    2 TIM(J+2,K1)=TIM(J+2,K1+1)                                                .
C     .........................................................................
      RETURN
      END
```

Fig. 9.21 Subroutine SERV for the simulator in Example 9.13.

```
                  SUBROUTINE CHANGE(J)
                  DIMENSION SYSN(10),REN(12),TIM(10,20),QLEN(10),AJOC(12),SYSNT(10)
                  DIMENSION AMU(10)
                  COMMON/BLOKA/IX
                  COMMON/BLOKB/TIM,TNEXT,NSYS,NMAX
                  COMMON/BLOKC/SYSN,QLEN
                  COMMON/BLOKE/A
                  COMMON/BLOKF/A1,A2
                  COMMON/BLOKG/REN,AJOC
                  COMMON/BLOKH/AMU
                  COMMON/BLOKK/JPOS
                  COMMON/BLOKL/IPER
          C       DETERMINE THE SUBSYSTEM POSSESSING THE LEAST NUMBER OF UNITS *************
                  APOS=JPOS                                                               *
                  JMIN=J                                                                  *
                  DO 1 J1=1,NSYS                                                          *
                  IF(J1.EQ.J) GO TO 1                                                     *
                  IF(SYSN(J1).GE.APOS-1.) GO TO 1                                         *
                  APOS=SYSN(J1)+1.                                                        *
                  JMIN=J1                                                                 *
                1 CONTINUE                                                                *
          C       ==***************************************************************
          C       DETERMINE WHETHER OR NOT THE UNIT RENIGS ................................
                  IF(RANDU(IX,IY).LE.1./(1.+A*APOS)) GO TO 5                              .
          C       ........................................................................
          C       INCREASE THE NUMBER OF UNITS RENIGING IN PERIOD IPER BY ONE -------------
                  REN(IPER)=REN(IPER)+1.                                                  -
          C       ------------------------------------------------------------------------
          C       DECREASE THE NUMBER OF UNITS REMAINING IN THE SUBSYSTEM WHICH THE
          C       RENIGING OR JOCKEYING UNIT LEFT BY ONE AND MOVE EACH REMAINING UNIT IN
          C       THAT SUBSYSTEM UP ONE POSITION IN THE WAITING LINE *********************
                2 APOS=JPOS                                                               *
                  IF(APOS.EQ.SYSN(J)) GO TO 4                                             *
                  K=SYSN(J)-1.                                                            *
                  DO 3 K1=JPOS,K                                                          *
                3 TIM(J+2,K1)=TIM(J+2,K1+1)                                               *
                4 SYSN(J)=SYSN(J)-1.                                                      *
                  QLEN(J)=QLEN(J)-1.                                                      *
          C       *************************************************************************
                  RETURN
          C       GENERATE THE TIME AT WHICH THE JOCKEYING UNIT BECOMES IMPATIENT AGAIN,  .
          C       INCREASE THE NUMBER IN THE SUBSYSTEM TO WHICH THE JOCKEYING UNIT MOVES BY .
          C       ONE, IF NECESSARY INCREASE THE NUMBER OF UNITS IN THE QUEUE ENTERED BY  .
          C       ONE, INCREASE THE NUMBER OF UNITS JOCKEYING BY ONE, AND IF THE SUBSYSTEM .
          C       ENTERED IS EMPTY PLACE THE JOCKEYING UNIT IN THE SERVICE CHANNEL AND    .
          C       GENERATE THE TIME OF THE NEXT SERVICE COMPLETION IN THAT CHANNEL ........
                5 K=APOS                                                                  .
                  TIM(JMIN+2,K)=TNEXT+TCHANG(A1,A2)                                       .
                  IF(J.NE.JMIN) AJOC(IPER)=AJOC(IPER)+1.                                  .
                  IF(J.EQ.JMIN) RETURN
                  SYSN(JMIN)=SYSN(JMIN)+1.                                                .
                  IF(SYSN(JMIN).EQ.1.) GO TO 6                                            .
                  QLEN(JMIN)=QLEN(JMIN)+1.                                                .
                  GO TO 2                                                                 .
                6 TIM(2,JMIN)=TNEXT-(1./AMU(JMIN))*ALOG(RANDU(IX,IY))                     .
                  GO TO 2                                                                 .
          C       .........................................................................
                  END
```

Fig. 9.22 Subroutine CHANGE for the simulator in Example 9.13.

Subroutine UPDATE is called after the time of the next event is determined but before the status of any subsystem is changed due to the next event. This routine first determines whether the next event occurs in the present 1-hr period or in the next. If the next event occurs in the next 1-hr period, then the indicator defining the period number is increased by one. Otherwise this indicator remains unchanged. If the next event occurs in the next 1-hr period, then the arrival rate changes and the next arrival is regenerated using the arrival rate for the next period. If the next event is the end of the simulation period for the last replication, subroutine UPDATE calculates and writes out the average number of balks, reneges, jockeys, arrivals, and services for each 1-hr period.

Function TCHANG generates the time interval a unit remains in the waiting line before becoming impatient. This random variable is normally distributed as defined in Eq. (9.69).

```
      SUBROUTINE UPDATE(I1)
      DIMENSION TIM(10,20),BALK(12),REN(12),AJOC(12),ARRV(12),ASERV(12)
      DIMENSION ALAM(12)
      COMMON/BLOKA/IX
      COMMON/BLOKB/TIM,TNEXT,NSYS,NMAX
      COMMON/BLOKD/ARRV,ASERV,NPER,NSIM,NEXT
      COMMON/BLOKG/REN,AJOC
      COMMON/BLOKI/ALAM
      COMMON/BLOKJ/BALK
      COMMON/BLOKL/IPER
C     DETERMINE WHETHER OR NOT THE NEXT EVENT OCCURS IN THE NEXT PERIOD ********
      IPR=IPER                                                                 *
      APER=NPER                                                                *
      TINT=TIM(NMAX+3,1)/APER                                                  *
      TIME=0                                                                   *
      IPER=0                                                                   *
      DO 1 I=1,NPER                                                            *
      TIME=TIME+TINT                                                           *
      IPER=IPER+1                                                              *
      IF(TIME.GT.TNEXT) GO TO 2                                                *
    1 CONTINUE                                                                 *
C     ****************************************************************************
C     IF THE NEXT EVENT OCCURS IN THE NEXT PERIOD, REGENERATE THE TIME OF THE
C     NEXT ARRIVAL SINCE THE ARRIVAL RATE CHANGES AS THE PERIOD CHANGES ........
    2 IF(IPR.LT.IPER) TIM(1,1)=TIME-TINT-(1./ALAM(IPER))*ALOG(RANDU(IX,I       .
     1Y))                                                                      .
C     ..........................................................................
C     IF THE NEXT EVENT IS THE END OF THE SIMULATION FOR PERIOD NSIM CALCULATE
C     AND WRITE OUT THE AVERAGE NUMBER OF BALKS, RENIGS, JOCKEYS, ARRIVALS, AND
C     SERVICES PER PERIOD ------------------------------------------------------
      IF(I1.LT.NSIM.OR.NEXT.LT.5) RETURN                                       -
      WRITE(6,100)                                                             -
      ASIM=NSIM                                                                -
      SUMB=0                                                                   -
      SUMR=0                                                                   -
      SUMJ=0                                                                   -
      SUMA=0                                                                   -
      SUMS=0                                                                   -
      DO 3 I=1,NPER                                                            -
      BALK(I)=BALK(I)/ASIM                                                     -
      REN(I)=REN(I)/ASIM                                                       -
      AJOC(I)=AJOC(I)/ASIM                                                     -
      ARRV(I)=ARRV(I)/ASIM                                                     -
      ASERV(I)=ASERV(I)/ASIM                                                   -
      SUMB=SUMB+BALK(I)                                                        -
      SUMR=SUMR+REN(I)                                                         -
      SUMJ=SUMJ+AJOC(I)                                                        -
      SUMA=SUMA+APRV(I)                                                        -
      SUMS=SUMS+ASERV(I)                                                       -
    3 WRITE(6,200) I, BALK(I),REN(I),AJOC(I),ARRV(I),ASERV(I)                  -
      WRITE(6,300) SUMB,SUMR,SUMJ,SUMA,SUMS                                    -
C     --------------------------------------------------------------------------
      RETURN
  100 FORMAT(1X,'PERIOD',3X,'AVG. NO. OF BALKS',3X,'AVG. NO. OF RENIGS',
     13X,'AVG. NO. JOCKEYING',6X,'ARRIVALS',6X,'SERVICES'////)
  200 FORMAT(1X,I5,7X,F10.4,8X,F10.4,14X,F10.4,7X,F10.4,4X,F10.4)
  300 FORMAT(//1X,'TOTAL',7X,F10.4,8X,F10.4,14X,F10.4,7X,F10.4,4X,F10.4)
      END
```

Fig. 9.23 Subroutine UPDATE for the simulator in Example 9.13.

The simulator described above can be used when each period of a replication of the simulation is not one hour. Specifically, the length of one replication may be divided into any desired number of periods, each of equal length. The variable NPER specifies the number of periods into which a replication is divided. The arrival rate can be changed from one period to the next. In addition, the points in time at which the number of service channels may be changed during a replication is not restricted to those points in time at which the arrival rate changes. The variable MPER defines the number of times the number of channels available changes during each replication and

```
      FUNCTION TCHANG(A1,A2)
      COMMON/BLOKA/IX
      TCHANG=A1+A2*SQRT(-2.*ALOG(RANDU(IX,IY)))*COS(6.23*RANDU(IX,IY))
      RETURN
      END
```

Fig. 9.24 Subroutine TCHANG for the simulator in Example 9.13.

NSER(I) is the number of channels available during the Ith period. The variable TSER(I) defines the point in time at which the number of channels available changes from NSER(I) to NSER(I + 1).

We first simulate the system for 15 periods of one day each with two channels available throughout the day. The input data for the simulation are as follows:

$$\text{NMAX} = 5$$

$$\text{IX} = \text{five-digit odd integer}$$

$$\text{NPER} = 8$$

$$\text{NSIM} = 15 \text{ (15 replications of one day's operation)}$$

$$\text{MPER} = 1$$

$$\text{IPRINT} = 0$$

$$\text{AMU(J)} = \begin{cases} 42.0, & J = 1 \\ 40.0, & J = 2 \\ 38.0, & J = 3 \\ 35.0, & J = 4 \\ 30.0, & J = 5 \end{cases}$$

$$A = 0.10$$

$$A1 = 0.20$$

$$A2 = 0.05$$

$$\text{TIM(NMAX} + 3, 1) = 1.0$$

$$\text{ALAM(I)} = \begin{cases} 50.0, & I = 1 \\ 30.0, & I = 2 \\ 220.0, & I = 3 \\ 70.0, & I = 4 \\ 200.0, & I = 5 \\ 10.0, & I = 6 \\ 50.0, & I = 7 \\ 10.0, & I = 8 \end{cases}$$

$$\text{TSER}(1) = 1.0$$

$$\text{NSER}(1) = 2$$

TABLE 9.15 Simulation of 15 days' operation of the system in Example 9.13, where two service channels are available throughout each day

Period	Average number of balks	Average number of reneges	Average number jockeying	Arrivals	Services
1	0.2667	0.0000	0.0000	5.5333	4.5333
2	0.2000	0.0000	0.0000	4.6000	4.0667
3	4.3333	0.0000	0.0000	21.8667	8.4667
4	2.4667	0.0667	0.0000	9.0667	9.4667
5	9.0000	0.5333	0.0000	24.8667	9.7333
6	0.5333	0.2667	0.1333	1.7333	9.4667
7	0.4667	0.4000	0.1333	5.6667	7.0667
8	0.1333	0.0000	0.0000	1.2667	2.8667
Total	17.4000	1.2667	0.2667	74.5999	55.6666

The results of this simulation are shown in Table 9.15, where the average number of lost customers per day is 18.6667. As seen in this table, the period between 0.250 and 0.625 hr accounted for a large portion of the lost customers. Thus we increase the number of channels available between 0.250 and 0.375 hr to four, increase the number between 0.375 and 0.500 hr to three, and increase the number between 0.500 and 0.625 hr to five. The system operates with two channels until 0.250 hr and two channels between 0.625 and 1.000 hr. The input data for this simulation run are identical to the first except that

$$\text{MPER} = 5$$

$$\text{NSER(I)} = \begin{cases} 2, & I = 1 \\ 4, & I = 2 \\ 3, & I = 3 \\ 5, & I = 4 \\ 2, & I = 5 \end{cases}$$

$$\text{TSER(I)} = \begin{cases} 0.250, & I = 1 \\ 0.375, & I = 2 \\ 0.500, & I = 3 \\ 0.625, & I = 4 \\ 1.000, & I = 5 \end{cases}$$

Multiple Channels in Parallel

TABLE 9.16 Simulation of 15 days' operation of the system in Example 9.13, where there are two channels available until 0.250 hr, four between 0.250 and 0.375 hr, three between 0.375 and 0.500 hr, five between 0.500 and 0.625 hr, and two between 0.625 and 1.000 hr

Period	Average number of balks	Average number of reneges	Average number jockeying	Arrivals	Services
1	0.1333	0.0000	0.0000	7.2000	5.4000
2	0.1333	0.0000	0.0000	4.4000	4.9333
3	1.9333	0.0000	0.0000	24.7333	13.9333
4	0.7333	0.0667	0.0000	7.2000	13.5333
5	1.8000	0.0667	0.0000	24.8667	17.4000
6	0.0000	0.0000	0.0000	1.2000	9.2667
7	0.1333	0.0000	0.0000	4.4000	3.1333
8	0.0000	0.0000	0.0000	1.2667	2.2000
Total	4.8667	0.1333	0.0000	75.2666	69.8000

The results of the second simulation are given in Table 9.16, where the average number of lost customers is 5.0000/day. Although this average number of lost customers/day is acceptable, we will attempt further allocations to reduce the total number of channel-hours operated per day.

Since the average number of lost customers is small until 0.250 hr, the number of channels available during this period will be reduced to one. In an attempt to reduce the average number of lost customers between 0.250 and 0.375 hr, the number of channels available will be increased to five. We will leave the number of channels available between 0.375 and 0.500 hr at three and the number available between 0.500 and 0.625 hr will remain five. However, the number of available channels between 0.625 and 1.000 hr will be reduced to one. The results of this simulation are shown in Table 9.17. The input data for this run are

$$\text{NSER(I)} = \begin{cases} 1, & I = 1 \\ 5, & I = 2 \\ 3, & I = 3 \\ 5, & I = 4 \\ 1, & I = 5 \end{cases}$$

The results shown in Table 9.17 indicate that the average number of lost customers per day is 5.0667. We will conduct one final run in an attempt to reduce the average number of lost customers. For this case we will increase the number of available channels between 0.000 and 0.125 hr to two. The previous allocation will remain unchanged for the period 0.125 to 1.000 hr.

TABLE 9.17 Simulation of 15 days' operation of the system in Example 9.13, where there is one channel available until 0.250 hr, five between 0.250 and 0.375 hr, three between 0.375 and 0.500 hr, five between 0.500 and 0.625 hr, and one between 0.625 and 1.000 hr

Period	Average number of balks	Average number of reneges	Average number jockeying	Arrivals	Services
1	0.4667	0.0000	0.0000	6.8000	3.8000
2	0.3333	0.0000	0.0000	4.0667	4.5333
3	1.333	0.0667	0.0000	27.2000	18.0000
4	0.4000	0.0000	0.1333	8.2000	13.2667
5	1.8667	0.0000	0.0667	25.2000	17.4667
6	0.0000	0.0000	0.0000	1.2000	10.2000
7	0.3333	0.0000	0.1333	3.6667	2.6667
8	0.2667	0.0000	0.0000	2.0667	2.9333
Total	5.0000	0.0667	0.3333	78.4000	72.8666

The input data for this run are the same as that for the second and third runs, except that

$$\text{MPER} = 6$$

$$\text{NSER}(I) = \begin{cases} 2, & I = 1 \\ 1, & I = 2 \\ 5, & I = 3 \\ 3, & I = 4 \\ 5, & I = 5 \\ 1, & I = 6 \end{cases}$$

$$\text{TSER}(I) = \begin{cases} 0.125, & I = 1 \\ 0.250, & I = 2 \\ 0.375, & I = 3 \\ 0.500, & I = 4 \\ 0.625, & I = 5 \\ 1.000, & I = 6 \end{cases}$$

The results of the final run are given in Table 9.18. Here the average number of lost customers is 4.4667/day.

Multiple Channels in Parallel

TABLE 9.18 Simulation of 15 days' operation of the system in Example 9.13, where there are two channels available until 0.125 hr, one between 0.125 and 0.250 hr, five between 0.250 and 0.375 hr, three between 0.375 and 0.500 hr, five between 0.500 and 0.625 hr, and one between 0.625 and 1.000 hr

Period	Average number of balks	Average number of reneges	Average number jockeying	Arrivals	Services
1	0.2000	0.0000	0.0000	7.6000	4.8000
2	0.6000	0.0000	0.0000	3.7333	4.4000
3	0.9333	0.0000	0.0000	24.9333	17.1333
4	0.5333	0.0000	0.0000	9.4667	13.4000
5	2.0667	0.0000	0.0667	24.1333	19.0667
6	0.0000	0.0000	0.0667	1.4000	7.7333
7	0.0667	0.0000	0.0000	3.4667	2.8000
8	0.0667	0.0000	0.0000	1.7333	2.3333
Total	4.4667	0.0000	0.1333	76.4666	71.6666

On the basis of the average number of lost customers/day, any one of the last three allocations is probably acceptable. However, since the last two allocations require less channel-hours/day, either of these allocations is likely to be preferable to the second.

Example 9.14 A job shop has M machines that are nearly always in use. However, these machines break down from time to time and must be repaired. The job shop has N repair crews to keep the machines operational. When a machine breaks down, one of the repair crews starts working on that machine immediately unless all N repair crews are busy working on other machines. In the latter case, the machine that has just gone down must wait for service, which is provided on a first come–first served basis. The density functions of time until failure and service time are machine dependent; that is, if f_i and f_j are the times until failure for the ith and jth machines, respectively, then f_i and f_j are independently but not necessarily identically distributed random variables. Similarly, if s_i and s_j are the service times for the ith and jth machines, then s_i and s_j are not necessarily identically distributed random variables, although they are independent.

Each repair crew costs C_c per time unit. When a machine fails, three costs arise. First, a fixed cost of C_{Ri} arises whenever a repair begins on the ith machine and excludes the cost of labor. The second cost, C_{f_2i}, is dependent on the time required to repair the ith machine. This expenditure includes the cost of lost production, overhead, and the materials consumed in the repair, but does not include labor. The final cost, C_{f_1i}, is that due to the ith machine waiting for service and is the cost of lost production. If n_i is the number of repairs started on the ith machine during a 1-yr period, r_i the time spent on repairs of the ith machine during a 1-yr period, and w_i the time spent waiting for service by the ith machine during a 1-yr period, then the total cost of failures and maintenance for the system, C_T, is given by

$$C_T = NC_c + \sum_{i=1}^{M} (n_i C_{Ri} + W_i C_{f_1 i} + r_i C_{f_2 i}) \tag{9.70}$$

Develop a simulator that will estimate the annual cost of maintaining the M machines in the job shop.

The system described above is a multiple–parallel-channel queueing system. However, the arrival population is finite as opposed to that considered in Examples 9.10–9.12, where the arrival population was implicitly assumed to be infinite. To account for this phenomenon in the simulator, we will generate the time until failure for a given machine at the start of the simulation and upon completion of a repair of that machine. For generality it will be assumed that the failure-time distributions for the M machines are different. As in our previous discussion of multiple–parallel-channel queueing simulators, service time for a particular machine will be generated when repairs begin on that machine. However, in this problem service time is assumed to be machine dependent rather than dependent on the channel (repair crew) providing the service. Thus, service time is generated from a distribution associated with each machine rather than each channel. The following variables will be used in connection with service and failure times:

$$\text{TIM}(I, J) = \begin{cases} \text{time of next failure of machine I,} & J = 1 \\ \text{time of completion of service for} \\ \text{machine I,} & J = 2 \end{cases}$$

$$A(I, 1), B(I, 1) = \begin{cases} \text{parameters of the failure time} \\ \text{distribution for machine I,} & J = 1 \\ \text{parameters of the service time} \\ \text{distribution for machine I,} & J = 2 \end{cases}$$

$$\text{DIST}(I, J) = \begin{cases} \text{distribution type for failure time} \\ \text{for machine I,} & J = 1 \\ \text{distribution type for service time} \\ \text{for machine I,} & J = 2 \end{cases}$$

In the course of the simulation, each machine may be in one of three states: operating, down and waiting for service, or down and under repair. Each of the repair crews may be in one of two states: idle or repairing a machine. These states will be represented by the following variables in the simulator:

$$STATEL(I) = \begin{cases} 0, & \text{machine I down and waiting for repair} \\ 1, & \text{machine I being repaired} \\ 2, & \text{machine I operating} \end{cases}$$

$$STATEC(I) = \begin{cases} 0, & \text{repair crew I idle} \\ 1, & \text{repair crew I repairing a machine} \end{cases}$$

To calculate the annual cost of the system, the simulator must keep track of the number of failures of each machine, NFAIL(I). The current number of machines down is given by NDOWN. In addition, the cumulative waiting time WAITIM(I) and the cumulative repair time RTIME(I) for each machine must be calculated. When the Ith machine fails, NDOWN and NFAIL(I) are incremented by one. To calculate machine waiting time in the system, we record the time of entry of the Ith machine to the service system, TWAIT(I). Waiting time is then given by the difference between the time at which repairs are started on machine I and TWAIT(I). When a machine fails it either enters the waiting line or goes into service. If machine I enters the waiting line, the waiting-line length NWAIT is increased by one, and the position of the Ith machine, IPOS(I), is set equal to NWAIT. When repairs begin on the Ith machine, IPOS(I) is reduced to zero and remains zero until that machine takes a position in the waiting line again.

If a machine fails and NDOWN is less than or equal to the number of repair crews N after that unit has failed, one of the idle repair crews starts working on the machine. The repair crew that services the machine that just failed is that crew J such that $STATEC(I) > 0$ for $I < J$ and $STATEC(J) = 0$. The variable LINE(J) is used to identify the machine on which the Jth repair crew is working. Thus, if repair crew J is servicing machine I, then $LINE(J) = I$. Idle time for crew J, TIDLE(J), is increased by $TNEXT - TIMID(J)$, where TIMID(J) is the time at which crew J last became idle, and TNEXT is present time. The state of the Jth crew is then set to busy, and the state of machine I is set to "in service," $STATEL(I) = 1$. Next, the time of service on machine I is generated, and the waiting time for machine I is accumulated. To accumulate total service time on machine I, the time at which the Ith machine moves into service, TSERV(I), is recorded. Service time on machine I is then given by the difference between TIM(I, 2) and TSERV(I) and is calculated when service is completed on machine I.

When service on machine I is completed, total down time on machine I, TDOWN(I), is accumulated by adding the quantity $TNEXT - TWAIT(I)$

```
C     N=NUMBER OF REPAIR CREWS
C     M=NUMBER OF MACHINES
C     IX=INITIAL RANDOM NUMBER
C     A(I,1),B(I,1)=PARAMETERS OF FAILURE DISTRIBUTION FOR MACHINE I
C     A(I,2),B(I,2)=PARAMETERS OF SERVICE DISTRIBUTION FOR MACHINE I
C     TIMLIM=TIME LIMIT ON THE SIMULATION
C     TIM(I,1)=TIME OF NEXT FAILURE OF MACHINE I
C     TIM(I,2)=TIME OF NEXT SERVICE COMPLETION OF MACHINE I
C     NWAIT=NUMBER IN THE WAITING LINE
C     NDOWN=NUMBER OF MACHINES DOWN (NUMBER IN THE SYSTEM)
C     NREP=CUMULATIVE NUMBER OF REPAIRS BY ALL REPAIR CREWS
C     IPOS(I)=0, ITH MACHINE IS EITHER OPERATING OR IS UNDER REPAIR
C            =K, ITH MACHINE IS IN THE KTH POSITION IN THE WAITING LINE, K
C                GREATER THAN OR EQUAL TO UNITY
C     NFAIL(I)=CUMULATIVE NUMBER OF FAILURES BY THE ITH MACHINE
C     WAITIM(I)=CUMULATIVE WAITING TIME FOR THE ITH MACHINE
C     TDOWN(I)=CUMULATIVE DOWN TIME FOR THE ITH MACHINE
C     RTIME(I)=CUMULATIVE REPAIR TIME FOR THE ITH MACHINE
C     STATEL(I)=0, ITH MACHINE IS IN THE WAITING LINE
C              =1, ITH MACHINE IS UNDERGOING REPAIR
C              =2, ITH MACHINE IS OPERATING
C     STATEC(J)=0, JTH REPAIR CREW IS IDLE
C              =1, JTH REPAIR CREW IS REPAIRING A MACHINE
C     TIDLE(J)=CUMULATIVE IDLE TIME FOR THE JTH REPAIR CREW
C     TREP(J)=CUMULATIVE REPAIR TIME FOR THE JTH REPAIR CREW
C     LINE(J)=MACHINE WHICH THE JTH REPAIR CREW IS PRESENTLY SERVICING
C     TIMID(J)=TIME AT WHICH THE JTH REPAIR CREW BECOMES IDLE
C     MREP(J)=CUMULATIVE NUMBER OF REPAIRS BY THE JTH REPAIR CREW
C     INEXT=NUMBER OF THE MACHINE INVOLVED IN THE NEXT EVENT
C     JNEXT=1, NEXT EVENT IS A MACHINE FAILURE
C          =2, NEXT EVENT IS THE COMPLETION OF THE REPAIR OF A MACHINE
C     TNEXT=TIME OF OCCURRENCE OF THE NEXT EVENT
C     NPRINT=0, STATUS OF THE SYSTEM IS NOT WRITTEN OUT AFTER EACH EVENT
C           =1, STATUS OF THE SYSTEM IS WRITTEN OUT AFTER EACH EVENT
C     CCREW=CREW COST PER YEAR
C     CFAIL1(I)=COST OF TIME SPENT BY THE ITH MACHINE WAITING FOR REPAIR
C     CFAIL2(I)=COST OF TIME SPENT FOR REPAIR OF THE ITH MACHINE
C               EXCLUDING LABOR COST
C     CREP(I)=FIXED PORTION OF THE COST OF REPAIRING THE ITH MACHINE
C     CO(I)=TOTAL COST OF FAILURE AND REPAIR (EXCLUDING LABOR COST) FOR THE
C           ENTIRE PERIOD OF THE SIMULATION FOR THE ITH MACHINE
C     CT(I)=AVERAGE ANNUAL COST OF FAILURES AND REPAIRS (EXCLUDING LABOR COST)
C           FOR THE ITH MACHINE
C     COST=TOTAL AND AVERAGE ANNUAL COST OF FAILURES, REPAIRS, AND LABOR FOR ALL
C          MACHINES AND REPAIR CREWS
C     Z(J)=TOTAL COST OF REPAIR CREW J FOR THE ENTIRE PERIOD OF THE SIMULATION
C     TUNIT(1),TUNIT(2)=UNITS OF TIME CARRIED IN THE SIMULATION (WEEKS,
C                 MONTHS,YEARS,ETC.) THESE VARIABLES ARE USED TO LABEL OUTPUT
C     TUNIT(3),TUNIT(4)=WEEKLY,MONTHLY,YEARLY,ETC. THESE VARIABLES ARE USED TO
C                 LABEL OUTPUT
      DIMENSION AMFAIL(20),ANFAIL(20)
      DIMENSION A(20,20),B(20,20),CFAIL1(20),CFAIL2(20),CREP(20)
      DIMENSION DIST(20,20),IPOS(20),NFAIL(20)
      DIMENSION TDOWN(20),RTIME(20),STATEL(20),TIM(20,2),WAITIM(20)
      DIMENSION TIDLE(20),TREP(20),LINE(20),TIMID(20),STATEC(20)
      DIMENSION TWAIT(20),TSERV(20),MREP(20)
      DIMENSION CO(20),CT(20),Z(20),T(20),TUNIT(4)
      COMMON/BLOKA/A,B,DIST
C READ INPUT DATA *********************************************************
      READ(5,100) N,M,IX,CCREW,TIMLIM,NPRINT,(TUNIT(I),I=1,4)            *
      WRITE(6,100) N,M,IX,CCREW,TIMLIM,NPRINT,(TUNIT(I),I=1,4)           *
      DO 1 I=1,M                                                         *
      READ(5,110) (A(I,J),B(I,J),DIST(I,J),J=1,2)                        *
      WRITE(6,110) (A(I,J),B(I,J),DIST(I,J),J=1,2)                       *
      READ(5,110) CFAIL1(I),CFAIL2(I),CREP(I)                            *
    1 WRITE(6,110) CFAIL1(I),CFAIL2(I),CREP(I)                           *
C **************************************************************************
C INITIALIZE COUNTERS, INDICES, ACCUMULATORS, AND EVENT TIMES ##############
      DO 2 I=1,M                                                         #
      IPOS(I)=0                                                          #
      NFAIL(I)=0                                                         #
      WAITIM(I)=0.                                                       #
      TDOWN(I)=0                                                         #
      RTIME(I)=0                                                         #
      TIM(I,2)=10.**30                                                   #
      STATEL(I)=2                                                        #
    2 TIM(I,1)=RNVAR(IX,I,1)                                             #
      NWAIT=0                                                            #
      NDOWN=0                                                            #
      NREP=0                                                             #
      DO 3 J=1,N                                                         #
      STATEC(J)=0                                                        #
      TIDLE(J)=0                                                         #
      TREP(J)=0                                                          #
      LINE(J)=0                                                          #
      TIMID(J)=0                                                         #
    3 MREP(J)=0                                                          #
      INEXT=0                                                            #
      JNEXT=0                                                            #
C ##########################################################################
```

Fig. 9.25 FORTRAN IV program for the machine failure and repair problem given in Example 9.14.

```
C DETERMINE THE NEXT EVENT TO OCCUR IN THE SIMULATION <<<<<<<<<<<<<<<<<<<<
    4 TNEXT=10.**30                                                     <
      DO 5 I=1,M                                                        <
      IF(NPRINT.EQ.0) GO TO 7                                           <
C WRITE OUT THE STATUS OF THE SYSTEM AT THE LAST EVENT *****************
      WRITE(6,120) I,IPOS(I),NFAIL(I),WAITIM(I),TDOWN(I),RTIME(I)     * <
    5 WRITE(6,130) (TIM(I,J),J=1,2),STATEL(I)                         * <
      WRITE(6,140) NWAIT,TNEXT,NDOWN,NREP,INEXT,JNEXT                 * <
      DO 6 J=1,N                                                        <
    6 WRITE(6,150) J,STATEC(J),TIDLE(J),TREP(J),LINE(J),TIMID(J),MREP(J) * <
C **********************************************************************<
    7 DO 8 I=1,M                                                        <
      DO 8 J=1,2                                                        <
      IF(TNEXT.LT.TIM(I,J)) GO TO 8                                     <
      TNEXT=TIM(I,J)                                                    <
      INEXT=I                                                           <
      JNEXT=J                                                           <
    8 CONTINUE                                                          <
C <<<<<<<<<<<<<<<<<<<<<<<<<<<<<<<<<<<<<<<<<<<<<<<<<<<<<<<<<<<<<<<<<<<<<<
C BRANCH TO NEXT EVENT ROUTINE ##########################################
      IF(TNEXT.GT.TIMLIM) GO TO 18                                      #
      IF(JNEXT.GT.1) GO TO 12                                           #
C ######################################################################
C MACHINE FAILURE ROUTINE ***********************************************
C INCREASE NUMBER OF MACHINES DOWN BY ONE, INCREASE NUMBER OF FAILURES BY *
C MACHINE INEXT BY ONE, SET STATE OF MACHINE INEXT TO DOWN (0), SET TIME OF *
C NEXT FAILURE OF MACHINE INEXT TO AN ARBITRARILY LARGE VALUE, AND SET TIME AT *
C WHICH MACHINE INEXT STARTS WAITING FOR SERVICE TO PRESENT TIME <<<<<<<<<<<< *
      NDOWN=NDOWN+1                                                   < *
      NFAIL(INEXT)=NFAIL(INEXT)+1                                     < *
      STATEL(INEXT)=0                                                 < *
      TIM(INEXT,1)=10.**30                                            < *
      TWAIT(INEXT)=TNEXT                                              < *
C <<<<<<<<<<<<<<<<<<<<<<<<<<<<<<<<<<<<<<<<<<<<<<<<<<<<<<<<<<<<<<<<<<<< *
C DETERMINE WHICH REPAIR CREW IS AVAILABLE FOR SERVICE IF ANY ############ *
      IF(NDOWN.GT.N) GO TO 10                                         # *
      DO 9 J1=1,N                                                     # *
      IF(STATEC(J1).GT.0.) GO TO 9                                    # *
      J=J1                                                            # *
      GO TO 11                                                        # *
    9 CONTINUE                                                        # *
C ####################################################################### *
      GO TO 4                                                           *
C INCREASE THE NUMBER OF MACHINES WAITING FOR SERVICE BY ONE AND PLACE MACHINE *
C INEXT IN THE LAST POSITION IN THE WAITING LINE <<<<<<<<<<<<<<<<<<<<<<<<<< *
   10 NWAIT=NWAIT+1                                                   < *
      IPOS(INEXT)=NWAIT                                               < *
C <<<<<<<<<<<<<<<<<<<<<<<<<<<<<<<<<<<<<<<<<<<<<<<<<<<<<<<<<<<<<<<<<<<< *
      GO TO 4                                                           *
C REPAIR CREW BEGINS SERVICE ON NEXT MACHINE WAITING OR ON MACHINE WHICH HAS *
C JUST FAILED ########################################################### *
   11 LINE(J)=INEXT                                                   # *
      STATEC(J)=1                                                     # *
      TIDLE(J)=TIDLE(J)+TNEXT-TIMID(J)                                # *
      TIM(INEXT,2)=TNEXT+RNVAR(IX,INEXT,2)                            # *
      WAITIM(INEXT)=WAITIM(INEXT)+TNEXT-TWAIT(INEXT)                  # *
      TSERV(INEXT)=TNEXT                                              # *
      STATEL(INEXT)=1                                                 # *
C ####################################################################### *
      GO TO 4                                                           *
C ***********************************************************************
C COMPLETION OF SERVICE ROUTINE #########################################
C UPDATE DOWNTIME AND REPAIR TIME FOR MACHINE INEXT, SET STATE OF MACHINE INEXT#
C TO ''OPERATING'' (2), REDUCE THE NUMBER OF MACHINES DOWN BY ONE, GENERATE  #
C TIME OF NEXT FAILURE FOR MACHINE INEXT, INCREASE THE NUMBER OF REPAIRS BY  #
C ONE, AND SET TIME OF NEXT SERVICE OF MACHINE INEXT TO ARBITRARILY LARGE    #
C VALUE *****************************************************************  #
   12 TDOWN(INEXT)=TDOWN(INEXT)+TNEXT-TWAIT(INEXT)                    * #
      RTIME(INEXT)=RTIME(INEXT)+TNEXT-TSERV(INEXT)                    * #
      STATEL(INEXT)=2                                                 * #
      NDOWN=NDOWN-1                                                   * #
      TIM(INEXT,1)=TNEXT+RNVAR(IX,INEXT,1)                            * #
      NREP=NREP+1                                                     * #
      TIM(INEXT,2)=10.**30                                            * #
C **********************************************************************  #
C DETERMINE WHICH REPAIR CREW WAS SERVICING MACHINE INEXT <<<<<<<<<<<<<<<<<  #
      DO 13 J1=1,N                                                    <  #
      IF(LINE(J1).NE.INEXT) GO TO 13                                  <  #
      J=J1                                                            <  #
      GO TO 14                                                        <  #
   13 CONTINUE                                                        <  #
C <<<<<<<<<<<<<<<<<<<<<<<<<<<<<<<<<<<<<<<<<<<<<<<<<<<<<<<<<<<<<<<<<<<<  #
      GO TO 4                                                            #
C SET STATE OF REPAIR CREW J TO ''IDLE'' (0), RECORD TIME AT WHICH REPAIR CREW *#
C J BECAME IDLE TO PRESENT TIME, AND UPDATE REPAIR TIME AND THE NUMBER OF    *
C REPAIRS FOR CREW J ****************************************************   *
   14 LINE(J)=0                                                       * #
      TIMID(J)=TNEXT                                                  * #
      STATEC(J)=0                                                     * #
      TREP(J)=TREP(J)+TNEXT-TSERV(INEXT)                              * #
      MREP(J)=MREP(J)+1                                               * #
C **********************************************************************
```

Fig. 9.25 (*continued*)

```
C DETERMINE THE NEXT MACHINE WAITING FOR SERVICE IF ANY <<<<<<<<<<<<<<<<<<
      DO 15 I=1,M
      IF(IPOS(I).NE.1) GO TO 15
      INEXT=I
      GO TO 16
   15 CONTINUE
C <<<<<<<<<<<<<<<<<<<<<<<<<<<<<<<<<<<<<<<<<<<<<<<<<<<<<<<<<<<<<<<<<<<<<
      GO TO 4
C MACHINE INEXT GOES INTO SERVICE, NUMBER OF MACHINES IN THE WAITING LINE IS
C REDUCED BY ONE, STATE OF REPAIR CREW J IS SET TO ''BUSY'' (1), AND EACH
C MACHINE WAITING FOR SERVICE IS MOVED UP ONE POSITION ******************
   16 IPOS(INEXT)=0
      NWAIT=NWAIT-1
      DO 17 I=1,M
      IF(IPOS(I).EQ.0) GO TO 17
      IPOS(I)=IPOS(I)-1
   17 CONTINUE
C ******************************************************************
      GO TO 11
C ##################################################################
C TERMINATION OF THE SIMULATION - CALCULATE AND WRITE OUT ALL STATISTICS ******
   18 WRITE(6,290)
      WRITE(6,160) TIMLIM,(TUNIT(I),I=1,2)
      WRITE(6,170) M,N
      WRITE(6,180)
      WRITE(6,190) (TUNIT(I),I=3,4)
      WRITE(6,200)
      COST=0
      DO 19 I=1,M
C COMPUTE COST OF DOWNTIME AND REPAIRS FOR MACHINE I, COMPUTE WAITING TIME,
C REPAIR TIME, AND DOWNTIME FOR MACHINE I <<<<<<<<<<<<<<<<<<<<<<<<<<<<<
      AMFAIL(I)=NFAIL(I)
      ANFAIL(I)=NFAIL(I)-1
      IF(STATEL(I).EQ.0.) WAITIM(I)=WAITIM(I)+TIMLIM-TWAIT(I)
      IF(STATEL(I).EQ.1.) RTIME(I)=RTIME(I)+TIMLIM-TSERV(I)
      TDOWN(I)=RTIME(I)+WAITIM(I)
      IF(STATEL(I).EQ.0.) CO(I)=ANFAIL(I)*CREP(I)+WAITIM(I)*CFAIL1(I)+RT
     1IME(I)*CFAIL2(I)
      IF(STATEL(I).GE.1.) CO(I)=AMFAIL(I)*CREP(I)+WAITIM(I)*CFAIL1(I)+RT
     1IME(I)*CFAIL2(I)
      CT(I)=CO(I)/TIMLIM
C <<<<<<<<<<<<<<<<<<<<<<<<<<<<<<<<<<<<<<<<<<<<<<<<<<<<<<<<<<<<<<<<<<<<<
      WRITE(6,210) I,NFAIL(I),TDOWN(I),RTIME(I),WAITIM(I),CO(I),CT(I)
      COST=COST+CO(I)
   19 CONTINUE
      WRITE(6,220)
      WRITE(6,230) (TUNIT(I),I=3,4)
      WRITE(6,240)
      DO 20 J=1,N
      Z(J)=CCREW*TIMLIM
      K=LINE(J)
      IF(STATEC(J).EQ.1.) TREP(J)=TREP(J)+TIMLIM-TSERV(K)
      TIDLE(J)=TIMLIM-TREP(J)
      WRITE(6,250) J,MREP(J),TREP(J),TIDLE(J),Z(J),CCREW
      COST=COST+Z(J)
   20 CONTINUE
      WRITE(6,260) COST
      COST=COST/TIMLIM
      WRITE(6,270) (TUNIT(I),I=3,4),COST
      WRITE(6,280) (TUNIT(I),I=1,2)
C ******************************************************************
      STOP
  100 FORMAT(1X3I6,2XF10.3,5X,F5.2,I5,4A4)
  110 FORMAT(1X,6F10.4)
  120 FORMAT(1X,'I=',I3,3X,'IPOS=',I5,3X,'NFAIL=',I5,3X,'WAITIM=',E14.7,
     13X,'TDOWN=',E14.7,3X,'RTIME=',E14.7)
  130 FORMAT(9X,'TIM(I,1)=',E14.7,3X,'TIM(I,2)=',E14.7,3X,'STATEL(I)=',E
     114.7)
  140 FORMAT(1X,'NWAIT=',I5,3X,'TNEXT=',E14.7,3X,'NDOWN=',I5,3X,'NREP=',
     1I5,3X,'INEXT=',I5,3X,'JNEXT=',I5)
  150 FORMAT(1X,'J=',I5,3X,'STATEC=',E14.7,3X,'TIDLE=',E14.7,3X,'TREP=',
     1E14.7/1X,'LINE=',I5,3X,'TIMID=',E14.7,3X,'MREP=',I5///)
  160 FORMAT(1X,F5.2,2X,A4,A4,1X,'SUMMARY')
  170 FORMAT(1X,I5,2X,'PRODUCTION LINES'/1X,I5,2X,'REPAIR CREWS')
  180 FORMAT(/////1X,' LINE ',2X,'    NUMBER',6X,'    TOTAL',7X,'    TOT
     1AL',7X,'    TOTAL',7X,'    TOTAL',7X,'  AVERAGE')
  190 FORMAT(1X,'NUMBER',2X,'    OF',8X,'    TIME',8X,'    REPAIR',6X,
     1'  WAITING',6X,'    COST',10X,2A4)
  200 FORMAT(9X,'  FAILURES',5X,'    DOWN',8X,'    TIME',7X,'    TIME
     1',23X,'    COST'///)
  210 FORMAT(2XI5,9XI5,6XF10.4,6XF10.4,6XF10.4,6XF10.2,6XF10.2)
  220 FORMAT(/////1X,'    REPAIR',6X,'    NUMBER',6X,'    TIME',7X,'
     1 IDLE',7X,'    TOTAL',7X,'  AVERAGE')
  230 FORMAT(1X,'    CREW',7X,'    OF',8X,'    IN',8X,'    TIME',7
     1X,'    COST',10X,2A4)
  240 FORMAT(1X,'    NUMBER',6X,'    REPAIRS',6X,'    SERVICE',37X,'
     1COST'///)
  250 FORMAT( 6X I5,11X I5,6XF10.4,6XF10.4,6XF10.2,6XF10.2)
  260 FORMAT(1X,'TOTAL COST =',F10.2)
  270 FORMAT(1X,'AVERAGE TOTAL'2A4,1X,'COST=',F10.2)
  280 FORMAT(1X,'* ALL TIMES IN ',2A4/3X,'ALL COSTS IN DOLLARS')
  290 FORMAT(1H1)
      END
```

Fig. 9.25 (continued)

```
       FUNCTION RNVAR(IX,I,J)
       DIMENSION A(20,20),B(20,20),DIST(20,20)
       COMMON/BLOKA/A,B,DIST
       A1=A(I,J)
       A2=B(I,J)
       M=DIST(I,J)
       GO TO (3,5,6,7,8),M
     3 IA=A2
       RNVAR=0
       DO 4 K=1,IA
     4 RNVAR=RNVAR-(1./A1)*ALOG(RANDU(IX,IY))
       RETURN
     5 R1=RANDU(IX,IY)
       R2=RANDU(IX,IY)
       RNVAR=A1+A2*SQRT(-2.*ALOG(R1))*COS(6.28*R2)
       RETURN
     6 R1=RANDU(IX,IY)
       R2=RANDU(IX,IY)
       A3=1./A1
       A4=1./A2
       RNVAR=(R1**A3)/(R1**A3+R2**A4)
       IF(R1**A3+R2**A4.GT.1.) GO TO 6
       RETURN
     7 RNVAR=A1+(A2-A1)*RANDU(IX,IY)
       RETURN
     8 RNVAR=A1
       RETURN
       END
```

Fig. 9.25 (*continued*)

to TDOWN(I). Repair time on machine I, RTIME(I), is accumulated by adding TNEXT − TSERVE(I) to RTIME(I). The state of machine I is set to "operating," STATEL(I) = 2, the number of machines down, NDOWN, is reduced by one, the time of the next failure on machine I is generated, and the accumulated number of repairs, NREP, is increased by one. Finally, the time of the next service on machine I is set to an arbitrarily large value.

When the repair of a machine is completed, the repair crew J that serviced the machine becomes idle and LINE(J) = 0, STATEC(J) = 0, and TIMID(J) = TNEXT. The number of repairs by crew J, MREP(J), is increased by one and the total repair time by crew J, TREP(J), is increased by TNEXT − TSERV(I), where I is the machine just serviced by crew J. If there are any machines waiting for service, the first machine in the waiting line, machine I, is selected and IPOS(I) = 0. The machines in the waiting line are moved up one position and the number of units in the waiting line, NWAIT, is reduced by one. At this point LINE(J) = I, STATEC(J) = 1, STATEL(I) = 1, service time for machine I is generated, waiting time is accumulated, and the time the Ith machine entered service, TSERV(I), is recorded.

The length of the simulation run is defined by TIMLIM. When the simulation is terminated, TNEXT = TIMLIM, waiting time, down time, and repair time for each machine are accumulated. Next, repair time and idle time for each repair crew are accumulated. Finally, all component costs and total average cost are calculated and the results of the simulation are written out.

An annotated FORTRAN IV simulation program for this system is given in Fig. 9.25. Included in the program are definitions of those variables already discussed as well as other variables that were not defined above.

Example 9.15 Consider a system of the type described in Example 9.14, where five machines are to be maintained by two repair crews. Each repair crew costs $600.00/week. The fixed cost of repairing any machine is $40.00, or $C_{Ri} = 40.00$, $i = 1, 2, 3, 4, 5$. The time-dependent costs of repair and waiting in dollars per week are

$$C_{f_{1i}} = \begin{cases} 2000.00, & i = 1 \\ 2200.00, & i = 2 \\ 2100.00, & i = 3 \\ 2400.00, & i = 4 \\ 2000.00, & i = 5 \end{cases} \quad C_{f_{2i}} = \begin{cases} 2700, & i = 1 \\ 3000, & i = 2 \\ 2400, & i = 3 \\ 3400, & i = 4 \\ 3800, & i = 5 \end{cases}$$

Let $f_i(t)$ be the density function of time until failure for the ith machine and let $g_i(s)$ be the density function of service time for the ith machine, where

$$f_i(t) = \begin{cases} 10e^{-10t}, & 0 < t < \infty, \quad i = 1 \\ \dfrac{1}{(2\pi)^{1/2}} \exp\left[-\dfrac{1}{2}(t-5)^2\right], & -\infty < t < \infty, \quad i = 2 \\ \dfrac{1}{0.3}, & 0.1 < t < 0.4, \quad i = 3 \\ 400te^{-20t}, & 0 < t < \infty, \quad i = 4 \\ 1600te^{-40t}, & 0 < t < \infty, \quad i = 5 \end{cases}$$

$$g_i(s) = \begin{cases} 30e^{-30s}, & 0 < s < \infty, \quad i = 1 \\ 40e^{-40s}, & 0 < s < \infty, \quad i = 2 \\ 2500se^{-50s}, & 0 < s < \infty, \quad i = 3 \\ 1667s^3e^{-100s}, & 0 < s < \infty, \quad i = 4 \\ 450s^2e^{-30s}, & 0 < s < \infty, \quad i = 5 \end{cases}$$

Based on the simulation of 20 weeks of operation of the system, estimate the average weekly cost of maintaining the five machines.

The input data for this problem for the simulator given in Fig. 9.25 are as follows:

$$N = 2$$
$$M = 5$$
$$IX = \text{five-digit odd integer}$$
$$CCREW = 600.00$$

TIMLIM = 20.00

NPRINT = 0

$$TUNIT(I) = \begin{cases} ----W, & I = 1 \\ EEKS, & I = 2 \\ --WE, & I = 3 \\ EKLY, & I = 4 \end{cases}$$

$$A(I, 1), B(I, 1), DIST(I, 1) = \begin{cases} 10.00, & 1.00, & 1.00, & I = 1 \\ 5.00, & 1.00, & 2.00, & I = 2 \\ 0.10, & 0.40, & 4.00, & I = 3 \\ 20.00, & 2.00, & 1.00, & I = 4 \\ 40.00, & 2.00, & 1.00, & I = 5 \end{cases}$$

$$A(I, 2), B(I, 2), DIST(I, 2) = \begin{cases} 30.00, & 1.00, & 1.00, & I = 1 \\ 40.00, & 1.00, & 1.00, & I = 2 \\ 50.00, & 2.00, & 1.00, & I = 3 \\ 100.00, & 4.00, & 1.00, & I = 4 \\ 30.00, & 3.00, & 1.00, & I = 5 \end{cases}$$

$$CFAIL(I) = \begin{cases} 2000.00, & I = 1 \\ 2200.00, & I = 2 \\ 2100.00, & I = 3 \\ 2400.00, & I = 4 \\ 2000.00, & I = 5 \end{cases}$$

$$CFAIL2(I) = \begin{cases} 2700.00, & I = 1 \\ 3000.00, & I = 2 \\ 2400.00, & I = 3 \\ 3400.00, & I = 4 \\ 3800.00, & I = 5 \end{cases}$$

$$CREP(I) = 40.00, \quad I = 1, 2, 3, 4, 5$$

The output from the 20-wk simulation is given in Fig. 9.26.

20.00 WEEKS SUMMARY
 5 PRODUCTION LINES
 2 REPAIR CREWS

LINE NUMBER	NUMBER OF FAILURES	TOTAL TIME DOWN	TOTAL REPAIR TIME	TOTAL WAITING TIME	TOTAL COST	AVERAGE WEEKLY COST
1	136	5.6245	4.7565	0.8680	20018.63	1000.93
2	4	0.1824	0.1456	0.0368	677.76	33.89
3	70	3.5118	2.8903	0.6215	11041.83	552.09
4	140	6.8663	5.5186	1.3477	27597.61	1379.82
5	128	13.2663	12.9239	0.3424	54915.56	2745.78

REPAIR CREW NUMBER	NUMBER OF REPAIRS	TIME IN SERVICE	IDLE TIME	TOTAL COST	AVERAGE WEEKLY COST
1	252	13.9970	6.0030	12000.00	600.00
2	224	12.2375	7.7625	12000.00	600.00

TOTAL COST = 138251.30
AVERAGE TOTAL WEEKLY COST = 6912.57
*ALL TIMES IN WEEKS
ALL COSTS IN DOLLARS

Fig. 9.26 Output from the simulation of 20 weeks of operation of the system described in Example 9.15.

SIMULATION OF NETWORKS OF QUEUES

While a complete mathematical analysis of a complex network of queueing systems often presents an insurmountable problem to the analyst, the analysis of such systems can be accomplished through simulation. For our purposes, a network of queues may be thought of as a collection of parallel-channel queueing systems such that each individual system in some way interacts with at least one other system. To illustrate, consider the following example: Parts are produced on manufacturing line A. After each unit is produced it must pass a sequence of three inspections at inspection stations I_1, I_2, I_3. The number of inspectors at station I_j is M_j. If a unit is rejected at station I_j it is reworked in production process B_j and returned to I_j for reinspection. The probability of rejecting a unit at I_j is p_j^{n+1}, where n is the number of times the unit was reworked at B_j. Rework at B_j continues until the unit is passed at I_j. Units of product leave manufacturing line A at a constant rate of λ/day. The time to inspect a unit at I_j is S_j and rework time at B_j is T_j, where S_j and T_j are random variables. A schematic representation of this network is shown in Fig. 9.27.

With the exception of arrivals, the operation of each inspection and rework station can be simulated using a parallel-channel queueing simulator such as that given in Fig. 9.15. Thus, the network of queueing systems given

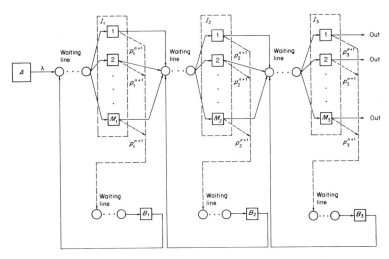

Fig. 9.27 Queueing network for three inspection stations with rework upon rejection at any station.

in Fig. 9.27 could be simulated by appropriate modification of the program given in Fig. 9.15. The first of these modifications would be to add a subscript to each variable in the program in Fig. 9.15 that records the status of the queueing system. For example, QLEN becomes QLEN(J), the length of the queue before the Jth system; TIM(I), $I \leq IC$, becomes TIM(I, J), the time of the next service in the Ith channel in the Jth system; and SYSN becomes SYSN(J), the total number of units in the Jth system. In a similar fashion, a subscript must be added to all accumulators and statistics so that measures of system effectiveness can be estimated for each individual queueing system. Moreover, since it is likely that statistics for the network as a whole would be desired, variables would have to be defined for their calculation.

As already mentioned, in addition to the modifications discussed above, a routine must be added to the basic multiple–parallel-channel simulator to take care of arrivals to one queueing system that emanate from another queueing system. Two possibilities exist when a unit leaves a given queueing system: Either the unit proceeds to another queueing system, becoming an arrival to that system, or the unit leaves the network. Thus, a routine must be programmed that defines the route each unit follows through the network.

When the origin of an arrival is outside the network, the mechanism that determines the initial queueing system that serves the unit must be defined. In some cases, all units entering the network proceed to the same system for initial processing. The network in Fig. 9.27 is an example of such a situation. However, the analyst will frequently encounter networks where different entering units receive initial processing from different queueing systems.

For the network described in Fig. 9.27, let us designate inspection stations 1, 2, and 3 as queueing systems 1, 3, and 5, respectively. Let the rework stations 1, 2, and 3 be designated as queueing systems 2, 4, and 6. Thus, each unit that enters the network proceeds to queueing system 1 for initial processing. After the unit enters the first system, we assign it a number j equal to one plus the number of units already in the network, record its position in the queue or a service channel, and record the number of rework periods for that unit n_j as zero.

Now consider what happens when unit number j leaves channel i. For $i = 1$ or 3, unit j proceeds to queueing system $i + 1$ with probability $p_i^{n_j+1}$ and to queueing system $i + 2$, with probability $1 - p_i^{n_j+1}$. If j proceeds to queueing system $i + 2$, n_j is set to zero to indicate that no rework has been carried out as a result of inspection at the inspection station the unit is about to enter. If $i = 2$, 4, or 6, unit j proceeds to queueing system $i - 1$ and n_j is increased by one. If $i = 5$, the unit leaves the network with probability $1 - p_i^{n_j+1}$ and enters queueing system 6 with probability $p_i^{n_j+1}$. When the jth unit in the network leaves the network, the total number of units in the network is reduced by one. In addition, the reference number for each unit currently in the network is reduced by one if its reference number is greater than j. In a similar fashion, all variables referencing a unit with number greater than j must be altered to refer to the unit with the preceding reference number.

The above discussion of the network given in Fig. 9.27 is intended to indicate some of the problems that may arise in attempting to simulate a network of queues. Of course, the complexities of network simulation are as varied as the networks that may arise, and a complete discussion of the topic is beyond the scope of this chapter.

In general, the simulation of the progress of a given unit once it is in a specific queueing system within the network does not present a serious problem. The major difficulties encountered in developing a network simulator usually arise in defining the links connecting the various queueing systems comprising the network and in keeping track of the location and stage of processing of the various units in the network at a given point in time.

Example 9.16 A job shop has N machine centers. The number of machines in the ith center is c_i, $i = 1, 2, \ldots, N$, where all machines in a given center perform the same function. The probability density function of service time for the jth machine in the ith center is given by $f_{ij}(s)$, where s is the service time. The probability density function of time between successive order arrivals to the job shop is $g(t)$, where t is the time between the arrival of successive orders.

All orders that arrive at the job shop do not necessarily require the same type of processing. Specifically, a given order may require processing through any one of L possible sequences of operations. For example, one order may require processing at machine centers 2, 5, 8, and 9 in the sequence 5-2-8-9. On the other hand, another order may require processing at machine centers 1, 2, 3, 5, and 7 in the sequence 7-5-2-3-2-1-3. In the latter case, the order requires processing at machine centers 1, 5, and 7 once, at machine center 2 twice, and at machine center 3 twice. As already indicated, there are L such sequences possible. The probability that an arriving order requires processing through the kth sequence is given by p_k, where

$$\sum_{k=1}^{L} p_k = 1$$

Each order processed at a given machine center is served on a first come–first served basis; that is, orders are processed at each center in the order in which they enter that center.

Develop a simulator for this queueing network. The simulator should calculate:

1. the probability mass function of the number of orders in each machine center and in the entire network,
2. the probability mass function of the number of orders waiting for service at each machine center and in the entire network,
3. the average number of orders in each machine center and in the entire network,
4. the average number of orders waiting for service at each machine center and in the entire network,
5. the average time spent by an order in each machine center (given that the order entered that machine center) and in the entire network,
6. the average time spent waiting for service by an order in each machine center (given that the order entered that machine center) and in the entire network, and
7. the utilization of each machine center and the total network.

In addition, the program should include an option whereby the analyst can obtain output on the status of each machine center after each event in the simulation.

The FORTRAN IV program for the network simulator is given in Fig. 9.28. The program is annotated to aid the reader in understanding the

```fortran
C     NPRINT=0, STATUS OF EACH SUBSYSTEM IS NOT WRITTEN OUT AFTER EACH EVENT
C            1, STATUS OF EACH SUBSYSTEM IS WRITTEN OUT AFTER EACH EVENT
C     NSYS=NUMBER OF SUBSYSTEMS
C     ISEQ=NUMBER OF POSSIBLE SEQUENCES
C     TIM(I,K)=TIME OF THE NEXT SERVICE IN THE KTH CHANNEL IN SUBSYSTEM I, FOR
C              I LESS THAN OR EQUAL TO NSYS
C             =TIME OF NEXT ARRIVAL FOR I=NSYS+1 AND K=1
C             =END OF THE SIMULATION FOR I=NSYS+1 AND K=2
C     TLAST(I)=TIME OF THE LAST EVENT IN THE ITH SUBSYSTEM
C     TNEXT=TIME OF THE NEXT EVENT IN THE SYSTEM
C     TTLAST=TIME OF LAST EVENT IN THE SYSTEM
C     INEXT=0,END OF SIMULATION IS THE NEXT EVENT
C           I,NEXT EVENT TAKES PLACE IN SUBSYSTEM I, I GREATER THAN 0
C     KNEXT=0,NEXT EVENT IS AN ARRIVAL
C           K,NEXT EVENT IS A SERVICE IN CHANNEL K, K GREATER THAN 0
C     IC(I)=NUMBER OF CHANNELS IN THE ITH SUBSYSTEM
C     IPOS(I,J)=NUMBER OF THE UNIT IN THE JTH POSITION IN THE ITH SUBSYSTEM
C     ISTAGE(I)=STAGE OF PROCESSING THE ITH UNIT IN THE SYSTEM IS
C               PRESENTLY IN
C     JSEQ(J)=SEQUENCE FOLLOWED BY THE JTH UNIT IN THE SYSTEM
C     P(J)=PROBABILITY THAT AN ARRIVING UNIT IS PROCESSED THROUGH THE
C          JTH SEQUENCE
C     NSEQ(J)=NUMBER OF PROCESSES IN THE JTH SEQUENCE
C     KSEQ(I,J)=ITH PROCESS IN THE JTH SEQUENCE
C     SYSN(I)=NUMBER OF UNITS IN THE ITH SUBSYSTEM
C     SYSNT(I)=NUMBER OF ENTRIES TO THE ITH SUBSYSTEM
C     QLEN(I)=QUEUE LENGTH AT THE ITH SUBSYSTEM
C     NUNIT=TOTAL NUMBER OF UNITS PRESENTLY IN THE ENTIRE SYSTEM
C     NOENT=CUMULATIVE NUMBER OF ENTRIES TO THE ENTIRE SYSTEM
C     PSYS(I,J)=PROBABILITY THAT THERE ARE J UNITS IN SUBSYSTEM I
C     PQU(I,J)=PROBABILITY THAT THERE ARE J UNITS IN THE WAITING LINE IN
C              SUBSYSTEM I
C     PTSYS(J)=PROBABILITY THAT THERE ARE J UNITS IN THE ENTIRE SYSTEM
C     PTQU(J)=PROBABILITY THAT THERE ARE J UNITS WAITING IN THE ENTIRE SYSTEM
C     TSYS(I)=CUMULATIVE UNIT-TOTAL TIME IN THE ITH SUBSYSTEM
C     TQU(I)=CUMULATIVE UNIT-WAITING TIME IN SUBSYSYEM I
C     ANSYS(I)=AVERAGE NUMBER OF UNITS IN SUBSYSTEM I
C     ANQU(I)=AVERAGE NUMBER OF UNITS WAITING IN SUBSYSTEM I
C     UTIL(I)=UTILIZATION OF SUBSYSTEM I
C     TTSYS=CUMULATIVE UNIT-TOTAL TIME IN THE ENTIRE SYSTEM
C     TTQU=CUMULATIVE UNIT-WAITING TIME IN THE ENTIRE SYSTEM
C     TANSYS=AVERAGE NUMBER OF UNITS IN THE ENTIRE SYSTEM
C     TANQU=AVERAGE NUMBER OF UNITS WAITING IN THE ENTIRE SYSTEM
C     TUTIL=UTILIZATION OF THE ENTIRE SYSTEM
C     DIST(I,J)=DISTRIBUTION TYPE FOR THE JTH CHANNEL IN THE ITH SUBSYSTEM, I
C               LESS THAN OR EQUAL TO NSYS
C              =DISTRIBUTION TYPE FOR INTERARRIVAL TIME, I=NSYS+1
C     A(I,K),B(I,K)=PARAMETERS OF THE SERVICE TIME DISTRIBUTION FOR THE
C                   KTH CHANNEL IN THE ITH SUBSYSTEM
C     C,D,=PARAMETERS OF THE INTERARRIVAL TIME DISTRIBUTION
      DIMENSION A(4,5),B(4,5),TIM(5,5),KSEQ(5,10),IPOS(4,50)
      DIMENSION PSYS(4,100),PQU(4,100),DIST(5,5)
      DIMENSION P(5),QLEN(4),SYSN(4),SYSNT(4),TLAST(4),IC(4)
      DIMENSION NSEQ(5),ISTAGE(500),JSEQ(100)
      DIMENSION TSYS(4),TQU(4),PTSYS(100),PTQU(100)
      COMMON/BLOKA/TIM,PSYS,PQU
      COMMON/BLOKB/KSEQ,IPOS
      COMMON/BLOKC/P,QLEN,SYSN,SYSNT,TLAST,TSYS,TQU,PTSYS,PTQU
      COMMON/BLOKD/IC,NSEQ,ISTAGE,JSEQ
      COMMON/BLOKE/TNEXT,NSYS,ISEQ,NUNIT,INEXT,KNEXT,NOENT,NPRINT
      COMMON/BLOKF/TTSYS,TTQU,TTLAST
      COMMON/BLOKG/A,B,DIST,C,D
C READ INPUT DATA AND INITIALIZE COUNTERS, INDICATORS, AND EVENT TIMES *********
      READ(5,100) IX,NSYS,ISEQ,C,D,NPRINT                                 *
      WRITE(6,100) IX,NSYS,ISEQ,C,D,NPRINT                                *
      READ(5,200) DIST(NSYS+1,1),TIM(NSYS+1,2)                            *
      WRITE(6,200) DIST(NSYS+1,1),TIM(NSYS+1,2)                           *
      DO 1 I=1,100                                                        *
      PTSYS(I)=0.                                                         *
    1 PTQU(I)=0.                                                          *
      TTSYS=0.                                                            *
      TTQU=0.                                                             *
      DO 3 I=1,NSYS                                                       *
      READ(5,100) IC(I)                                                   *
      WRITE(6,100) IC(I)                                                  *
      QLEN(I)=0                                                           *
      SYSN(I)=0                                                           *
      SYSNT(I)=0                                                          *
      TLAST(I)=0                                                          *
      TSYS(I)=0                                                           *
      TQU(I)=0                                                            *
      DO 2 J=1,100                                                        *
      PSYS(I,J)=0                                                         *
    2 PQU(I,J)=0                                                          *
      K=IC(I)                                                             *
      DO 3 J=1,K                                                          *
      READ(5,200) A(I,J),B(I,J),DIST(I,J)                                 *
    3 WRITE(6,200) A(I,J),B(I,J),DIST(I,J)                                *
C SET TIME OF NEXT SERVICE COMPLETION IN CHANNEL J OF SUBSYSTEM I TO AN   *
C ARBITRARILY LARGE QUANTITY <<<<<<<<<<<<<<<<<<<<<<<<<<<<<<<<<<<<<<<<<<   *
      TIM(I,J)=10.**30                                                   < *
C <<<<<<<<<<<<<<<<<<<<<<<<<<<<<<<<<<<<<<<<<<<<<<<<<<<<<<<<<<<<<<<<<<<<   *
```

Fig. 9.28 FORTRAN IV program for the network simulator in Example 9.16.

```
      3 CONTINUE                                                        *
        DO 4 J=1,ISEQ                                                   *
        READ(5,300) P(J),NSEQ(J)                                        *
        WRITE(6,300) P(J),NSEQ(J)                                       *
        K=NSEQ(J)                                                       *
        DO 4 I=1,K                                                      *
        READ(5,100) KSEQ(I,J)                                           *
        WRITE(6,100) KSEQ(I,J)                                          *
      4 CONTINUE                                                        *
        NOENT=0                                                         *
        NUNIT=0                                                         *
        TTLAST=0                                                        *
C GENERATE THE TIME OF THE FIRST ARRIVAL TO THE SYSTEM ################  *
        TIM(NSYS+1,1)=RNVAR(IX,NSYS+1,1)                              # *
C ####################################################################  *
C ********************************************************************
C DETERMINE THE NEXT EVENT AND BRANCH TO THE APPROPRIATE EVENT ROUTINE <<<<<<<<
      5 INEXT=0                                                          <
        KNEXT=0                                                          <
        TNEXT=10.**30                                                    <
        NSYS1=NSYS+1                                                     <
        DO 7 I=1,NSYS1                                                   <
        K=2                                                              <
        IF(I.EQ.NSYS1) GO TO 6                                           <
        K=IC(I)                                                          <
      6 DO 7 J=1,K                                                       <
        IF(TNEXT.LT.TIM(I,J)) GO TO 7                                    <
        TNEXT=TIM(I,J)                                                   <
        INEXT=I                                                          <
        KNEXT=J                                                          <
      7 CONTINUE                                                         <
        IF(INEXT.EQ.NSYS+1.AND.KNEXT.EQ.1) GO TO 8                       <
        IF(INEXT.EQ.NSYS+1.AND.KNEXT.EQ.2) GO TO 9                       <
C <<<<<<<<<<<<<<<<<<<<<<<<<<<<<<<<<<<<<<<<<<<<<<<<<<<<<<<<<<<<<<<<<<<<<
C THE NEXT EVENT IS A SERVICE COMPLETION. CALL THE SERVICE ROUTINE #############
        CALL SERVE(IX)                                                  #
        GO TO 5                                                         #
C #####################################################################
C THE NEXT EVENT IS AN ARRIVAL. CALL THE ARRIVAL ROUTINE ********************
      8 CALL ARRIV(IX)                                                  *
        GO TO 5                                                         *
C *********************************************************************
C THE NEXT EVENT IS THE END OF THE SIMULATION. CALL SUBROUTINE UPDATE WHICH WILL
C UPDATE ALL ACCUMULATORS, COMPUTE ALL STATISTICS, AND WRITE OUT FINAL
C INFORMATION <<<<<<<<<<<<<<<<<<<<<<<<<<<<<<<<<<<<<<<<<<<<<<<<<<<<<<<<
      9 CALL UPDATE(INEXT)                                               <
C <<<<<<<<<<<<<<<<<<<<<<<<<<<<<<<<<<<<<<<<<<<<<<<<<<<<<<<<<<<<<<<<<<<<<
        STOP
    100 FORMAT(1X,3I5,2F10.4,I5)
    200 FORMAT(1X,3F10.4)
    300 FORMAT(1X,F10.6,I5)
        END

        SUBROUTINE ARRIV(IX)
        DIMENSION TIM(5,5),KSEQ(5,10),IPOS(4,50),PSYS(4,100),PQU(4,100)
        DIMENSION P(5),QLEN(4),SYSN(4),SYSNT(4),TLAST(4),IC(4)
        DIMENSION NSEQ(5),ISTAGE(500),JSEQ(100)
        DIMENSION TSYS(4),TQU(4),PTSYS(100),PTQU(100)
        COMMON/BLOKA/TIM,PSYS,PQU
        COMMON/BLOKB/KSEQ,IPOS
        COMMON/BLOKC/P,QLEN,SYSN,SYSNT,TLAST,TSYS,TQU,PTSYS,PTQU
        COMMON/BLOKD/IC,NSEQ,ISTAGE,JSEQ
        COMMON/BLOKE/TNEXT,NSYS,ISEQ,NUNIT,INEXT,KNEXT,NOENT,NPRINT
        COMMON/BLOKF/TTSYS,TTQU,TTLAST
C INCREASE THE CUMULATIVE NUMBER OF ENTRIES TO THE SYSTEM BY ONE ***************
        NOENT=NOENT+1                                                   *
C *****************************************************************************
C DETERMINE THE SEQUENCE OF OPERATIONS FOLLOWED BY THE ARRIVING UNIT, INCREASE
C THE NUMBER PRESENTLY IN THE SYSTEM BY ONE, RECORD THE PRESENT STAGE OF
C PROCESSING OF THE ENTERING UNIT, RECORD THE SEQUENCE OF OPERATIONS FOLLOWED
C BY THE ENTERING UNIT, RECORD THE SUBSYSTEM ENTERED BY THE ARRIVING UNIT, CALL
C SUBROUTINE SYSENT, AND GENERATE THE TIME OF THE NEXT ARRIVAL TO THE SYSTEM <<<
        R=RANDU(IX,IY)                                                   <
        Q=0                                                              <
        DO 1 J=1,ISEQ                                                    <
        Q=Q+P(J)                                                         <
        IF(R.GT.Q) GO TO 1                                               <
        NUNIT=NUNIT+1                                                    <
        ISTAGE(NUNIT)=1                                                  <
        JSEQ(NUNIT)=J                                                    <
        ISYS=KSEQ(1,J)                                                   <
C UPDATE STATISTICS FOR SUBSYSTEM ISYS AND FOR THE TOTAL SYSTEM (FOR THE
C PERIOD PRIOR TO THE INCREASE IN NUNIT) ################################ <
        NUNIT=NUNIT-1                                                   # <
        CALL UPDATE(ISYS)                                               # <
        NUNIT=NUNIT+1                                                   # <
C ##################################################################### <
        CALL SYSENT(ISYS,NUNIT,IX)                                       <
        TIM(NSYS+1,1)=TNEXT+RNVAR(IX,NSYS+1,1)                           <
        RETURN                                                           <
      1 CONTINUE                                                         <
C <<<<<<<<<<<<<<<<<<<<<<<<<<<<<<<<<<<<<<<<<<<<<<<<<<<<<<<<<<<<<<<<<<<<<
        RETURN
        END
```

Fig. 9.28 (*continued*)

```
      SUBROUTINE SYSENT(I,J,IX)
      DIMENSION TIM(5,5),KSEQ(5,10),IPOS(4,50),PSYS(4,100),PQU(4,100)
      DIMENSION P(5),QLEN(4),SYSN(4),SYSNT(4),TLAST(4),IC(4)
      DIMENSION NSEQ(5),ISTAGE(500),JSEQ(100)
      DIMENSION TSYS(4),TQU(4),PTSYS(100),PTQU(100)
      COMMON/BLOKA/TIM,PSYS,PQU
      COMMON/BLOKB/KSEQ,IPOS
      COMMON/BLOKC/P,QLEN,SYSN,SYSNT,TLAST,TSYS,TQU,PTSYS,PTQU
      COMMON/BLOKD/IC,NSEQ,ISTAGE,JSEQ
      COMMON/BLOKE/TNEXT,NSYS,ISEQ,NUNIT,INEXT,KNEXT,NOENT,NPRINT
      COMMON/BLOKF/TTSYS,TTQU,TLAST
C INCREASE THE CURRENT NUMBER IN SUBSYSTEM I AND THE CUMULATIVE NUMBER OF
C ENTRIES TO SUBSYSTEM I BY ONE <<<<<<<<<<<<<<<<<<<<<<<<<<<<<<<<<<<<<<<
      SYSN(I)=SYSN(I)+1.                                              <
      SYSNT(I)=SYSNT(I)+1.                                            <
C <<<<<<<<<<<<<<<<<<<<<<<<<<<<<<<<<<<<<<<<<<<<<<<<<<<<<<<<<<<<<<<<<<<<
C DETERMINE WHETHER THE CURRENT NUMBER OF UNITS IN SUBSYSTEM I IS GREATER THAN
C THE NUMBER OF CHANNELS PROVIDED BY SUBSYSTEM I ######################
      S=IC(I)                                                         #
      IF(SYSN(I).LE.S) GO TO 1                                        #
C ####################################################################
C THE NUMBER OF UNITS IN SUBSYSTEM I IS GREATER THAN THE NUMBER OF CHANNELS
C PROVIDED BY THAT SUBSYSTEM. INCREASE THE QUEUE LENGTH OF SUBSYSTEM I BY ONE
C AND RECORD THAT THE UNIT IN THE KTH POSITION IN SUBSYSTEM I IS UNIT NUMBER J
C CURRENTLY IN THE TOTAL SYSTEM **************************************
      QLEN(I)=QLEN(I)+1.                                              *
      K=SYSN(I)                                                       *
      IPOS(I,K)=J                                                     *
      RETURN                                                          *
C ********************************************************************
C THE NUMBER OF UNITS CURRENTLY IN SUBSYSTEM I IS LESS THAN OR EQUAL TO THE
C NUMBER OF CHANNELS PROVIDED BY THAT SUBSYSTEM. PLACE UNIT NUMBER J IN THE
C FIRST AVAILABLE CHANNEL IN SUBSYSTEM I AND GENERATE THE TIME OF SERVICE
C COMPLETION FOR UNIT NUMBER J #######################################
    1 IS=IC(I)                                                        #
      DO 2 K=1,IS                                                     #
      IF(TIM(I,K).LT.10.**30) GO TO 2                                 #
      IPOS(I,K)=J                                                     #
      TIM(I,K)=TNEXT+RNVAR(IX,I,K)                                    #
      RETURN                                                          #
    2 CONTINUE                                                        #
C ####################################################################
      RETURN
      END

      SUBROUTINE SERVE(IX)
      DIMENSION TIM(5,5),KSEQ(5,10),IPOS(4,50),PSYS(4,100),PQU(4,100)
      DIMENSION P(5),QLEN(4),SYSN(4),SYSNT(4),TLAST(4),IC(4)
      DIMENSION NSEQ(5),ISTAGE(500),JSEQ(100)
      DIMENSION TSYS(4),TQU(4),PTSYS(100),PTQU(100)
      COMMON/BLOKA/TIM,PSYS,PQU
      COMMON/BLOKB/KSEQ,IPOS
      COMMON/BLOKC/P,QLEN,SYSN,SYSNT,TLAST,TSYS,TQU,PTSYS,PTQU
      COMMON/BLOKD/IC,NSEQ,ISTAGE,JSEQ
      COMMON/BLOKE/TNEXT,NSYS,ISEQ,NUNIT,INEXT,KNEXT,NOENT,NPRINT
      COMMON/BLOKF/TTSYS,TTQU,TLAST
C UPDATE STATISTICS FOR SUBSYSTEM INEXT AND FOR THE TOTAL SYSTEM ****************
      CALL UPDATE(INEXT)                                              *
C ********************************************************************
      I=INEXT
      K=KNEXT
C REDUCE THE CURRENT NUMBER OF UNITS IN THE ITH SUBSYSTEM BY ONE <<<<<<<<<<<<<<
      SYSN(I)=SYSN(I)-1.                                              <
C <<<<<<<<<<<<<<<<<<<<<<<<<<<<<<<<<<<<<<<<<<<<<<<<<<<<<<<<<<<<<<<<<<<<
C RECORD J AS THE UNIT IN THE KTH POSITION OF SUBSYSTEM I, SET J1 AS THE NUMBER
C OF THE NEXT STAGE OF PROCESSING FOR UNIT NUMBER J, AND RECORD J2 AS THE
C SEQUENCE FOLLOWED BY UNIT NUMBER J ##################################
      J=IPOS(I,K)                                                     #
      J1=ISTAGE(J)+1                                                  #
      J2=JSEQ(J)                                                      #
C ####################################################################
C DETERMINE WHETHER THE UNIT JUST SERVED LEAVES THE ENTIRE SYSTEM OR PROCEEDS
C TO ANOTHER SUBSYSTEM FOR FURTHER PROCESSING *************************
      IF(J1.GT.NSEQ(J2)) GO TO 5                                      *
C ********************************************************************
C UNIT JUST SERVED PROCEEDS FOR FURTHER PROCESSING. RECORD THE SUBSYSTEM TO
C WHICH THE SERVED UNIT PROCEEDS AS ISYS, SET THE STAGE OF PROCESSING OF THE
C UNIT JUST SERVED TO J1, AND CALL SUBROUTINE SYSENT <<<<<<<<<<<<<<<<<<<<<
      ISYS=KSEQ(J1,J2)                                                <
      ISTAGE(J)=J1                                                    <
C UPDATE STATISTICS FOR SUBSYSTEM ISYS ********************************  <
      CALL UPDATE(ISYS)                                             * <
C ********************************************************************  <
      CALL SYSENT(ISYS,J,IX)                                          <
C <<<<<<<<<<<<<<<<<<<<<<<<<<<<<<<<<<<<<<<<<<<<<<<<<<<<<<<<<<<<<<<<<<<<
C DETERMINE WHETHER OR NOT THERE IS ANOTHER UNIT WAITING FOR SERVICE IN
C SUBSYSTEM I #########################################################
    1 IF(QLEN(I).GT.0.) GO TO 2                                       #
C ####################################################################
C THE QUEUE FOR THE ITH SUBSYSTEM IS EMPTY. SET TIME OF THE NEXT SERVICE IN
C THE KTH CHANNEL (THE CHANNEL JUST VACATED) OF SUBSYSTEM I TO 10.**30 *********
      TIM(I,K)=10.**30                                                *
      GO TO 10                                                        *
C ********************************************************************
```

Fig. 9.28 (*continued*)

```
C THE QUEUE FOR THE ITH SUBSYSTEM IS NOT EMPTY. PLACE THE UNIT IN THE FIRST
C POSITION OF THE QUEUE IN CHANNEL K AND GENERATE THE TIME OF THE NEXT SERVICE
C COMPLETION IN CHANNEL K <<<<<<<<<<<<<<<<<<<<<<<<<<<<<<<<<<<<<<<<<<<<<<<
    2 K1=IC(I)+1                                                              <
      IQ=QLEN(I)                                                              <
      IPOS(I,K1)=IPOS(I,K1)                                                   <
      TIM(I,K)=TNEXT+RNVAR(IX,I,K)                                            <
C <<<<<<<<<<<<<<<<<<<<<<<<<<<<<<<<<<<<<<<<<<<<<<<<<<<<<<<<<<<<<<<<<<<<<<<
C DETERMINE WHETHER THE LENGTH OF THE QUEUE FOR THE ITH SUBSYSTEM IS ONE OR
C LESS ###################################################################
      IF(QLEN(I).LE.1.) GO TO 4                                               #
C ########################################################################
C THE QUEUE FOR THE ITH SUBSYSTEM WAS OF LENGTH GREATER THAN ONE. MOVE EACH UNIT
C IN THE QUEUE UP ONE POSITION *******************************************
      DO 3 K2=K1,IQ                                                           *
    3 IPOS(I,K2)=IPOS(I,K2+1)                                                 *
C ************************************************************************
C REDUCE THE QUEUE LENGTH FOR THE ITH SUBSYSTEM BY ONE <<<<<<<<<<<<<<<<<<<<<<
    4 QLEN(I)=QLEN(I)-1.                                                      <
C <<<<<<<<<<<<<<<<<<<<<<<<<<<<<<<<<<<<<<<<<<<<<<<<<<<<<<<<<<<<<<<<<<<<<<<
      RETURN
C THE UNIT JUST SERVED LEAVES THE ENTIRE SUSTEM. REDUCE THE CURRENT NUMBER IN
C THE SYSTEM BY ONEAND DETERMINE WHETHER THE UNIT JUST SERVED WAS THE LAST UNIT
C TO ENTER THE SYSTEM OR NOT #############################################
    5 K1=NUNIT-1                                                              #
      NUNIT=K1                                                                #
      IF(K1.LT.J) GO TO 1                                                     #
C ########################################################################
C THE UNIT JUST SERVED WAS NOT THE LAST UNIT TO ENTER THE SYSTEM. MOVE EACH UNIT
C IN THE SYSTEM FROM THE J+1 ST ON UP ONE POSITION IN THE LIST OF UNITS
C PRESENTLY IN THE TOTAL SYSTEM *******************************************
      DO 6 K2=J,K1                                                            *
      ISTAGE(K2)=ISTAGE(K2+1)                                                 *
    6 JSEQ(K2)=JSEQ(K2+1)                                                     *
C ************************************************************************
C FOR EACH SUBSYSTEM, REDUCE THE NUMBER OF THE UNIT IN EACH POSITION BY ONE FOR
C ALL UNITS FROM THE J+1 ST ON <<<<<<<<<<<<<<<<<<<<<<<<<<<<<<<<<<<<<<<<<<<<<
      DO 9 K3=1,NSYS                                                          <
      IF(SYSN(K3).LE.0.) GO TO 9                                              <
      K5=SYSN(K3)                                                             <
      IF(K3.EQ.I) K5=K5+1                                                     <
      IF(K5.LT.IC(K3)) K5=IC(K3)                                              <
      DO 8 K4=1,K5                                                            <
      IF(K4.GT.IC(K3)) GO TO 7                                                <
      IF(TIM(K3,K4).GE.10.**30) GO TO 8                                       <
    7 IF(J.LT.IPOS(K3,K4)) IPOS(K3,K4)=IPOS(K3,K4)-1                          <
    8 CONTINUE                                                                <
    9 CONTINUE                                                                <
C <<<<<<<<<<<<<<<<<<<<<<<<<<<<<<<<<<<<<<<<<<<<<<<<<<<<<<<<<<<<<<<<<<<<<<<
      GO TO 1
   10 RETURN
      END

      SUBROUTINE UPDATE(I)
      DIMENSION TIM(5,5),KSEQ(5,10),IPOS(4,50),PSYS(4,100),PQU(4,100)
      DIMENSION P(5),QLEN(4),SYSN(4),SYSNT(4),TLAST(4),IC(4)
      DIMENSION NSEQ(5),ISTAGE(500),JSEQ(100),ISTOP(100)
      DIMENSION TSYS(4),TQU(4),ANSYS(4),ANQU(4),UTIL(4)
      DIMENSION PTSYS(100),PTQU(100)
      COMMON/BLOKA/TIM,PSYS,PQU
      COMMON/BLOKB/KSEQ,IPOS
      COMMON/BLOKC/P,QLEN,SYSN,SYSNT,TLAST,TSYS,TQU,PTSYS,PTQU
      COMMON/BLOKD/IC,NSEQ,ISTAGE,JSEQ
      COMMON/BLOKE/TNEXT,NSYS,ISEQ,NUNIT,INEXT,KNEXT,NOENT,NPRINT
      COMMON/BLOKF/TTSYS,TTQU,TTLAST
C DETERMINE WHETHER OUTPUT OF THE STATUS OF EACH SUBSYSTEM IS DESIRED AFTER EACH
C EVENT ******************************************************************
      IF(NPRINT.LE.0) GO TO 6                                                 *
C ************************************************************************
C WRITE STATUS OF EACH SUBSYSTEM <<<<<<<<<<<<<<<<<<<<<<<<<<<<<<<<<<<<<<<<<<
      WRITE(6,1300) TIM(NSYS+1,1),TNEXT,TTLAST                                <
      WRITE(6,1200) NUNIT,NOENT                                               <
      DO 3 I1=1,NSYS                                                          <
      WRITE(6,700) I1,QLEN(I1),SYSN(I1),SYSNT(I1),TLAST(I1)                   <
      K=IC(I1)                                                                <
      KQ=QLEN(I1)+IC(I1)                                                      <
      IF(QLEN(I1).LE.0.) GO TO 2                                              <
      KC=IC(I1)+1                                                             <
      DO 1 K8=KC,KQ                                                           <
    1 WRITE(6,800) K8,IPOS(I1,K8)                                             <
    2 CONTINUE                                                                <
      DO 3 J1=1,K                                                             <
      WRITE(6,900) J1,TIM(I1,J1)                                              <
      IF(TIM(I1,J1).GE.10.**30) GO TO 3                                       <
      WRITE(6,1000) IPOS(I1,J1)                                               <
    3 CONTINUE                                                                <
      IF(NUNIT.LE.0) GO TO 5                                                  <
      DO 4 I1=1,NUNIT                                                         <
    4 WRITE(6,1100) I1,ISTAGE(I1),JSEQ(I1)                                    <
    5 CONTINUE                                                                <
C <<<<<<<<<<<<<<<<<<<<<<<<<<<<<<<<<<<<<<<<<<<<<<<<<<<<<<<<<<<<<<<<<<<<<<<
```

Fig. 9.28 (*continued*)

```
C ACCUMULATE TOTAL UNIT-TIME IN THE SUBSYSTEM, TOTAL UNIT-TIME IN THE QUEUE FOR
C THE SUBSYSTEM, THE TIME FOR WHICH THERE WERE J1-1 UNITS IN THE SUBSYSTEM
C (PSYS(J,J1)), AND THE TIME FOR WHICH THERE WERE J2-1 UNITS IN THE QUEUE FOR
C THE SUBSYSTEM (PQU(J,J2)) ##################################################
    6 DO 9 J=1,NSYS
C DETERMINE WHETHER OR NOT THE PRESENT UPDATE OCCURS AT THE END OF THE
C END OF THE SIMULATION ***************************************************
      IF(I.NE.NSYS+1) GO TO 8
C ****************************************************************************
C THE PRESENT UPDATE OCCURS AT THE END OF THE SIMULATION. ACCUMULATE STATISTICS
C FOR ALL SUBSYSTEMS <<<<<<<<<<<<<<<<<<<<<<<<<<<<<<<<<<<<<<<<<<<<<<<<<<<<<
    7 DELT=TNEXT-TLAST(J)
      TLAST(J)=TNEXT
      TSYS(J)=TSYS(J)+SYSN(J)*DELT
      TQU(J)=TQU(J)+QLEN(J)*DELT
      J1=SYSN(J)+1.
      J2=QLEN(J)+1.
      IF(J1.GT.100) J1=100
      IF(J2.GT.100) J2=100
      PSYS(J,J1)=PSYS(J,J1)+DELT
      PQU(J,J2)=PQU(J,J2)+DELT
      GO TO 9
C <<<<<<<<<<<<<<<<<<<<<<<<<<<<<<<<<<<<<<<<<<<<<<<<<<<<<<<<<<<<<<<<<<<<<<<<
C THE PRESENT UPDATE DOES NOT OCCUR AT THE END OF THE SIMULATION. DETERMINE
C WHETHER OR NOT THE CURRENT VALUE OF J IS THE NUMBER OF THE SUBSYSTEM FOR
C WHICH AN UPDATE SHOULD BE CARRIED OUT. IF AN UPDATE IS TO BE CARRIED OUT FOR
C SUBSYSTEM J, GO TO STATEMENT 7 AND ACCUMULATE STATISTICS FOR SUBSYSTEM J.
    8 IF(J.EQ.I) GO TO 7
C ****************************************************************************
    9 CONTINUE
C ############################################################################
C ACCUMULATE TOTAL UNIT-TIME IN THE ENTIRE SYSTEM, TOTAL UNIT-TIME WAITING IN
C THE ENTIRE SYSTEM, TOTAL TIME FOR WHICH THERE WERE J1-1 UNITS IN THE ENTIRE
C SYSTEM (PTSYS(J1)), AND TOTAL TIME FOR WHICH THERE WERE J2-1 UNITS WAITING IN
C THE ENTIRE SYSTEM (PTQU(J2)). <<<<<<<<<<<<<<<<<<<<<<<<<<<<<<<<<<<<<<<<<<<
      DELT=TNEXT-TTLAST
      TTLAST=TNEXT
      SS=NUNIT
      SQ=0
      DO 10 J=1,NSYS
   10 SQ=SQ+QLEN(J)
      TTSYS=TTSYS+SS*DELT
      TTQU=TTQU+SQ*DELT
      J1=SS+1.
      J2=SQ+1.
      IF(J1.GT.100) J1=100
      IF(J2.GT.100) J2=100
      PTSYS(J1)=PTSYS(J1)+DELT
      PTQU(J2)=PTQU(J2)+DELT
C <<<<<<<<<<<<<<<<<<<<<<<<<<<<<<<<<<<<<<<<<<<<<<<<<<<<<<<<<<<<<<<<<<<<<<<<
C DETERMINE WHETHER OR NOT THE PRESENT UPDATE OCCURS AT THE END OF THE
C SIMULATION. IF NOT RETURN. ##############################################
      IF(I.NE.NSYS+1) RETURN
C ############################################################################
C THE PRESENT UPDATE OCCURS AT THE END OF THE SIMULATION. CALCULATE AVERAGE
C TNIT-TIME IN THE
C UNIT-TIME IN EACH SUBSYSTEM AND THE TOTAL SYSTEM, AVERAGE UNIT-TIME IN THE
C WAITING LINE FOR EACH SUBSYSTEM AND THE TOTAL SYSTEM, AVERAGE NUMBER OF UNITS
C IN EACH SUBSYSTEM AND THE TOTAL SYSTEM, AVERAGE NUMBER OF UNITS WAITING FOR
C EACH SUBSYSTEM AND THE TOTAL SYSTEM, AND THE PROBABILITY MASS FUNCTIONS FOR
C THE NUMBER OF UNITS IN EACH SUBSYSTEM AND THE TOTAL SYSTEM AND FOR THE NUMBER
C OF UNITS WAITING IN EACH SUBSYSTEM AND THE TOTAL SYSTEM. WRITE OUT ALL
C COMPUTED STATISTICS. *****************************************************
      DO 12 I=1,NSYS
      TSYS(I)=TSYS(I)/SYSNT(I)
      TQU(I)=TQU(I)/SYSNT(I)
      ANSYS(I)=0
      ANQU(I)=0
      DO 11 J=1,100
      PSYS(I,J)=PSYS(I,J)/TIM(NSYS+1,2)
      PQU(I,J)=PQU(I,J)/TIM(NSYS+1,2)
      IF(PSYS(I,J).NE.0.) ISTOP(I)=J
      AJ=J-1
      ANSYS(I)=ANSYS(I)+AJ*PSYS(I,J)
   11 ANQU(I)=ANQU(I)+AJ*PQU(I,J)
   12 UTIL(I)=1.-PSYS(I,1)
      DO 14 I=1,NSYS
      K=ISTOP(I)
      WRITE(6,100) I
      DO 13 J=1,K
      K1=J-1
   13 WRITE(6,200) K1,PSYS(I,J),PQU(I,J)
   14 WRITE(6,300) ANSYS(I),ANQU(I),TSYS(I),TQU(I),UTIL(I)
      TANSYS=0.
      TANQU=0.
      SS=NOENT
      TTSYS=TTSYS/SS
      TTQU=TTQU/SS
      DO 15 J=1,100
      PTSYS(J)=PTSYS(J)/TIM(NSYS+1,2)
      PTQU(J)=PTQU(J)/TIM(NSYS+1,2)
      IF(PTSYS(J).NE.0.) JSTOP=J
```

Fig. 9.28 (*continued*)

Simulation of Networks of Queues

```
              AJ=J-1                                                             *
              TANSYS=TANSYS+AJ*PTSYS(J)                                          *
           15 TANQU=TANQU+AJ*PTQU(J)                                             *
              TUTIL=1.-PTSYS(1)                                                  *
              WRITE(6,500)                                                       *
              DO 16 J=1,JSTOP                                                    *
              K1=J-1                                                             *
           16 WRITE(6,200) K1,PTSYS(J),PTQU(J)                                   *
              WRITE(6,600) TANSYS,TANQU,TTSYS,TTQU,TUTIL                         *
        *****************************************************************************
              RETURN
          100 FORMAT(1X,'SUBSYSTEM NUMBER',I4///1X,'NUMBER',10X,'PROBABILITY OF
             1NUMBER IN SYSTEM',6X,'PROBABILITY OF NUMBER IN QUEUE'/)
          200 FORMAT(1X,I6,17X,E14.7,23X,E14.7)
          300 FORMAT(//1X,'AVERAGE NUMBER IN SUBSYSTEM=',F10.4/1X,'AVERAGE NUMBE
             1R IN QUEUE=',F10.4/1X,'AVERAGE TIME IN SUBSYSTEM=',F10.4/1X,'AVERA
             2GE TIME IN QUEUE=',F10.4/1X,'UTILIZATION=',F10.4///)
          500 FORMAT(1X,'TOTAL SYSTEM'///1X,'NUMBER',10X,'PROBABILITY OF NUMBER
             1IN SYSTEM',6X,'PROBABILITY OF NUMBER IN QUEUE'/)
          600 FORMAT(1X,'AVERAGE NUMBER IN SYSTEM=',F10.4/1X,'AVERAGE NUMBER IN
             1QUEUE=',F10.4/1X,'AVERAGE TIME IN SYSTEM=',F10.4/1X,'AVERAGE TIME
             2IN QUEUE=',F10.4/1X,'UTILIZATION=',F10.4///)
          700 FORMAT(1X,'SYSTEM=',I3,1X,'QLEN=',E14.7,1X,'SYSN=',E14.7,1X,'SYSNT
             1=',E14.7,1X,'TLAST=',E14.7)
          800 FORMAT(1X,'QUEUE POS=',I3,1X,'UNIT IN POS=',I4)
          900 FORMAT(1X,'CHANNEL NO.=',I3,1X,'SERVICE TIME=',E14.7)
         1000 FORMAT(17X,'UNIT IN CHANNEL=',I5)
         1100 FORMAT(1X,'UNIT NO.=',I5,1X,'STAGE=',I5,1X,'SEQUENCE=',I5///)
         1200 FORMAT(1X,'NUNIT=',I5/1X,'NOENT=',I5)
         1300 FORMAT(1X,'TIME OF NEXT ARRIVAL=',E14.7,3X,'TIME OF NEXT EVENT=',E
             114.7,3X,'TIME OF LAST EVENT=',E14.7)
              END

              FUNCTION RNVAR(IX,I,J)
              DIMENSION A(4,5),B(4,5),DIST(5,5)
              COMMON/BLOKE/TNEXT,NSYS,ISEQ,NUNIT,INEXT,KNEXT,NOENT,NPRINT
              COMMON/BLOKG/A,B,DIST,C,D
              M=DIST(I,J)
              IF(I.GT.NSYS) GO TO 1
              A1=A(I,J)
              A2=B(I,J)
              GO TO 2
            1 A1=C
              A2=D
            2 GO TO (3,5,6,7,8),M
            3 IA=A2
              RNVAR=0
              DO 4 L=1,IA
            4 RNVAR=RNVAR-(1./A1)*ALOG(RANDU(IX,IY))
              RETURN
            5 R1=RANDU(IX,IY)
              R2=RANDU(IX,IY)
              RNVAR=A1+A2*SQRT(-2.*ALOG(R1))*COS(6.28*R2)
              RETURN
            6 R1=RANDU(IX,IY)
              R2=RANDU(IX,IY)
              A3=1./A1
              A4=1./A2
              RNVAR=(R1**A3)/(R1**A3+R2**A4)
              IF(R1**A3+R2**A4.GT.1.) GO TO 6
              RETURN
            7 RNVAR=A1+(A2-A1)*RANDU(IX,IY)
              RETURN
            8 RNVAR=A1
              RETURN
              END
```

Fig. 9.28 (continued)

logic of the simulator. The program is broken into five modules plus the process generator RNVAR and the random number generator. The main program reads all input data, initializes counters, indicators, and event times, determines the time of occurrence of the next event, identifies the next event as an arrival, service, or termination of the simulation, and calls the appropriate subroutine, which alters the status of the network in accordance with the character of the next event.

When the next event is an arrival to the system, subroutine ARRIV is called. This subroutine increases the cumulative number of arrivals to the network, determines the sequence of operations necessary to process the arriving order, and calls subroutine UPDATE, which updates all statistical accumulators. Finally subroutine SYSENT is called, which places the arriving order in the appropriate machine center (queueing system).

As already mentioned, subroutine SYSENT places an order in the appropriate queueing system upon arrival of that order to the system. Subroutine SYSENT is also called when processing of an order at a given machine center is completed and that order is sent to another machine center for further processing. Specifically, this subroutine increases the current number at the appropriate machine center and the total number of entries to that center. The entering order is either placed in the waiting line at the machine center or is placed in the first available service channel, in which case the service time for that order is generated.

Subroutine SERVE is called whenever the next event to occur in the network is the completion of processing of an order at a given machine center. First, subroutine UPDATE is called to update all statistical accumulators. The order on which processing was just completed is then removed from the machine center. If further processing is necessary, the next stage of processing is determined and subroutine SYSENT is called. If the order has completed the sequence of operations for full processing, that order is removed from the network. The number at the machine center where the service occurred is then reduced by one, and if the length of the waiting line at that center is greater than zero the first unit in the waiting line is placed in the vacated service channel and each unit remaining in the waiting line is moved up one position.

Subroutine UPDATE writes out the status of each machine center prior to the occurrence of each event when called for. In addition, this subroutine accumulates unit-time in the system, unit-time in the queue, and the time for which a given number of units were in the system and in the queue for the machine center where the current event takes place. In addition, similar accumulators are brought up to date for the network as a whole.

When subroutine UPDATE is called at the termination of the simulation, the accumulators mentioned above are updated for each machine center as well as for the entire network. Subroutine UPDATE then calculates and writes out estimates of the average time per unit in each machine center, the average time per unit in each queue, the average number in each machine center, the average number in each queue, the average utilization for each machine center, and the probability mass functions of the number in each machine center and in each queue. The same estimates are calculated and written out for the network as a whole.

The above description of the network simulator presented here highlights the salient operations of the simulation program. However, the comments included in the program listing should explain each of the operations carried out in the simulator in detail. In addition, all variables used in the program are defined at the beginning of the main program. In the definition of variables, a machine center is referred to as a subsystem and the network is referred to as the system.

Example 9.17 The operation of a network of the type described in Example 9.16 is to be simulated for a period of twelve months. The job shop consists of three machine centers and production orders may follow one of five different sequences of operations for complete processing. The five sequences of machine center operations and the associated probability of an arriving order requiring each sequence are given in Table 9.19.

TABLE 9.19

Sequence of machine center operations required	Probability that an arriving order requires the sequence
1-2-1	0.10
2-3	0.20
1-2-1-3	0.30
2	0.20
3-2-3-1	0.20

Orders arrive at the job shop in an exponential fashion at an average rate of 40 orders/month. The first machine center consists of one machine, the second consists of two machines in parallel, and the third consists of three machines in parallel. Let $f_{ij}(t)$ be the density function of service time, t in months, for the jth machine at the ith machine center, where

$$f_{11}(t) = 10{,}000te^{-100t}, \qquad 0 < t < \infty$$

$$f_{2j}(t) = \begin{cases} \dfrac{1}{0.001(2\pi)^{1/2}} \exp\left[\dfrac{-(t-0.01)^2}{0.000001}\right], & -\infty < t < \infty, \quad j = 1 \\ 100t^{99}, & 0 < t < 1, \quad j = 2 \end{cases}$$

$$f_{3j}(t) = \begin{cases} 50.0, & 0 < t < 0.02, \quad j = 1 \\ 33.3, & 0 < t < 0.03, \quad j = 2 \\ 25.0, & 0 < t < 0.04, \quad j = 3 \end{cases}$$

The program given in Fig. 9.28 is to be used to simulate the operation of the job shop and the status of each machine center is to be written out at each event. Plot the number in each machine center, the total number in the job shop, and the total number waiting for service in the job shop at each event for the first month.

The input data for the network simulation are as follows:

$$IX = \text{five-digit odd integer}$$
$$NSYS = 3$$
$$ISEQ = 5$$
$$C = 40.$$
$$D = 1.$$
$$NPRINT = 1$$

$$IC(I) = \begin{cases} 1, & I = 1 \\ 2, & I = 2 \\ 3, & I = 3 \end{cases}$$

$$A(I, J), B(I, J), DIST(I, J) = \begin{cases} 100.00, & 2.000, & 1., & I = 1, J = 1 \\ 0.01, & 0.001, & 2., & I = 2, J = 1 \\ 100.00, & 1.000, & 3., & I = 2, J = 2 \\ 0.00, & 0.020, & 4., & I = 3, J = 1 \\ 0.00, & 0.030, & 4., & I = 3, J = 2 \\ 0.00, & 0.040, & 4., & I = 3, J = 3 \end{cases}$$

$$DIST(4, 1) = 1.$$
$$TIM(4, 2) = 12.$$

$$P(J) = \begin{cases} 0.10, & J = 1 \\ 0.20, & J = 2 \\ 0.30, & J = 3 \\ 0.20, & J = 4 \\ 0.20, & J = 5 \end{cases} \qquad NSEQ(J) = \begin{cases} 3, & J = 1 \\ 2, & J = 2 \\ 4, & J = 3 \\ 1, & J = 4 \\ 4, & J = 5 \end{cases}$$

Simulation of Networks of Queues

$$KSEQ(I, J) = \begin{cases} 1, & I = 1, \ J = 1 \\ 2, & I = 2, \ J = 1 \\ 1, & I = 3, \ J = 1 \\ 2, & I = 1, \ J = 2 \\ 3, & I = 2, \ J = 2 \\ 1, & I = 1, \ J = 3 \\ 2, & I = 2, \ J = 3 \\ 1, & I = 3, \ J = 3 \\ 3, & I = 4, \ J = 3 \\ 3, & I = 1, \ J = 4 \\ 3, & I = 1, \ J = 5 \\ 2, & I = 2, \ J = 5 \\ 3, & I = 3, \ J = 5 \\ 1, & I = 4, \ J = 5 \end{cases}$$

The variation of the number at each machine center as a function of time is shown graphically for a period of one month in Fig. 9.29. The variation of

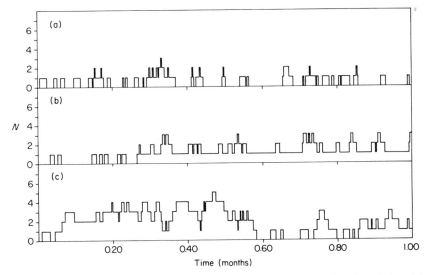

Fig. 9.29 Variation of the number N at each machine center as a function of time. (a) Machine center III, (b) machine center II, (c) machine center I.

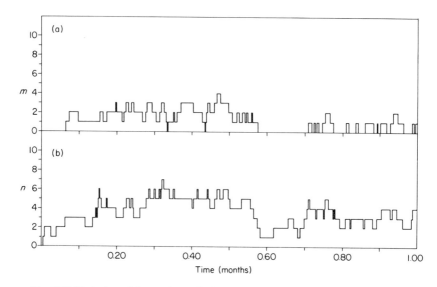

Fig. 9.30 Variation of the number N in the network and the number waiting for service in the network as a function of time. (a) Number waiting in the job shop, (b) number in the job shop.

SUBSYSTEM NUMBER 1

NUMBER	PROBABILITY OF NUMBER IN SYSTEM	PROBABILITY OF NUMBER IN QUEUE
0	0.2022031E 00	0.4077392E 00
1	0.2055360E 00	0.1717480E 00
2	0.1717480E 00	0.1317535E 00
3	0.1317535E 00	0.9709024E−01
4	0.9709024E−01	0.5720621E−01
5	0.5720621E−01	0.3340761F−01
6	0.3340761E−01	0.2421148E−01
7	0.2421148E−01	0.1585976E−01
8	0.1585976E−01	0.1730505E−01
9	0.1730505E−01	0.2364556E−01
10	0.2364556E−01	0.1685723E−01
11	0.1685723E−01	0.3176053E−02
12	0.3176053E−02	0.0000000E 00

AVERAGE NUMBER IN SUBSYSTEM = 2.7312
AVERAGE NUMBER IN QUEUE = 1.9334
AVERAGE TIME IN SUBSYSTEM = 0.0697
AVERAGE TIME IN QUEUE = 0.0494
UTILIZATION = 0.7978

Fig. 9.31 Statistical output for machine center I.

Simulation of Networks of Queues

SUBSYSTEM NUMBER 2

NUMBER	PROBABILITY OF NUMBER IN SYSTEM	PROBABILITY OF NUMBER IN QUEUE
0	0.7744360E−01	0.9530317E−00
1	0.6499201E 00	0.4542469E−01
2	0.2256678E 00	0.1543601E−02
3	0.4542469E−01	0.0000000E 00
4	0.1543601E−02	0.0000000E 00

AVERAGE NUMBER IN SUBSYSTEM = 1.2437
AVERAGE NUMBER IN QUEUE = 0.0485
AVERAGE TIME IN SUBSYSTEM = 0.0393
AVERAGE TIME IN QUEUE = 0.0015
UTILIZATION = 0.9226

Fig. 9.32 Statistical output for machine center II.

SUBSYSTEM NUMBER 3

NUMBER	PROBABILITY OF NUMBER IN SYSTEM	PROBABILITY OF NUMBER IN QUEUE
0	0.6172701E 00	0.9979523E 00
1	0.2937649E 00	0.1822551E−02
2	0.7657975E−01	0.2250671E−03
3	0.1033752E−01	0.0000000E 00
4	0.1822551E−02	0.0000000E 00
5	0.2250671E−03	0.0000000E 00

AVERAGE NUMBER IN SUBSYSTEM = 0.4864
AVERAGE NUMBER IN QUEUE = 0.0023
AVERAGE TIME IN SUBSYSTEM = 0.0114
AVERAGE TIME IN QUEUE = 0.0001
UTILIZATION = 0.3827

Fig. 9.33 Statistical output for machine center III.

TOTAL SYSTEM

NUMBER	PROBABILITY OF NUMBER IN SYSTEM	PROBABILITY OF NUMBER IN QUEUE
0	0.7825501E−02	0.3880212E 00
1	0.8117867E−01	0.1847246E 00
2	0.1732179E 00	0.1306126E 00
3	0.1816528E 00	0.9718519E−01
4	0.1414138E 00	0.6171034E−01
5	0.1283242E 00	0.3389748E−01
6	0.9869623E−01	0.2517239E−01
7	0.5753179E−01	0.1716296E−01
8	0.3547350E−01	0.1761603E−01
9	0.2122045E−01	0.2386371E−01
10	0.2579021E−01	0.1685723E−01
11	0.2288103E−01	0.3176053E−02
12	0.2014430E−01	0.0000000E 00
13	0.4649401E−02	0.0000000E 00

AVERAGE NUMBER IN SYSTEM = 4.4613
AVERAGE NUMBER IN QUEUE = 1.9842
AVERAGE TIME IN SYSTEM = 0.1120
AVERAGE TIME IN QUEUE = 0.0498
UTILIZATION = 0.9922

Fig. 9.34 Statistical output for the job shop as a whole.

the total number of orders in the job shop and the total number of orders waiting for processing as a function of time is shown in Fig. 9.30. The statistical output from the simulator is given in Figs. 9.31–9.34.

SUMMARY

Our attempt has been to present simulation as complementary to mathematical modeling in the analysis of queueing systems; that is, we do not suggest that simulation be used where mathematical methods are appropriate, but rather where the complexity of the system studied is such that meaningful mathematical analysis is not possible or beyond the capability of the analyst. However, as we have attempted to point out, simulation does offer the analyst a relatively simple and yet versatile tool for treating a wide range of complex queueing situations.

REFERENCES

Abramowitz, M., and Stegun, I. A. (eds.), *Handbook of Mathematical Functions.* Washington: National Bureau of Standards, 1964.

Berman, M. B., *Generating Gamma Distributed Variates for Computer Simulation Models.* Santa Monica, California: Rand Corporation, 1971.

Elmaghraby, S. E., *The Design of Production Systems.* New York: Reinhold, 1966.

Morris, W. T., *Analysis for Materials Handling Management.* Homewood, Illinois: Irwin, 1962.

Rand Corporation, *A Million Random Digits with 100,000 Normal Deviates.* Glencoe, Illinois: Free Press, 1955.

Saaty, T. L., *Elements of Queueing Theory.* New York: McGraw-Hill, 1961.

Schmidt, J. W., and Taylor, R. E., *Simulation and Analysis of Industrial Systems.* Homewood, Illinois: Irwin, 1970.

PROBLEMS

1. The following data have been collected on the sale price per square foot of living area for residential property in a small community during the year 1972:

Sale price/ft^2 ($)	Sale price/ft^2 ($)	Sale price/ft^2 ($)
8.50	12.02	9.02
10.20	11.41	14.31
16.30	12.21	15.11
9.80	6.82	13.14
4.70	7.21	8.31
8.90	8.91	9.81
7.30	10.05	7.61
9.20	10.82	11.04
10.75	8.94	8.77
8.21	6.72	10.94

Problems

Sale price/ft² ($)	Sale price/ft² ($)	Sale price/ft² ($)
7.71	11.21	10.04
8.91	10.42	17.21
10.77	16.84	13.05
14.31	11.91	13.41
12.72	9.05	12.82
11.11	6.21	6.09
18.05	6.64	14.89
6.81	12.89	17.79
9.31	8.68	9.44
14.81	8.98	10.96

Develop an empirical process generator for sale price per square foot based on these data.

2. The density function of the random variable X is given by

$$f(x) = \tfrac{1}{9}x^2, \quad 0 < x < 3$$

Develop a process generator for this random variable.

3. Let

$$f(x) = b(1-x)^{b-1}, \quad 0 < x < 1$$

be the density function of the random variable X, where b is a positive integer. Develop a process generator for X.

4. If X is a random variable having the density function given by

$$f(x) = \begin{cases} \tfrac{1}{2}e^x, & -\infty < x < 0 \\ \tfrac{1}{2}e^{-x}, & 0 < x < \infty \end{cases}$$

find a process generator for X.

5. The random variable X has the density function given by

$$f(x) = \begin{cases} 30x - 3, & 0.1 < x \le 0.2 \\ \dfrac{13 - 2x}{3}, & 0.2 < x \le 0.5 \\ 2(1 - x), & 0.5 < x \le 1 \end{cases}$$

Develop a process generator for X.

6. The discrete random variable X has the cumulative distribution function given by

$$F(x) = \frac{x(x+1)(2x+1)}{n(n+1)(2n+1)}, \quad x = 1, 2, 3, \ldots, n$$

Develop a process generator for X.

7. The discrete random variable X has the probability mass function given by

$$p(x) = \frac{p(1-p)^{x-1}}{1-(1-p)^n}, \quad x = 1, 2, 3, \ldots, n$$

Develop a process generator for X.

8. Develop an algorithm for generating values of the hypergeometric random variable.

9. The discrete random variable X has the probability mass function given by

$$p(x) = \frac{2x}{k(k+1)}, \quad x = 1, 2, \ldots, k$$

Develop a process generator for X.

10. Develop a process generator for the discrete random variable X with probability mass function given by

$$p(x) = \begin{cases} \dfrac{x}{k(k+1)}, & x = 1, 2, \ldots, k \\ \dfrac{2k+1-x}{k(k+1)}, & x = k+1, k+2, \ldots, 2k \end{cases}$$

11. Show that

$$x = c - \frac{(b-c)}{a} \ln(r)$$

is a process generator for the Weibull random variable, where r is a uniformly distributed random number on the interval (0, 1).

12. Show that

$$x = \begin{cases} -\dfrac{1}{\lambda_1} \ln(r_2), & r_1 \le p \\ -\dfrac{1}{\lambda_2} \ln(r_2), & r_1 > p \end{cases}$$

is a valid process generator for the hyperexponential random variable, where r_1 and r_2 are independently distributed random numbers each on the interval (0, 1), λ_1 and λ_2 are positive constants, and $0 < p < 1$.

13. Arrivals occur to a single-service channel in a Poisson fashion with mean rate 50/hr. Units are served on a first come–first served basis. Service time is gamma distributed with parameters $n = 10$, $\lambda = 1000$. The cost of serving a customer C_s, is given by

$$C_s = C_w W + C_t S$$

where W is the waiting time of the unit in hours, S the service time in hours, $C_w = \$1000.00$, and $C_t = \$4000.00$. Develop a simulator that will estimate the average cost of serving a unit.

14. In Problem 13, suppose that an arriving unit will not join the waiting line if it contains more than two units. In addition, an income of $10.00 is derived from each unit served by the system. Using a 90% confidence interval, estimate the hourly profit resulting from operation of the system based on 30 days of simulation.

15. Arrivals to a service center occur in a Poisson fashion with mean rate 30 units/day. The service center is comprised of 5 service channels arranged in parallel. Service time in the ith channel is exponentially distributed with service rate μ_i per day, $i = 1, 2, \ldots, 5$, where

$$\mu_1 = 4, \quad \mu_2 = 12, \quad \mu_3 = 15, \quad \mu_4 = 5, \quad \mu_5 = 8$$

Problems

If more than one service channel is available for service when an arrival occurs, the probability that the given available channel is chosen is proportional to the service rate for that channel; that is, if channels 1, 3, and 5 are available for service when an arrival occurs, then

$$P(\text{unit enters channel } i \mid 1, 3, 5 \text{ available}) = \frac{\mu_i}{\mu_1 + \mu_3 + \mu_5}, \quad i = 1, 3, 5$$

Develop a simulator for this system; simulate the system for a period of 30 days; and determine (a) the mean utilization for each service channel, (b) the mean utilization for the entire system, and, (c) the mean number of units serviced per day for each channel.

16. A single-channel queueing system services arrivals in an exponential fashion with mean rate μ. The number of arrivals occurring in a fixed period of time is Poisson distributed with mean rate λ. Units are served in an order based on service time; that is, the time required to serve a unit is determined when the arrival enters the system. When the service channel becomes empty, the time required to service each unit in the waiting line, if any, is checked and the unit having the minimum service time enters the service channel and service begins. Develop a simulator for this system. Let

$$\lambda = 100, \quad \mu = 200$$

Determine the mean time in the system and the waiting line per unit for this system.

17. Given the service system described in Problem 16, compare the mean time in the system and the waiting line for that system and for a similar queueing system where the service discipline is first come–first served. Let

$$\lambda = 100, \quad \mu = 200$$

for both systems.

18. A single exponential service channel operates at a mean rate of μ units/week. Arrivals to the service channel occur from 5 populations. The distribution of interarrival time for each population is given as follows.

Population number	Interarrival-time distribution	Parameters	
1	Normal	$m = 20$,	$\sigma^2 = 9$
2	Gamma	$\lambda = 0.10$,	$k = 2$
3	Exponential	$\lambda = 0.10$	
4	Normal	$m = 40$,	$\sigma^2 = 4$
5	Gamma	$\lambda = 0.20$,	$k = 4$

All interarrival times are in weeks. Develop a simulator for this system. Let $\mu = 190$. Based upon the simulation of 100 weeks of operation of the system, estimate (a) the expected number of units in the system, and (b) the expected total time a unit spends in the system.

19. Arrivals to a single-channel service system are Poisson distributed with mean rate λ given by

$$\lambda = a + b \sin\left(2\pi \frac{t}{365}\right)$$

where t is measured in days; that is, if the time of the next arrival is generated at time t_0, then the parameter of the exponential variable generated is given by

$$\lambda = a + b \sin\left(2\pi \frac{t_0}{365}\right)$$

and the generated interarrival time is measured in days. Let

$$a = 20, \quad b = 5, \quad \mu = 30$$

Compare the operation of this system with that of a corresponding single-channel queueing system with a constant arrival rate of 20 units/day and a constant service rate of 30 units/day. Compare the two systems on the basis of expected total time in the system per unit.

20. For the system given in Problem 18, estimate the probability that there will be four or more units in the system at an arbitrary point in time.

21. Arrivals to a single-channel service system occur in a Poisson fashion with mean rate 100/day. Consider the following service-time distributions:

Service-time distribution	Parameters	
Exponential	$\mu = 300$	
Gamma	$\mu = 600$,	$n = 2$
Gamma	$\mu = 900$,	$n = 3$
Gamma	$\mu = 1200$,	$n = 4$
Gamma	$\mu = 2100$,	$n = 7$

Note that in each case the mean service time is $1/300$. However, the variance of service time is given by n/λ^2 for each distribution, where $n = 1$ in the case of the exponential distribution. Determine the effect of change in the variance of service time (while keeping the mean constant) on the expected time a unit spends in the system and the variance of total time a unit spends in the system.

22. Units arrive to a single exponential service channel in a Poisson fashion at a mean rate of 10/day. The service rate is 15/day. When a unit arrives it is assigned one of three priorities. Units having a priority of 1 are serviced first, and the probability of receiving this priority is 0.20. After all units with priority 1 are served, units with priority 2 are served. The probability that a unit is given a priority of 2 is 0.50. When all units having priority 1 and 2 are served, units with priority 3 are served and the probability that a unit has priority 3 is 0.30. Units are served on a first come–first served basis within each priority class. Priorities are not preemptive. Simulate this system for 200 days and estimate the mean time a unit spends in the system for units within each priority class.

23. Consider the problem examined in Example 9.9. Determine and plot $P_0(t)$ when arrivals or Poisson distributed with mean rate 49/day but service time is:
   (a) Erlang distributed with $\mu = 51$, $n = 4$;
   (b) Normally distributed with $\mu = 1/51$, $\sigma = 0.001$.

24. For the problem in Example 9.9, assume that service time is exponentially distributed with mean rate 51/day but interarrival time is:
   (a) Erlang distributed with $\lambda = 49$, $n = 10$;
   (b) Normally distributed with $\mu = 1/49$, $\sigma = 0.001$.

25. A nurse schedules patients to see one of three available doctors in a Poisson fashion with mean rate 12/hr. The time each doctor spends with a patient is exponentially distributed with a

mean of 10 min. Occasionally a doctor is called out of the office on an emergency. The time between successive emergencies is normally distributed with a mean of 6 hr and a standard deviation of 1 hr. The time required to handle the emergency is exponentially distributed with a mean of 30 min. Patients are scheduled between 9:00 A.M. and 3:00 P.M., but the doctors remain in the office until all patients have been taken care of. If the probability of a patient having to stand for any length of time is to be less than 0.05, how many seats should be available in the waiting room? Estimate the mean time each doctor spends in the office per day including emergencies.

26. Jobs to be run on a digital computer arrive in a Poisson fashion with mean rate 100/hr. Processing time per job is Erlang distributed with $\mu = 0.5$ min and $n = 3.0$. However, when jobs arrive, they are assigned one of the three nonpreemptive priorities 1, 2, 3. Jobs with priority 1 are served first and those with priority 3 are served last. However, if a job with priority 2 is in the waiting line for more than 15 min it is automatically moved up to priority 1. In a similar fashion, jobs with priority 3 are moved up to priority 2 after waiting 21 min. The probability that a job is assigned a priority of 1 is 0.20, that a job is assigned a priority of 2 is 0.35, and that a job is assigned a priority of 3 is 0.45. Develop a simulator for this system and estimate the average waiting time per job in each priority class using 95% confidence intervals.

27. In order to produce the required volume of product, it is necessary that a manufacturer operate the equivalent of 3 production lines continuously (without interruption) for a period of 6 months. However, each production line assigned to the manufacture of this product may fail periodically. The density function of time until failure for each production line is exponential with a mean of 3 weeks. The time required to repair a production line is uniformly distributed between 1 and 2 weeks. There are a total of 2 repair crews available to service production lines when they fail. How many production lines should be operated if the manufacturer is to be 99% certain that the required volume of product will be produced during the 1-yr period?

28. In Problem 26 suppose that you were submitting a job and could select the priority you wished upon entry to the system. However, if you select priority 1, the cost of running the job is $600.00/hr; for priority 2, the cost is $200.00/hr, and for priority 3, the cost is $50.00/hr. In addition, the cost of the time you spend waiting for the job from the time you submit it until you get it back is $50.00/hr. If your job will require 30 min of running time, what priority should you select to minimize the total cost of completing the run?

29. A production process operates continuously except during periods when it is down for repairs. The distribution of time until process failure is normally distributed with a mean of 0.01 yr and a standard deviation of 0.001 yr. When the process fails, repairs commence immediately and repair time is normally distributed with a mean of 0.002 yr and a standard deviation of 0.0001 yr. In addition, when the process fails a number of production units are usually damaged and must be repaired before the production process starts up again. The number of units damaged when the process fails is Poisson distributed with a mean of 10 units. The time to repair a single unit is exponentially distributed with a mean of 0.0002 yr. Hence, down time, as a result of a process failure, is the maximum of the time to repair the process and the time required to repair units damaged when the failure occurs. Develop a simulator for this system and estimate the percentage of down time per year for the process.

30. The time between successive arrivals to a single service channel is uniformly distributed on the interval 1 to 3 hr. Service time is hyperexponentially distributed with $\lambda_1 = 0.50$, $\lambda_2 = 0.80$. $p = 0.30$. When an arrival occurs, it may or may not join the system. Specifically, the probability that an arrival joins the system is dependent on the number of units in the system when the arrival occurs and is given by

$$P(\text{arrival joins system} \mid n \text{ in system}) = \frac{1}{n+1}$$

Simulate the operation of this system for a period of 500 hr and estimate:
   (a) the proportion of arrivals that join the system,
   (b) the average time the unit spends in the system,
   (c) the average length of the waiting line,
   (d) the distribution of facility idle time.

31. In Problem 30, suppose that parallel service channels can be added to the system to reduce the number of lost customers. If the distribution of service time is identically distributed for all service channels, how many channels are necessary to reduce the probability of losing an arrival to 0.10 or less?

32. Arrivals to a single-channel queueing system are Poisson distributed with a mean of 0.04/min. Service time is exponentially distributed with a mean of 20 min. No unit will remain in the waiting line for more than 60 min. Simulate the operation of the system for a period of 100 hr and estimate the average number of units leaving the waiting line per hour using a 95% confidence interval.

33. Units arriving to a service system require service at 3 different centers. The time between successive arrivals is normally distributed with a mean of 100 min and a standard deviation of 4 min. Service time at the first center is normally distributed with mean service time of 43 min and standard deviation 8 min. Service time at the second center is exponentially distributed with a mean of 14 min. Service time at the third center is uniformly distributed between 13 and 32 min. There is no restriction on the size of the waiting line before any service channel. Simulate this system for a period of 50 days. Estimate the mean and standard deviation of the total time a unit spends in the system and the mean and standard deviation of the number of units in the system.

34. For the system described in Problem 33, suppose that the maximum number of units allowed before the first, second, and third service centers is 10, 5, and 3, respectively. If a unit arrives to the system and there are 10 units in the waiting line before the first channel, that unit may not enter the system. If a unit completes service in the first channel and there are 5 units in the second waiting line, the unit remains in the first channel until a service is completed in the second channel and, hence, blocks entry to the first channel. In a similar manner, if there are 3 units in the third waiting line when service is completed in the second channel, the unit remains in the second channel until the waiting line before the third channel is reduced. Using the data given in Problem 33, estimate the mean and standard deviation of total time in the system per unit and the total number in the system.

35. Develop a simulator for the system described in Fig. 9.27.

36. Orders for production of a certain product arrive in a Poisson fashion at an average rate of 10/day. The number of units required per order is geometrically distributed with $p = 0.20$. Each unit of product is manufactured through a sequence of four production processes. Production time on process 1 is exponentially distributed with a mean of 0.005 days/unit of product. Production time on process 2 is normally distributed with a mean of 0.001 days and a standard deviation of 0.0001 days/unit. Production time on process 3 is uniformly distributed per unit on the interval 0.001 to 0.002 days. Production time on process 4 is normally distributed with a mean of 0.007 days and a standard deviation of 0.002 days/unit. Orders are placed in the waiting line for production as they are received, and units are produced one at a time on a first come–first served basis. Orders are to be delivered one day after they are placed and a complete order must be filled before it can be delivered. Simulate the operation of this system for 100 days and estimate the percentage of orders that are delivered late.

37. Consider the system described in Problem 36. The cost of each order received is $5.00. The cost of operating time on processes 1, 2, 3, and 4 is $100.00, $200.00, $400.00, and $800.00/day,

respectively. The selling price of each unit produced is $50.00. However, a penalty of $15.00/unit is incurred for each day for which delivery is late. Through simulation of this system, determine the average profit per unit.

38. For the system given in Problem 36, suppose that each unit is inspected after each production process. If the unit is defective, it is immediately reworked on that process. Otherwise, the unit continues to the next process. The probability that a unit is defective after completion of the ith process, p_i, is given by

$$p_i = \begin{cases} 0.05, & i = 1 \\ 0.10, & i = 2 \\ 0.14, & i = 3 \\ 0.12, & i = 4 \end{cases}$$

Based on simulation of the system for a period of 50 days, estimate the probability of a late order delivery.

39. Solve Problem 37 with the modification to the system described in Problem 34.

Appendix

TABLES

TABLE A Cumulative distribution function $F(z)$ of the standard normal random variable Z[a]

Z	$F(z)$	Z	$F(z)$	Z	$F(z)$	Z	$F(z)$
−4.000	0.0000	−3.750	0.0001	−3.500	0.0002	−3.250	0.0006
−3.990	0.0000	−3.740	0.0001	−3.490	0.0002	−3.240	0.0006
−3.980	0.0000	−3.730	0.0001	−3.480	0.0003	−3.230	0.0006
−3.970	0.0000	−3.720	0.0001	−3.470	0.0003	−3.220	0.0006
−3.960	0.0000	−3.710	0.0001	−3.460	0.0003	−3.210	0.0007
−3.950	0.0000	−3.700	0.0001	−3.450	0.0003	−3.200	0.0007
−3.940	0.0000	−3.690	0.0001	−3.440	0.0003	−3.190	0.0007
−3.930	0.0000	−3.680	0.0001	−3.430	0.0003	−3.180	0.0007
−3.920	0.0000	−3.670	0.0001	−3.420	0.0003	−3.170	0.0008
−3.910	0.0001	−3.660	0.0001	−3.410	0.0003	−3.160	0.0008
−3.900	0.0001	−3.650	0.0001	−3.400	0.0003	−3.150	0.0008
−3.890	0.0001	−3.640	0.0001	−3.390	0.0004	−3.140	0.0009
−3.880	0.0001	−3.630	0.0001	−3.380	0.0004	−3.130	0.0009
−3.870	0.0001	−3.620	0.0002	−3.370	0.0004	−3.120	0.0009
−3.860	0.0001	−3.610	0.0002	−3.360	0.0004	−3.110	0.0009
−3.850	0.0001	−3.600	0.0002	−3.350	0.0004	−3.100	0.0010
−3.840	0.0001	−3.590	0.0002	−3.340	0.0004	−3.090	0.0010
−3.830	0.0001	−3.580	0.0002	−3.330	0.0004	−3.080	0.0010
−3.820	0.0001	−3.570	0.0002	−3.320	0.0005	−3.070	0.0011
−3.810	0.0001	−3.560	0.0002	−3.310	0.0005	−3.060	0.0011
−3.800	0.0001	−3.550	0.0002	−3.300	0.0005	−3.050	0.0012
−3.790	0.0001	−3.540	0.0002	−3.290	0.0005	−3.040	0.0012
−3.780	0.0001	−3.530	0.0002	−3.280	0.0005	−3.030	0.0012
−3.770	0.0001	−3.520	0.0002	−3.270	0.0005	−3.020	0.0013
−3.760	0.0001	−3.510	0.0002	−3.260	0.0006	−3.010	0.0013

TABLE A—*Continued*

Z	F(z)	Z	F(z)	Z	F(z)	Z	F(z)
−3.000	0.0014	−2.500	0.0062	−2.000	0.0227	−1.500	0.0668
−2.990	0.0014	−2.490	0.0064	−1.990	0.0233	−1.490	0.0681
−2.980	0.0014	−2.480	0.0066	−1.980	0.0239	−1.480	0.0695
−2.970	0.0015	−2.470	0.0067	−1.970	0.0244	−1.470	0.0708
−2.960	0.0015	−2.460	0.0069	−1.960	0.0250	−1.460	0.0722
−2.950	0.0016	−2.450	0.0071	−1.950	0.0256	−1.450	0.0735
−2.940	0.0016	−2.440	0.0073	−1.940	0.0262	−1.440	0.0750
−2.930	0.0017	−2.430	0.0075	−1.930	0.0268	−1.430	0.0764
−2.920	0.0018	−2.420	0.0078	−1.920	0.0274	−1.420	0.0778
−2.910	0.0018	−2.410	0.0080	−1.910	0.0281	−1.410	0.0793
−2.900	0.0019	−2.400	0.0082	−1.900	0.0287	−1.400	0.0808
−2.890	0.0019	−2.390	0.0084	−1.890	0.0294	−1.390	0.0823
−2.880	0.0020	−2.380	0.0086	−1.880	0.0301	−1.380	0.0838
−2.870	0.0021	−2.370	0.0089	−1.870	0.0307	−1.370	0.0854
−2.860	0.0021	−2.360	0.0091	−1.860	0.0314	−1.360	0.0869
−2.850	0.0022	−2.350	0.0094	−1.850	0.0322	−1.350	0.0885
−2.840	0.0023	−2.340	0.0096	−1.840	0.0329	−1.340	0.0901
−2.830	0.0023	−2.330	0.0099	−1.830	0.0336	−1.330	0.0918
−2.820	0.0024	−2.320	0.0102	−1.820	0.0344	−1.320	0.0934
−2.810	0.0025	−2.310	0.0104	−1.810	0.0352	−1.310	0.0951
−2.800	0.0026	−2.300	0.0107	−1.800	0.0359	−1.300	0.0968
−2.790	0.0026	−2.290	0.0110	−1.790	0.0367	−1.290	0.0985
−2.780	0.0027	−2.280	0.0113	−1.780	0.0375	−1.280	0.1003
−2.770	0.0028	−2.270	0.0116	−1.770	0.0384	−1.270	0.1021
−2.760	0.0029	−2.260	0.0119	−1.760	0.0392	−1.260	0.1038
−2.750	0.0030	−2.250	0.0122	−1.750	0.0401	−1.250	0.1057
−2.740	0.0031	−2.240	0.0125	−1.740	0.0409	−1.240	0.1075
−2.730	0.0032	−2.230	0.0129	−1.730	0.0418	−1.230	0.1094
−2.720	0.0033	−2.220	0.0132	−1.720	0.0427	−1.220	0.1112
−2.710	0.0034	−2.210	0.0135	−1.710	0.0436	−1.210	0.1132
−2.700	0.0035	−2.200	0.0139	−1.700	0.0446	−1.200	0.1151
−2.690	0.0036	−2.190	0.0143	−1.690	0.0455	−1.190	0.1170
−2.680	0.0037	−2.180	0.0146	−1.680	0.0465	−1.180	0.1190
−2.670	0.0038	−2.170	0.0150	−1.670	0.0475	−1.170	0.1210
−2.660	0.0039	−2.160	0.0154	−1.660	0.0485	−1.160	0.1230
−2.650	0.0040	−2.150	0.0158	−1.650	0.0495	−1.150	0.1251
−2.640	0.0041	−2.140	0.0162	−1.640	0.0505	−1.140	0.1272
−2.630	0.0043	−2.130	0.0166	−1.630	0.0516	−1.130	0.1293
−2.620	0.0044	−2.120	0.0170	−1.620	0.0526	−1.120	0.1314
−2.610	0.0045	−2.110	0.0174	−1.610	0.0537	−1.110	0.1335
−2.600	0.0047	−2.100	0.0179	−1.600	0.0548	−1.100	0.1357
−2.590	0.0048	−2.090	0.0183	−1.590	0.0559	−1.090	0.1379
−2.580	0.0049	−2.080	0.0188	−1.580	0.0571	−1.080	0.1401
−2.570	0.0051	−2.070	0.0192	−1.570	0.0582	−1.070	0.1423
−2.560	0.0052	−2.060	0.0197	−1.560	0.0594	−1.060	0.1446
−2.550	0.0054	−2.050	0.0202	−1.550	0.0606	−1.050	0.1469
−2.540	0.0055	−2.040	0.0207	−1.540	0.0618	−1.040	0.1492
−2.530	0.0057	−2.030	0.0212	−1.530	0.0630	−1.030	0.1515
−2.520	0.0059	−2.020	0.0217	−1.520	0.0643	−1.020	0.1539
−2.510	0.0060	−2.010	0.0222	−1.510	0.0655	−1.010	0.1563

TABLE A—*Continued*

Z	F(z)	Z	F(z)	Z	F(z)	Z	F(z)
−1.000	0.1587	−0.500	0.3086	0.000	0.5000	0.500	0.6915
−0.990	0.1611	−0.490	0.3121	0.010	0.5040	0.510	0.6950
−0.980	0.1636	−0.480	0.3157	0.020	0.5080	0.520	0.6985
−0.970	0.1660	−0.470	0.3192	0.030	0.5120	0.530	0.7020
−0.960	0.1685	−0.460	0.3228	0.040	0.5160	0.540	0.7054
−0.950	0.1711	−0.450	0.3264	0.050	0.5200	0.550	0.7089
−0.940	0.1736	−0.440	0.3300	0.060	0.5240	0.560	0.7123
−0.930	0.1762	−0.430	0.3336	0.070	0.5279	0.570	0.7157
−0.920	0.1788	−0.420	0.3373	0.080	0.5319	0.580	0.7191
−0.910	0.1814	−0.410	0.3409	0.090	0.5359	0.590	0.7224
−0.900	0.1841	−0.400	0.3446	0.100	0.5399	0.600	0.7258
−0.890	0.1867	−0.390	0.3483	0.110	0.5438	0.610	0.7291
−0.880	0.1894	−0.380	0.3520	0.120	0.5478	0.620	0.7324
−0.870	0.1922	−0.370	0.3557	0.130	0.5517	0.630	0.7357
−0.860	0.1949	−0.360	0.3595	0.140	0.5557	0.640	0.7389
−0.850	0.1977	−0.350	0.3632	0.150	0.5596	0.650	0.7422
−0.840	0.2005	−0.340	0.3670	0.160	0.5636	0.660	0.7454
−0.830	0.2033	−0.330	0.3707	0.170	0.5675	0.670	0.7486
−0.820	0.2061	−0.320	0.3745	0.180	0.5714	0.680	0.7518
−0.810	0.2090	−0.310	0.3783	0.190	0.5754	0.690	0.7549
−0.800	0.2119	−0.300	0.3821	0.200	0.5793	0.700	0.7581
−0.790	0.2148	−0.290	0.3859	0.210	0.5832	0.710	0.7612
−0.780	0.2177	−0.280	0.3898	0.220	0.5871	0.720	0.7643
−0.770	0.2207	−0.270	0.3936	0.230	0.5910	0.730	0.7673
−0.760	0.2236	−0.260	0.3975	0.240	0.5949	0.740	0.7704
−0.750	0.2266	−0.250	0.4013	0.250	0.5987	0.750	0.7734
−0.740	0.2297	−0.240	0.4052	0.260	0.6026	0.760	0.7764
−0.730	0.2327	−0.230	0.4091	0.270	0.6064	0.770	0.7794
−0.720	0.2358	−0.220	0.4130	0.280	0.6103	0.780	0.7823
−0.710	0.2389	−0.210	0.4169	0.290	0.6141	0.790	0.7853
−0.700	0.2420	−0.200	0.4208	0.300	0.6179	0.800	0.7882
−0.690	0.2451	−0.190	0.4247	0.310	0.6217	0.810	0.7911
−0.680	0.2483	−0.180	0.4286	0.320	0.6255	0.820	0.7939
−0.670	0.2515	−0.170	0.4325	0.330	0.6293	0.830	0.7968
−0.660	0.2547	−0.160	0.4365	0.340	0.6331	0.840	0.7996
−0.650	0.2579	−0.150	0.4404	0.350	0.6368	0.850	0.8024
−0.640	0.2611	−0.140	0.4444	0.360	0.6406	0.860	0.8051
−0.630	0.2644	−0.130	0.4483	0.370	0.6443	0.870	0.8079
−0.620	0.2677	−0.120	0.4523	0.380	0.6480	0.880	0.8106
−0.610	0.2710	−0.110	0.4562	0.390	0.6517	0.890	0.8133
−0.600	0.2743	−0.100	0.4602	0.400	0.6554	0.900	0.8160
−0.590	0.2776	−0.090	0.4642	0.410	0.6591	0.910	0.8186
−0.580	0.2810	−0.080	0.4681	0.420	0.6628	0.920	0.8212
−0.570	0.2844	−0.070	0.4721	0.430	0.6664	0.930	0.8238
−0.560	0.2878	−0.060	0.4761	0.440	0.6700	0.940	0.8264
−0.550	0.2912	−0.050	0.4801	0.450	0.6737	0.950	0.8290
−0.540	0.2946	−0.040	0.4841	0.460	0.6773	0.960	0.8315
−0.530	0.2981	−0.030	0.4881	0.470	0.6808	0.970	0.8340
−0.520	0.3016	−0.020	0.4920	0.480	0.6844	0.980	0.8365
−0.510	0.3051	−0.010	0.4960	0.490	0.6879	0.990	0.8389

TABLE A—*Continued*

Z	F(z)	Z	F(z)	Z	F(z)	Z	F(z)
1.000	0.8414	1.500	0.9332	2.000	0.9773	2.500	0.9938
1.010	0.8438	1.510	0.9345	2.010	0.9778	2.510	0.9940
1.020	0.8462	1.520	0.9357	2.020	0.9783	2.520	0.9941
1.030	0.8485	1.530	0.9370	2.030	0.9788	2.530	0.9943
1.040	0.8509	1.540	0.9382	2.040	0.9793	2.540	0.9945
1.050	0.8532	1.550	0.9394	2.050	0.9798	2.550	0.9946
1.060	0.8554	1.560	0.9406	2.060	0.9803	2.560	0.9948
1.070	0.8577	1.570	0.9418	2.070	0.9808	2.570	0.9949
1.080	0.8599	1.580	0.9429	2.080	0.9812	2.580	0.9951
1.090	0.8622	1.590	0.9441	2.090	0.9817	2.590	0.9952
1.100	0.8643	1.600	0.9452	2.100	0.9821	2.600	0.9953
1.110	0.8665	1.610	0.9463	2.110	0.9826	2.610	0.9955
1.120	0.8687	1.620	0.9474	2.120	0.9830	2.620	0.9956
1.130	0.8708	1.630	0.9484	2.130	0.9834	2.630	0.9957
1.140	0.8729	1.640	0.9495	2.140	0.9838	2.640	0.9959
1.150	0.8749	1.650	0.9505	2.150	0.9842	2.650	0.9960
1.160	0.8770	1.660	0.9515	2.160	0.9846	2.660	0.9961
1.170	0.8790	1.670	0.9525	2.170	0.9850	2.670	0.9962
1.180	0.8810	1.680	0.9535	2.180	0.9854	2.680	0.9963
1.190	0.8830	1.690	0.9545	2.190	0.9857	2.690	0.9964
1.200	0.8849	1.700	0.9554	2.200	0.9861	2.700	0.9965
1.210	0.8869	1.710	0.9564	2.210	0.9865	2.710	0.9966
1.220	0.8888	1.720	0.9573	2.220	0.9868	2.720	0.9967
1.230	0.8907	1.730	0.9582	2.230	0.9871	2.730	0.9968
1.240	0.8925	1.740	0.9591	2.240	0.9875	2.740	0.9969
1.250	0.8944	1.750	0.9599	2.250	0.9878	2.750	0.9970
1.260	0.8962	1.760	0.9608	2.260	0.9881	2.760	0.9971
1.270	0.8980	1.770	0.9616	2.270	0.9884	2.770	0.9972
1.280	0.8997	1.780	0.9625	2.280	0.9887	2.780	0.9973
1.290	0.9015	1.790	0.9633	2.290	0.9890	2.790	0.9974
1.300	0.9032	1.800	0.9641	2.300	0.9893	2.800	0.9974
1.310	0.9049	1.810	0.9648	2.310	0.9896	2.810	0.9975
1.320	0.9066	1.820	0.9656	2.320	0.9898	2.820	0.9976
1.330	0.9082	1.830	0.9664	2.330	0.9901	2.830	0.9977
1.340	0.9099	1.840	0.9671	2.340	0.9904	2.840	0.9977
1.350	0.9115	1.850	0.9678	2.350	0.9906	2.850	0.9978
1.360	0.9131	1.860	0.9686	2.360	0.9909	2.860	0.9979
1.370	0.9147	1.870	0.9693	2.370	0.9911	2.870	0.9979
1.380	0.9162	1.880	0.9699	2.380	0.9914	2.880	0.9980
1.390	0.9177	1.890	0.9706	2.390	0.9916	2.890	0.9981
1.400	0.9192	1.900	0.9713	2.400	0.9918	2.900	0.9981
1.410	0.9207	1.910	0.9719	2.410	0.9920	2.910	0.9982
1.420	0.9222	1.920	0.9726	2.420	0.9922	2.920	0.9982
1.430	0.9236	1.930	0.9732	2.430	0.9925	2.930	0.9983
1.440	0.9251	1.940	0.9738	2.440	0.9927	2.940	0.9984
1.450	0.9265	1.950	0.9744	2.450	0.9929	2.950	0.9984
1.460	0.9279	1.960	0.9750	2.460	0.9931	2.960	0.9985
1.470	0.9292	1.970	0.9756	2.470	0.9932	2.970	0.9985
1.480	0.9306	1.980	0.9762	2.480	0.9934	2.980	0.9986
1.490	0.9319	1.990	0.9767	2.490	0.9936	2.990	0.9986

TABLE A—*Continued*

Z	F(z)	Z	F(z)	Z	F(z)	Z	F(z)
3.000	0.9986	3.250	0.9994	3.500	0.9998	3.750	0.9999
3.010	0.9987	3.260	0.9994	3.510	0.9998	3.760	0.9999
3.020	0.9987	3.270	0.9995	3.520	0.9998	3.770	0.9999
3.030	0.9988	3.280	0.9995	3.530	0.9998	3.780	0.9999
3.040	0.9988	3.290	0.9995	3.540	0.9998	3.790	0.9999
3.050	0.9988	3.300	0.9995	3.550	0.9998	3.800	0.9999
3.060	0.9989	3.310	0.9995	3.560	0.9998	3.810	0.9999
3.070	0.9989	3.320	0.9995	3.570	0.9998	3.820	0.9999
3.080	0.9990	3.330	0.9996	3.580	0.9998	3.830	0.9999
3.090	0.9990	3.340	0.9996	3.590	0.9998	3.840	0.9999
3.100	0.9990	3.350	0.9996	3.600	0.9998	3.850	0.9999
3.110	0.9991	3.360	0.9996	3.610	0.9998	3.860	0.9999
3.120	0.9991	3.370	0.9996	3.620	0.9998	3.870	0.9999
3.130	0.9991	3.380	0.9996	3.630	0.9999	3.880	0.9999
3.140	0.9991	3.390	0.9996	3.640	0.9999	3.890	0.9999
3.150	0.9992	3.400	0.9997	3.650	0.9999	3.900	0.9999
3.160	0.9992	3.410	0.9997	3.660	0.9999	3.910	0.9999
3.170	0.9992	3.420	0.9997	3.670	0.9999	3.920	1.0000
3.180	0.9993	3.430	0.9997	3.680	0.9999	3.930	1.0000
3.190	0.9993	3.440	0.9997	3.690	0.9999	3.940	1.0000
3.200	0.9993	3.450	0.9997	3.700	0.9999	3.950	1.0000
3.210	0.9993	3.460	0.9997	3.710	0.9999	3.960	1.0000
3.220	0.9994	3.470	0.9997	3.720	0.9999	3.970	1.0000
3.230	0.9994	3.480	0.9997	3.730	0.9999	3.980	1.0000
3.240	0.9994	3.490	0.9998	3.740	0.9999	3.990	1.0000
						4.000	1.0000

[a] Table taken from Schmidt, J. W., *Mathematical Foundations for Management Science and Systems Analysis.* New York: Academic Press, 1974.

TABLE B Percentiles of the t-distribution[a]

d	$P = 0.6$[b]	0.75	0.9	0.95	0.975	0.99	0.995	0.9975	0.999	0.9995
1	0.325	1.000	3.078	6.314	12.706	31.821	63.657	127.32	318.31	636.62
2	0.289	0.816	1.886	2.920	4.303	6.965	9.925	14.089	22.327	31.598
3	0.277	0.765	1.638	2.353	3.182	4.541	5.841	7.453	10.214	12.924
4	0.271	0.741	1.533	2.132	2.776	3.747	4.604	5.598	7.173	8.610
5	0.267	0.727	1.476	2.015	2.571	3.365	4.032	4.773	5.893	6.869
6	0.265	0.718	1.440	1.943	2.447	3.143	3.707	4.317	5.208	5.959
7	0.263	0.711	1.415	1.895	2.365	2.998	3.499	4.029	4.785	5.408
8	0.262	0.706	1.397	1.860	2.306	2.896	3.355	3.833	4.501	5.041
9	0.261	0.703	1.383	1.833	2.262	2.821	3.250	3.690	4.297	4.781
10	0.260	0.700	1.372	1.812	2.228	2.764	3.169	3.581	4.144	4.587
11	0.260	0.697	1.363	1.796	2.201	2.718	3.106	3.497	4.025	4.437
12	0.259	0.695	1.356	1.782	2.179	2.681	3.055	3.428	3.930	4.318
13	0.259	0.694	1.350	1.771	2.160	2.650	3.012	3.372	3.852	4.221
14	0.258	0.692	1.345	1.761	2.145	2.624	2.977	3.326	3.787	4.140
15	0.258	0.691	1.341	1.753	2.131	2.602	2.947	3.286	3.733	4.073
16	0.258	0.690	1.337	1.746	2.120	2.583	2.921	3.252	3.686	4.015
17	0.257	0.689	1.333	1.740	2.110	2.567	2.898	3.222	3.646	3.965
18	0.257	0.688	1.330	1.734	2.101	2.552	2.878	3.197	3.610	3.922
19	0.257	0.688	1.328	1.729	2.093	2.539	2.861	3.174	3.579	3.883
20	0.257	0.687	1.325	1.725	2.086	2.528	2.845	3.153	3.552	3.850
21	0.257	0.686	1.323	1.721	2.080	2.518	2.831	3.135	3.527	3.819
22	0.256	0.686	1.321	1.717	2.074	2.508	2.819	3.119	3.505	3.792
23	0.256	0.685	1.319	1.714	2.069	2.500	2.807	3.104	3.485	3.767
24	0.256	0.685	1.318	1.711	2.064	2.492	2.797	3.091	3.467	3.745
25	0.256	0.684	1.316	1.708	2.060	2.485	2.787	3.078	3.450	3.725
26	0.256	0.684	1.315	1.706	2.056	2.479	2.779	3.067	3.435	3.707
27	0.256	0.684	1.314	1.703	2.052	2.473	2.771	3.057	3.421	3.690
28	0.256	0.683	1.313	1.701	2.048	2.467	2.763	3.047	3.408	3.674
29	0.256	0.683	1.311	1.699	2.045	2.462	2.756	3.038	3.396	3.659
30	0.256	0.683	1.310	1.697	2.042	2.457	2.750	3.030	3.385	3.646
40	0.255	0.681	1.303	1.684	2.021	2.423	2.704	2.971	3.307	3.551
60	0.254	0.679	1.296	1.671	2.000	2.390	2.660	2.915	3.232	3.460
120	0.254	0.677	1.289	1.658	1.980	2.358	2.617	2.860	3.160	3.373
∞	0.253	0.674	1.282	1.645	1.960	2.326	2.576	2.807	3.090	3.291

[a] From Pearson, E. S., and Hartley, H. O., *Tables for Statisticians*, Vol. 1, 3rd ed. London: Biometrika, 1966.

[b] $P = F_t(t_p(d))$ is the cumulative distribution for d degrees of freedom.

TABLE C Percentiles of the χ^2 distributions[a]

df	$P_{0.5}$	P_{01}	$P_{02.5}$	P_{05}	P_{10}	P_{90}	P_{95}	$P_{97.5}$	P_{99}	$P_{99.5}$
1	0.000039	0.00016	0.00098	0.0039	0.0158	2.71	3.84	5.02	6.63	7.88
2	0.0100	0.0201	0.0506	0.1026	0.2107	4.61	5.99	7.38	9.21	10.60
3	0.0717	0.115	0.216	0.352	0.584	6.25	7.81	9.35	11.34	12.84
4	0.207	0.297	0.484	0.711	1.064	7.78	9.49	11.14	13.28	14.86
5	0.412	0.554	0.831	1.15	1.61	9.24	11.07	12.83	15.09	16.75
6	0.676	0.872	1.24	1.64	2.20	10.64	12.59	14.45	16.81	18.55
7	0.989	1.24	1.69	2.17	2.83	12.02	14.07	16.01	18.48	20.28
8	1.34	1.65	2.18	2.73	3.49	13.36	15.51	17.53	20.09	21.96
9	1.73	2.09	2.70	3.33	4.17	14.68	16.92	19.02	21.67	23.59
10	2.16	2.56	3.25	3.94	4.87	15.99	18.31	20.48	23.21	25.19
11	2.60	3.05	3.82	4.57	5.58	17.28	19.68	21.92	24.73	26.76
12	3.07	3.57	4.40	5.23	6.30	18.55	21.03	23.34	26.22	28.30
13	3.57	4.11	5.01	5.89	7.04	19.81	22.36	24.74	27.69	29.82
14	4.07	4.66	5.63	6.57	7.79	21.06	23.68	26.12	29.14	31.32
15	4.60	5.23	6.26	7.26	8.55	22.31	25.00	27.49	30.58	32.80
16	5.14	5.81	6.91	7.96	9.31	23.54	26.30	28.85	32.00	34.27
18	6.26	7.01	8.23	9.39	10.86	25.99	28.87	31.53	34.81	37.16
20	7.43	8.26	9.59	10.85	12.44	28.41	31.41	34.17	37.57	40.00
24	9.89	10.86	12.40	13.85	15.66	33.20	36.42	39.36	42.98	45.56
30	13.79	14.95	16.79	18.49	20.60	40.26	43.77	46.98	50.89	53.67
40	20.71	22.16	24.43	26.51	29.05	51.81	55.76	59.34	63.69	66.77
60	35.53	37.48	40.48	43.19	46.46	74.40	79.08	83.30	88.38	91.95
120	83.85	86.92	91.58	95.70	100.62	140.23	146.57	152.21	158.95	163.64

[a] For large values of degrees of freedom, the approximate formula

$$\chi_\alpha^2 = v\left[1 - \frac{2}{9v} + z_\alpha\left(\frac{2}{9v}\right)^{1/2}\right]^3$$

where z_α is the normal deviate and v the number of degrees of freedom, may be used. For example $\chi_{0.99}^2 = 60[1 = 0.00370 + (2.326 \times 0.06086)]^3 = 60(1.1379)^3 = 88.4$ for the 99th percentile for 60 degrees of freedom. (From Dixon, W. J., and Massey, F. J., Jr., *Introduction to Statistical Analysis*, copyright 1969, McGraw-Hill, New York. Used with permission of McGraw-Hill Book Company.)

TABLE D Critical values D_n^{α} of the maximum absolute difference between sample and population cumulative distributions[a]

Sample size (n)	Level of significance (α)				
	0.20	0.15	0.10	0.05	0.01
1	0.900	0.925	0.950	0.975	0.995
2	0.684	0.726	0.776	0.842	0.929
3	0.565	0.597	0.642	0.708	0.828
4	0.494	0.525	0.564	0.624	0.733
5	0.446	0.474	0.510	0.565	0.669
6	0.410	0.436	0.470	0.521	0.618
7	0.381	0.405	0.438	0.486	0.577
8	0.358	0.381	0.411	0.457	0.543
9	0.339	0.360	0.388	0.432	0.514
10	0.322	0.342	0.368	0.410	0.490
11	0.307	0.326	0.352	0.391	0.468
12	0.295	0.313	0.338	0.375	0.450
13	0.284	0.302	0.325	0.361	0.433
14	0.274	0.292	0.314	0.349	0.418
15	0.266	0.283	0.304	0.338	0.404
16	0.258	0.274	0.295	0.328	0.392
17	0.250	0.266	0.286	0.318	0.381
18	0.244	0.259	0.278	0.309	0.371
19	0.237	0.252	0.272	0.301	0.363
20	0.231	0.246	0.264	0.294	0.356
25	0.21	0.22	0.24	0.27	0.32
30	0.19	0.20	0.22	0.24	0.29
35	0.18	0.19	0.21	0.23	0.27
> 35	$\dfrac{1.07}{\sqrt{n}}$	$\dfrac{1.14}{\sqrt{n}}$	$\dfrac{1.22}{\sqrt{n}}$	$\dfrac{1.36}{\sqrt{n}}$	$\dfrac{1.63}{\sqrt{n}}$

[a] Values of D_n^{α} such that $P[\max|S_n(x) - F_0(x)| > D_n^{\alpha}] = \alpha$, where $F_0(x)$ is the theoretical cumulative distribution and $S_n(x)$ an observed cumulative distribution for a sample of n. (From Massey, F. J., Jr., Kolmogorov–Smirnov test for goodness of fit, *J. Amer. Statist. Assoc.* **46**, 70, 1951.)

TABLE E Laplace transform pairs $L(s) = \mathscr{L}[f(t)]$

$f(t)$	$L(s)$
$\sum_{i=1}^{n} a_i f_i(t)$	$\sum_{i=1}^{n} a_i L_i(s)$
$\dfrac{d}{dt} f(t)$	$sL(s) - f(0)$
$\dfrac{d^n}{dt^n} f(t)$	$s^n L(s) - s^{n-1} f(0) - s^{n-2} f'(0) - \cdots - f^{(n-1)}(0)$
$\int_0^t f(x)\, dx$	$\dfrac{L(s)}{s}$
$\int_0^t f(x) g(t-x)\, dx$	$\mathscr{L}[f(t)] \mathscr{L}[g(t)]$
$e^{\lambda t}$	$\dfrac{1}{s - \lambda}$
$\lambda e^{-\lambda t}$	$\left(1 + \dfrac{s}{\lambda}\right)^{-1}$
$e^{\lambda t} f(t)$	$L(s - \lambda)$
$p\lambda_1 \exp(-\lambda_1 t) + (1-p)\lambda_2 \exp(-\lambda_2 t)$	$\dfrac{(\lambda_1 - \lambda_2)ps + \lambda_2(\lambda_1 + s)}{(s + \lambda_1)(s + \lambda_2)}$
$\dfrac{f(t)}{t}$	$\int_s^\infty L(x)\, dx$
$\delta(t)$	1
$U(t - a)$	$-\dfrac{e^{-sa}}{s}$
λ	$\dfrac{\lambda}{s}$
t	$\dfrac{1}{s^2}$
$\dfrac{t^{n-1}}{\Gamma(n)},\quad n > 0$	$\dfrac{1}{s^n}$
$\dfrac{(k\lambda)^k}{(k-1)!} t^{k-1} e^{-k\lambda t}$	$\left(1 + \dfrac{s}{k\lambda}\right)^{-k}$
$\sin(\lambda t)$	$\dfrac{\lambda}{s^2 + \lambda^2}$

TABLE E (*continued*)

$f(t)$	$L(s)$
$\cos(\lambda t)$	$\dfrac{s}{s^2 + \lambda^2}$
$\sinh(\lambda t)$	$\dfrac{\lambda}{s^2 - \lambda^2}$
$\cosh(\lambda t)$	$\dfrac{s}{s^2 - \lambda^2}$
$\dfrac{e^{\lambda t} - e^{\mu t}}{\lambda - \mu}, \quad \lambda \neq \mu$	$\dfrac{1}{(s - \lambda)(s - \mu)}$
$\dfrac{\lambda e^{\lambda t} - \mu e^{\mu t}}{\lambda - \mu}, \quad \mu \neq \lambda$	$\dfrac{s}{(s - \lambda)(s - \mu)}$
$t \sin(\lambda t)$	$\dfrac{2\lambda s}{(s^2 + \lambda^2)^2}$
$t \cos(\lambda t)$	$\dfrac{s^2 - \lambda^2}{(s^2 + \lambda^2)^2}$
$t \sinh(\lambda t)$	$\dfrac{2\lambda s}{(s^2 - \lambda^2)^2}$
$t \cosh(\lambda t)$	$\dfrac{s^2 + \lambda^2}{(s^2 - \lambda^2)^2}$
$\dfrac{e^{\lambda t} - e^{\mu t}}{t}$	$\ln\left(\dfrac{s - \mu}{s - \lambda}\right)$
$\ln(t)$	$-\dfrac{\gamma + \ln(s)}{s}$*

*$\gamma = 0.5772$, Euler's constant.

TABLE F Some generating functions

Function $f(x)$, $x = 0, 1, 2, \ldots$	Generating function $\mathscr{G}[f(x)] = \sum_{x=0}^{\infty} f(x)z^x$
$\delta(x) = \begin{cases} 1, & x = 0 \\ 0, & x > 0 \end{cases}$	1
$U(k) = \begin{cases} 0, & x < k \\ 1, & x \geq k \end{cases}$	$\dfrac{z^k}{1-z}$
$U(0) = 1$	$\dfrac{1}{1-z}$
c	$\dfrac{c}{1-z}$
x	$\dfrac{z}{(1-z)^2}$
$1 + x$	$\dfrac{1}{(1-z)^2}$
x^2	$\dfrac{z(1+z)}{(1-z)^3}$
x^3	$\dfrac{z(1+4z+z^2)}{(1-z)^4}$
c^x	$\dfrac{1}{1-cz}$
xc^x	$\dfrac{cz}{(1-cz)^2}$
$x^2 c^x$	$\dfrac{cz(1+cz)}{(1-cz)^3}$
$ag(x) + bh(x)$	$a\mathscr{G}[g(x)] + b\mathscr{G}[h(x)]$
$g(x + 1)$	$\dfrac{\mathscr{G}[g(x)] - g(0)}{z}$
$g(x + k)$	$\{\mathscr{G}[g(x)] - g(0) - g(1)z$ $\quad - \cdots - g(k-1)z^{k-1}\}z^{-k}$

TABLE F (*continued*)

Function $f(x)$, $x = 0, 1, 2, \ldots$	Generating function $\mathscr{G}[f(x)] = \sum_{x=0}^{\infty} f(x) z^x$
$g(x-k)U(x-k)$	$z^k \mathscr{G}[g(x)]$
$\sum_{k=0}^{x} g(k) h(x-k)$	$\mathscr{G}[g(x)] \mathscr{G}[h(x)]$
$x g(x)$	$z \dfrac{d \mathscr{G}[g(x)]}{dz}$
$\sum_{k=0}^{x} g(k)$	$\dfrac{\mathscr{G}[g(x)]}{1-z}$
$a^x g(x)$	$\mathscr{G}[g(x)] \bigg|_{z=az}$
$\dfrac{c^x}{x}, \quad x > 0$	$-\ln(1 - cz)$
$\dfrac{e^{-\lambda} \lambda^x}{x!}$	$e^{-\lambda(1-z)}$
$\dfrac{1}{b-a+1}, \quad a \leq x \leq b$	$\dfrac{z^a - z^{b+1}}{(b-a+1)(1-z)}$
$\binom{n}{x} p^x (1-p)^{n-x}, \quad x = 0, 1, \ldots, n$	$[pz + (1-p)]^n$
$\binom{x-1}{x-n} p^n (1-p)^{x-n}, \quad x = n, n+1, \ldots$	$\left(\dfrac{pz}{1 - z(1-p)} \right)^n$
$\sin(cx)$	$\dfrac{z \sin c}{z^2 + 1 - 2z \cos c}$
$\cos(cx)$	$\dfrac{1 - z \cos c}{z^2 + 1 - 2z \cos c}$
$\lim_{z \to 0} \mathscr{G}[f(x)]$	Initial-value theorem
$\lim_{z \to 1} (1-z) \mathscr{G}[f(x)]$	Final-value theorem

TABLE G Some results for the $(M/M/c); (GD/K/K)$ queue[a]

K	c	D	F	K	c	D	F	K	c	D	F
		$X = 0.05$		12	1	0.879	0.764	10	1	0.987	0.497
4	1	0.149	0.992		2	0.361	0.970		2	0.692	0.854
5	1	0.198	0.989		3	0.098	0.996		3	0.300	0.968
6	1	0.247	0.985		4	0.019	0.999		4	0.092	0.994
	2	0.023	0.999	14	1	0.946	0.690		5	0.020	0.999
7	1	0.296	0.981		2	0.469	0.954	12	1	0.998	0.416
	2	0.034	0.999		3	0.151	0.992		2	0.841	0.778
8	1	0.343	0.977		4	0.036	0.999		3	0.459	0.940
	2	0.046	0.999	16	1	0.980	0.618		4	0.180	0.986
9	1	0.391	0.972		2	0.576	0.935		5	0.054	0.997
	2	0.061	0.998		3	0.214	0.988	14	2	0.934	0.697
10	1	0.437	0.967		4	0.060	0.998		3	0.619	0.902
	2	0.076	0.998	18	1	0.994	0.554		4	0.295	0.973
12	1	0.528	0.954		2	0.680	0.909		5	0.109	0.993
	2	0.111	0.996		3	0.285	0.983		6	0.032	0.999
14	1	0.615	0.939		4	0.092	0.997	16	2	0.978	0.621
	2	0.151	0.995		5	0.024	0.999		3	0.760	0.854
	3	0.026	0.999	20	1	0.999	0.500		4	0.426	0.954
16	1	0.697	0.919		2	0.773	0.878		5	0.187	0.987
	2	0.195	0.993		3	0.363	0.975		6	0.066	0.997
	3	0.039	0.999		4	0.131	0.995		7	0.019	0.999
18	1	0.772	0.895		5	0.038	0.999	18	2	0.994	0.555
	2	0.243	0.991	25	2	0.934	0.776		3	0.868	0.797
	3	0.054	0.999		3	0.572	0.947		4	0.563	0.928
20	1	0.837	0.866		4	0.258	0.987		5	0.284	0.977
	2	0.293	0.988		5	0.096	0.997		6	0.118	0.993
	3	0.073	0.988		6	0.030	0.999		7	0.040	0.998
25	1	0.950	0.771	30	2	0.991	0.664	20	2	0.999	0.500
	2	0.429	0.978		3	0.771	0.899		3	0.938	0.736
	3	0.132	0.997		4	0.421	0.973		4	0.693	0.895
	4	0.032	0.999		5	0.187	0.993		5	0.397	0.963
30	1	0.992	0.663		6	0.071	0.998		6	0.187	0.988
	2	0.571	0.963			$X = 0.20$			7	0.074	0.997
	3	0.208	0.994	4	1	0.549	0.862		8	0.025	0.999
	4	0.060	0.999		2	0.108	0.988	25	3	0.996	0.599
		$X = 0.10$			3	0.008	0.999		4	0.920	0.783
4	1	0.294	0.965	5	1	0.689	0.801		5	0.693	0.905
	2	0.028	0.999		2	0.194	0.976		6	0.424	0.963
5	1	0.386	0.950		3	0.028	0.998		7	0.221	0.987
	2	0.054	0.997	6	1	0.801	0.736		8	0.100	0.995
6	1	0.475	0.932		2	0.291	0.961		9	0.039	0.999
	2	0.086	0.995		3	0.060	0.995	30	4	0.991	0.665
7	1	0.559	0.912	7	1	0.883	0.669		5	0.905	0.814
	2	0.123	0.992		2	0.395	0.941		6	0.693	0.913
	3	0.016	0.999		3	0.105	0.991		7	0.446	0.963
8	1	0.638	0.889		4	0.017	0.999		8	0.249	0.985
	2	0.165	0.989	8	1	0.937	0.606		9	0.123	0.995
	3	0.027	0.999		2	0.499	0.916		10	0.054	0.998
9	1	0.711	0.862		3	0.162	0.985		11	0.021	0.999
	2	0.210	0.985		4	0.035	0.998				
	3	0.040	0.998	9	1	0.970	0.548				
10	1	0.776	0.832		2	0.599	0.887				
	2	0.258	0.981		3	0.227	0.978				
	3	0.056	0.998		4	0.060	0.996				

[a] Adapted from Peck, L. G., and Hazelwood, R. N., *Finite Queueing Tables*, New York: Wiley, 1958 (with permission of the publisher).

TABLE H Critical values for Kolmogorov–Smirnov test for the exponential distribution

Sample size N	Kolmogorov–Smirnov level of significance α				
	0.20	0.15	0.10	0.05	0.01
3	0.451	0.479	0.511	0.551	0.600
4	0.396	0.422	0.449	0.487	0.548
5	0.359	0.382	0.406	0.442	0.504
6	0.331	0.351	0.375	0.408	0.470
7	0.309	0.327	0.350	0.382	0.442
8	0.291	0.308	0.329	0.360	0.419
9	0.277	0.291	0.311	0.341	0.399
10	0.263	0.277	0.295	0.325	0.380
11	0.251	0.264	0.283	0.311	0.365
12	0.241	0.254	0.271	0.298	0.351
13	0.232	0.245	0.261	0.287	0.338
14	0.224	0.237	0.252	0.277	0.326
15	0.217	0.229	0.224	0.269	0.315
16	0.211	0.222	0.236	0.261	0.306
17	0.204	0.215	0.229	0.253	0.297
18	0.199	0.210	0.223	0.246	0.289
19	0.193	0.204	0.218	0.239	0.283
20	0.188	0.199	0.212	0.234	0.278
25	0.170	0.180	0.191	0.210	0.247
30	0.155	0.164	0.174	0.192	0.226
>30	$\dfrac{0.86}{\sqrt{N}}$	$\dfrac{0.91}{\sqrt{N}}$	$\dfrac{0.96}{\sqrt{N}}$	$\dfrac{1.06}{\sqrt{N}}$	$\dfrac{1.25}{\sqrt{N}}$

INDEX

A

Alternative hypothesis, 361
Aspiration level models, 226-231
Asymptotic distribution, 310
Asymptotic mean, 310
Asymptotic normal distribution, 315
Asymptotic sample properties, 304
  consistent estimator, 311
  efficient estimator, 312
  squared-error consistent estimator, 312
  unbiased estimator, 310
Asymptotic variance, 310

B

Bayes's theorem, 24
Birth-death equations, 93
Bulk arrivals
  $(M^{(b)}|M|1) : (GD|\infty|\infty)$ results, 121-124
  random bulk quantity, 138, 141
Busy period, 252, 253, 292

C

Carried load, 101
Central limit theorem, 309
Conditional density function, 41
Conditional expectation, 46
Conditional probability mass function, 41

Confidence interval, 302, 345
  approximate intervals, 353
  mean of normal distribution, 347, 349
  standard deviation of normal distribution, 352
Continuous probability distributions, 30-39
  beta, 32-33
  Erlang, 32-33, 35
  exponential, 32-34
  gamma, 32-35
  hyperexponential, 32-33, 35-36
  normal, 32-33, 38
  tables of, 32-33
  uniform, 32-33, 38-39
Control charts, 16, 395-409
Cost determination, 231-238
Cost models, 207-226
  $(M|M|c) : (GD|\infty|\infty)$, determine c, 207-210
  $(M|G|c) : (GD|c|\infty)$, determine c, 210-211
  $(M|M|c) : (GD|N|\infty)$, determine c and N, 211-212
  $(M|M|c) : (GD|K|K)$, determine c, 212-213
  $(M|M|1) : (GD|K|K)$, determine K, 213-217
  $(M|M|1) : (GD|\infty|\infty)$, determine μ, 218-220

$(M^{(b)}|M|1) : (GD|\infty|\infty)$, determine b, 220-221
$(M|M|c) : (GD|\infty|\infty)$, determine λ and c, 221-226
Cramer-Rao inequality, 306
Cumulative distribution function, 24

D

Data collection, 298
Difference, 192
Differential-difference equations, 87, 92, 133, 143
Discrete probability distributions, 24-30
   Bernoulli, 27-29
   binomial, 27-29
   geometric, 28-30
   hypergeometric, 28-29
   negative binomial, 28-30
   Poisson, 27-29
   rectangular, 28-29
   tables of, 28-29
Distribution identification, 298

E

Effective arrival rate, 96, 98, 102
Entering customer's distribution, 97
Erlang's loss formula, 111-112
Expectation, 45
Extreme point
   definition, 184
   interior, 184
   maximum, 184, 188
   minimum, 184, 188
   necessary conditions, 185, 188, 190
   sufficient conditions, 186, 190

F

Finite sample properties, 304
   efficient estimator, 305
   invariance, 318
   minimum variance unbiased estimator, 306-307
   unbiased estimator, 304

G

Gauss-Seidel method, *see* Method of successive displacements

General balance equations, 92, 124, 131, 144
General birth-death process, 91
Generating function, 69-77
   convolutions, 73
   random sums of random variables, 74
   sums of random variables, 73
   tables of, 525-526
Geometric transform, *see* Generating function
Goodness-of-fit tests, 330
   Chi-square test, 332, 521
   Cramer-Smirnov-Von Mises test, 340
   Kolmogorov-Smirnov test, 330, 522
   Lilliefors test, 340
   Poisson tests, 338
   see also Statistical tests

H

Hessian, 190

I

Interval estimation, 302, 344

J

Joint distributions, 40

L

Laplace transforms, 51-69
   convolutions, 56
   differential equations, 61
   Dirac delta function, 58
   moments, 67
   partial fraction expansion, 63
   tables of, 523-524
   unit step function, 57
Least-squares regression, *see* Regression
Linear partial differential equations, 266

M

Marginal density functions, 41
Maximum, *see* Extreme point
Maximum likelihood estimation, 316
   Erlang distribution, 320
   exponential distribution, 318
   Poisson distribution, 319

Index

Mean, 43
Mean squared error, 303
Method of moments, 314
   exponential distribution, 315
   Poisson distribution, 315
Method of stages, 150–174
   $(E_2|M|1) : (GD|N|\infty)$, 160
   $(E_2|M|2) : (GD|N|\infty)$, 162
   $(HE_2|E_2|1) : (GD|1|\infty)$, 172
   $(HE_2|M|1) : (GD|N|\infty)$, 170
   $(HE_2|M|2) : (GD|N|\infty)$, 171
   $(M|E_K|1) : (GD|\infty|\infty)$, 153
   $(M|E_2|2) : (GD|2|\infty)$, 158
   $(M|HE_2|1) : (GD|\infty|\infty)$, 164
   $(M|HE_2|2) : (GD|2|\infty)$, 168
Method of successive displacements, 174
Method of successive overrelaxation, 181
Method of successive underrelaxation, 181
Minimum, *see* Extreme point
Models, 11–14
   analog, 11
   descriptive, 12, 85, 89, 134
   heuristic, 11
   iconic, 11
   prescriptive (normative), 12, 13, 17, 207–226
   sensitivity, 14
   simulation, 11, 17, 415–418
   symbolic, 11
   tradeoffs, 12

N

Network of queues, 124
Nonnormality, effects, 393–394
Non-Poisson queues, 143–178
   $(M|G|1) : (GD|\infty|\infty)$, 246, 251
   $(M|E_K|1) : (GD|\infty|\infty)$ 153, 250
   $(GI|M|1) : (GD|\infty|\infty)$, 255, 257
   $(E_K|M|1) : (GD|\infty|\infty)$, 257
   $(D|M|1) : (GD|\infty|\infty)$, 257
   $(M|G|1) : (NPRP|\infty|\infty)$, 258
   see also Method of stages
Null hypothesis, 361
Numerical solution of steady-state balance equations, *see* Method of successive displacements

O

Offered load, 101–102

Optimization
   aspiration level models, 226–231
   constrained, 192
   continuous variables, 184
   cost models, 207–226
   discrete variables, 192
Outside customer's distribution, 97

P

Point estimation, 302
Point of inflection, 185
Poisson process, 47–50, 85–86
Poisson process identification, 341
Poisson queueing models, 85–133
   $(M|M|c) : (FCFS|N|\infty)$ results, 107
   $(M|M|c) : (FCFS|\infty|\infty)$ results, 110
   $(M|M|1) : (FCFS|\infty|\infty)$, 252
   $(M|M|c) : (GD|N|\infty)$ results, 99
   $(M|M|c) : (GD|N|\infty)$, 89–91, 99
   $(M|M|c) : (GD|\infty|\infty)$, 90, 105
   $(M|M|c) : (NPRP|\infty|\infty)$, 258–263
   $(M|M|1) : (GD|N|\infty)$, 90
   $(M|M|1) : (GD|\infty|\infty)$, 90, 250
   operating characteristics, 90, 94–96, 99, 103–105, 116, 134, 144, 146
Pollaczek-Khintchine formula, 143–146
Population size, 85
Probability mass function, 24
Probability theory, 18–50
   basic concepts, 18
   dependent event, 22
   event, 20
   independent event, 21
   outcome, 18
   random variable, 19
   sample space, 19
Process generators, 422–435
   Bernoulli, 426–427
   beta, 425
   binomial, 427
   chi-square, 425
   continuous, 422–425
   discrete, 426–428
   empirical, 428–435
   Erlang, 424
   exponential, 422
   geometric, 427–428
   normal, 424
   Poisson, 428

tables of, 430–431
uniform, 425
Weibull, 422–423
Pure birth process, 267
Pure death process, 268

Q

Quasi-random imput, 115
Queueing system, 2–16
  classification, 8
  elements, 2
  Poisson, 85

R

Random number generation, 421–422
Random variable
  continuous, 30–40
  defined, 19
  discrete, 24–30
Rate out equals rate in, 130–131, 158
Regression, 322, 394–395
Ridge, 201
Runge-Kutta methods, 277

S

Saddle point, 189
Sample size
  relation to error, 362–363
  test for means, 369, 372–373, 389–391
  test for variances, 378–379, 383–385
Search techniques, 193–206
  one-factor-at-a-time, 195–201
  pattern, 201–206
Series of queues, 124, 130
Server occupancy, *see* Server utilization
Server utilization, 101
Service discipline, 8
  first come-first served, 9, 107–111, 118–120
  general, 9, 89–107, 111–118, 120–133, 144–177, 246–257, 265–291
  last come-first served, 9
  priority, 257–265
    nonpreemptive, 258–263
    preemptive, 263–265
  random, 9
  shortest processing time, 9
Simulation, 414–506
  modeling, 11, 17, 415–418

Monte Carlo method, 418–421
multiple-parallel channel queues, 450–490
networks, 490–506
single-channel queue, 435–450
Standard error, 303
Stationary point, 185
Statistical tests
  binomial, 402–405
  chi-square, 375–380, 521
  F, 381–386
  lack of fit, 394
  means, known variances, 363–369
  means, unknown variances, 369–375 386–393
  normal, 363–369, 515–519
  runs, 405–409
  t, 369, 375, 386–393, 520
  variance, 375–380, 381–386
  see also Goodness-of-fit tests
Steady-state solution, 92
Systems analysis
  black-box approach, 4, 9
  steps, 2–17

T

Time-dependent parameter estimation, 321
Total time distribution
  $(M|M|c) : (FCFS|N|\infty)$ results, 107–111
  $(M|M|c) : (FCFS|\infty|\infty)$ results, 110–111
  $(M|M|c) : (FCFS|K|K)$ results, 118–120
Traffic intensity, 101–102
Transform methods, 50–77
Transient analysis
  $(M|M|\infty) : (GD|\infty|\infty)$, 272
  $(M|M|1) : (GD|\infty|\infty)$, 273
  machine-release model, 279
Two-state process, 276
Type I error, 361–362
Type II error, 361–362

V

Variance, 43

W

Waiting-time distribution, *see* Total-time distribution

Z

z transform, *see* Generating function